Investigating the Social World

W9-DFW-376

7

To Julia Ellen Schutt

Investigating the Social World

The Process and Practice of Research

RUSSELL K. SCHUTT

University of Massachusetts Boston

Los Angeles | London | New Delhi
Singapore | Washington DC

Los Angeles | London | New Delhi
Singapore | Washington DC

FOR INFORMATION:

SAGE Publications, Inc.
2455 Teller Road
Thousand Oaks, California 91320
E-mail: order@sagepub.com

SAGE Publications Ltd.
1 Oliver's Yard
55 City Road
London EC1Y 1SP
United Kingdom

SAGE Publications India Pvt. Ltd.
B 1/I 1 Mohan Cooperative Industrial Area
Mathura Road, New Delhi 110 044
India

SAGE Publications Asia-Pacific Pte. Ltd.
33 Pekin Street #02-01
Far East Square
Singapore 048763

Acquisitions Editor: Jerry Westby
Associate Editor: Megan Krattli
Editorial Assistant: Erim Sarbuland
Production Editor: Eric Garner
Copy Editor: Gretchen Treadwell
Typesetter: C&M Digitals (P) Ltd.
Proofreader: Susan Schon
Indexer: Molly Hall
Cover Designer: Bryan Fishman
Marketing Manager: Erica DeLuca
Permissions Editor: Karen Ehrmann

Copyright © 2012 by SAGE Publications, Inc.

All rights reserved. No part of this book may be reproduced or utilized in any form or by any means, electronic or mechanical, including photocopying, recording, or by any information storage and retrieval system, without permission in writing from the publisher.

Printed in Canada

Library of Congress Cataloging-in-Publication Data

Schutt, Russell K.
Investigating the social world : the process and practice of research / Russell K. Schutt.—7th ed.

p. cm.
Includes bibliographical references and index.

ISBN 978-1-4129-9980-9 (pbk.)
ISBN 978-1-4129-9981-6 (pbk. w/SPSS CD)

1. Social problems—Research. 2. Social sciences—Research. I. Title.

HN29.S34 2012
361.1072—dc23   2011031279

This book is printed on acid-free paper.

11 12 13 14 15 10 9 8 7 6 5 4 3 2 1

Contents

On the Study Site

Detailed Contents

On the Study Site

About the Author

Russell K. Schutt, PhD, is Professor and Chair of Sociology at the University of Massachusetts, Boston, and Lecturer on Sociology in the Department of Psychiatry (Beth Israel-Deaconess Medical Center) at the Harvard Medical School. He completed his BA, MA, and PhD (1977) degrees at the University of Illinois at Chicago and was a postdoctoral fellow in the Sociology of Social Control Training Program at Yale University (1977–1979). In addition to *Investigating the Social World: The Process and Practice of Research,* now in its seventh edition, *Making Sense of the Social World* (with Dan Chambliss), and adaptations for the fields of psychology (with Paul G. Nestor), social work (with Ray Engel), criminology/criminal justice (with Ronet Bachman), and education (with Joseph Check), he is the author of *Homelessness, Housing, and Mental Illness* and *Organization in a Changing Environment*, coeditor of *The Organizational Response to Social Problems* and coauthor of *Responding to the Homeless: Policy and Practice* . He has authored and coauthored almost 50 peer-reviewed journal articles as well as many book chapters and research reports on homelessness, service preferences and satisfaction, mental health, organizations, law, and teaching research methods. His current funded research is a mixed-methods investigation of a coordinated care program in the Massachusetts Department of Public Health, while his prior research projects include a study of community health workers and recruitment for cancer clinical trials (National Cancer Institute), a randomized evaluation of housing alternatives for homeless persons diagnosed with severe mental illness (National Institute of Mental Health), a large translational research project for the Massachusetts Department of Public Health's Women's Health Network, and evaluations of case management programs (Massachusetts Department of Public Health and Massachusetts Department of Mental Health). His publications in peer-reviewed journals range in focus from the effect of social context on cognition, satisfaction, and functioning; to the service preferences of homeless persons and service personnel; the admission practices of craft unions; and the social factors in legal decisions.

Preface

How do women and men handle the conflicting demands of work and family and other relationships in the always-connected social world of the 21st century? How often do parents turn away from their children to check their BlackBerry devices for the latest communication? Does it make life easier to work on a computer from home, or does it prevent full engagement with family routines? Are friends distracted from their conversations by beeps signaling the arrival of a new text message? Is the result of our electronic connectedness more fulfilling social lives or more conflict and anxiety? "Why are so many working mothers haunted by constant guilt?" asked Francie Latour (2011:K1) in a *Boston Sunday Globe* article.

It was concerns like these that encouraged University of Toronto sociologist Scott Schieman and graduate students Paul Glavin and Sarah Reid (2011) to investigate the psychological impact of "boundary-spanning work demands." For their investigation, they used data collected in the 2005 Work, Stress, and Health survey in the United States. As illustrated in Exhibit P.1, they found that frequent work-related contacts by email, phone, and text messaging outside of work resulted in women feeling guilty—but not men. Scott Schieman (cited in Latour, 2011) explained, "For men, the answer is, it doesn't seem to affect the response much at all. For women, when these things are occurring, the guilt starts taking a greater and greater toll" (p. K2).

Does this finding resonate with your own experiences or observations? Do you find yourself wondering how Glavin, Schieman and Reid determined whether the people who were surveyed felt guilty? Are you curious how people were selected for participation in the Work, Stress, and Health survey? And, does Exhibit P.1 convince you that it was the higher levels of work contact that were responsible for the more frequent feelings of guilt among the women surveyed? We all react to stories like Francie Latour's based on our own experiences, but before we can determine how confident we can be in social science research findings like these, we also need to answer these other questions about the research methods that lie behind the findings.

Of course, I have only presented a little bit of the evidence in the Glavin, Schieman, and Reid (2011) research article and just one of its many conclusions—there are many related issues to consider and important questions about the evidence to ask. If you're interested, the complete article appears on the book's study site, **www.sagepub.com/schuttisw7e**. In any case, this research provides a good starting point for introducing a text on research methods, for it illustrates both how research methods help us understand pressing social questions and why our own impressions of the social world can be misleading: Neither our own limited perceptions nor a few facts gleaned from even reputable sources provide us with a trustworthy basis for understanding the social world. We need systematic methods for investigating our social world that enable us to chart our course through the passions of the moment and to see beyond our own personal experience.

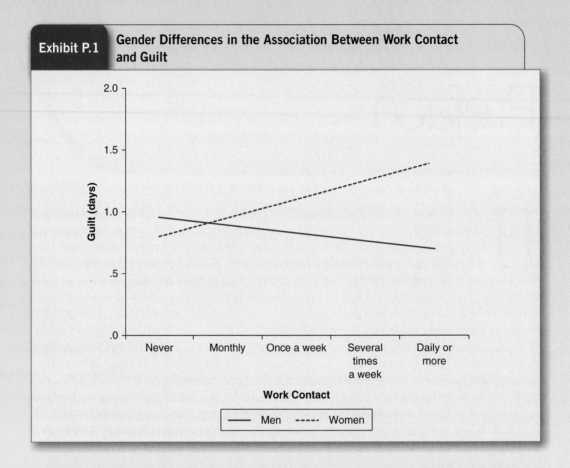

Exhibit P.1 **Gender Differences in the Association Between Work Contact and Guilt**

🖿 Teaching and Learning Goals

If you see the importance of pursuing answers to questions about the social world the way that Paul Glavin, Scott Schieman, and Sarah Reid did, you can understand the importance of investigating the social world. One purpose of this book is to introduce you to social science research methods like the ones involved in the Work, Stress, and Health survey and show how they improve everyday methods of answering our questions about the social world. Each chapter integrates instruction in research methods with investigation of interesting aspects of the social world, such as the use of social networking; the police response to domestic violence; and influences on crime, homelessness, work organizations, health, patterns of democratization, and likelihood of voting.

Another purpose of this book is to give you the critical skills necessary to evaluate research. Just "doing research" is not enough. Just reading that some conclusions are "based on a research study" is not sufficient. You must learn to ask many questions before concluding that research-based conclusions are appropriate. What did the researchers set out to investigate? How were people selected for study? What information was collected, and how was it analyzed? Throughout this book, you will learn what questions to ask when critiquing a research study and how to evaluate the answers. You can begin to sharpen your critical teeth on the illustrative studies throughout the book.

Another goal of this book is to train you to actually do research. Substantive examples will help you see how methods are used in practice. Exercises at the end of each chapter give you ways to try different methods alone or in a group. A checklist for research proposals will chart a course when you plan more ambitious studies. But research methods cannot be learned by rote and applied mechanically. Thus, you will learn the benefits and liabilities of each major approach to research and why employing a combination of them is often preferable. You will come to appreciate why the results of particular research studies must always be interpreted within the context of prior research and through the lens of social theory.

▣ Organization of the Book

The way the book is organized reflects my beliefs in making research methods interesting, teaching students how to critique research, and viewing specific research techniques as parts of an integrated research strategy. The first three chapters introduce the why and how of research in general. Chapter 1 shows how research has helped us understand the impact of social networking and changes in social ties. It also introduces some alternative approaches to social research. Chapter 2 illustrates the basic stages of research with a series of experiments on the police response to domestic violence, it emphasizes the role of theory in guiding research, and it introduces the decisions that researchers must make during an entire research project. Chapter 3 highlights issues of research ethics by taking you inside of Stanley Milgram's research on obedience to authority. It also introduces you to different research philosophies. The next three chapters discuss how to evaluate the way researchers design their measures, draw their samples, and justify their statements about causal connections. As you learn about these procedures, you will also read about research on substance abuse and gangs, homelessness, and the causes of violence.

Chapters 7, 8, and 9 present the three most important primary methods of data collection: experiments, surveys, and qualitative methods (including participant observation, intensive interviews, and focus groups), and Chapter 10 extends the coverage of qualitative methods with a focus on methods of analyzing qualitative data. The substantive studies in these chapters show how these methods have been used to improve our understanding of responses to prejudice, the relation between education and health, dealing with disaster, and understanding social interaction. Studying Chapters 9 and 10 together will provide a firm foundation for further use of a range of qualitative methods.

Chapters 11 and 12 present methodologies that can combine the primary methods reviewed in the preceding four chapters. Evaluation research, the focus of Chapter 11, can employ experiments, surveys, or qualitative methods to learn about the effects of and the need for social and other types of programs. This chapter begins with an overview of evaluation research on drug abuse prevention programs. Chapter 12 focuses on historical and comparative methodologies, which can use data obtained with one or more of the primary methods to study processes at the regional and societal levels over time and between units; research examples focus on the process of democratization and the bases of social revolutions.

Chapter 13 introduces the techniques and challenges of secondary data analysis and content analysis. The online availability of thousands of data sets from social science studies has helped make secondary data analysis—the use of previously collected data to investigate new research questions—the method of choice in many investigations. Content analysis—the use of text or pictures as the source of data about the social world—opens many possibilities; the examples in this chapter primarily involve research on gender roles. Chapter 14 works through an analysis of voting patterns in the 2008 U.S. presidential election to see how statistics can be used to investigate the likelihood of voting. Chapter 15 finishes up with an overview of the process of and techniques for reporting research results, an introduction to meta-analysis—a special tool for studying prior research—and some research on ethical problems in writing.

Distinctive Features of the Seventh Edition

The success of this book has been due in no small measure to the availability in the research literature of so many excellent examples of how social scientists investigate interesting and important questions about the social world. As a result, the first change that I make in each new edition of this text is to add examples of some of the newest and best research. But you will also find many other innovations in approach, coverage, and organization in the seventh edition:

Examples of social research as it occurs in real-world settings. Interesting studies of social ties, domestic violence, crime, and other social issues have been updated and extended from the sixth edition. New examples have been introduced from research on social networking, aggression and victimization, job satisfaction, and international social policies. The examples demonstrate that the exigencies and complexities of real life shape the application of research methods. This book also highlights the value of using multiple methods to study a research question.

Recent developments in research methods. Research methods continue to change in response to the spread of cell phones and the use of the Internet, the growth of interdisciplinary scholarship, and concerns with community impact. Some researchers have begun to explore the Internet with qualitative techniques. Reflecting these changes, I have expanded my coverage of web surveys and related issues, and added sections on Internet-based forms of qualitative research, added a section on mixed methods, and expanded my coverage of participatory action research.

Streamlined introduction to research. I have consolidated material from the sixth edition's appendix on searching the literature with the guidance for reviewing the literature in Chapter 2, thus providing a comprehensive overview of the process for finding and using the results of prior research.

Focus on international research. I have expanded my use of research conducted in countries around the globe, as well as continuing my focus on issues involving diversity in race, ethnicity, gender, and culture within the United States and in other countries.

Expanded statistics coverage. In response to instructor requests, I have expanded coverage of inferential statistics and multiple regression analysis.

Web-based instructional aids. The book's study site includes new interactive exercises that link directly to original research articles, published by SAGE, on each major research topic. It is important to spend enough time with these exercises to become very comfortable with the basic research concepts presented. The interactive exercises allow you to learn about research on a range of interesting topics as you practice using the language of research. The linked articles now allow the opportunity to learn more about the research that illustrates particular techniques.

Updated end-of-chapter exercises. Exercises using websites have been updated and those involving IBM SPSS Statistics* software have been revised for use with the new 2010 General Social Survey (GSS) data set (a subset of this data set is available on the website). The HyperRESEARCH program for qualitative analysis is also available online to facilitate qualitative analyses presented in exercises for Chapter 10.

---

*IBM SPSS® Statistics was formerly called PASW® Statistics.

Aids to effective study. The many effective study aids included in the previous editions have been updated, as needed. Appendix F, which you will find on the book's website, presents an annotated list of useful websites. The website also includes an online reader with articles from SAGE journals that illustrate topics in the text.

It is a privilege to be able to share with so many students the results of excellent social science investigations of the social world. If *Investigating the Social World* communicates the excitement of social research and the importance of evaluating carefully the methods we use in that research, then I have succeeded in representing fairly what social scientists do. If this book conveys accurately the latest developments in research methods, it demonstrates that social scientists are themselves committed to evaluating and improving their own methods of investigation. I think it is fair to say that we practice what we preach.

Now you're the judge. I hope that you and your instructor enjoy learning how to investigate the social world and perhaps do some investigating along the way. And, I hope you find that the knowledge and (dare I say it?) enthusiasm you develop for social research in this course will serve you well throughout your education, in your career, and in your community.

▣ A Note About Using SPSS

To carry out the SPSS exercises at the end of each chapter and in Appendix E (on the website), you must already have SPSS on your computer. The exercises use a subset of the 2010 GSS data set (included on the study site). This data set includes variables on topics such as work, family, gender roles, government institutions, race relations, and politics. Appendix E will get you up and running with SPSS (Statistical Package for the Social Sciences), and you can then spend as much time as you like exploring characteristics and attitudes of Americans. Just download the GSS2010 files (gss2010x and gss2010chapter14) and save them on your computer, then start SPSS on your PC, open one of the GSS2010 files, and begin with the first SPSS exercise in Chapter 1. (If you purchased the Student Version of SPSS with your text, you must first install it on your PC—simply insert the CD-ROM and follow the instructions). If you can use the complete SPSS program, you can do all the analyses in the book using the gss2010vars71 file.

The website also includes a subset of the 2002 International Social Survey Program data set. In addition, the GSS website listed subsequently contains documentation files for the GSS2010 and the ISSP2002, as well as the complete GSS2010 data set. You can use the complete data set if you have access to the full version of SPSS. You can also carry out analyses of the GSS at the University of California, Berkeley, website: http://sda.berkeley .edu/archive.htm) or at the National Opinion Research Center site: www.norc.uchicago.edu/GSS+Website/.

Acknowledgments

M y thanks first to Jerry Westby, publisher for SAGE Publications. Jerry's consistent support and exceptional vision have made it possible for this project to flourish, while his good cheer and collegiality have even made it all rather fun. Editor extraordinaire Elise Caffee also contributed vitally to the success of this edition, with the assistance of Erim Sarbuland and Megan Krattli. Book production was managed with great expertise and good cheer by Eric J. Garner, while Gretchen Treadwell proved herself to be one of publishing's most conscientious and effective copy editors. Rachael Leblond assisted with book ancillaries. And, I continue to be grateful to work with what has become the world's best publisher in social science, SAGE Publications, and their exceptional marketing manager, Erica DeLuca.

I also am indebted to the first-rate social scientists Jerry Westby recruited to critique the sixth edition. Their thoughtful suggestions and cogent insights have helped improve every chapter. They are:

Robyn Brown, DePaul University

Jennifer Bulanda, Miami University

Jerry Daday, Western Kentucky University

Marvin Dawkins, University of Miami

Patricia Drentea, University of Alabama-Birmingham

Kenneth Fernandez, University of Nevada at Las Vegas

Elizabeth Monk-Turner, Old Dominion University

David Sanders, Angelo State University

Jimmy Kazaara Tindigarukayo, University of the West Indies

Susan Wurtzburg, University of Hawai'i

The quality of *Investigating the Social World* benefits increasingly from the wisdom and creativity of my coauthors on adaptations for other markets and disciplines, as well as from the pleasure of being part of the support group that we provide each other. My profound gratitude to my SAGE coauthors: Ronet Bachman (University of Delaware), Dan Chambliss (Hamilton College), Joe Check (University of Massachusetts, Boston), Ray Engel (University of Pittsburgh), and Paul Nestor (University of Massachusetts, Boston). Thanks are also due to Charles DiSogra, Karen Hacker, Robert J. Sampson, Sunshine Hillygus, and Catherine Kohler Riessman for sharing with me and my coauthors their insights about recent developments in social research at one of our biennial coauthor meetings.

Kate Russell, one of my talented graduate students, provided indispensable, timely assistance for the seventh edition, checking websites, updating SPSS exercises and the SPSS appendix, adding bibliographic entries, reviewing exhibits and developing new interactive exercises for this edition. My thanks for her exceptional dedication, consistent good humor, and impressive skills.

Reviewers for the sixth edition were

Von Bakanic, College of Charleston

Marvin Dawkins, University of Miami

Carol Erbes, Old Dominion University

Isaac Heacock, Indiana University, Bloomington

Kenneth E. Fernandez, University of Nevada, Las Vegas

Edward Lascher, California State University, Sacramento

Quan Li, University of Central Florida

Steve McDonald, North Carolina State University

Kevin A. Yoder, University of North Texas

Reviewers for the fifth edition were

James David Ballard, California State University, Northridge

Carl Bankston, Tulane University

Diana Bates, The College of New Jersey

Sandy Cook-Fon, University of Nebraska at Kearny

Christopher Donoghue, William Paterson University

Tricia Mein, University of California at Santa Barbara

Jeanne Mekolichick, Radford University

Kevin Mulvey, George Washington University

Jennifer Parker-Talwar, Pennsylvania State University at Lehigh Valley

Nicholas Parsons, Washington State University

Michael J. Scavio, University of California at Irvine

Shaihid M. Shahidullah, Virginia State University

Tabitha Sharp, Texas Woman's University

John Talmage, University of North Florida

Bill Tillinghas, San Jose State University

Fourth edition reviewers were

Marina A. Adler, University of Maryland, Baltimore

Diane C. Bates, Sam Houston State University

Andrew E. Behrendt, University of Pennsylvania

Robert A. Dentler, University of Massachusetts, Boston (Chapter 10)

David H. Folz, University of Tennessee

Christine A. Johnson, Oklahoma State University

Carolyn Liebler, University of Washington

Carol D. Miller, University of Wisconsin-La Crosse

Dan Olson, Indiana University, South Bend

Brian J. Stults, University of Florida

John R. Warren, University of Washington

Ken Wilson, East Carolina University

Third edition reviewers were

Emmanuel N. Amadi, Mississippi Valley State University

Doug Anderson, University of Southern Maine

Robert B. Arundale, University of Alaska, Fairbanks

Hee-Je Bak, University of Wisconsin, Madison

Marit Berntson, University of Minnesota

Deborah Bhattacharayya, Wittenbert University

Karen Bradley, Central Missouri State University

Cynthia J. Buckley, The University of Texas, Austin

J. P. Burnham, Cumberland College

Gerald Charbonneau, Madonna University

Hugh G. Clark, Texas Woman's University

Mark E. Comadena, Illinois State University

John Constantelos, Grand Valley State University

Mary T. Corrigan, Binghamton University

John Eck, University of Cincinnati

Kristin Espinosa, University of Wisconsin, Milwaukee

Kimberly Faust, Fitchburg State College

Kenneth Fidel, DePaul University

Jane Hood, University of New Mexico

Christine Johnson, Oklahoma State University

Joseph Jones, Taylor University

Sean Keenan, Utah State University

Debra Kelley, Longwood College

Kurt Kent, University of Florida

Jan Leighley, Texas A&M University

Joel Lieberman, University of Nevada, Las Vegas

Randall MacIntosh, California State University, Sacramento

Peter J. May, University of Washington

Michael McQuestion, University of Wisconsin, Madison

Bruce Mork, University of Minnesota

Jennifer R. Myhre, University of California, Davis

Zeynep Özgen, Arizona State University

Norah Peters-Davis, Beaver College

Ronald Ramke, High Point University

Adinah Raskas, University of Missouri

Akos Rona-Tas, University of California, San Diego

Pamela J. Shoemaker, Syracuse University

Therese Seibert, Keene State College

Mark A. Shibley, Southern Oregon University

Herbert L. Smith, University of Pennsylvania

Paul C. Smith, Alverno College

Glenna Spitze, State University of New York, Albany

Beverly L. Stiles, Midwestern State University

Carolina Tolbert, Kent State University

Tim Wadsworth, University of Washington

Charles Webb, Freed-Hardeman University

Adam Weinberg, Colgate University

Special thanks to Barbara Costello, University of Rhode Island; Nancy B. Miller, University of Akron; and Gi-Wook Shin, University of California, Los Angeles, for their contributions to the third edition.

Second edition reviewers were

Nasrin Abdolali, Long Island University, C. W. Post

Lynda Ames, State University of New York, Plattsburgh

Matthew Archibald, University of Washington

Karen Baird, State University of New York, Purchase

Kelly Damphousse, Sam Houston State University

Ray Darville, Stephen F. Austin State University

Jana Everett, University of Colorado, Denver

Virginia S. Fink, University of Colorado, Colorado Springs

Jay Hertzog, Valdosta State University

Lin Huff-Corzine, University of Central Florida

Gary Hytrek, University of California, Los Angeles

Debra S. Kelley, Longwood College

Manfred Kuechler, Hunter College (CUNY)

Thomas Linneman, College of William & Mary

Andrew London, Kent State University

Stephanie Luce, University of Wisconsin, Madison

Ronald J. McAllister, Elizabethtown College

Kelly Moore, Barnard College, Columbia University

Kristen Myers, Northern Illinois University

Michael R. Norris, University of Texas, El Paso

Jeffrey Prager, University of California, Los Angeles

Liesl Riddle, University of Texas, Austin

Janet Ruane, Montclair State University

Josephine A. Ruggiero, Providence College

Mary Ann Schwartz, Northeastern Illinois University

Mildred A. Schwartz, University of Illinois, Chicago (Chapter 11)

Gi-Wook Shin, University of California, Los Angeles

Howard Stine, University of Washington

William J. Swart, The University of Kansas

Guang-zhen Wang, Russell Sage College

Shernaaz M. Webster, University of Nevada, Reno

Karin Wilkins, University of Texas, Austin

Keith Yanner, Central College

First edition reviewers were

Catherine Berheide, Skidmore College

Terry Besser, University of Kentucky

Lisa Callahan, Russell Sage College

Herbert L. Costner, formerly of University of Washington

Jack Dison, Arkansas State University

Sandra K. Gill, Gettysburg College

Gary Goreham, North Dakota State University

Barbara Keating, Mankato State University

Bebe Lavin, Kent State University

Scott Long, Indiana University

Elizabeth Morrissey, Frostburg State University

Chandra Muller, University of Texas

G. Nanjundappa, California State University, Fullerton

Josephine Ruggiero, Providence College

Valerie Schwebach, Rice University

Judith Stull, Temple University

Robbyn Wacker, University of Northern Colorado

Daniel S. Ward, Rice University

Greg Weiss, Roanoke College

DeeAnn Wenk, University of Oklahoma

I am also grateful for Kathy Crittenden's support on the first three editions, for the contributions of Herbert L. Costner and Richard Campbell to the first edition, and to Steve Rutter, whose vision and enthusiasm launched the whole project on a successful journey.

The interactive exercises on the website began with a series of exercises that I developed in a project at the University of Massachusetts, Boston. They were expanded for the second edition by Tom Linneman and a team of graduate students he directed at the University of Washington—Mark Edwards, Lorella Palazzo, and Tim Wadsworth—and tested by Gary Hytrek and Gi-Wook Shin at UCLA. My format changes in the exercises for the third edition were tested by my daughter, Julia Schutt. Diane Bates and Matthew Archibald helped revise material for instructors and Judith Richlin-Klonsky revised some examples in Chapter 9, for the third edition. Kate Russell developed a new set of exercises for this edition.

Reef Youngreen and Phil Kretsedemas provided helpful feedback on, respectively, chapters 3 and 4. Several former faculty, staff, and graduate students at the University of Massachusetts, Boston, made important contributions to earlier editions: Jeffrey Xavier, Megan Reynolds, Kathryn Stoeckert, Tracey Newman, Heather Johnson (Northeastern University), Chris Gillespie (University of Massachusetts, Boston, and Brandeis University), Heather Albertson, Ra'eda Al-Zubi, Anne Foxx, Bob Dentler and students in his 1993–94 graduate research methods class. I continue to be indebted to the many students I have had the opportunity to teach and mentor, at both the undergraduate and graduate levels. In many respects, this book could not have been so successful without the ongoing teaching experiences we have shared. I also share a profound debt to the many social scientists and service professionals with whom I have collaborated in social science research projects.

No scholarly book project can succeed without good library resources, and for these I continue to incur a profound debt to the Harvard University library staff and their extraordinary collection. I also have benefited from the resources maintained by the excellent librarians at the University of Massachusetts, Boston.

Again, most important, I thank my wife for her love and support and our daughter for the joy she brings to our lives and the good she has done in the social world.

—Russell K. Schutt

CHAPTER 1

Science, Society, and Social Research

F acebook and other online social networking services added a new dimension to the social world in the early years of the new millennium. If you saw the film, *The Social Network*, you know that Mark Zuckerberg started Facebook in 2004, as a service for college students like himself, and that he didn't stop there. By the end of 2010, Facebook had more than 550,000,000 members—one out of every 12 people in the world and almost half the United States population—and was adding about 700,000 new members each day (Grossman 2010). With so many people connecting through Facebook and similar services, you might think that the social world will never again be the same.

Do you know what impact online social networking has had? Has it helped you to keep in touch with your friends? To make new friends? Does it distract you from reading your textbooks and paying attention in class? Is it reducing your face-to-face interactions with other people? And, is your experience with Facebook similar to that of other people, including those in other countries? Overall, are computer-mediated forms of communication enriching your social ties or impoverishing them?

You have probably asked yourself some of these questions and you may have thought about answers to them. That's where social researchers begin, with a question about the social world and a desire to find an answer. What makes social research different from the ordinary process of thinking about our experiences is a focus on broader questions that involve people outside our immediate experience, and the use of systematic research methods in order to answer those questions. Keith Hampton, Oren Livio, and Lauren Sessions Goulet (2010) asked whether people who make wireless Internet connections in public spaces reduce their engagement with others in those spaces. Rich Ling and Gitte Stald (2010) asked whether different technology-mediated forms of communication like cell phones and social networking differ in their effects on social ties. Kevin Lewis and others (2008) asked if some types of students were more likely to use Facebook than others. Although their research methods ranged from observing others (Hampton & Gupta, 2008) to conducting a survey on the web (Ling & Stald 2010), and analyzing Facebook records (Lewis et al. 2008), each of them carefully designed a research project about their question and presented their research findings in a published report.

In this chapter, we often return to questions about technology-based forms of communication and their effects on social ties. As we do so, I hope to convince you that the use of research methods to investigate questions about the social world results in knowledge that can be more important, more trustworthy, and more useful than personal opinions or individual experiences. You will learn how investigations such as by Ling and Stald (2010) are helpful in answering questions about social ties and the impact of technology on these ties. You will also learn about the challenges that researchers confront. By the chapter's end, you should know what is "scientific" in social science and appreciate how the methods of science can help us understand the problems of society.

▣ Learning About the Social World

Merely one research question about the social world raises so many more questions. Let's think about several more of the questions. Take a few minutes to read each of the following questions and jot down your answers. Don't ruminate about the questions or worry about your responses: *This is not a test;* there are no "wrong" answers.

1. Does social networking software help you stay in touch with your friends?

2. What percentage of Americans are connected to the Internet?

3. How many close friends does an average American have?

4. Does wireless access (Wi-Fi) in public places like Starbucks decrease social interaction among customers?

5. Do both cell phones and e-mail tend to hinder the development of strong social ties?

6. How does Internet use vary across social groups?

I'll bet you didn't have any trouble answering the first question, about your own experiences. But the second question and the others concern *the social world*—the experiences and orientations of people in addition to yourself. To answer questions such as these, we need to combine the answers of many different people and perhaps other sources. If you're on your toes, you also recognize that your answers to these other questions will be shaped in part by your answer to the first question—that is, what we think about the social world will be shaped by our own experiences. Of course, this means that other people, with different experiences, will often come up with different answers to the same questions. Studying research methods will help you learn what criteria to apply when evaluating these different answers and what methods to use when seeking to develop your own answers.

Are you convinced? Let's compare your answers to Questions 2 through 6 with findings from research using social science methods.

2. The 2009 Current Population Survey by the U.S. Bureau of the Census of approximately 54,000 households revealed that 63.5% of the U.S. population had a broadband connection to the Internet at home (a total of 68.7% were connected to the Internet at home). This percentage has increased rapidly since broadband first came into use in 2000 (see Exhibit 1.1).

3. In 2004, the average American had 2.08 discussion partners, or "confidants" (see Exhibit 1.2). This average had declined from 2.94 in 1985, so you might speculate that it is even lower by now (McPherson et al. 2006:358).

4. After observing Internet use in coffee shops with wireless access in two cities, Hampton and Gupta (2008) concluded that there were two types of Wi-Fi users: some who used their Internet connection to create a secondary work office and others who used their Internet connection as a tool for meeting others in the coffee shop.

5. Based on surveys in Norway and Denmark, Ling and Stald (2010) concluded that mobile phones increase social ties among close friends and family members, while e-mail communication tends to decrease the intensity of our focus on close friends and family members.

6. Internet use differs dramatically between social groups. As indicated in Exhibit 1.3, in 2007 Internet use ranged from as low as 32% among those with less than a high school education to 88% among those with at least a bachelor's degree—although Internet has increased for all education levels since 1997 (Strickling 2010). Internet use also increases with family income and is higher among non-Hispanic whites and Asian Americans than among Hispanic Americans and non-Hispanic black Americans (Cooper & Gallager 2004:Appendix, Table 1). People who are under 30 are most likely to use the Internet (93% of those age 12 to 29), as compared to those who are middle aged (81% among those age 30 to 49 and 70% of those age 50 to 64) or older (38% of those age 65 and older) (Pew Internet 2010). In their study at one college, Lewis et al. (2008) found that black students had more Facebook friends than whites, while female students posted more Facebook pictures than male students.

How do these answers compare with the opinions you recorded earlier? Do you think your personal experiences led you to different estimates than what others might have given? You have just learned that about two-thirds of U.S. households have an Internet connection at home and that those with more education use the Internet more than those with less education. Does this variability lead you to be cautious about using your own experience as a basis for estimating the average level of Internet use (Question 2)? How about estimating the number of intimate social ties (Question 3)? Were your opinions based primarily on what you have experienced or observed? How well do your experiences represent those of residents of the United States as a whole (Question 3)? Do you see how different people can come to such different conclusions about the effect of technology-based communication (Questions 4, 5, 6)?

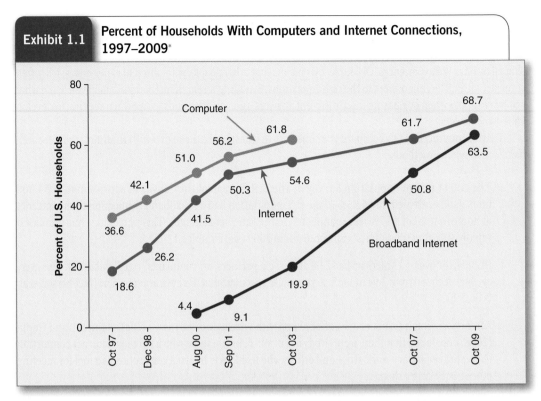

Exhibit 1.1 Percent of Households With Computers and Internet Connections, 1997–2009*

*Note: 2001, 2003, 2007, and 2009 Census-based weights and earlier years use 1990 Census-based weights.

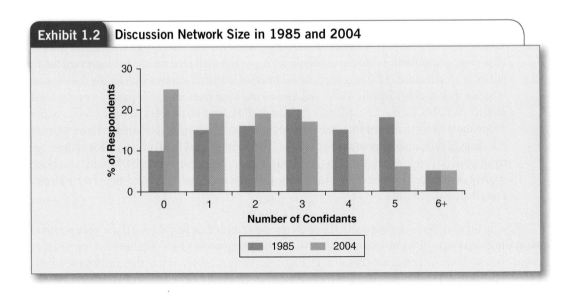

Exhibit 1.2 Discussion Network Size in 1985 and 2004

We cannot avoid asking questions about our complex social world or trying to make sense of our position in it. In fact, the more that you begin to "think like a social scientist," the more such questions will come to mind—and that's a good thing! But as you've just seen, in our everyday reasoning about the social world, our own prior experiences and orientations can have a major influence on what we perceive and how we interpret these perceptions. As a result, one person may see a person posting a message on

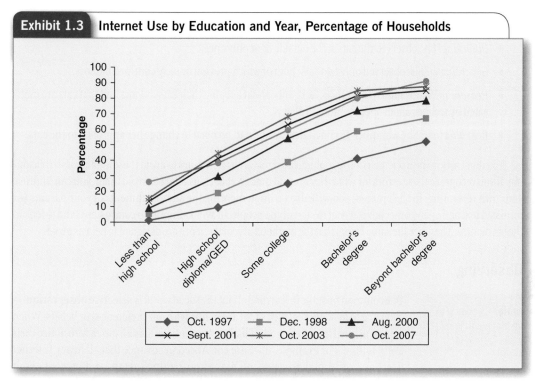

Exhibit 1.3   **Internet Use by Education and Year, Percentage of Households**

Note: Reports available online at www.census.gov/population/www/socdemo//computer/html

Facebook as being typical of what's wrong with modern society, while another person may see the same individual as helping people to "get connected" with others.

🖳 Avoiding Errors in Reasoning About the Social World

How can we avoid errors rooted in the particularities of our own backgrounds and improve our reasoning about the social world? First, let's identify the different processes involved in learning about the social world and the types of errors that can result as we reason about the social world.

When we think about the social world, we engage in one or more of four processes: (1) "*observing*" through our five senses (seeing, hearing, feeling, tasting, or smelling); (2) *generalizing* from what we have observed to other times, places, or people; (3) *reasoning* about the connections between different things that we have observed; and (4) *reevaluating* our understanding of the social world on the basis of these processes. It is easy to make mistakes in each of these processes.

My favorite example of the errors in reasoning that occur in the nonscientific, unreflective discourse about the social world that we hear on a daily basis comes from a letter to Ann Landers. The letter was written by someone who had just moved with her two cats from the city to a house in the country. In the city, she had not let her cats outside and felt guilty about confining them. When they arrived in the country, she threw her back door open. Her two cats cautiously went to the door and looked outside for a while, then returned to the living room and lay down. Her conclusion was that people shouldn't feel guilty about keeping their cats indoors—even when they have the chance, cats don't really want to play outside.

Do you see this person's errors in her approach to

- *Observing?* She observed the cats at the outside door only once.
- *Generalizing?* She observed only two cats, both of which previously were confined indoors.
- *Reasoning?* She assumed that others feel guilty about keeping their cats indoors and that cats are motivated by feelings about opportunities to play.
- *Reevaluating?* She was quick to conclude that she had no need to change her approach to the cats.

You don't have to be a scientist or use sophisticated research techniques to avoid these four errors in reasoning. If you recognize these errors for what they are and make a conscious effort to avoid them, you can improve your own reasoning. In the process, you will also be implementing the admonishments of your parents (or minister, teacher, or any other adviser) not to stereotype people, to avoid jumping to conclusions, and to look at the big picture. These are the same errors that the methods of social science are designed to help us avoid.

Observing

Selective observation Choosing to look only at things that are in line with our preferences or beliefs.

Inaccurate observation An observation based on faulty perceptions of empirical reality.

One common mistake in learning about the social world is **selective observation**—choosing to look only at things that are in line with our preferences or beliefs. When we are inclined to criticize individuals or institutions, it is all too easy to notice their every failure. For example, if we are convinced in advance that all heavy Internet users are antisocial, we can find many confirming instances. But what about elderly people who serve as Internet pen pals for grade-school children? Doctors who exchange views on medical developments? Therapists who deliver online counseling? If we acknowledge only the instances that confirm our predispositions, we are victims of our own selective observation.

Our observations can also simply be inaccurate. If, after a quick glance around the computer lab, you think there are 14 students present, when there are actually 17, you have made an **inaccurate observation**. If you hear a speaker say that "for the oppressed, the flogging never really stops," when what she said was, "For the obsessed, the blogging never really stops" (Hafner 2004), you have made an inaccurate observation.

Such errors occur often in casual conversation and in everyday observation of the world around us. In fact, our perceptions do not provide a direct window onto the world around us, for what we think we have sensed is not necessarily what we have seen (or heard, smelled, felt, or tasted). Even when our senses are functioning fully, our minds have to interpret what we have sensed (Humphrey 1992). The optical illusion in Exhibit 1.4, which can be viewed as either two faces or a vase, should help you realize that perceptions involve interpretations. Different observers may perceive the same situation differently because they interpret it differently.

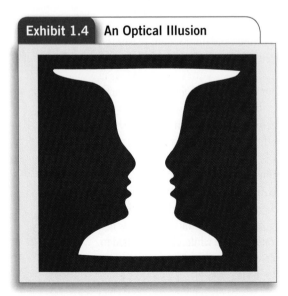

Exhibit 1.4 An Optical Illusion

Generalizing

Overgeneralization occurs when we conclude that what we have observed or what we know to be true for some cases is true for all or most cases (Exhibit 1.5). We are always drawing conclusions about people and social processes from our own interactions with them and perceptions of them, but sometimes we forget that our

Exhibit 1.5 | **The Difference Between Selective Observation and Overgeneralization**

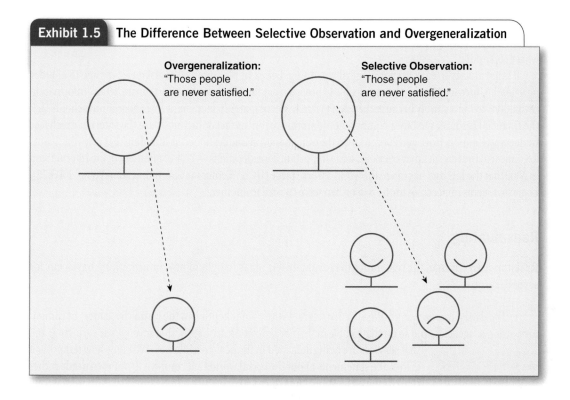

experiences are limited. The social (and natural) world is, after all, a complex place. We have the ability (and inclination) to interact with just a small fraction of the individuals who inhabit the social world, especially within a limited span of time. Thanks to the Internet and the practice of "blogging" (i.e., posting personal ruminations on web sites), we can easily find many examples of overgeneralization in people's thoughts about the social world. Here's one posted by a frequent blogger who was called for jury duty (www.tonypierce .com/blog/bloggy.htm, posted on June 17, 2005):

> **Overgeneralization** Occurs when we unjustifiably conclude that what is true for some cases is true for all cases.

> yesterday i had to go to jury duty to perform my civil duty. *unlike most people* i enjoy jury duty because i find the whole legal process fascinating, especially when its unfolding right in front of you and you get to help decide yay or nay.

Do you know what the majority of people think about jury duty? According to a Harris Poll, 75% of Americans consider jury service to be a privilege (Grey 2005), so the blogger's generalization about "most people" is not correct. Do you ever find yourself making a quick overgeneralization like that?

Reasoning

When we prematurely jump to conclusions or argue on the basis of invalid assumptions, we are using **illogical reasoning**. An Internet blogger posted a conclusion about the cause of the tsunami wave that devastated part of Indonesia in 2004 (cited in Schwartz 2005):

> **Illogical reasoning** When we prematurely jump to conclusions or argue on the basis of invalid assumptions.

> Since we know that the atmosphere has become contaminated by all the atomic testing, space stuff, electronic stuff, earth pollutants, etc., is it logical to wonder if: Perhaps the "bones" of our earth where this earthquake spawned have also been affected?

Is that logical? Another blogger soon responded with an explanation of plate tectonics: "The floor of the Indian Ocean slid over part of the Pacific Ocean" (Schwartz 2005:A9). The earth's crust moves no matter what people do!

It is not always so easy to spot illogical reasoning. For example, about 63% of Americans age 18 or older now use the Internet. Would it be reasonable to propose that the 37% who don't participate in the "information revolution" avoid it simply because they don't want to participate? In fact, many low-income households lack the financial resources to buy a computer or maintain an online account and so use the Internet much less frequently (Rainie & Horrigan 2005:63). On the other hand, an unquestioned assumption that everyone wants to connect to the Internet may overlook some important considerations—17% of nonusers of the Internet said in 2002 that the Internet has made the world a worse place (UCLA Center for Communication Policy 2003:78). Logic that seems impeccable to one person can seem twisted to another.

Reevaluating

Resistance to change, the reluctance to reevaluate our ideas in light of new information, may occur for several reasons:

- *Ego-based commitments.* We all learn to greet with some skepticism the claims by leaders of companies, schools, agencies, and so on that people in their organization are happy, that revenues are growing, and that services are being delivered in the best possible way. We know how tempting it is to make statements about the social world that conform to our own needs rather than to the observable facts. It can also be difficult to admit that we were wrong once we have staked out a position on an issue. Barry Wellman (Boase et al. 2006:1) recalls a call from a reporter after the death of four "cyber addicts." The reporter was already committed to the explanation that computer use had caused the four deaths; now, he just wanted an appropriate quote from a computer-use expert, such as Wellman. But the interview didn't last long.

> **Resistance to change** The reluctance to change our ideas in light of new information.

> The reporter lost interest when Wellman pointed out that other causes might be involved, that "addicts" were a low percentage of users, and that no one worries about "neighboring addicts" who chat daily in their front yards. (Boase et al. 2006:1)

- *Excessive devotion to tradition.* Some degree of devotion to tradition is necessary for the predictable functioning of society. Social life can be richer and more meaningful if it is allowed to flow along the paths charted by those who have preceded us. Some skepticism about the potential for online learning once served as a healthy antidote to unrealistic expectations of widespread student enthusiasm (Bray 1999). But too much devotion to tradition can stifle adaptation to changing circumstances. When we distort our observations or alter our reasoning so that we can maintain beliefs that "were good enough for my grandfather, so they're good enough for me," we hinder our ability to accept new findings and develop new knowledge. Of course, there was nothing "traditional" about maintaining social ties through e-mail when this first became possible in the late 20th century. Many social commentators assumed that the result of increasing communication by e-mail would be fewer social ties maintained through phone calls and personal contact. As a result, it was claimed, the social world would be impoverished. But subsequent research indicated that people who used e-mail more also kept in touch with others more in person and by phone (Benkler 2006:356). As you are learning in this chapter, debate continues about the long-term effects of our new forms of technology-based communication.

- *Uncritical agreement with authority.* If we do not have the courage to evaluate critically the ideas of those in positions of authority, we will have little basis for complaint if they exercise their authority

Research in the News

TWITTER USERS FLOCK TOGETHER

Like "birds of a feather that flock together," Twitter users tend to respond to others who express sentiments similar to their own. Johan Bolen from the University of Indiana reviewed 6-months-worth of tweets from 102,009 active Twitter users. Twitter users who expressed happy moods tended to retweet or reply to others who were happy, while those who indicated loneliness engaged more with other lonely Twitterers.

Source: Bilton, Nick. 2011. "Twitter Users Flock Together." *The New York Times,* March 21:B7.

over us in ways we don't like. And, if we do not allow new discoveries to call our beliefs into question, our understanding of the social world will remain limited. Was it in part uncritical agreement with computer industry authorities that led so many to utopian visions for the future of the Internet? "Entrepreneurs saw it as a way to get rich, policy makers thought it could remake society, and business people hoped that online sales would make stock prices soar. Pundits preached the gospel of the new Internet millennium" (Wellman 2004:25).

Now take just a minute to reexamine the opinions about social ties and Internet use that you recorded earlier. Did you grasp at a simple explanation even though reality is far more complex? Did your own ego and feelings about your similarities to or differences from others influence your beliefs? Did you weigh carefully the opinions of authorities who decry the decline of "community"? Could knowledge of research methods help improve your own understanding of the social world? Do you see some of the challenges social science faces?

Science and Social Science

The scientific approach to answering questions about the natural world and the social world is designed to reduce greatly these potential sources of error in everyday reasoning. **Science** relies on logical and systematic methods to answer questions, and it does so in a way that allows others to inspect and evaluate its methods. In this way, scientific research develops a body of knowledge that is continually refined, as beliefs are rejected or confirmed on the basis of testing empirical evidence.

Exhibit 1.6 shows one example of the use of scientific methods: The rapid increase in transportation speeds as scientific knowledge in the past two centuries has fueled transportation technologies.

Social science relies on scientific methods to investigate individuals, societies, and social processes. It is important to realize that when we apply scientific methods to understanding ourselves, we often engage in activities—asking questions, observing social groups, and/or counting people—that are similar to things we do

> **Science** A set of logical, systematic, documented methods for investigating nature and natural processes; the knowledge produced by these investigations.

> **Social science** The use of scientific methods to investigate individuals, societies, and social processes; the knowledge produced by these investigations.

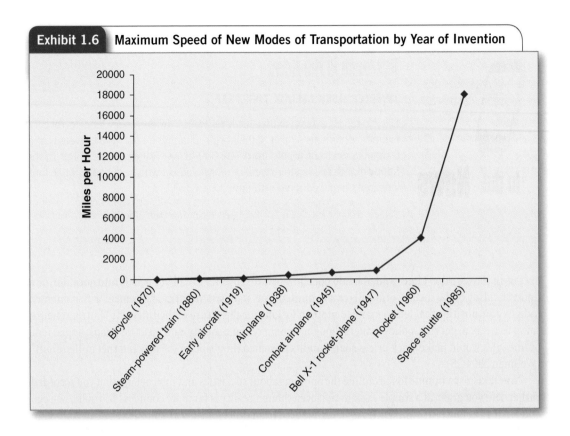

Exhibit 1.6 Maximum Speed of New Modes of Transportation by Year of Invention

in our everyday lives. However, social scientists develop, refine, apply, and report their understanding of the social world more systematically, or "scientifically," than Joanna Q. Public does:

- Social science research methods can reduce the likelihood of overgeneralization by using systematic procedures for selecting individuals or groups to study that are representative of the individuals or groups to which we wish to generalize.

- To avoid illogical reasoning, social researchers use explicit criteria for identifying causes and for determining whether these criteria are met in a particular instance.

- Social science methods can reduce the risk of selective or inaccurate observation by requiring that we measure and sample phenomena systematically.

- Because they require that we base our beliefs on evidence that can be examined and critiqued by others, scientific methods lessen the tendency to develop answers about the social world from ego-based commitments, excessive devotion to tradition, and/or unquestioning respect for authority.

Even as you learn to appreciate the value of social science methods, however, you shouldn't forget that *social* scientists face three specific challenges:

1. The objects of our research are people like us, so biases rooted in our personal experiences and relationships are more likely to influence our conclusions.

2. Those whom we study can evaluate us, even as we study them. As a result, subjects' decisions to "tell us what they think we want to hear" or, alternatively, to refuse to cooperate in our investigations can produce misleading evidence.

3. In physics or chemistry, research subjects (objects and substances) may be treated to extreme conditions and then discarded when they are no longer useful. However, social (and medical) scientists must concern themselves with the way their human subjects are treated in the course of research (much could also be said about research on animals, but this isn't the place for that).

We must never be so impressed with the use of scientific methods in investigations of the social world that we forget to evaluate carefully the quality of the resulting evidence. And we cannot ignore the need always to treat people ethically, even when that involves restrictions on the manipulations in our experiments, the questions in our surveys, or the observations in our field studies.

We must also be on guard against our natural tendency to be impressed with knowledge that is justified with what sounds like scientific evidence, but which has not really been tested. **Pseudoscience** claims are not always easy to identify, and many people believe them.

Are you surprised that more than half of Americans believe in astrology, with all its charts and numbers and references to stars and planets, even though astrological predictions have been tested and found baseless? (Shermer 1997:26). Are any of your beliefs based on pseudoscience?

> **Pseudoscience** Claims presented so that they appear scientific even though they lack supporting evidence and plausibility. (Shermer 1997:33)

Motives for Social Research

Similar to you, social scientists have friends and family, observe other persons' social ties, and try to make sense of what they experience and observe. For most, that's the end of it. But for some social scientists, the quality and impact of social ties has become a major research focus. What motivates selection of this or any other particular research focus? Usually, it's one or more of the following reasons:

- *Policy motivations.* Many government agencies, elected officials, and private organizations seek better descriptions of social ties in the modern world so they can identify unmet strains in communities, deficits in organizations, or marketing opportunities. Public officials may need information for planning zoning restrictions in residential neighborhoods. Law enforcement agencies may seek to track the connections between criminal gangs and the effect of social cohesion on the crime rate. Military leaders may seek to strengthen unit cohesion. These policy guidance and program management needs can stimulate numerous research projects. As Cooper and Victory (2002a) said in their foreword to a U.S. Department of Commerce report on the Census Bureau's survey of Internet use,

> this information will be useful to a wide variety of policymakers and service providers . . . help all of us determine how we can reach Americans more effectively and take maximum advantage of the opportunities available through new information technologies. (p. 3)

- *Academic motivations.* Questions about changing social relations have stimulated much academic social science. One hundred years ago, Emile Durkheim (1951) linked social processes stemming from urbanization and industrialization to a higher rate of suicide. Fifty years ago, David Reisman (1950/1969) considered whether the growing role of the mass media, among other changes, was leading Americans to become a "lonely crowd." Similar to this earlier research, contemporary investigations of the effect of computers and the Internet are often motivated by a desire to understand influences on the strength and meaning of social bonds.

Example: Can Internet resources help elderly persons manage heart conditions? The Internet provides a "space where disparate individuals can find mutual solace and exchange information within a common community of interest" (Loader et al. 2002:53). It is easy to understand why these features of the Internet "space" have made it a popular medium for individuals seeking help for health problems. Too often, however, elderly persons who grew up without computers do not benefit from this potentially important resource.

British social scientists Sally Lindsay, Simon Smith, Frances Bell, and Paul Bellaby (2007) were impressed with the potential of Internet-based health resources and wondered how access to those resources might help elderly persons manage heart conditions. They decided to explore this question by introducing a small group of older men to computers and the Internet and then letting them discuss their experiences with using the Internet for the next 3 years. Through the Internet, participants sought support from others with similar health problems, they helped others to cope, and they learned more about their condition.

Sally Lindsay and her colleagues read through transcripts of interviews and a guided group discussion with their participants. They then identified different themes and categorized text passages in terms of the themes and their interrelations. Two researchers read each transcript and compared their classifications of themes. These two researchers also discussed their interpretations of what they learned with their coauthors as well as with two of the elderly interviewees. For example, the researchers categorized one passage as showing *how the Internet could help reduce fear about participants' heart conditions:* "There's a lot of information there. It makes you feel a lot better. It takes a lot of the fear away. It's a horrible feeling once you've had a heart attack" (Lindsay et al. 2007:103).

In general, 3 years after being introduced to the Internet, "the majority were more informed and confident about managing their health and had developed strategies for meeting their specific informational needs and making better informed decisions" (Lindsay et al. 2007:107).

The Internet provided these new users with both more knowledge and greater social support in dealing with their health problems.

Explanatory Research

Explanatory research Seeks to identify causes and effects of social phenomena and to predict how one phenomenon will change or vary in response to variation in some other phenomenon.

Many consider explanation the premier goal of any science. **Explanatory research** seeks to identify the causes and effects of social phenomena and to predict how one phenomenon will change or vary in response to variation in some other phenomenon. Internet researchers adopted explanation as a goal when they began to ask such questions as "Does the internet increase, decrease, or supplement social capital?" (Wellman et al. 2001). "Do students who meet through Internet interaction like each other more than those who meet face-to-face"? (Bargh, McKenna, & Fitzsimons 2002:41). And "how [does] the Internet affect the role and use of the traditional media?" (Nie & Erbring 2002:276). I focus on ways of identifying causal effects in Chapter 6. Explanatory research often involves experiments (see Chapter 7) or surveys (see Chapter 8), both of which are most likely to use quantitative methods.

Example: What effect does Internet use have on social relations? Norman H. Nie and Lutz Erbring (2002), political scientists at the Stanford Institute for the Quantitative Study of Society, designed a large, innovative survey of Americans to answer this and other questions. They drew a random sample of 4,113 adults in 2,689 households across the United States and then gave every member of the sample a free web TV, which was then connected to the Internet, also free of charges. The survey was conducted on the Internet, with respondents answering questions directly on their web TVs.

The first study report focused on survey respondents who had already been using the Internet when they were contacted for the study. These respondents were questioned about their Internet usage, their personal characteristics and orientations, and the impact of Internet usage on their lives. Their answers suggested adverse effects of Internet use on social relations. The more time people spent using the Internet, the less time they spent for other social activities, even talking on the phone to friends and family. The heavier Internet users reported an increase in time spent working both at home and at the office. Nie and Erbring also found what some might view as positive effects: less time watching TV, shopping in stores, and commuting in traffic.

However, Nie and Erbring (2000) were troubled by the results.

> E-mail is a way to stay in touch, but you can't share a coffee or a beer with somebody on e-mail or give them a hug. . . . The Internet could be the ultimate isolating technology that further reduces our participation in communities even more than television did before it. (p. 19)

But as more research evidence on Internet use has accumulated, it seems that the Internet can be "a catalyst for creating and maintaining friendships and family relationships" (UCLA Center for Communication Policy 2001:8).

Evaluation Research

Evaluation research seeks to determine the effects of programs, policies, or other efforts to affect social patterns, whether by government agencies, private non-profits, or for-profit businesses. This is a type of explanatory research, because it deals with cause and effect, but it differs from other forms of explanatory research because evaluation research focuses on one type of cause: programs, policies, and other conscious efforts to create change (Lewis-Beck, Bryman, & Liao 2004:337). This focus raises some issues that are not relevant in other types of explanatory research. Concern over the potential impact of alternative policies regarding the Internet provided an impetus for new evaluation research. Chapter 11 introduces evaluation research.

> **Evaluation research** Research that describes or identifies the impact of social policies and programs.

Example: Does high-speed Internet access change community life? Netville's developers connected all homes in this new suburban Toronto community with a high-speed cable and appropriate devices for Internet access. Sociologists Barry Wellman and Keith Hampton (1999) used this arrangement to evaluate the impact of Internet access on social relations. They surveyed Netville residents who were connected to the Internet and compared them with residents who had not activated their computer connections. Hampton actually lived in Netville for 2 years, participating in community events and taking notes on social interaction.

It proved to be difficult to begin research in a rapidly developing community (Hampton & Wellman 1999), but a combination of household surveys and participant observation, supplemented by analysis of postings to the community e-mail list and special group discussions (focus groups), resulted in a comprehensive investigation of the role of the computer network in community social life (Hampton & Wellman 2000).

Hampton and Wellman found that Internet access increased social relations of residents ("Ego" in Exhibit 1.8) with other households, resulting in a larger and less geographically concentrated circle of friends. E-mail was used to set up face-to-face social events rather than as a substitute for them. Information about home repair and other personal and community topics and residents' service needs were exchanged over the Internet. Sensitive personal topics, however, were discussed offline. In fact, while wired residents knew more people within Netville by name and talked to more people on a regular basis than did the non-wired residents, they were not more likely to actually visit other residents (Hampton 2003:422). They also found that being wired into the computer network enabled residents to maintain more effectively their relations with friends and relatives elsewhere. Overall, community ties were enriched and extracommunity social ties were strengthened (Hampton & Wellman 2001).

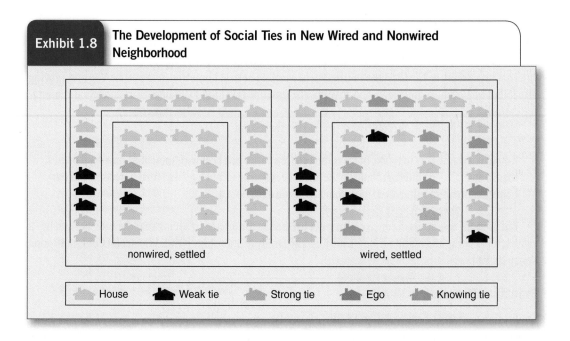

| Exhibit 1.8 | The Development of Social Ties in New Wired and Nonwired Neighborhood |

nonwired, settled

wired, settled

House   Weak tie   Strong tie   Ego   Knowing tie

Alternative Research Orientations

In addition to deciding on the type of research they will conduct, social researchers also must choose among several alternative orientations to research. Some researchers always adopt the same orientation in their research, but others vary their orientation based on the research particulars. It's also possible to combine these alternative orientations in different ways. I introduce alternative orientations in this chapter that represent answers to two important questions that must be considered when you begin a research project: (1) Will the research use primarily quantitative or qualitative methods, or some mixture? (2) Is the goal to accumulate new knowledge (basic science) or to make a practical contribution (applied research), or to do both? You will learn more about these alternatives in Chapter 2. In Chapter 3, I introduce ethical principles and alternative research philosophies that should guide an entire research project.

Quantitative and Qualitative Methods

Did you notice the difference between the types of data used in the studies about social ties? The primary data used in the descriptive social ties survey were counts of the number of people who had particular numbers of social ties and particular kinds of social ties, as well as their age, education, and other characteristics (McPherson et al. 2006:363). These data were numerical, so we say that this study used **quantitative methods.** The Bureau of the Census survey (Strickling 2010), the Lewis research (Lewis et al. 2008), and the Ling and Stald (2010) research also used quantitative methods—they reported their findings as percentages and other statistics that summarized the relationship between Internet usage and various aspects of social relations. In contrast, Hampton and Gupta (2008) observed Wi-Fi users in public spaces. Because they recorded their

Quantitative methods Methods such as surveys and experiments that record variation in social life in terms of quantities. Data that are treated as quantitative are either numbers or attributes that can be ordered in terms of magnitude.

actual observations and did not attempt to quantify what they were studying, we say that Hampton and Gupta (2008) used **qualitative methods.**

The distinction between quantitative and qualitative methods involves more than just the type of data collected. Quantitative methods are most often used when the motives for research are explanation, description, or evaluation. Exploration is more often the motive for using qualitative methods, although researchers also use these methods for descriptive, explanatory, and evaluative purposes. I highlight several other differences between quantitative and qualitative methods in the next two chapters. Chapters 9 and 10 present qualitative methods in much more detail. Chapter 3 introduces the alternative research philosophies that often lie behind the preference for quantitative or qualitative methods.

> **Qualitative methods** Methods such as participant observation, intensive interviewing, and focus groups that are designed to capture social life as participants experience it rather than in categories predetermined by the researcher. These methods rely on written or spoken words or observations that do not often have a direct numerical interpretation and typically involve exploratory research questions, an orientation to social context, and the meanings attached by participants to events and to their lives.

Important as it is, I don't want to place too much emphasis on the distinction between quantitative and qualitative orientations or methods. Social scientists often combine these methods to enrich their research. For example, Hampton and Wellman (2000) used surveys to generate counts of community network usage and other behaviors in Netville, but to help interpret these behaviors, they also observed social interaction and recorded spoken comments. In this way, qualitative data about social settings can be used to understand patterns in quantitative data better (Campbell & Russo 1999:141).

The use of multiple methods to study one research question is called **triangulation**. The term suggests that a researcher can get a clearer picture of the social reality being studied by viewing it from several different perspectives. Each will have some liabilities in a specific research application, and all can benefit from a combination of one or more other methods (Brewer & Hunter 1989; Sechrest & Sidani 1995).

> **Triangulation** The use of multiple methods to study one research question. Also used to mean the use of two or more different measures of the same variable.

The distinction between quantitative and qualitative data is not always sharp. Qualitative data can be converted to quantitative data, when we count the frequency of particular words or phrases in a text or measure the time elapsed between different observed behaviors. Surveys that collect primarily quantitative data may also include questions asking for written responses, and these responses may be used in a qualitative, textual analysis. Qualitative researchers may test explicit explanations of social phenomena using textual or observational data. We consider a *mixed-method* strategy in more detail in Chapter 10 and we examine particular combinations of methods in most other chapters.

Basic Science or Applied Research

You know that social scientists seek to describe and explain how society works. McPherson et al. (2006) sought to answer questions such as, "How do social ties vary between people or societies?" and "Why do some people, groups, or societies have more social ties than others?" Other researchers have investigated the meaning people attach to social ties and the consequences of having fewer social ties. The effort to figure out what the world is like and why it works as it does—academic motivations—is the goal of *basic science* (Hammersley 2008:50).

Social research may also have more immediate, practical concerns. Evaluation research like that conducted by Keith Hampton and Barry Wellman (1999) seeks to determine whether one program or policy has a more desirable impact than another. This knowledge can then lead to practical changes, such as increasing community members' access to the Internet so that their possibilities for social relations will expand. Evaluation research and other social research motivated by practical concerns are termed applied research.

Do you think that doing applied research would be good for society as well as for social researchers? Or do you think that a focus on how to improve society might lead social researchers to distort their understanding of how society works? Whether you think you would prefer a basic or applied orientation in social research, you have lots of company. In the 19th century, sociologist Lester Frank Ward (who subsequently became the American Sociological Society's first president) endorsed applied research: "The real object of science is to benefit man. A science which fails to do this, however agreeable its study, is lifeless" (Ward 1897:xxvii).

But in 1929, the American Sociological Society President William Fielding Ogburn urged sociologists to be guided by a basic research orientation: "Sociology as a science is not interested in making the world a better place to live. . . . Science is interested directly in one thing only, to wit, discovering new knowledge" (Ogburn 1930:300–301).

Tension between basic and applied research orientations has continued ever since these early disputes. Lynn Smith-Lovin (2007), who collaborated with Miller McPherson in the "social isolation" study, has argued recently for the importance of the basic science orientation: "I would, indeed, argue for knowledge for knowledge's sake" (p. 127).

In contrast, Robert Bellah, and his *Habits of the Heart* coauthors (1985) urged social scientists to focus explicit attention on achieving a more just society:

> Social science . . . whether it admits it or not, makes assumptions about good persons and a good society and considers how far these conceptions are embodied in our actual society . . . By probing the past as well as the present, by looking at "values" as much as at "facts," such a social science [as "public philosophy"] is able to make connections that are not obvious and to ask difficult questions. (p. 301)

You will encounter examples of basic and applied research throughout this book. By the time you finish *Investigating the Social World,* I know you'll have a good understanding of the difference between these orientations, but I can't predict whether you'll decide one is preferable. Maybe you'll conclude that they both have some merit.

▣ Strengths and Limitations of Social Research

Using social scientific research methods to develop answers to questions about the social world reduces the likelihood of making everyday errors in reasoning. The various projects that we have reviewed in this chapter illustrate this point:

- A clear definition of the population of interest in each study increased the researchers' ability to draw conclusions without overgeneralizing findings to groups to which they did not apply. Selection of a data set based on a broad, representative sample of the population enabled McPherson et al. (2006) to describe social ties throughout the United States rather than among some unknown set of their friends or acquaintances. The researchers' explicit recognition that persons who do not speak English were not included in their data set helps prevent overgeneralization to groups that were not actually studied (McPherson et al. 2006:356).

- The use of surveys in which each respondent was asked the same set of questions reduced the risk of selective or inaccurate observation, as did careful attention to a range of measurement issues (McPherson et al. 2006:355–356).

- The risk of illogical reasoning was reduced by carefully describing each stage of the research, clearly presenting the findings, and carefully testing the bases for cause-and-effect conclusions. For example, Ling and Stald (2010) test to see whether age or gender, rather than cell phone use, might have increased the tightness of social group ties in Norway.

- Resistance to change was reduced by providing free computers to participants in the Internet health study (Lindsay et al. 2007:100). The publications by all the researchers help other researchers critique and learn from their findings as well as inform the general public.

Nevertheless, I would be less than honest if I implied that we enter the realm of truth and light when we conduct social research or when we rely solely on the best available social research. Research always has some limitations and some flaws (as does any human endeavor), and our findings are always subject to differing interpretations. Social research permits us to see more, to observe with fewer distortions, and to describe more clearly to others what our opinions are based on, but it will not settle all arguments. Others will always have differing opinions, and some of those others will be social scientists who have conducted their own studies and drawn different conclusions.

Although Nie and Erbring (2000) concluded that the use of the Internet diminished social relations, their study at Stanford was soon followed by the Pew Internet & American Life Project (2000) and another Internet survey by the UCLA Center for Communication Policy (2001). These two studies also used survey research methods, but their findings suggested that the use of the Internet does *not* diminish social relations. Psychologist Robert Kraut's early research suggested that Internet use was isolating, but his own more recent research indicates more positive effects (Kraut et al. 2002). To what extent are different conclusions due to differences in research methods, to different perspectives on similar findings, or to rapid changes in the population of Internet users?

It's not easy to answer such questions, so one research study often leads to another, and another, each one improving on previous research or examining a research question from a somewhat different angle. Part of becoming a good social researcher is learning that we have to evaluate critically each research study and weigh carefully the entire body of research about a research question before coming to a conclusion. And, we have to keep an open mind about alternative interpretations and the possibility of new discoveries. The social phenomena we study are often complex, so we must take this complexity into account when we choose methods to study social phenomena and when we interpret the results of these studies.

However, even in the areas of research that are fraught with controversy, where social scientists differ in their interpretations of the evidence, the quest for new and more sophisticated research has value. What is most important for improving understanding of the social world is not the result of any particular study but the accumulation of evidence from different studies of related issues. By designing new studies that focus on the weak points or controversial conclusions of prior research, social scientists contribute to a body of findings that gradually expands our knowledge about the social world and resolves some of the disagreements about it.

Whether you plan to conduct your own research projects, read others' research reports, or just think about and act in the social world, knowing about research methods has many benefits. This knowledge will give you greater confidence in your own opinions; improve your ability to evaluate others' opinions; and encourage you to refine your questions, answers, and methods of inquiry about the social world.

Conclusions

I hope this first chapter has given you an idea of what to expect from the rest of the book. My aim is to introduce you to social research methods by describing what social scientists have learned about the social world as well as how they have learned it. The substance of social science is inevitably more interesting than its methods, but the methods become more interesting when they're linked to substantive investigations. I have focused attention in this chapter on research about social ties; in the subsequent chapters, I introduce research examples from other areas.

Chapter 2 continues to build the foundation for investigating the social world. I review how social scientists select research questions for investigation, how they orient themselves to those questions with social theories, and how they review related prior research. Most of the chapter focuses on the steps involved in the overall research process and the criteria that researchers use to assess the quality of their answers to the original research questions. Several studies of domestic violence illustrate the research process in Chapter 2. I also introduce in this chapter the process of writing research proposals, which I then continue in the end-of-chapter exercises throughout the book. Chapter 3, on research ethics and research philosophies, completes the foundation for our study of social research. I emphasize in this chapter and in the subsequent end-of-chapter exercises the importance of ethical treatment of human subjects in research. I also introduce in Chapter 3 alternative philosophies and guidelines that should be considered throughout a research project.

Chapters 4, 5, and 6 focus on the issues in measurement, sampling, and research design that must be considered in any social research project. In Chapter 4, I discuss the concepts we use to think about the social world and the measures we use to collect data about those concepts. This chapter begins with the example of research on student substance abuse, but you will find throughout this chapter a range of examples from contemporary research. In Chapter 5, I use research on homelessness to exemplify the issues involved in sampling cases to study. In Chapter 6, I use research on violence to illustrate how research can be designed to answer causal research questions such as "What causes violence?" I also explain in this chapter the decisions that social researchers must make about two related research design issues: (1) whether to use groups or individuals as their units of analysis and (2) whether to use a cross-sectional or longitudinal research design.

Chapters 7, 8, and 9 introduce the three primary methods of data collection. Experimental studies, the subject of Chapter 7, are favored by many psychologists, social psychologists, and policy evaluation researchers. Survey research is the most common method of data collection in sociology, so in Chapter 8, I describe the different types of surveys and explain how researchers design survey questions. I highlight in this chapter the ways in which the Internet and cell phones are changing the nature of survey research. Qualitative methods have long been the method of choice in anthropology, but they also have a long tradition in American sociology and a growing number of adherents around the world. Chapter 9 shows how qualitative techniques can uncover aspects of the social world that we are likely to miss in experiments and surveys and can sometimes result in a different perspective on social processes.

Chapter 10 continues my overview of qualitative methods but with a focus on the logic and procedures of analyzing qualitative data. If you read Chapters 9 and 10 together, you will obtain a richer understanding of qualitative methods. In these chapters, you will learn about research on disasters such as Hurricane Katrina, on work organizations, psychological distress, gender roles, and classroom behavior.

Chapters 11, 12, and 13 introduce data collection approaches that can involve several methods. Evaluation research, the subject of Chapter 11, is conducted to identify the impact of social programs or to clarify social processes involving such programs. Evaluation research often uses experimental methods, but survey research and qualitative methods can also be helpful in evaluation research projects. Historical and comparative methods, the subject of Chapter 12, may involve either quantitative or qualitative methods that are used to compare societies and groups at one point in time and to analyze their development over time. We will see how these different approaches have been used to learn about political change in transitional societies. Chapter 13 reviews the methods of secondary data analysis and content analysis. In this chapter, you will learn how to obtain previously collected data and to investigate important social issues such as poverty dynamics. I think that by the time you finish Chapter 13, you will realize why secondary methods and content analysis often provide researchers with the best options for investigating important questions about the social world.

Chapter 14 gives you a good idea of how to use statistics when analyzing research data and how to interpret statistics in research reports. This single chapter is not a substitute for an entire course in statistics, but it provides the basic tools you can use to answer most research questions. To make this chapter realistic, I walk you through an analysis of quantitative data on voting in the 2008 presidential election. You can replicate this analysis with data on the book's study site (if you have access to the SPSS statistical analysis program). You can also learn more about statistics with the SPSS exercises at the end of most chapters and with the web site's tutorials.

Plan to read Chapter 15 carefully. Our research efforts are only as good as the attention given to our research reports, so my primary focus in this chapter is on writing research reports. I also review the strengths and weaknesses of the different major research methods we have studied. In addition, I introduce meta-analysis—a statistical technique for assessing many research studies about a particular research question. By the end of the chapter, you should have a broader perspective on how research methods can improve understanding of the social world (as well as an appreciation for how much remains to be done).

Each chapter ends with several helpful learning tools. Lists of key terms and chapter highlights will help you review the ideas that have been discussed. Discussion questions and practice exercises will help you apply and deepen your knowledge. Special exercises guide you in developing your first research proposal, finding information on the World Wide Web, grappling with ethical dilemmas and conducting statistical analyses. The Internet study site for this book provides interactive exercises and quizzes for reviewing key concepts, as well as research articles to review, web sites to visit, and data to analyze.

Key Terms

Descriptive research 12	Inaccurate observation 6	Resistance to change 8
Evaluation research 15	Overgeneralization 6	Science 9
Explanatory research 14	Pseudoscience 11	Selective observation 6
Exploratory research 13	Qualitative methods 17	Social science 9
Illogical reasoning 7	Quantitative methods 16	Triangulation 17

Highlights

- Empirical data are obtained in social science investigations from either direct experience or others' statements.

- Four common errors in reasoning are (1) overgeneralization, (2) selective or inaccurate observation, (3) illogical reasoning, and (4) resistance to change. These errors result from the complexity of the social world, subjective processes that affect the reasoning of researchers and those they study, researchers' self-interestedness, and unquestioning acceptance of tradition or of those in positions of authority.

- Social science is the use of logical, systematic, documented methods to investigate individuals, societies, and social processes, as well as the knowledge produced by these investigations.

- Social research cannot resolve value questions or provide permanent, universally accepted answers.

- Social research can be motivated by policy guidance and program management needs, academic concerns, and charitable impulses.

- Social research can be descriptive, exploratory, explanatory, or evaluative—or some combination of these.

- Quantitative and qualitative methods structure research in different ways and are differentially appropriate for diverse research situations. They may be combined in research projects.

- Research seeking to contribute to basic science focuses on expanding knowledge and providing results to the other researchers. Applied research seeks to have an impact on social practice and to share results with a wide audience.

STUDENT STUDY SITE

To assist in completing the web exercises, please access the study site at **www.sagepub.com/schuttisw7e**, where you will find the web exercises with accompanying links. You'll find other useful study materials such as self-quizzes and e-flash-cards for each chapter, along with a group of carefully selected articles from research journals that illustrate the major concepts and techniques presented in the book.

Discussion Questions

1. Select a social issue that interests you, such as Internet use or crime. List at least four of your beliefs about this phenomenon. Try to identify the sources of each of these beliefs.

2. Does the academic motivation to do the best possible job of understanding how the social world works conflict with policy and/or personal motivations? How could personal experiences with social isolation or with Internet use shape research motivations? In what ways might the goal of influencing policy about social relations shape a researcher's approach to this issue?

3. Pick a contemporary social issue of interest to you. Describe different approaches to research on this issue that would involve descriptive, exploratory, explanatory, and evaluative approaches.

4. Review each of the research alternatives. Do you find yourself more attracted to a quantitative or a qualitative approach? Or, to doing research to contribute to basic knowledge or shape social policy? What is the basis of your preferences? Would you prefer to take a mixed-methods approach? What research questions do you think are most important to pursue in order to improve government policies?

Practice Exercises

1. Read the abstracts (initial summaries) of each article in a recent issue of a major social science journal. (Ask your instructor for some good journal titles.) On the basis of the abstract only, classify each research project represented in the articles as primarily descriptive, exploratory, explanatory, or evaluative. Note any indications that the research focused on other types of research questions.

2. Find a report of social science research in an article in a daily newspaper. What are the motives for the research? How much information is provided about the research design? What were the major findings? What additional evidence would you like to see in the article to increase your findings in the research conclusions?

3. Review "Types of Research" from the Interactive Exercises link on the study site. To use these lessons, choose one of the four "Types of Research" exercises from the opening menu. About 10 questions are presented in each version of the lesson. After reading each question, choose one answer from the list presented. The program will evaluate your answers. If an answer is correct, the program will explain why you were right and go on to the next question. If you have made an error, the program will explain the error to you and give you another chance to respond.

4. Now, select the Learning From Journal Articles link (www.sagepub.com/schuttisw7e). Open one article and read its abstract. Identify the type of research (descriptive, exploratory, or evaluative) that appeared to be used. Now scan the article and decide whether the approach was quantitative or qualitative (or both) and whether it included any discussion of policy implications.

Ethics Questions

Throughout the book, we will discuss the ethical challenges that arise in social research. At the end of each chapter, you are asked to consider some questions about ethical issues related to that chapter's focus. I introduce this critical topic formally in Chapter 3, but we will begin here with some questions for you to ponder.

1. The chapter refers to research on social isolation. What would *you* do if you were interviewing elderly persons in the community and found that one was very isolated and depressed or even suicidal, apparently as a result of his or her isolation? Do you believe that social researchers have an obligation to take action in a situation like this? What if you discovered a similar problem with a child? What guidelines would you suggest for researchers?

2. Would you encourage social researchers to announce their findings in press conferences about topics such as the impact of the Internet on social *ties*, and to encourage relevant agencies to adopt policies aimed to lessen social isolation? Are there any advantages to studying research questions only in order to contribute to academic knowledge? Do you think there is a fundamental conflict between academic and policy motivations? Do social researchers have an ethical obligation to recommend policies that their research suggests would help other people?

Web Exercises

1. The research on social ties by McPherson and his colleagues was publicized in a *Washington Post* article that also included comments by other sociologists. Read the article at www .washingtonpost.com/wp-dyn/content/article/2006/06/22/ AR2006062201763_pf.html and continue the commentary. Do your own experiences suggest that there is a problem with social ties in your community? Does it seem, as Barry Wellman suggests in the *Washington Post* article, that a larger number of social ties can make up for the decline in intimate social ties that McPherson found?

2. Scan one of the publications about the Internet and society at the Berkman Center for Internet & Society website, http://cyber.law.harvard.edu/. Describe one of the projects discussed: its goals, methods, and major findings. What do the researchers conclude about the impact of the Internet on social life in the United States? Next, repeat this process with a report from the Pew Internet Project at www.pewinternet .org, or with the Digital Future report from the University of Southern California's Center for the Digital Future site, www .digitalcenter.org. What aspects of the methods, questions, or findings might explain differences in their conclusions? Do you think the researchers approached their studies with different perspectives at the outset? If so, what might these perspectives have been?

SPSS Exercises

As explained in the Preface, the SPSS Exercises at the end of each chapter focus on support for the death penalty. A portion of the 2010 GSS survey data is available on the study site as well as on the CD-ROM packaged with this book (if you purchased the SPSS version). You will need this portion of the 2010 GSS to carry out these exercises. You will begin your empirical investigation by thinking a bit about the topic and the data you have available for study.

1. What personal motivation might you have for studying support for the death penalty? What might motivate other people to conduct research on this topic? What policy and academic motives might be important?

2. Open the GSS2010x file containing the 2010 GSS data. In the SPSS menu, click on File, then Open and Data, and then on the name of the data file on the CD-ROM drive, or on the C: drive if GSS2010x was copied there. How many respondents are there in this subset of the complete GSS file? (Scroll down to the bottom of the data set.) How many variables were measured? (Scroll down to the bottom of

the Variable View in SPSS v. 13–19, or click on Utilities, then Variable List in earlier versions.)

3. What would you estimate as the level of support for capital punishment in the United States in 2008? Now for your first real research experience in this text: Describe the distribution of support for capital punishment. Obtaining the relevant data is as simple as "a, b, c, d, e."

   a. Click on Graphs.

   b. Click on Legacy Dialogs > Bar

   c. Select "Simple" and "Summaries for groups of cases" under Data in Chart Area > Define

   d. Place the CAPPUN variable in the box below "Category Axis:" and select "% of cases" under "Bar Represent."

   e. Click OK.

Now describe the distribution of support for capital punishment. What percentage of the population supported capital punishment in the United States in 2010?

Developing a Research Proposal

Will you develop a research proposal in this course? If so, you should begin to consider your alternatives.

1. What topic would you focus on, if you could design a social research project without any concern for costs? What are your motives for studying this topic?

2. Develop four questions that you might investigate about the topic you just selected. Each question should reflect a different research motive: description, exploration, explanation, or evaluation. Be specific.

Which question most interests you? Would you prefer to attempt to answer that question with quantitative or qualitative methods? Do you seek to contribute to basic science or to applied research?

The Process and Problems of Social Research

Domestic violence is a major problem in countries around the world. An international survey by the World Health Organization of 24,000 women in 10 countries estimated lifetime physical or sexual abuse ranging from a low of 15% in Japan to 71% in rural Ethiopia (WHO 2005:6) (Exhibit 2.1). In a United States survey of 16,000 men and women sponsored by the Department of Justice and the Centers for Disease Control and Prevention, 25% of women and 7.6% of men said they had been raped and/or physically assaulted by a current or former spouse, cohabiting partner, or date at some time in their lives (Tjaden & Thoennes 2000:iii). And, most partners seem to get away with the abuse: only one-fifth of all rapes and one-quarter of all physical assaults perpetrated against female respondents by intimates were reported to the police (Tjaden & Thoennes 2000:v).

What can be done to reduce this problem? In 1981, the Police Foundation and the Minneapolis Police Department began an experiment to determine whether arresting accused spouse abusers on the spot would deter repeat incidents. The study's results, which were widely publicized, indicated that arrests did have a deterrent effect. In part because of this, the percentage of urban police departments that made arrest the preferred response to complaints of domestic violence rose from 10% in 1984 to 90% in 1988 (Sherman 1992:14). Researchers in six other cities then conducted similar experiments to determine whether changing the location or other research procedures would result in different outcomes (Sherman 1992; Sherman & Berk 1984). The Minneapolis Domestic Violence Experiment, the studies modeled after it, and the controversies arising from it will provide good examples for our systematic overview of the social research process.

Exhibit 2.1 **International Prevalence of Lifetime Physical and Sexual Violence by an Intimate Partner, Among Ever-Partnered Women, by Site**

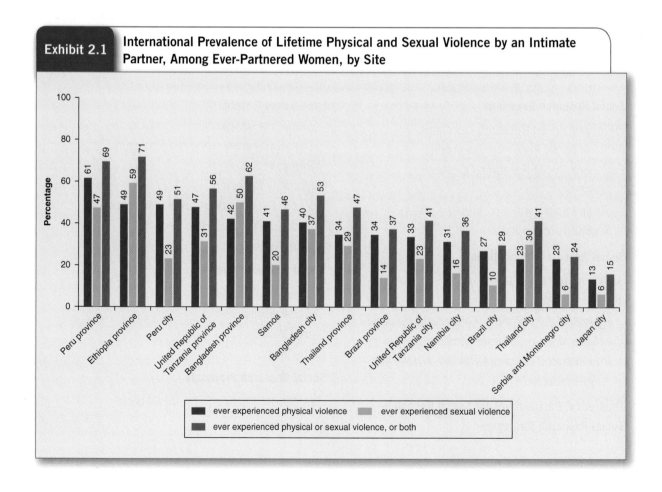

In this chapter, we will examine different social research strategies. We will also consider in some detail the techniques required to begin the research process: formulating research questions, finding information, reviewing prior research, and writing a research proposal. You will also find more details in the end-of-chapter exercises on proposal writing, and in Appendixes A and B about reviewing the literature. I will use the Minneapolis experiment and the related research to illustrate the different research strategies and some of the related techniques. The chapter also explains the role of social theories in developing research questions and guiding research decisions. By the chapter's end, you should be ready to formulate a research question, design a general strategy for answering this question, critique previous studies that addressed this question, and begin a proposal for additional research on the question. You can think of Chapter 1 as having introduced the *why* of social research; Chapter 2 introduces the *how*.

▣ Social Research Questions

A **social research question** is a question about the social world that one seeks to answer through the collection and analysis of firsthand, verifiable, empirical data. It is not a question about who did what to whom, but a question about people in groups, about general social processes, or about tendencies in community change such as the following: What distinguishes Internet users from other persons? Does community policing reduce the crime rate? What influences the likelihood of spouse abuse? How do people react to social isolation? So many research questions are possible that it is more of a challenge to specify what does not qualify as a social research question than to specify what does.

> **Social research question**
> A question about the social world that is answered through the collection and analysis of firsthand, verifiable, empirical data.

But that doesn't mean it is easy to specify a research question. In fact, formulating a good research question can be surprisingly difficult. We can break the process into three stages: (1) identifying one or more questions for study, (2) refining the questions, and then, (3) evaluating the questions.

Identifying Social Research Questions

Social research questions may emerge from your own experience—from your "personal troubles," as C. Wright Mills (1959) put it. One experience might be membership in a church, another could be victimization by crime, and yet another might be moving from a dorm to a sorority house. You may find yourself asking a question such as "In what ways do people tend to benefit from church membership?" "Does victimization change a person's trust in others?" or "How do initiation procedures influence group commitment?" What other possible research questions can you develop based on your own experiences in the social world?

The research literature is often the best source for research questions. Every article or book will bring new questions to mind. Even if you're not feeling too creative when you read the literature, most research articles highlight unresolved issues and end with suggestions for additional research. For example, Lawrence Sherman and Douglas Smith (1992), with their colleagues, concluded an article on some of the replications of the Minneapolis experiment on police responses to spouse abuse by suggesting that "deterrence may be effective for a substantial segment of the offender population. . . . However, the underlying mechanisms remain obscure" (p. 706). A new study could focus on these mechanisms: Why does the arrest of offenders deter some of them from future criminal acts? Any research article in a journal in your field is likely to have comments that point toward unresolved issues.

Many social scientists find the source of their research questions in social theory. Some researchers spend much of their careers conducting research intended to refine an answer to one research question that is critical for a particular social theory. For example, you may have concluded that labeling theory can explain much social deviance, so you may ask whether labeling theory can explain how spouse abusers react to being arrested.

Finally, some research questions have very pragmatic sources. You may focus on a research question someone else posed because it seems to be to your advantage to do so. Some social scientists conduct research on specific questions posed by a funding source in what is termed an RFP, a request for proposals. (Sometimes the acronym RFA is used, meaning request for applications.) The six projects to test the conclusions of the Minneapolis Domestic Violence Experiment were developed in response to such a call for proposals from the National Institute of Justice. Or, you may learn that the social workers in the homeless shelter where you volunteer need help with a survey to learn about client needs, which becomes the basis for another research question.

Refining Social Research Questions

It is even more challenging to focus on a problem of manageable size than it is to come up with an interesting question for research. We are often interested in much more than we can reasonably investigate with limited time and resources. In addition, researchers may worry about staking a research project (and thereby a grant or a grade) on a single problem, and so they may address several research questions at once. Also, it might seem risky to focus on a research question that may lead to results discrepant with our own cherished assumptions about the social world. The prospective commitment of time and effort for some research questions may seem overwhelming, resulting in a certain degree of paralysis.

The best way to avoid these problems is to develop the research question gradually. Don't keep hoping that the perfect research question will just spring forth from your pen. Instead, develop a list of possible research questions as you go along. At the appropriate time, look through this list for the research questions that appear more than once. Narrow your list to the most interesting, most workable candidates. Repeat this process as long as it helps improve your research question.

Evaluating Social Research Questions

In the third stage of selecting a research question, we evaluate the best candidate against the criteria for good social research questions: feasibility, given the time and resources available; social importance; and scientific relevance (King, Keohane, & Verba 1994).

Feasibility

We must be able to conduct any study within the time and resources available. If time is short, questions that involve long-term change may not be feasible. Another issue is to what people or groups we can expect to gain access. Observing social interaction in corporate boardrooms may be taboo. Next, we must consider whether we will have any additional resources, such as research funds or other researchers with whom to collaborate. Remember that there are severe limits on what one person can accomplish. On the other hand, we may be able to piggyback our research onto a larger research project. We also must take into account the constraints we face due to our schedules and other commitments and our skill level.

The Minneapolis Domestic Violence Experiment shows how ambitious a social research question can be when a team of seasoned researchers secures the backing of influential groups. The project required hundreds of thousands of dollars, the collaboration of many social scientists and criminal justice personnel, and the volunteer efforts of 41 Minneapolis police officers. Of course, for this reason, the Sherman and Berk (1984) question would not be a feasible one for a student project. You might instead ask the question "Do students think punishment deters spouse abuse?" Or, perhaps you could work out an arrangement with a local police department to study the question, "How satisfied are police officers with their treatment of domestic violence cases?"

Social Importance

Social research is not a simple undertaking, so it's hard to justify the expenditure of effort and resources unless we focus on a substantive area that is important. Besides, you need to feel motivated to carry out the study. Nonetheless, "importance" is relative, so for a class assignment, student reactions to dormitory rules or something similar might be important enough.

For most research undertakings, we should consider whether the research question is important to other people. Will an answer to the research question make a difference for society or for social relations? Again, the Minneapolis Domestic Violence Experiment is an exemplary case. But the social sciences are not wanting for important research questions. The April 1984 issue of the *American Sociological Review,* which contained the first academic article on the Minneapolis experiment, also included articles reporting research on elections, school tracking, discrimination, work commitment, school grades, organizational change, and homicide. All these articles addressed research questions about important social issues, and all raised new questions for additional research.

Scientific Relevance

Every research question should be grounded in the social science literature. Whether we formulate a research question because we have been stimulated by an academic article or because we want to investigate a current social problem, we should first turn to the social science literature to find out what has already been learned about this question. You can be sure that some prior study is relevant to almost any research question you can think of.

The Minneapolis experiment was built on a substantial body of contradictory theorizing about the impact of punishment on criminality (Sherman & Berk 1984). Deterrence theory predicted that arrest would deter individuals from repeat offenses; labeling theory predicted that arrest would make repeat offenses more likely. Only one prior experimental study of this issue was about juveniles, and studies among adults had yielded inconsistent findings. Clearly, the Minneapolis researchers had good reason for another study. Any new research question should be connected in this way to past research.

回 Social Research Foundations

How do we find prior research on questions of interest? You may already know some of the relevant material from prior coursework or your independent reading, but that won't be enough. When you are about to launch an investigation of a new research question, you must apply a very different standard than when you are studying for a test or just seeking to "learn about domestic violence." You need to find reports of previous investigations that sought to answer the same research question that you wish to answer, not just those that were about a similar topic. If there have been no prior studies of exactly the same research question on which you wish to focus, you should seek to find reports from investigations of very similar research questions. Once you have located reports from prior research similar to the research that you wish to conduct, you may expand your search to include investigations about related topics or studies that used similar methods. You want to be able to explain what your proposed study adds to prior research as well as how it takes into account what has already been learned about your research question.

Although it's most important when you're starting out, reviewing the literature is also important at later stages of the research process. Throughout a research project, you will uncover new issues and encounter unexpected problems; at each of these times, you should search the literature to locate prior research on these issues and to learn how others responded to similar problems. Published research that you ignored when you were seeking to find other research on domestic violence might become very relevant when you have to decide which questions to ask people about their attitudes toward police and other authorities.

Searching the Literature

Conducting a thorough search of the research literature and then reviewing critically what you have found lays an essential foundation for any research project. Fortunately, much of this information can be identified online, without leaving your desktop, and an increasing number of published journal articles can be downloaded directly onto your own computer (depending on your particular access privileges). But, just because there's a lot available online doesn't mean that you need to find it *all*. Keep in mind that your goal is to find relevant reports of prior research investigations. The type of reports you should focus on are those that have been screened for quality through critique by other social scientists prior to publication. Scholarly journals, or *refereed journals* that publish *peer reviewed articles*, manage this review process. Most often, editors of refereed journals send articles that authors submit to them to three or more other social scientists for anonymous review. Based on the reviewers' comments, the journal editor then decides whether to accept or reject the article, or to invite the author to "revise and resubmit." This process results in the rejection of most articles (top journals like the *American Sociological Review* or the *American Journal of Sociology* may reject about 90% of the articles submitted), while those that are ultimately accepted for publication normally have to be revised and resubmitted first. This helps to ensure a much higher-quality standard, although journals vary in the rigor of their review standards, and of course, different reviewers may be impressed by different types of articles; you thus always have to make your own judgment about article quality.

Newspaper and magazine articles may raise important issues or summarize social science research investigations, but they are not an acceptable source for understanding the research literature. The web offers much useful material, including research reports from government and other sources, sites that describe social programs, and even indexes of the published research literature. You may find copies of particular rating scales, reports from research in progress, papers that have been presented at professional conferences, and online discussions of related topics. Web search engines will also find academic journal articles that you can access directly online, although usually for a fee. Most of the published research literature will be available to you online only if you go through the website of your college or university library. The library pays a fee to companies that provide online journals so that you can retrieve this information without paying anything extra yourself. Of course, no library can afford to pay for every journal, so if you can't find a particular issue of a particular journal that you need online, you will have to order the article that you need through interlibrary loan or, if the hard copy of the journal is available, walk over to your library to read it.

As with any part of the research process, your method for searching the literature will affect the quality of your results. Your search method should include the following steps:

Specify your research question. Your research question should be neither so broad that hundreds of articles are judged relevant nor so narrow that you miss important literature. "Is informal social control effective?" is probably too broad. "Does informal social control reduce rates of burglary in large cities?" is probably too narrow. "Is informal social control more effective in reducing crime rates than policing?" provides about the right level of specificity.

Identify appropriate bibliographic databases to search. Sociological Abstracts or SocINDEX may meet many of your needs, but if you are studying a question about social factors in illness, you should also search in Medline, the database for searching the medical literature. If your focus is on mental health, you'll also want to include a search in the online Psychological Abstracts database, PsycINFO, or the version that also contains the full text of articles, PsycARTICLES. Search Criminal Justice Abstracts if your topic is in the area of criminology or criminal justice. You might also find relevant literature in EconLit, which indexes the economic literature, and in ContempWomenIss, which indexes literature on contemporary women's issues. It will save you a lot of time in the long run if you ask a librarian to teach you the best techniques for retrieving the most relevant articles to answer your questions.

To find articles that refer to a previous publication, like Sherman and Berk's study of the police response to domestic violence, the Social Science Citation Index (SSCI) will be helpful. SSCI is an extremely useful tool for tracing the cumulative research in an area across the social sciences. SSCI has a unique "citation searching" feature that allows you to look up articles or books, see who else has cited them in their work, and find out which articles and books have had the biggest impact in a field.

Create a tentative list of search terms. List the parts and subparts of your research question and any related issues that you think are important: "informal social control," "policing," "influences on crime rates," and perhaps "community cohesion and crime." List the authors of relevant studies. Specify the most important journals that deal with your topic.

Narrow your search. The sheer number of references you find can be a problem. For example, searching for "social capital" in March 2011 resulted in 4,863 citations in SocINDEX. Depending on the database you are working with and the purposes of your search, you may want to limit your search to English-language publications, to journal articles rather than conference papers or dissertations (both of which are more difficult to acquire), and to materials published in recent years. If your search yields too many citations, try specifying the search terms more precisely. If you have not found much literature, try using more general terms. Whatever terms you search first, don't consider your search complete until you have tried several different approaches and have seen how many articles you find. A search for "domestic violence" in SocINDEX on March 22, 2011, yielded 7,548 hits; by adding "effects" *or* "influences" as required search terms and limiting the search to peer reviewed articles published since 2000, the number of hits dropped to 368.

Use Boolean search logic. It's often a good idea to narrow down your search by requiring that abstracts contain combinations of words or phrases that include more of the specific details of your research question. Using the Boolean connector *and* allows you to do this, while using the connector *or* allows you to find abstracts containing different words that mean the same thing (see Exhibit 2.2).

Use appropriate subject descriptors. Once you have found an article that you consider to be appropriate, take a look at the "descriptors" field in the citation. You can then redo your search after requiring that the articles be classified with some or all of these descriptor terms.

Check the results. Read the titles and abstracts you have found and identify the articles that appear to be most relevant. If possible, click on these article titles and generate a list of their references. See if you find more articles that are relevant to your research question but that you have missed so far. You will be surprised (I always am) at how many important articles your initial online search missed.

Locate the articles. Whatever database you use, the next step after finding your references is to obtain the articles themselves. You will probably find the full text of many articles available online, but this will be determined by what journals your library subscribes to and the period for which it pays for online access.

The most recent issues of some journals may not be available online. Keep in mind that your library will not have anywhere near all the journals (and books) that you run across in your literature search, so you will have to add another step to your search: checking the "holdings" information.

If an article that appears to be important for your topic isn't available from your own library, nor online, you may be able to request a copy online through your library site or by asking a member of the library staff. You can also check http://worldcat.org to see what other libraries have the journal, and http://publist.com to find out if you can purchase the article.

Be sure to take notes on each article you read, organizing your notes into standard sections: theory, methods, findings, conclusions. In any case, write your review of the literature so that it contributes to your study in some concrete way; don't feel compelled to discuss an article just because you have read it. Be judicious. You are conducting only one study of one issue and it will only obscure the value of your study if you try to relate it to every tangential point in related research.

Don't think of searching the literature as a one-time-only venture—something that you leave behind as you move on to your *real* research. You may encounter new questions or unanticipated problems as you conduct your research or as you burrow deeper into the literature. Searching the literature again to determine what others have found in response to these questions or what steps they have taken to resolve these problems can yield substantial improvements in your own research. There is so much literature on so many topics that it often is not possible to figure out in advance every subject for which you should search the literature, or what type of search will be most beneficial.

Another reason to make searching the literature an ongoing project is that the literature is always growing. During the course of one research study, whether it takes only one semester or several years, new findings will be published and relevant questions will be debated. Staying attuned to the literature and checking it at least when you are writing up your findings may save your study from being outdated as soon as it is finished.

Reviewing Research

Your literature review will suggest specific research questions for further investigation and research methods with which to study those questions. Sherman and Berk (1984) learned from their literature review that there had been little empirical research about the impact of arrest policies in domestic violence cases. What prior research had been conducted did not use very rigorous research designs. There was thus potential value in conducting new research using a rigorous design. Subsequent researchers questioned whether Sherman and Berk's results would be replicated in other cities and whether some of their methods could be improved. When the original results did not replicate, researchers designed more investigations to test explanations for the different findings. In this way, reviewing the literature identifies unanswered questions and contradictory evidence.

Effective review of the prior research is thus an essential step in building the foundation for new research. You must assess carefully the quality of each research study, consider the implications of each article for your own plans, and expand your thinking about your research question to take account of new perspectives and alternative arguments. It is through reviewing the literature and using it to extend and sharpen your own ideas and methods that you become a part of the social science community. Instead of being just one individual studying an issue that interests you, you are building on an ever-growing body of knowledge that is being constructed by the entire community of scholars.

Sometimes you'll find that someone else has already searched the literature on your research question and discussed what they found in a special review article or book chapter. For example, Rosemary Chalk and Joel H. Garner (2001) published an excellent review of the research on arrest and domestic violence in the journal *New Directions for Evaluation.* Most of the research articles that you find will include a short literature review on the specific focus of the research. These reviews can help a lot, but they are no substitute for searching the literature yourself, selecting the articles and other sources that are most pertinent to your research question, and then reviewing what you have found. No one but you can decide what is relevant for your research question and the research circumstances you will be facing—the setting you will study, the timing of your study, the new issues that you want to include in your study, and your specific methods. And, you can't depend on any published research review for information on the most recent works. New research results about many questions appear continually in scholarly journals and books, in research reports from government agencies and other organizations, and on websites all over the world; you'll need to check for new research like this yourself.

Caveat emptor (buyer beware) is the watchword when you search the web, but the published scholarly journal literature can be identified in databases such as Sociological Abstracts, SocINDEX, and Psychological Abstracts. Because these literature databases follow a more standard format and use a careful process to decide what literature to include, they are the sources on which you should focus. This section concentrates on the procedures you should use for reviewing the articles you find in a search of the scholarly literature. These procedures can also be applied to reviews of research monographs—books that provide more information from a research project than that which can be contained in a journal article.

Reviewing the literature is really a two-stage process. In the first stage, you must assess each article separately. This assessment should follow a standard format such as that represented by the "Questions to Ask About a Research Article" in Appendix B. However, you should keep in mind that you can't adequately understand a research study if you just treat it as a series of discrete steps, involving a marriage of convenience among separate techniques. Any research project is an integrated whole, so you must be concerned with how each component of the research design influenced the others—for example, how the measurement approach might have affected the causal validity of the researcher's conclusions and how the sampling strategy might have altered the quality of measures.

The second stage of the review process is to assess the implications of the entire set of articles (and other materials) for the relevant aspects of your research question and procedures, and then to write an integrated review that highlights these implications. Although you can find literature reviews that consist simply of assessments of one published article after another—that never get beyond the first stage in the review process—your understanding of the literature and the quality of your own work will be much improved if you make the effort to write an integrated review.

In the next two sections, I show how you might answer many of the questions in Appendix A as I review a research article about domestic violence. I will then show how the review of a single article can be used within an integrated review of the body of prior research on this research question. Because at this early point in the text you won't be familiar with all the terminology used in the article review, you might want to read through the more elaborate article review in Appendix B later in the course.

A Single-Article Review: Formal and Informal Deterrents to Domestic Violence

Antony Pate and Edwin Hamilton at the national Police Foundation designed one of the studies funded by the U.S. Department of Justice to replicate the Minneapolis Domestic Violence Experiment. In this section, we will examine the article that resulted from that replication, which was published in the *American Sociological Review* (Pate & Hamilton 1992). The numbers in square brackets refer to the article review questions in Appendix A.

The research question. Like Sherman and Berk's (1984) original Minneapolis study, Pate and Hamilton's (1992) Metro-Dade spouse assault experiment sought to test the deterrent effect of arrest in domestic violence cases, but with an additional focus on the role of informal social control [1]. The purpose of the study was explanatory, because the goal was to explain variation in the propensity to commit spouse abuse [2]. Deterrence theory provided the theoretical framework for the study, but this framework was broadened to include the proposition by Williams and Hawkins (1986) that informal sanctions such as stigma and the loss of valued relationships augment the effect of formal sanctions such as arrest [4]. Pate and Hamilton's (1992) literature review referred, appropriately, to the original Sherman and Berk (1984) research, to the other studies that attempted to replicate the original findings, and to research on informal social control [3].

Exhibit 2.2 shows what Pate and Hamilton might have entered on their computer if they searched Sociological Abstracts to find research on "informal social control" and "police" or "arrest."

There is no explicit discussion of ethical guidelines in the article, although reference is made to a more complete unpublished report [6]. Clearly, important ethical issues had to be considered, given the experimental intervention in the police response to serious assaults, but the adherence to standard criminal justice procedures suggests attention to the welfare of victims as well as to the rights of suspects. We will consider these issues in more detail later in this chapter.

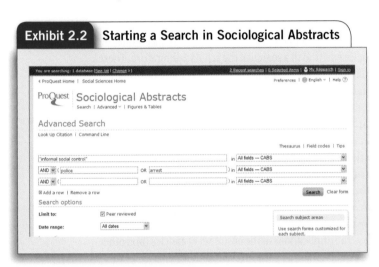

Exhibit 2.2 Starting a Search in Sociological Abstracts

The research design. Developed as a follow-up to the original Minneapolis experiment, the Metro-Dade experiment exemplifies the guidelines for scientific research that I present in Chapter 3 [5]. It was designed systematically, with careful attention to specification of terms and clarification of assumptions, and focused on the possibility of different outcomes rather than certainty about one preferred outcome. The major concepts in the study, formal and informal deterrence, were defined clearly [9] and then measured with straightforward indicators—arrest or nonarrest for formal deterrence and marital status and employment status for informal deterrence. However, the specific measurement procedures for marital and employment status were not discussed, and no attempt was made to determine whether they captured adequately the concept of informal social control [9, 10].

Three hypotheses were stated and also related to the larger theoretical framework and prior research [7]. The study design focused on the behavior of individuals [13] and collected data over time, including records indicating subsequent assault up to 6 months after the initial arrest [14]. The project's experimental design was used appropriately to test for the causal effect of arrest on recidivism [15, 17]. The research project involved all eligible cases, rather than a sample of cases, but there were a number of eligibility criteria that narrow down the ability to generalize these results to the entire population of domestic assault cases in the Metro-Dade area or elsewhere [11]. There is a brief discussion of the 92 eligible cases that were not given the treatment to which they were assigned, but it does not clarify the reasons for the misassignment [15].

The research findings and conclusion. Pate and Hamilton's (1992) analysis of the Metro-Dade experiment was motivated by concern with the effect of social context, because the replications in other cities of the original Minneapolis Domestic Violence Experiment had not had consistent results [19]. Their analysis gave strong support to the expectation that informal social control processes are important: As they had hypothesized, arrest had a deterrent effect on suspects who were employed, but not on those who were unemployed (Exhibit 2.3). However, marital status had no such effect [20]. The subsequent discussion of these findings gives no attention to the implications of the lack of support for the effect of marital status [21], but the study

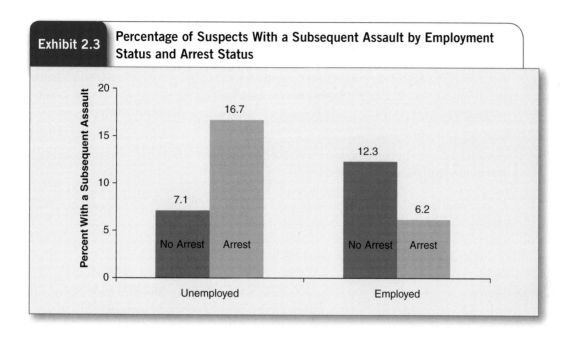

Exhibit 2.3 Percentage of Suspects With a Subsequent Assault by Employment Status and Arrest Status

represents an important improvement over earlier research that had not examined informal sanctions [22]. The need for additional research is highlighted and the importance of the findings for social policy are discussed: Pate and Hamilton suggest that their finding that arrest deters only those who have something to lose (e.g., a job) must be taken into account when policing policies are established [23].

Overall, the Pate and Hamilton (1992) study represents an important contribution to understanding how informal social control processes influence the effectiveness of formal sanctions such as arrest. Although the use of a population of actual spouse assault cases precluded the use of very sophisticated measures of informal social control, the experimental design of the study and the researchers' ability to interpret the results in the context of several other comparable experiments distinguishes this research as exceptionally worthwhile. It is not hard to understand why these studies continue to stimulate further research and ongoing policy discussions.

An Integrated Literature Review: When Does Arrest Matter?

The goal of the second stage of the literature review process is to integrate the results of your separate article reviews and develop an overall assessment of the implications of prior research. The integrated literature review should accomplish three goals: (1) summarize prior research, (2) critique prior research, and (3) present pertinent conclusions (Hart 1998:186–187). I'll discuss each of these goals in turn.

1. *Summarize prior research.* Your summary of prior research must focus on the particular research questions that you will address, but you may also need to provide some more general background. Carolyn Hoyle and Andrew Sanders (2000:14) begin their *British Journal of Criminology* research article about mandatory arrest policies in domestic violence cases with what they term a "provocative" question: What is the point of making it a crime for men to assault their female partners and ex-partners? They then review the different theories and supporting research that has justified different police policies: the "victim choice" position, the "pro-arrest" position, and the "victim empowerment" position. Finally, they review the research on the "controlling behaviors" of men that frames the specific research question on which they focus: how victims view the value of criminal justice interventions in their own cases (Hoyle & Sanders 2000:15).

Ask yourself three questions about your summary of the literature:

a. Have you been selective? If there have been more than a few prior investigations of your research question, you will need to narrow your focus to the most relevant and highest-quality studies. Don't cite a large number of prior articles "just because they are there."

b. Is the research up-to-date? Be sure to include the most recent research, not just the "classic" studies.

c. Have you used direct quotes sparingly? To focus your literature review, you need to express the key points from prior research in your own words. Use direct quotes only when they are essential for making an important point (Pyrczak 2005:51–59).

2. *Critique prior research.* Evaluate the strengths and weaknesses of the prior research. In addition to all the points that you develop as you answer the article review questions in Appendix A, you should also select articles for review that reflect work published in peer-reviewed journals and written by credible authors who have been funded by reputable sources. Consider the following questions as you decide how much weight to give each article:

a. How was the report reviewed prior to its publication or release? Articles published in academic journals go through a rigorous review process, usually involving careful criticism and revision. Top refereed journals may accept only 10% of the submitted articles, so they can be very selective. Dissertations go through a lengthy process of criticism and revision by a few members of the dissertation writer's home institution. A report released directly by a research organization is likely to have had only a limited review, although some research organizations maintain a rigorous internal review process. Papers presented at professional meetings may have had little prior review. Needless to say, more confidence can be placed in research results that have been subject to a more rigorous review.

b. What is the author's reputation? Reports by an author or team of authors who have published other work on the research question should be given somewhat greater credibility at the outset.

c. Who funded and sponsored the research? Major federal funding agencies and private foundations fund only research proposals that have been evaluated carefully and ranked highly by a panel of experts. They also often monitor closely the progress of the research. This does not guarantee that every such project report is good, but it goes a long way toward ensuring some worthwhile products. On the other hand, research that is funded by organizations that have a preference for a particular outcome should be given particularly close scrutiny (Locke, Silverman, & Spirduso 1998:37–44).

3. *Present pertinent conclusions.* Don't leave the reader guessing about the implications of the prior research for your own investigation. Present the conclusions you draw from the research you have reviewed. As you do so, follow several simple guidelines:

a. Distinguish clearly your own opinion of prior research from the conclusions of the authors of the articles you have reviewed.

b. Make it clear when your own approach is based on the theoretical framework that you use and not on the results of prior research.

c. Acknowledge the potential limitations of any empirical research project. Don't emphasize problems in prior research that you can't avoid (Pyrczak 2005:53–56).

d. Explain how the unanswered questions raised by prior research or the limitations of methods used in prior research make it important for you to conduct your own investigation (Fink 2005:190–192).

A good example of how to conclude an integrated literature review is provided by an article based on the replication in Milwaukee of the Minneapolis Domestic Violence Experiment. For this article, Raymond Paternoster et al. (1997) sought to determine whether police officers' use of fair procedures when arresting assault suspects would lessen the rate of subsequent domestic violence. Paternoster et al. (1997) conclude that there has been a major gap in the prior literature: "Even at the end of some seven experiments and millions of dollars, then, there is a great deal of ambiguity surrounding the question of how arrest impacts future spouse assault" (p. 164). Specifically, they note that each of the seven experiments focused on the effect of arrest itself, but ignored the possibility that "particular kinds of police procedure might inhibit the recurrence of spouse assault" (p. 165).

So, Paternoster and his colleagues (1997) ground their new analysis in additional literature on procedural justice and conclude that their new analysis will be "the first study to examine the effect of fairness judgments regarding a punitive criminal sanction (arrest) on serious criminal behavior (assaulting one's partner)" (p. 172).

Theoretical Perspectives for Social Research

As you review the research literature surrounding your research question, you will find that these publications often refer to one or more theories that have guided their research. Of course, you have already learned about social theories in other social science courses, but we need to give special attention to the role of social theory in social research.

Neither domestic violence nor police policies exist in a vacuum, set apart from the rest of the social world. We can understand the particular behaviors and orientations better if we consider how they reflect broader social patterns. Do abusive men keep their wives in positions of subservience? Are community members law abiding? Our answers to general questions such as these will help shape the research questions that we ask and the methods that we use. If we are responsible in our literature reviews, we will give special attention to the general orientation that researchers took to the research question they studied. If we are rigorous in our research methods, we will consider how the general orientation that we favor influences our selection of particular methods.

Although everyone has general notions about "how things work," "what people are like," and so on, social scientists draw on more formal sets of general ideas—social theories—to guide their research (Collins 1994). A **theory** is a logically interrelated set of propositions that helps us make sense of many interrelated phenomena and predict behavior or attitudes that are likely to occur when certain conditions are met. Theory helps social scientists decide which questions are important to ask about the social world and which are just trivial pursuits. Theory focuses a spotlight on the particular features of the social world where we should look to get answers for these questions and suggests other features that can be ignored. Building and evaluating theory is therefore one of the most important objectives of social science.

Lawrence Sherman and Richard Berk's (1984) domestic violence experiment tested predictions derived from **rational choice theory**, which can be viewed as taking a functionalist approach to explaining crime. Rational choice theory assumes that people's behavior is shaped by practical cost-benefit calculations

> **Theory** A logically interrelated set of propositions about empirical reality.

> **Rational choice theory** A social theory that explains individual action with the principle that actors choose actions that maximize their gains from taking that action.

(Coleman 1990:14). *Specific deterrence theory* applies rational choice theory to crime and punishment (Lempert & Sanders 1986:86–87). It states that arresting spouse abusers will lessen their likelihood of reoffending by increasing the costs of reoffending. Crime "doesn't pay" (as much) for these people (see Exhibit 2.4).

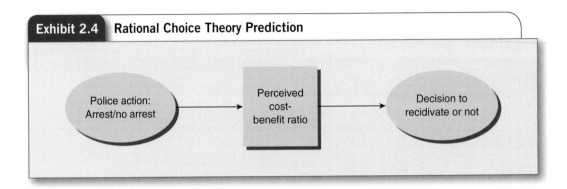

Exhibit 2.4 **Rational Choice Theory Prediction**

Do these concepts interest you? Do these propositions strike you as reasonable ones? If so, you might join a long list of researchers who have attempted to test, extend, and modify various aspects of rational choice theory.

Conflict theory has its origins in the way in which Karl Marx and Friedrich Engels (1961:13–16) explained the social consequences of the Industrial Revolution. They focused on social classes as the key groupings in society and believed that conflict between social classes was not only the norm but also the "engine" of social change.

> **Conflict theory** Identifies conflict between social groups as the primary force in society. Understanding the bases and consequences of the conflict is key to understanding social processes.

Although different versions of conflict theory emphasize different bases for conflict, they focus attention on the conflicting interests of groups rather than on the individuals' concerns with maximizing their self-interest.

Do these concepts strike a responsive chord with you? Can you think of instances when propositions of conflict theory might help explain social change?

Paternoster et al. (1997) concluded that rational choice theory—in particular, specific deterrence theory—did not provide an adequate framework for explaining how citizens respond to arrest. They turned to a type of conflict theory—procedural justice theory—for a very different prediction. *Procedural justice theory* predicts that people will obey the law from a sense of obligation that flows from seeing legal authorities as moral and legitimate (Tyler 1990). From this perspective, individuals who are arrested will be less likely to reoffend if they are treated fairly, irrespective of the outcome of their case, because fair treatment will enhance their view of legal authorities as moral and legitimate. Procedural justice theory expands our view of the punishment process by focusing attention on *how* authorities treat subjects rather than just on *what* decisions they make.

Some sociologists attempt to understand the social world by looking inward, at the meaning people attach to their interactions. They focus on the symbolic nature of social interaction—how social interaction conveys meaning and promotes socialization. Herbert Blumer developed these ideas into **symbolic interaction theory** (Turner, Beeghley, & Powers 1995:460).

> **Symbolic interaction theory** Focuses on the symbolic nature of social interaction—how social interaction conveys meaning and promotes socialization.

Labeling theory uses a symbolic interactionist approach to explain deviance as an "offender's" reaction to the application of rules and sanctions (Becker 1963:9; Scull 1988:678). Sherman and Berk (1984) recognized that a labeling process might influence offenders' responses to arrest in domestic violence cases. Once the offender is labeled as a deviant by undergoing arrest, other people treat the offender as deviant, and he or she is then more likely to act in a way that is consistent with the label *deviant*. Ironically, the act of punishment stimulates more of the very behavior that it was intended to eliminate. This theory suggests that persons arrested for domestic assault are more likely to reoffend than those who are not punished, which is the reverse of the deterrence theory prediction.

Do you find yourself thinking of some interesting research foci when you read about this labeling theory of deviance? If so, consider developing your knowledge of symbolic interaction theory and use it as a guide in your research.

As a social researcher, you may work with one of these theories, seeking to extend it, challenge it, or specify it. You may test alternative implications of the different theories against each other. If you're feeling ambitious, you may even seek to combine some aspects of the different perspectives. Maybe you'll come up with a different theoretical perspective altogether. Or, you may find that you lose sight of the larger picture in the midst of a research project; after all, it is easier to focus on accumulating particular findings rather than considering how those findings fit into a more general understanding of the social world. But you'll find that in any area of research, developing an understanding of relevant theories will help you ask important questions, consider reasonable alternatives, and choose appropriate research procedures.

▣ Social Research Strategies

With a research question formulated, a review of the pertinent literature taking shape, and a theoretical framework in mind, we are ready to consider the process of conducting our research.

When we conduct social research, we are attempting to connect theory with empirical data—the evidence we obtain from the social world. Researchers may make this connection by starting with a social theory and then testing some of its implications with data. This is the process of deductive research; it is most often the strategy used in quantitative methods. Alternatively, researchers may develop a connection between social theory and data by first collecting the data and then developing a theory that explains the patterns in the data (see Exhibit 2.5). This inductive research process is more often the strategy used in qualitative methods. As you'll see, a research project can draw on both deductive and inductive strategies.

Research in the News

INVESTIGATING CHILD ABUSE DOESN'T REDUCE IT

Congress intended the 1974 Child Abuse Prevention and Treatment Act to increase documentation of and thereby reduce the prevalence of child abuse. However, a review of records of 595 high-risk children nationwide from the ages of 4 to 8 found that those children whose families were investigated were not doing any better than those whose families were not investigated—except that mothers in investigated families had more depressive symptoms than mothers in uninvestigated families. Whatever services families were offered after being investigated failed to reduce the risk of future child abuse.

Source: Bakalar, Nicholas. 2010. "Child Abuse Investigations Didn't Reduce Risk, a Study Finds." *The New York Times,* October 12:D3.

Exhibit 2.5   The Links Between Theory and Data

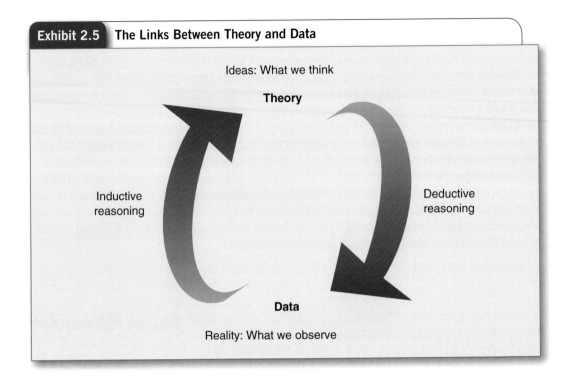

Ideas: What we think

Theory

Inductive reasoning

Deductive reasoning

Data

Reality: What we observe

Social theories do not provide the answers to the questions we pose as topics for research. Instead, social theories suggest the areas on which we should focus and the propositions that we should consider for a test. Exhibit 2.6 summarizes how the two theories that guided Sherman and Berk's (1984) research and the theory that guided Paternoster et al.'s (1997) re-analysis relate to the question of whether or not to arrest spouse abusers. By helping us make such connections, social theory makes us much more sensitive to the possibilities, and thus helps us design better research and draw out the implications of our results. Before, during, and after a research investigation, we need to keep thinking theoretically.

Explanatory Research

The process of conducting research designed to test explanations for social phenomena involves moving from theory to data and then back to theory. This process can be characterized with a **research circle** (Exhibit 2.7).

Deductive Research

As Exhibit 2.7 shows, in **deductive research**, a specific expectation is deduced from a general theoretical premise and then tested with data that have been collected for this purpose. We call the specific expectation deduced from the more general theory a **hypothesis**. It is the hypothesis that researchers actually test, not the complete theory itself. A hypothesis proposes a relationship between two or more **variables**—characteristics or properties that can vary.

Variation in one variable is proposed to predict, influence, or cause variation in the other. The proposed influence is the **independent variable**; its effect

Research circle A diagram of the elements of the research process, including theories, hypotheses, data collection, and data analysis.

Deductive research The type of research in which a specific expectation is deduced from a general premise and is then tested.

Hypothesis A tentative statement about empirical reality, involving a relationship between two or more variables.

Example of a hypothesis: The higher the poverty rate in a community, the higher the percentage of community residents who are homeless.

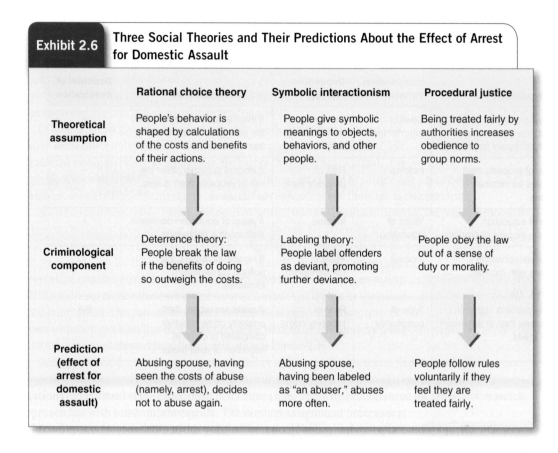

Exhibit 2.6	Three Social Theories and Their Predictions About the Effect of Arrest for Domestic Assault		
	Rational choice theory	**Symbolic interactionism**	**Procedural justice**
Theoretical assumption	People's behavior is shaped by calculations of the costs and benefits of their actions.	People give symbolic meanings to objects, behaviors, and other people.	Being treated fairly by authorities increases obedience to group norms.
Criminological component	Deterrence theory: People break the law if the benefits of doing so outweigh the costs.	Labeling theory: People label offenders as deviant, promoting further deviance.	People obey the law out of a sense of duty or morality.
Prediction (effect of arrest for domestic assault)	Abusing spouse, having seen the costs of abuse (namely, arrest), decides not to abuse again.	Abusing spouse, having been labeled as "an abuser," abuses more often.	People follow rules voluntarily if they feel they are treated fairly.

or consequence is the **dependent variable**. After the researchers formulate one or more hypotheses and develop research procedures, they collect data with which to test the hypothesis.

Hypotheses can be worded in several different ways, and identifying the independent and dependent variables is sometimes difficult. When in doubt, try to rephrase the hypothesis as an *if-then* statement: "*If* the independent variable increases (or decreases), *then* the dependent variable increases (or decreases)." Exhibit 2.8 presents several hypotheses with their independent and dependent variables and their if-then equivalents.

Exhibit 2.8 demonstrates another feature of hypotheses: **direction of association**. When researchers hypothesize that one variable increases as the other variable increases, the direction of association is positive (Hypotheses 1 and 4). When one variable decreases as the other variable decreases, the direction of association is also positive (Hypothesis 3). But when one variable increases as the other decreases, or vice versa, the direction of association is negative, or inverse (Hypothesis 2).

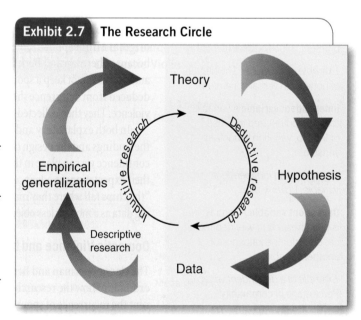

Exhibit 2.7	The Research Circle

Theory

Deductive research

Hypothesis

Data

Descriptive research

Inductive research

Empirical generalizations

follow-up data from several cities in an attempt to explain the discrepant results, thereby starting around the research circle once again (Berk et al. 1992; Pate & Hamilton 1992; Sherman & Smith 1992).

Inductive Research

Inductive research The type of research in which general conclusions are drawn from specific data.

In contrast to deductive research, **inductive research** begins with specific data, which are then used to develop (induce) a general explanation (a theory) to account for the data. One way to think of this process is in terms of the research circle: Rather than starting at the top of the circle with a theory, the inductive researcher starts at the bottom of the circle with data and then develops the theory. Another way to think of this process is represented in Exhibit 2.10. In deductive research, reasoning from specific premises results in a conclusion that a theory is supported, while in inductive research, the identification of similar empirical patterns results in a generalization about some social process.

Inductive reasoning enters into deductive research when we find unexpected patterns in the data we have collected for testing a hypothesis. We may call these patterns **serendipitous findings** or **anomalous findings**. Whether we begin by doing inductive research or add an inductive element later, the result of the inductive process can be new insights and provocative questions. However, the adequacy of an explanation formulated after the fact is necessarily less certain than an explanation presented prior to the collection of data. Every phenomenon can always be explained in *some* way. Inductive explanations are thus more trustworthy if they are tested subsequently with deductive research.

Exhibit 2.10   Deductive and Inductive Reasoning

Deductive

Premise 1:   *All unemployed spouse abusers recidivate.*

Premise 2:   *Joe is an unemployed spouse abuser.*

Conclusion: ***Joe will recidivate.***

Inductive

Evidence 1:   *Joe, an unemployed spouse abuser, recidivated.*

Evidence 2:   *Harold, an unemployed spouse abuser, recidivated.*

Evidence 3:   *George, an employed spouse abuser, didn't recidivate.*

Conclusion: ***All unemployed spouse abusers recidivate.***

Serendipitous or anomalous findings Unexpected patterns in data, which stimulate new ideas or theoretical approaches.

An inductive approach to explaining domestic violence. The domestic violence research took an inductive turn when Sherman and the other researchers began trying to make sense of the differing patterns in the data collected in the different cities. Could systematic differences in the samples or in the implementation of arrest policies explain the differing outcomes? Or was the problem an inadequacy in the theoretical basis of their research? Was deterrence theory really the best way to explain the patterns in the data they were collecting?

As you learned in my review of the Pate and Hamilton (1992) study, the researchers had found that individuals who were married and employed were deterred from repeat offenses by arrest, but individuals who were unmarried and unemployed were actually more likely to commit repeat offenses if they were arrested. What could explain this empirical pattern? The researchers turned to control theory, which predicts that having a "stake in conformity" (resulting from inclusion in social networks at work or in the community) decreases a person's likelihood of committing crimes (Toby 1957). The implication is that people who are employed and married are more likely to be deterred by the threat of arrest than those without such stakes in conformity. And this is indeed what the data revealed.

Now, the researchers had traversed the research circle almost three times, a process perhaps better described as a spiral (see Exhibit 2.11). The first two times, the researchers had traversed the research circle in a deductive, hypothesis-testing way. They started with theory and then deduced and tested hypotheses. The third time, they were more inductive: They started with empirical generalizations from the data they

Exhibit 2.11 **The Research Spiral: Domestic Violence Experiment**

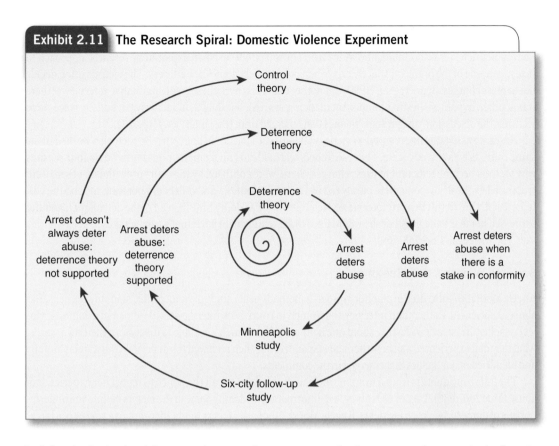

had already obtained and then turned to a new theory to account for the unexpected patterns in the data. At this point, they believed that deterrence theory made correct predictions, given certain conditions, and that another theory, control theory, might specify what these conditions were.

This last inductive step in their research made for a more complex, but also conceptually richer, picture of the impact of arrest on domestic violence. The researchers seemed to have come closer to understanding how to inhibit domestic violence. But they cautioned us that their initial question—the research problem—was still not completely answered. Employment status and marital status do not solely measure the strength of social attachments; they are also related to how much people earn and the social standing of victims in court. So, maybe social ties are not really what make arrest an effective deterrent to domestic violence. The real deterrent may be cost-benefit calculations ("If I have a higher income, jail is more costly for me") or perceptions about the actions of authorities ("If I am a married woman, judges will treat my complaint more seriously"). Additional research was needed (Berk et al. 1992).

Exploratory Research

Qualitative research is often exploratory and, hence, inductive: The researchers begin by observing social interaction or interviewing social actors in depth and then developing an explanation for what has been found. The researchers often ask questions such as "What is going on here?" "How do people interpret these experiences?" or "Why do people do what they do?" Rather than testing a hypothesis, the researchers are trying to make sense of some social phenomenon. They may even put off formulating a research question until after they begin to collect data—the idea is to let the question emerge from the situation itself (Brewer & Hunter 1989:54–58).

Victim Responses to Police Intervention

Lauren Bennett, Lisa Goodman, and Mary Ann Dutton (1999) used an exploratory research approach to investigate one of the problems that emerge when police arrest domestic batterers: The victims often decide not to press charges. Bennett et al. (1999) did not set out to test hypotheses with qualitative interviews (there was another, hypothesis-testing component in their research), but sought, inductively, to "add the voice of the victim to the discussion" and present "themes that emerged from [the] interviews" (p. 762).

Research assistants interviewed 49 victims of domestic violence in one court; Bennett also worked in the same court as a victim advocate. The researchers were able to cull from their qualitative data four reasons why victims became reluctant to press charges. Some were confused by the court procedures, others were frustrated by the delay, some were paralyzed by fear of retribution, and others did not want to send the batterer to jail. One victim Bennett interviewed felt that she "was doing time instead of the defendant"; another expressed her fear by saying that she would like "to keep him out of jail if that's what it takes to keep my kids safe" (Bennett et al. 1999:768–769).

Battered Women's Help Seeking

Angela Moe (2007) also used exploratory research methods in her study of women's decisions to seek help after abuse experiences. Rather than interviewing women in court, Moe interviewed 19 women in a domestic violence shelter. In interviews lasting about one hour each, the women were able to discuss, in their own words, what they had experienced and how they had responded. She then reviewed the interview transcripts carefully and identified major themes that emerged in the comments.

The following quote is from a woman who had decided not to call the police to report her experience of abuse (Moe 2007:686). We can use this type of information to identify some of the factors behind the underreporting of domestic violence incidents. Angela Moe or other researchers might then design a survey of a larger sample to determine how frequently each basis for underreporting occurs.

> I tried the last time to call the police and he ripped both the phones out of the walls. . . .
> That time he sat on my upper body and had his thumbs in my eyes and he was just squeezing.
> He was going, "I'll gouge your eyes out. I'll break every bone in your body. Even if they do find you alive, you won't know to tell them who did it to you because you'll be in intensive care for so long you'll forget." (Terri)

Both the Moe (2007) and Bennett et al. (1999) examples illustrate how the research questions that serve as starting points for qualitative data analyses do not simply emerge from the setting studied, but are shaped by the investigator. As Harry Wolcott (1995) explains,

> [The research question] is not embedded within the lives of those whom we study, demurely waiting to be discovered. Quite the opposite: *We instigate the problems we investigate.* There is no point in simply sitting by, passively waiting to see what a setting is going to "tell" us or hoping a problem will "emerge." (p. 156)

My focus on the importance of the research question as a tool for guiding qualitative data analyses should not obscure the iterative nature of the analytic process. The research question can change, narrow, expand, or multiply throughout the processes of data collection and analysis.

Explanations developed inductively from qualitative research can feel authentic because we have heard what people have to say in their own words, and we have tried to see the social world as they see it. Explanations derived from qualitative research will be richer and more finely textured than they often are in quantitative research, but they are likely to be based on fewer cases from a limited area. We cannot assume that the people studied in this setting are like others, or that other researchers will develop explanations similar to ours to make sense of what was observed or heard. Because we do not initially set up a test of a hypothesis according to some specific rules, another researcher cannot come along and conduct the same test.

Descriptive Research

You learned in Chapter 1 that some social research is purely descriptive. Such research does not involve connecting theory and data, but it is still a part of the research circle—it begins with data and proceeds only to the stage of making empirical generalizations based on those data (refer to Exhibit 2.7).

Valid description is important in its own right—in fact, it is a necessary component of all investigations. Before they began an investigation of differences in arrests for domestic violence in states with and without mandatory arrest laws, David Hirschel, Eve Buzawa, April Pattavina, and Don Faggiani (2008) carefully described the characteristics of incidents reported to the police (see Exhibit 2.12). Describing the prevalence of intimate partner violence is an important first step for societies that seek to respond to this problem (refer to Exhibit 2.1). Government agencies and nonprofit organizations frequently sponsor research that is primarily descriptive: How many poor people live in this community? Is the health of the elderly improving? How frequently do convicted criminals return to crime? Simply put, good description of data is the cornerstone of the scientific research process and an essential component for understanding the social world.

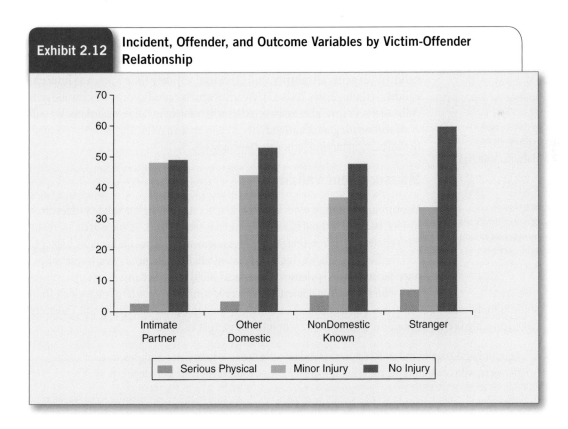

Exhibit 2.12 Incident, Offender, and Outcome Variables by Victim-Offender Relationship

Good descriptive research can also stimulate more ambitious deductive and inductive research. The Minneapolis Domestic Violence Experiment was motivated, in part, by a growing body of descriptive research indicating that spouse abuse is very common: 572,000 cases of women victimized by a violent partner each year; 1.5 million women (and 500,000 men) requiring medical attention each year due to a domestic assault (Buzawa & Buzawa 1996:1–3).

▣ Social Research Goals

> **Validity** The state that exists when statements or conclusions about empirical reality are correct.

Social science research can improve our understanding of empirical reality—the reality we encounter firsthand. We have achieved the goal of **validity** when our conclusions about this empirical reality are correct. I look out my window and observe that it is raining—a valid observation, if my eyes and ears are to be trusted. I pick up the newspaper and read that the rate of violence may be climbing after several years of decline. I am less certain of the validity of this statement, based as it is on an interpretation of some trends in crime indicators obtained through some process that isn't explained. As you learned in this chapter, many social scientists who have studied the police response to domestic violence came to the conclusion that arrest deters violence—that there is a valid connection between this prediction of rational choice theory and the data obtained in research about these processes.

If validity sounds desirable to you, you're a good candidate for becoming a social scientist. If you recognize that validity is often a difficult goal to achieve, you may be tough enough for social research. In any case, the goal of social science is not to come up with conclusions that people will like or conclusions that suit our own personal preferences. The goal is to figure out how and why the social world—some aspect of it—operates as it does. In *Investigating the Social World,* we are concerned with three aspects of validity: (1) **measurement validity**, (2) **generalizability**, and (3) **causal validity** (also known as **internal validity**) (Hammersley 2008:43). We will learn that invalid measures, invalid generalizations, or invalid causal inferences will result in invalid conclusions. We will also focus on the goal of **authenticity**, a concern with reflecting fairly the perspectives of participants in a setting that we study.

> **Measurement validity** Exists when a measure measures what we think it measures.
>
> **Generalizability** Exists when a conclusion holds true for the population, group, setting, or event that we say it does, given the conditions that we specify.
>
> **Causal validity (internal validity)** Exists when a conclusion that A leads to or results in B is correct.
>
> **Authenticity** When the understanding of a social process or social setting is one that reflects fairly the various perspectives of participants in that setting.

Measurement Validity

Measurement validity is our first concern in establishing the validity of research results, because without having measured what we think we measured, we really don't know what we're talking about. Measurement validity is the focus of Chapter 4.

The first step in achieving measurement validity is to specify clearly what it is we intend to measure. Tjaden and Thoennes (2000) identified this as one of the problems with research on domestic violence: "definitions of the term vary widely from study to study, making comparisons difficult" (p. 5). In order to avoid this problem, Tjaden and Thoennes (2000) presented a clear definition of what they meant by intimate partner violence:

> rape, physical assault, and stalking perpetrated by current and former dates, spouses, and cohabiting partners, with cohabiting meaning living together at least some of the time as a couple. (p. 5)

They also provided a measure of each type of violence. For example, "'physical assault' is defined as behaviors that threaten, attempt, or actually inflict physical harm" (Tjaden & Thoennes 2000:5).

With this definition in mind, Tjaden and Thoennes (2000:6) then specified the set of questions they would use to measure intimate partner violence (the questions pertaining to physical assault):

> Not counting any incidents you have already mentioned, after you became an adult, did any other adult, male or female, ever:

—Throw something at you that could hurt?
—Push, grab, or shove you?
—Pull your hair?
—Slap or hit you?
—Kick or bite you?
—Choke or attempt to drown you?
—Hit you with some object?
—Beat you up?
—Threaten you with a gun?
—Threaten you with a knife or other weapon?
—Use a gun on you?
—Use a knife or other weapon on you?

Do you believe that answers to these questions provide a valid measure of having been physical assaulted? Do you worry that some survey respondents might not report all the assaults they have experienced? Might some respondents make up some incidents? Issues like these must be considered when we evaluate measurement validity. Suffice it to say that we must be very careful in designing our measures and in subsequently evaluating how well they have performed. Chapter 4 introduces several different ways to test measurement validity. We cannot just *assume* that measures are valid.

Generalizability

The generalizability of a study is the extent to which it can be used to inform us about persons, places, or events that were not studied. Generalizability is the focus of Chapter 5.

You have already learned in this chapter that Sherman and Berk's findings in Minneapolis about the police response to domestic violence simply did not hold up in several other cities: The initial results could not be generalized. As you know, this led to additional research to figure out what accounted for the different patterns in different cities.

If every person or community we study were like every other one, generalizations based on observations of a small number would be valid. But that's not the case. We are on solid ground if we question the generalizability of statements about research based on the results of a restricted sample of the population or in just one community or other social context.

Generalizability has two aspects. **Sample generalizability** refers to the ability to generalize from a sample, or subset, of a larger population to that population itself. This is the most common meaning of generalizability. **Cross-population generalizability** refers to the ability to generalize from findings about one group, population, or setting to other groups, populations, or settings (see Exhibit 2.13). Cross-population generalizability can also be referred to as **external validity**. (Some social scientists equate the term *external validity* to *generalizability,* but in this book I restrict its use to the more limited notion of cross-population generalizability.)

Sample generalizability is a key concern in survey research. Political pollsters may study a sample of likely voters, for example, and then generalize their findings to the entire population of likely voters. No one would be interested in the results of political polls if they represented only the relatively tiny sample that actually was surveyed rather than the entire population. The procedures for the National Violence Against Women Survey that Tjaden and Thoennes (2000) relied on were designed to maximize sample generalizability.

Sample generalizability Exists when a conclusion based on a sample, or subset, of a larger population holds true for that population.

Cross-population generalizability (external validity) Exists when findings about one group, population, or setting hold true for other groups, populations, or settings.

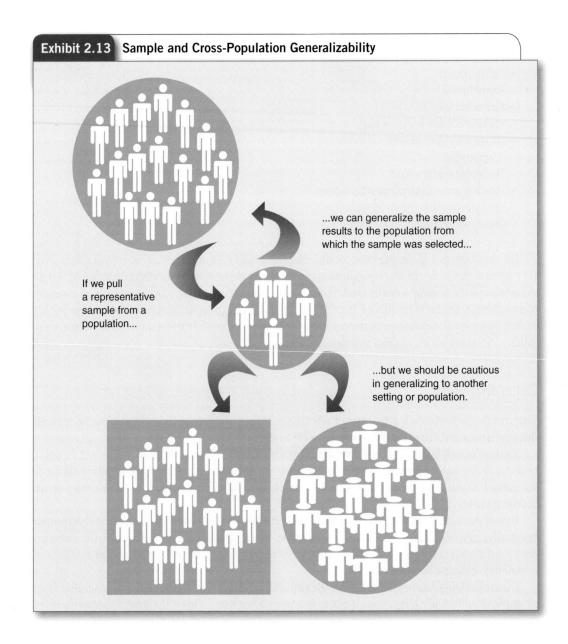

Exhibit 2.13 Sample and Cross-Population Generalizability

If we pull a representative sample from a population...

...we can generalize the sample results to the population from which the sample was selected...

...but we should be cautious in generalizing to another setting or population.

Cross-population generalizability occurs to the extent that the results of a study hold true for multiple populations; these populations may not all have been sampled, or they may be represented as subgroups within the sample studied. This was the problem with Sherman and Berk's (1984) results: persons in Minneapolis who were arrested for domestic violence did not respond in the same way as persons arrested for the same crime in several other cities. The conclusions from Sherman and Berk's (1984) initial research in Minneapolis were not "externally valid."

Generalizability is a key concern in research design. We rarely have the resources to study the entire population that is of interest to us, so we have to select cases to study that will allow our findings to be generalized to the population of interest. Chapter 5 reviews alternative approaches to selecting cases so that findings can be generalized to the population from which the cases were selected. Nonetheless, because we can never be sure that our findings will hold under all conditions, we should be cautious in generalizing to populations or periods that we did not actually sample.

Causal Validity

Causal validity, also known as internal validity, refers to the truthfulness of an assertion that A causes B. It is the focus of Chapter 6.

Most research seeks to determine what causes what, so social scientists frequently must be concerned with causal validity. Sherman and Berk (1984) were concerned with the effect of arrest on the likelihood of recidivism by people accused of domestic violence. To test their causal hypothesis, they designed their experiment so that some accused persons were arrested and others were not. Of course, it may seem heavy-handed for social scientists to influence police actions for the purpose of a research project, but this step reflects just how difficult it can be to establish causally valid understandings about the social world. It was only because police officials did not know whether arrest caused spouse abusers to reduce their level of abuse that they were willing to allow an experiment to test the effect of different policies. David Hirschel and his collaborators (2008) used a different approach to investigate the effect of mandatory arrest laws on police decisions to arrest: They compared the rate of arrest for domestic violence incidents in jurisdictions with and without mandatory arrest laws.

Which of these two research designs gives you more confidence in the causal validity of the conclusions? Chapter 6 will give you much more understanding of how some features of a research design can help us evaluate causal propositions. However, you will also learn that the solutions are neither easy nor perfect: We always have to consider critically the validity of causal statements that we hear or read.

Authenticity

The goal of authenticity is stressed by researchers who focus attention on the subjective dimension of the social world. An authentic understanding of a social process or social setting is one that reflects fairly the various perspectives of participants in that setting (Gubrium & Holstein 1997). Authenticity is one of several different standards proposed by some as uniquely suited to qualitative research; it reflects a belief that those who study the social world should focus first and foremost on how participants view that social world, not on developing a unique social scientists' interpretation of that world. Rather than expecting social scientists to be able to provide a valid mirror of reality, this perspective emphasizes the need for recognizing that what is understood by participants as reality is a linguistic and social construction of reality (Kvale 2002:306).

Angela Moe (2007) explained her basis for considering the responses of women she interviewed in the domestic violence shelter to be authentic:

> members of marginalized groups are better positioned than members of socially dominant groups to describe the ways in which the world is organized according to the oppressions they experience. (p. 682)

Moe's (2007) assumption was that "battered women serve as experts of their own lives" (p. 682). Adding to her assessment of authenticity, Moe (2007) found that the women "exhibited a great deal of comfort through their honesty and candor" as they produced "a richly detailed and descriptive set of narratives" (p. 683). You will learn more about how authenticity can be achieved in qualitative methods in Chapter 9.

🔲 Social Research Proposals

Be grateful to those people or groups who require you to write a formal research proposal (as hard as that seems), and be even more grateful to those who give you constructive feedback. Whether your proposal is written for a professor, a thesis committee, an organization seeking practical advice, or a government agency

---

Exhibit 2.14 (Continued)

Checkpoint 3

Alternatives:

- Continue as planned.
- Modify the plan.
- Stop. Abandon the plan.

Data Analysis (Chapters 10, 14)

18. Choosing an analytic approach:

- Statistics and graphs for describing data
- Identifying relationships between variables
- Deciding about statistical controls
- Testing for interaction effects
- Evaluating inferences from sample data to the population
- Developing a qualitative analysis approach

Checkpoint 4

Alternatives:

- Continue as planned.
- Modify the plan.
- Stop. Abandon the plan.

Reporting Research (Chapter 15)

19. Clarifying research goals and prior research findings

20. Identifying the intended audience

21. Searching the literature and the web

22. Organizing the text

23. Reviewing research limitations

Checkpoint 5

Alternatives:

- Continue as planned.
- Modify the plan.
- Stop. Abandon the plan.

Institute of Mental Health (NIMH) with colleagues from the University of Massachusetts Medical School. The Research Plan is limited by NIH guidelines to 25 pages. It must be preceded by an abstract (which I have excerpted), a proposed budget, biographical sketches of project personnel, and a discussion of the available resources for the project. Appendixes may include research instruments, prior publications by the authors, and findings from related work.

As you can see from the excerpts, our proposal (Schutt et al. 1992b) was to study the efficacy of a particular treatment approach for homeless mentally ill persons who abuse substances. The proposal included a procedure for recruiting subjects in two cities, randomly assigning half the subjects to a recently developed treatment program and measuring a range of outcomes. The NIMH review committee (composed of social scientists with expertise in substance abuse treatment programs and related methodological areas) approved the project for funding but did not rate it highly enough so that it actually was awarded funds (it often takes several resubmissions before even a worthwhile proposal is funded). The committee members recognized the proposal's strengths but also identified several problems that they believed had to be overcome before the proposal could

Exhibit 2.15 A Grant Proposal to the National Institute of Mental Health

Relapse Prevention for Homeless Dually Diagnosed
Abstract

This project will test the efficacy of shelter-based treatment that integrates Psychosocial Rehabilitation with Relapse Prevention techniques adapted for homeless mentally ill persons who abuse substances. Two hundred and fifty homeless persons, meeting . . . criteria for substance abuse and severe and persistent mental disorder, will be recruited from two shelters and then randomly assigned to either an experimental treatment condition . . . or to a control condition.

For one year, at the rate of three two-hour sessions per week, the treatment group ($n = 125$) will participate for the first six months in "enhanced" Psychosocial Rehabilitation . . . , followed by six months of Relapse Prevention training. . . . The control group will participate in a Standard Treatment condition (currently comprised of a twelve-step peer-help program along with counseling offered at all shelters). . . .

Outcome measures include substance abuse, housing placement and residential stability, social support, service utilization, level of distress. . . . The integrity of the experimental design will be monitored through a process analysis. Tests for the hypothesized treatment effects . . . will be supplemented with analyses to evaluate the direct and indirect effects of subject characteristics and to identify interactions between subject characteristics and treatment condition. . . .

Research Plan

1. **Specific Aims**

   The research demonstration project will determine whether an integrated clinical shelter-based treatment intervention can improve health and well-being among homeless persons who abuse alcohol and/or drugs and who are seriously and persistently ill—the so-called "dually diagnosed." . . . We aim to identify the specific attitudes and behaviors that are most affected by the integrated psychosocial rehabilitation/relapse prevention treatment, and thus to help guide future service interventions.

2. **Background and Significance**

   Relapse is the most common outcome in treating the chronically mentally ill, including the homeless. . . . Reviews of the clinical and empirical literature published to date indicate that treatment interventions based on social learning experiences are associated with more favorable outcomes than treatment interventions based on more traditional forms of psychotherapy and/or chemotherapy. . . . However, few tests of the efficacy of such interventions have been reported for homeless samples.

3. **Progress Report/Preliminary Studies**

   Four areas of Dr. Schutt's research help to lay the foundation for the research demonstration project here proposed. . . . The 1990 survey in Boston shelters measured substance abuse with selected ASI [Addiction Severity Index] questions. . . . About half of the respondents evidenced a substance abuse problem.

   Just over one-quarter of respondents had ever been treated for a mental health problem. . . . At least three-quarters were interested in help with each of the problems mentioned other than substance abuse. Since help with benefits, housing, and AIDS prevention will each be provided to all study participants in the proposed research demonstration project, we project that this should increase the rate of participation and retention in the study. . . . Results [from co-investigator Dr. Walter Penk's research] . . . indicate that trainers were more successful in engaging the dually diagnosed in Relapse Prevention techniques. . . .

4. **Research Design and Methods**

   Study Sample.

   Recruitment. The study will recruit 350 clients beginning in month 4 of the study and running through month 28 for study entry. The span of treatment is 12 months and is followed by 12 months of follow-up. . . .

(Continued)

Exhibit 2.15 **(Continued)**

Study Criteria.

Those volunteering to participate will be screened and declared eligible for the study based upon the following characteristics:

1. Determination that subject is homeless using criteria operationally defined by one of the accepted definitions summarized by . . .

Attrition.

Subject enrollment, treatment engagement, and subject retention each represent potentially significant challenges to study integrity and have been given special attention in all phases of the project. Techniques have been developed to address engagement and retention and are described in detail below. . . .

Research Procedures.

All clients referred to the participating shelters will be screened for basic study criteria. . . . Once assessment is completed, subjects who volunteer are then randomly assigned to one of two treatment conditions—RPST or Standard Treatment. . . .

Research Variables and Measures.

Measures for this study . . . are of three kinds: subject selection measures, process measures, and outcome measures. . . .

5. **Human Subjects**

Potential risks to subjects are minor. . . . Acute problems identified . . . can be quickly referred to appropriate interventions. Participation in the project is voluntary, and all subjects retain the option to withdraw . . . at any time, without any impact on their access to shelter care or services regularly offered by the shelters. Confidentiality of subjects is guaranteed. . . . [They have] . . . an opportunity to learn new ways of dealing with symptoms of substance abuse and mental illness.

be funded. The problems were primarily methodological, stemming from the difficulties associated with providing services to, and conducting research on, this particular segment of the homeless population.

The proposal has many strengths, including the specially tailored intervention derived from psychiatric rehabilitation technology developed by Liberman and his associates and relapse prevention methods adapted from Marlatt. This fully documented treatment . . . greatly facilitates the generalizability and transportability of study findings. . . . The investigative team is excellent . . . also attuned to the difficulties entailed in studying this target group. . . . While these strengths recommend the proposal . . . eligibility criteria for inclusion of subjects in the study are somewhat ambiguous. . . . This volunteer procedure could substantially underrepresent important components of the shelter population. . . . The projected time frame for recruiting subjects . . . also seems unrealistic for a three-year effort. . . . Several factors in the research design seem to mitigate against maximum participation and retention. (Services Research Review Committee 1992:3–4)

If you get the impression that researchers cannot afford to leave any stone unturned in working through the procedures in an NIMH proposal, you are right. It is very difficult to convince a government agency that a research project is worth spending a lot of money on (we requested about $2 million). And that is as it should be: Your tax dollars should be used only for research that has a high likelihood of yielding findings that are valid and useful. But even when you are proposing a smaller project to a more generous funding source—or just presenting a proposal to your professor—you should scrutinize the proposal carefully before submission and ask others to comment on it. Other people will often think of issues you neglected to consider, and you should allow yourself time to think about these issues and to reread and redraft the proposal. Also, you will get no credit for having thrown together a proposal as best you could in the face of an impossible submission deadline.

Let's review the issues identified in Exhibit 2.14 as they relate to the NIMH relapse prevention proposal. The research question concerned the effectiveness of a particular type of substance abuse treatment in a shelter for homeless persons—an evaluation research question [Question 1]. This problem certainly was suitable for social research, and it was one that could have been handled for the money we requested [2]. Prior research demonstrated clearly that our proposed treatment had potential and also that it had not previously been tried with homeless persons [3]. The treatment approach was connected to psychosocial rehabilitation theory [4] and, given prior work in this area, a deductive, hypothesis-testing stance was called for [5]. Our review of research guidelines continued up to the point of submission, and we felt that our proposal took each into account [6]. So it seemed reasonable to continue to develop the proposal (Checkpoint 1).

Measures were to include direct questions, observations by field researchers, and laboratory tests (of substance abuse) [7]. The proposal's primary weakness was in the area of generalizability [8]. We proposed to sample persons in only two homeless shelters in two cities, and we could offer only weak incentives to encourage potential participants to start and stay in the study. The review committee believed that these procedures might result in an unrepresentative group of initial volunteers beginning the treatment and perhaps an even less representative group continuing through the entire program. The problem was well suited to an experimental design [9] and was best addressed with longitudinal data [10], involving individuals [11]. Our design controlled for many *sources of invalidity,* but several sources of causal invalidity remained [12]. Clearly, we should have modified the proposal with some additional recruitment and retention strategies—although it may be that the research could not actually be carried out without some major modification of the research question (Checkpoint 2).

A randomized experimental design was preferable because this was to be a treatment-outcome study, but we did include a field research component so that we could evaluate treatment implementation [13, 14]. Because the effectiveness of our proposed treatment strategy had not been studied before among homeless persons, we could not propose doing a secondary data analysis or meta-analysis [15]. We sought only to investigate causation from a *nomothetic* perspective, without attempting to show how the particular experiences of each participant may have led to their outcome [6]. Because participation in the study was to be voluntary and everyone received *something* for participation, the research design seemed ethical—and it was approved by the University of Massachusetts Medical School's **institutional review board (IRB)** and by the state mental health agency's human subjects committee [17]. We planned several statistical tests, but the review committee remarked that we should have been more specific on this point [18]. Our goal was to use our research as the basis for several academic articles, and we expected that the funding agency would also require us to prepare a report for general distribution [19, 20]. We had reviewed the research literature carefully [21], but as is typical in most research proposals, we did not develop our research reporting plans any further [22, 23].

> **Institutional Review Board (IRB)**
> A group of organizational and community representatives required by federal law to review the ethical issues in all proposed research that is federally funded, involves human subjects, or has any potential for harm to subjects.

If your research proposal will be reviewed competitively, it must present a compelling rationale for funding. It is not possible to overstate the importance of the research problem that you propose to study (see the first section of this chapter). If you propose to test a hypothesis, be sure that it is one for which there are plausible alternatives. You want to avoid focusing on a "boring hypothesis"—one that has no credible alternatives, even though it is likely to be correct (Dawes 1995:93).

A research proposal also can be strengthened considerably by presenting results from a pilot study of the research question. This might have involved administering the proposed questionnaire to a small sample, conducting a preliminary version of the proposed experiment with a group of students, or making observations over a limited period of time in a setting like that proposed for a qualitative study. Careful presentation of the methods used in the pilot study and the problems that were encountered will impress anyone who reviews the proposal.

Don't neglect the procedures for the protection of human subjects. Even before you begin to develop your proposal, you should find out what procedure your university's IRB requires for the review of student research proposals. Follow these procedures carefully, even if they require that you submit your proposal for an IRB review. No matter what your university's specific requirements are, if your research involves human subjects, you will need to include in your proposal a detailed statement that describes how you will adhere to these requirements. I discuss the key issues in the next chapter.

You have learned in this chapter how to formulate a research question, review relevant literature, consider social theory, and identify some possible limitations, so you are now ready to begin proposing new research. If you plan to do so, you can use the proposal exercises at the end of each of the subsequent chapters to incorporate more systematically the research elements discussed in those chapters. By the book's end, in Chapter 15, you will have attained a much firmer grasp of the various research decisions outlined in Exhibit 2.14.

Conclusions

Selecting a worthy research question does not guarantee a worthwhile research project. The simplicity of the research circle presented in this chapter belies the complexity of the social research process. In the following chapters, I focus on particular aspects of the research process. Chapter 4 examines the interrelated processes of conceptualization and measurement, arguably the most important part of research. Measurement validity is the foundation for the other two aspects of validity. Chapter 5 reviews the meaning of generalizability and the sampling strategies that help us achieve this goal. Chapter 6 introduces causal validity and illustrates different methods for achieving it. Most of the remaining chapters then introduce different approaches to data collection—experiments, surveys, participant observation and intensive interviewing, evaluation research, comparative historical research, secondary data analysis, and content analysis—that help us, in different ways, achieve results that are valid.

Of course, our answers to research questions will never be complete or entirely certain. We always need to ground our research plans and results in the literature about related research. Our approach should be guided by explicit consideration of a larger theoretical framework. When we complete a research project, we should evaluate the confidence that can be placed in our conclusions, point out how the research could be extended, and consider the implications for social theory. Recall how the elaboration of knowledge about deterrence of domestic violence required sensitivity to research difficulties, careful weighing of the evidence, identification of unanswered questions, and consideration of alternative theories.

Owning a large social science toolkit is no guarantee for making the right decisions about which tools to use and how to use them in the investigation of particular research problems, but you are now forewarned about, and thus hopefully forearmed against, some of the problems that social scientists face in their work. I hope that you will return often to this chapter as you read the subsequent chapters, when you criticize the research literature and when you design your own research projects. To be conscientious, thoughtful, and responsible—this is the mandate of every social scientist. If you formulate a feasible research problem, ask the right questions in advance, try to adhere to the research guidelines, and steer clear of the most common difficulties, you will be well along the road to fulfilling this mandate.

Key Terms

Anomalous findings 44
Authenticity 48
Causal validity (internal validity) 48
Conflict theory 38
Cross-population generalizability
    (external validity) 49
Deductive research 40
Dependent variable 41
Direction of association 41
Empirical generalization 43

External validity (cross-population
    generalizability) 49
Generalizability 48
Hypothesis 40
Independent variable 40
Inductive research 44
Institutional
    Review Board (IRB) 57
Internal validity (causal validity) 48
Measurement validity 48

Rational choice theory 37
Replications 43
Research circle 40
Sample generalizability 49
Serendipitous findings 44
Social research question 27
Symbolic interaction theory 38
Theory 37
Validity 48
Variable 40

Highlights

- Research questions should be feasible (within the time and resources available), socially important, and scientifically relevant.

- Building social theory is a major objective of social science research. Relevant theories should be investigated before starting social research projects, drawing out the theoretical implications of research findings.

- Rational choice theory focuses attention on the rational bases for social exchange and explains most social phenomena in terms of these motives.

- Conflict theory focuses attention on the bases of conflict between social groups and uses these conflicts to explain most social phenomena.

- Symbolic interaction theory focuses attention on the meanings that people attach to and gain from social interaction and explains most social phenomena in terms of these meanings.

- The type of reasoning in most research can be described as primarily deductive or inductive. Research based on deductive reasoning proceeds from general ideas, deduces specific expectations from these ideas, and then tests the ideas with empirical data. Research based on inductive reasoning begins with specific data and then develops general ideas or theories to explain patterns in the data.

- It may be possible to explain unanticipated research findings after the fact, but such explanations have less credibility than those that have been tested with data collected for the purpose of the study.

- The scientific process can be represented as circular, with a path from theory to hypotheses, to data, and then to empirical generalizations. Research investigations may begin at different points along the research circle and traverse different portions of it. Deductive research begins at the point of theory, inductive research begins with data but ends with theory, and descriptive research begins with data and ends with empirical generalizations.

- Replications of a study are essential to establishing its generalizability in other situations. An ongoing line of research stemming from a particular research question should include a series of studies that, collectively, traverse the research circle multiple times.

- Writing a research proposal is an important part of preparing for research. Key decisions can be viewed as checkpoints that will shape subsequent stages.

STUDENT STUDY SITE

To assist in completing the web exercises, please access the study site at **www.sagepub.com/schuttisw7e**, where you will find the web exercises with accompanying links. You'll find other useful study materials such as self-quizzes and e-flashcards for each chapter, along with a group of carefully selected articles from research journals that illustrate the major concepts and techniques presented in the book.

Discussion Questions

1. Pick a social issue about which you think research is needed. Draft three research questions about this issue. Refine one of the questions and evaluate it in terms of the three criteria for good research questions.

2. Identify variables that are relevant to your three research questions. Now formulate three related hypotheses. Which are the independent and which the dependent variables in these hypotheses?

3. If you were to design research about domestic violence, would you prefer an inductive approach or a deductive approach? Explain your preference. What would be the advantages and disadvantages of each approach? Consider in your answer the role of social theory, the value of searching the literature, and the goals of your research.

4. Sherman and Berk's (1984) study of the police response to domestic violence tested a prediction derived from rational choice theory. Propose hypotheses about the response to domestic violence that are consistent with conflict and symbolic interactionist theories. Which theory seems to you to provide the best framework for understanding domestic violence and how to respond to it?

5. Review my description of the research projects in the section "Types of Social Research" in Chapter 1. Can you identify the stages of each project corresponding to the points on the research circle? Did each project include each of the four stages? Which theory (or theories) seem applicable to each of these projects?

6. The research on victim responses to police intervention and the study of conflicts among high school students used an exploratory research approach. Why do you think the researchers adopted this approach in these studies? Do you agree with their decision? Propose a research project that would address issues in one of these studies with a deductive approach.

7. Critique the Sherman and Berk (1984) research on the police response to domestic violence from the standpoint of measurement validity, generalizability, and causal validity. What else would you like to know about this research in order to strengthen your critique? What does consideration of the goal of authenticity add to your critique?

Practice Exercises

1. Pair up with one other student and select one of the research articles available on the book's study site, at www.sagepub.com/schuttisw7e. Evaluate the research article in terms of its research strategy. Be generally negative but not unreasonable in your criticisms. The student with whom you are working should critique the article in the same way but from a generally positive standpoint, defending its quality. Together, write a summary of the study's strong and weak points, or conduct a debate in the class.

2. Research problems posed for explanatory studies must specify variables and hypotheses, which need to be stated properly and need to correctly imply any hypothesized causal relationship. The "Variables and Hypotheses" lessons, found in the Interactive Exercises on the study site, will help you learn how to do this.

3. To use these lessons, choose one of the four sets of "Variables and Hypotheses" exercises from the opening menu. About 10 hypotheses are presented in the lesson. After reading each hypothesis, you must name the dependent and independent variables and state the direction (positive or negative) of the relationship between them. In this Interactive Exercise, you must write in your own answer, so type carefully. The program will evaluate your answers. If an answer is correct, the program will present its version of the correct answer and go on to the next question. If you have made an error, the program will explain the error to you and give you another chance to respond. If your answer is unrecognizable, the program will instruct you to check your spelling and try again.

4. Now choose another article from the Learning From Journal Articles option on the study site. Read one article based on empirical research and diagram the process of research that it reports. Your diagram should have the structure of the research circle in Exhibit 2.7. How well does the process of research in this study seem to match the process symbolized in Exhibit 2.7? How much information is provided about each step in that process?

5. Review the section in this chapter on literature searching. Now choose a topic for investigation and search the social science literature for prior research on this topic. You will

need to know how to use a database such as Sociological Abstracts at your own library as well as how to retrieve articles you locate (those that are available through your library). Try to narrow your search so that most of the articles you find are relevant to your topic (or broaden your search, if you don't find many relevant articles). Report on your search terms and the results of your search with each term or combination of terms.

Ethics Questions

1. Sherman and Berk (1984) and those who replicated their research on the police response to domestic violence assigned persons accused of domestic violence by chance (randomly) to be arrested or not. Their goal was to ensure that the people who were arrested were similar to those who were not arrested. Based on what you now know, do you feel that this random assignment procedure was ethical? Why or why not?

2. Concern with how research results are used is one of the hallmarks of ethical researchers, but deciding what form that concern should take is often difficult. You learned in this chapter about the controversy that occurred after Sherman and Berk (1984) encouraged police departments to adopt a pro-arrest policy in domestic abuse cases, based on findings from their Minneapolis study. Do you agree with the researchers' decision to suggest policy changes to police departments based on their study, in an effort to minimize domestic abuse? Several replication studies failed to confirm the Minneapolis findings. Does this influence your evaluation of what the researchers should have done after the Minneapolis study was completed? What about Larry Sherman's argument that failure to publicize the Omaha study finding of the effectiveness of arrest warrants resulted in some cases of abuse that could have been prevented?

Web Exercises

1. You can brush up on a range of social theorists at www.sociologyprofessor.com. Pick a theorist and read some of what you find. What social phenomena does this theorist focus on? What hypotheses seem consistent with his or her theorizing? Describe a hypothetical research project to test one of these hypotheses.

2. You've been assigned to write a paper on domestic violence and the law. To start, you can review relevant research on the American Bar Association's website at *www.americanbar.org/groups/domestic_violence/resources/statistics.html*. What does the research summarized at this site suggest about the prevalence of domestic violence, its distribution about social groups, and its causes and effects? Write your answers in a one- to two-page report.

SPSS Exercises

1. Formulate four research questions about support for capital punishment—one question per research purpose: (1) exploratory, (2) descriptive, (3) explanatory, and (4) evaluative. You should be able to answer two of these questions with the GSS2010x data. Highlight these two.

2. Now, to develop some foundation from the literature, check the bibliography of this book for the following articles that drew on the GSS: Aguirre and Baker (1993); Barkan and Cohn (1994); Borg (1997, 1998); Warr (1995); and Young (1992). How have social scientists used social theory to explain support for capital punishment? What potential influences on capital punishment have been tested? What influences could you test again with the 2010 GSS?

3. State four hypotheses in which support for capital punishment (CAPPUN) is the dependent variable and another variable in the GSS2010x is the independent variable. Justify each hypothesis in a sentence or two.

4. Test at least one hypothesis. Marian Borg (1997) suggests that region might be expected to influence support for the death

penalty. Test this as follows (after opening the GSS2010x file, as explained in Chapter 1, SPSS Exercise 3):

a. Click on Analyze/Descriptive Statistics/Crosstabs.

b. Highlight CAPPUN and click on the arrow so that it moves into the Rows box; highlight REGION and click on the arrow to move it into the Columns box.

c. Click on Cells, click off Counts-Observed, and click on Percentages-Column.

d. Click Continue and then OK. Inspect the table.

5. Does support for capital punishment vary by region? Scroll down to the percentage table (in which regions appear across the top) and compare the percentages in the Favor row for each region. Describe what you have found.

6. Now you can go on to test your other hypotheses in the same way, if you have the time. Due to space constraints, I can't give you more guidance, but I will warn you that there could be some problems at this point (e.g., if your independent variable has lots of values). Proceed with caution!

Developing a Research Proposal

Now it's time to start writing the proposal. The following exercises pertain to the very critical first steps identified in Exhibit 2.14:

1. State a problem for research (Exhibit 2.14, #1, #2). If you have not already identified a problem for study, or if you need to evaluate whether your research problem is doable, a few suggestions should help get the ball rolling and keep it on course.

    a. Jot down questions that have puzzled you in some area having to do with people and social relations, perhaps questions that have come to mind while reading textbooks or research articles or even while hearing news stories. Don't hesitate to jot down many questions, and don't bore yourself—try to identify questions that really interest you.

    b. Now take stock of your interests, your opportunities, and the work of others. Which of your research questions no longer seem feasible or interesting? What additional research questions come to mind? Pick out a question that is of interest and seems feasible and that your other coursework suggests has been the focus of some prior research or theorizing.

    c. Write out your research question in one sentence, and elaborate on it in one paragraph. List at least three reasons why it is a good research question for you to investigate. Then present your proposal to your classmates and instructor for discussion and feedback.

2. Search the literature (and the web) on the research question you identified (Exhibit 2.14, #3). Refer to the section on searching the literature for more guidance on conducting the search. Copy down at least 10 citations to articles (with abstracts from Sociological Abstracts or Psychological Abstracts) and five websites reporting research that seems highly relevant to your research question; then look up at least five of these articles and three of the sites. Inspect the article bibliographies and the links on the website, and identify at least one more relevant article and website from each source.

    Write a brief description of each article and website you consulted, and evaluate its relevance to your research question. What additions or changes to your thoughts about the research question do the sources suggest?

3. Which general theoretical perspective do you believe is most appropriate to guide your proposed research (Exhibit 2.14, #4)? Write two paragraphs in which you (1) summarize the major tenets of the theoretical perspective you choose and (2) explain the relevance of this perspective to your research problem.

4. Propose at least two hypotheses that pertain to your research question. Justify these hypotheses in terms of the literature you have read.

Research Ethics and Philosophies

L et's begin with a thought experiment (or a trip down memory lane, depending on your earlier exposure to this example). One spring morning as you are drinking coffee and reading the newspaper, you notice a small ad for a psychology experiment at the local university. "Earn money and learn about yourself," it says. Feeling a bit bored with your job as a high school teacher, you call and schedule an evening visit to the lab.

WE WILL PAY YOU $45 FOR ONE HOUR OF YOUR TIME

Persons Needed for a Study of Memory

You arrive at the assigned room at the university, ready for an interesting hour or so, and are impressed immediately by the elegance of the building and the professional appearance of the personnel. In the waiting room, you see a man dressed in a lab technician's coat talking to another visitor—a middle-aged fellow dressed in casual attire. The man in the lab coat turns and introduces himself and explains that as a psychologist, he is interested in the question of whether people learn things better when they are punished for making a mistake. He quickly convinces you that this is a very important question for which there has been no adequate answer; he then explains that his experiment on punishment and learning will help answer this question. Then he announces, "I'm going to ask one of you to be the teacher here tonight and the other one to be the learner."

"The experimenter" [as we'll refer to him from now on] says he will write either *teacher* or *learner* on small identical slips of paper and then asks both of you to draw out one. Yours says teacher.

The experimenter now says, in a matter-of-fact way, "All right. Now the first thing we'll have to do is to set the learner up so that he can get some type of punishment."

He leads you both behind a curtain, sits the learner down, attaches a wire to his left wrist and straps both his arms to the chair so that he cannot remove the wire (see Exhibit 3.1). The wire is connected to a console with 30 switches and a large dial on the other side of the room. When you ask what the wire is for, the experimenter says he will demonstrate. He then asks you to hold the end of the wire, walks back to the control console, flips several switches and focuses his attention on the dial. You hear a clicking noise, see the dial move, and then feel an electric shock in your hand. The shock increases and the dial registers more current when the experimenter flips the next switch on the console.

"Oh, I see," you say. "This is the punishment. Couldn't it cause injury?" The experimenter explains that the machine is calibrated so that it will not cause permanent injury, but acknowledges that when it is turned up all the way it is very, very painful and can result in severe, although momentary, discomfort.

Now you walk back to the other side of the room (so that the learner is behind the curtain) and sit before the console. The experimental procedure has four simple steps: (1) You read aloud a series of word pairs, such as *blue box, nice day, wild duck,* and so on. (2) You read one of the first words from those pairs and a set of

Exhibit 3.1   Learner Strapped in Chair With Electrodes

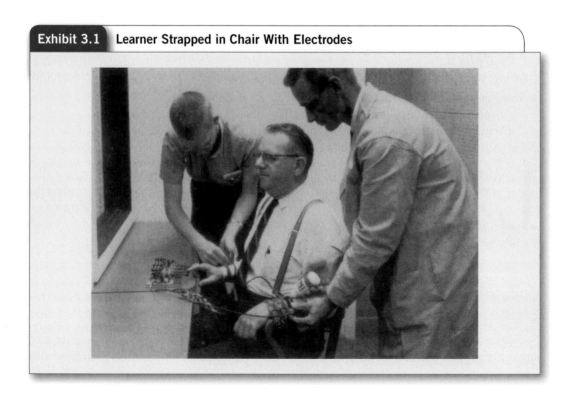

four words, one of which contains the original paired word. For example, you might say, "blue: sky ink box lamp." (3) The learner states the word that he thinks was paired with the first word you read ("blue"). If he gives a correct response, you compliment him and move on to the next word. If he makes a mistake, you flip a switch on the console. This causes the learner to feel a shock on his wrist. (4) After each mistake, you are to flip the next switch on the console, progressing from left to right. You note that there is a label corresponding to every fifth mark on the dial, with the first mark labeled *slight shock,* the fifth mark labeled *moderate shock,* the tenth *strong shock,* and so on through *very strong shock, intense shock, extreme intensity shock,* and *danger: severe shock.*

You begin. The learner at first gives some correct answers, but then he makes a few errors. Soon you are beyond the fifth mark (moderate shock) and are moving in the direction of more and more severe shocks.

You recall having heard about this experiment and so you know that as you turn the dial, the learner's responses increase in intensity from a grunt at the tenth mark (strong shock) to painful groans at higher levels, anguished cries to "get me out of here" at the extreme intensity shock levels, to a deathly silence at the highest level. You also know that as you proceed and indicate your discomfort at administering the stronger shocks, the experimenter will inform you that "The experiment requires that you continue," and occasionally, "It is absolutely essential that you continue." Now, please note on the meter in Exhibit 3.2 the most severe shock that you would agree to give to the learner.

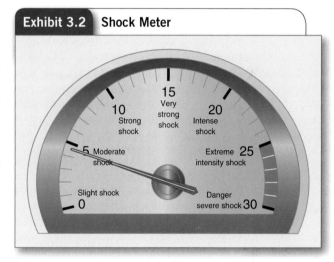

Exhibit 3.2 **Shock Meter**

You may very well recognize that this thought experiment is a slightly simplified version of **Milgram's obedience experiments,** begun at Yale University in 1960. Did you know that Stanley Milgram also surveyed Yale undergraduates and asked them to indicate at what level they would terminate their "shocks"? The average (mean) maximum shock level predicted by the Yale undergraduates was 9.35, corresponding to a strong shock. Only one student predicted that he would provide a stimulus above that level, but only barely so, for he said he would stop at the very strong level. Responses were similar from nonstudent groups who were asked the same question.

What was the actual average level of shock administered by the 40 New Haven adults who volunteered for the experiment? A shock level of 24.53, or a level higher than extreme intensity shock and just short of danger: severe shock. Of Milgram's original 40 subjects, 25 (62.5%) complied with the experimenter's demands, all the way to the top of the scale (originally labeled simply as *XXX*). And lest you pass this result off as simply the result of the subjects having thought that the experiment wasn't "real," we hasten to point out that there is abundant evidence from the subjects' own observed high stress and their subsequent reports that they really believed that the learner was receiving actual, hurtful shocks.

Are you surprised by the subjects' responses? By the Yale undergraduates' predictions of so many compassionate responses? By your own response? (I leave it to you to assess how accurately you predicted the response you would have given if you had been an actual subject.)

Of course, my purpose in introducing this small "experiment" is not to focus attention on the prediction of obedience to authority; instead, I want to introduce the topic of research ethics by encouraging you to think about research from the standpoint of the people who are the subjects of behavioral research. I will refer to Stanley Milgram's (1963) famous research on obedience throughout this chapter, since it is fair to say that this research ultimately had as profound an influence on the way that social scientists think about research ethics as it had on the way that they understand obedience to authority.

Every social scientist needs to consider how to practice their discipline ethically. Whenever we interact with other people as social scientists, we must give paramount importance to the rational concerns and emotional needs that will shape their responses to our actions. It is here that ethical research practice begins, with the recognition that our research procedures involve people who deserve as much respect for their well-being as we do for ours.

Historical Background

Concern with ethical practice in relation to people who are in some respect dependent, whether as patients or research subjects, is not a new idea. Ethical guidelines for medicine trace back to Hippocrates in 5 BC Greece (World Medical Association 2009:11), and the American Medical Association (AMA) adopted the world's first formal professional ethics code in medicine in 1847 (AMA, 2011). Current AMA ethical principles include respecting patient rights, maintaining confidentiality, and regarding "responsibility to the patient as paramount" (AMA, 2011). Yet the history of medical practice makes it clear that having an ethics code is not sufficient to ensure ethical practice, at least when there are clear incentives to do otherwise.

A defining event occurred in 1946, when the **Nuremberg War Crime Trials** exposed horrific medical experiments conducted by Nazi doctors and others in the name of "science." However, as late as 1972, Americans learned from news reports that researchers funded by the U.S. Public Health Service had followed 399 low-income African American men with syphilis (and some without the disease) since the 1930s, collecting data to study the "natural" course of the illness (Exhibit 3.3). At the time, there was no effective treatment for the disease, but the men were told they were being treated for "bad blood," whether they had syphilis or not. Participants received free medical exams, meals, and burial insurance, but were not asked for their consent to be studied. What made this research study, known as the Tuskegee Syphilis Experiment, so shocking was that many participants were not informed of their illness and, even after penicillin was recognized as an effective treatment in 1945 and in large-scale use by 1947, the study participants were not treated. The research was only ended after the study was exposed. In 1973, congressional hearings began and in 1974 an out-of-court settlement of $10 million was reached; it was not until 1997 that President Bill Clinton made an official apology (CDC 2009).

Research in the News

SYPHILIS EXPERIMENTS IN GUATEMALA

The U.S. government has asked the Institute of Medicine and an international panel of specialists to investigate an unethical experiment in Guatemala in the 1940s. In the experiment, U.S. scientists, with the support of Guatemalan health authorities, infected hundreds of Guatemalans with syphilis and gonorrhea, without their consent, in order to test the efficacy of new treatments. At least one infected person died, although most were treated.

Source: Smith, Stephen. 2010. "Wellesley Professor Unearths a Horror: Syphilis Experiments in Guatemala." *The New York Times,* October 2:A1, A7.

In the News

- Research should cause no harm to subjects.

- Participation in research should be voluntary, and therefore subjects must give their informed consent to participate in the research.

- Researchers should fully disclose their identity.

- Anonymity or confidentiality must be maintained for individual research participants unless it is voluntarily and explicitly waived.

- Benefits from a research project should outweigh any foreseeable risks.

Each of these standards became a focus of debate about Milgram's experiments, so we will return frequently to that debate in order to keep our discussion realistic. We will also refer frequently to the ASA code in order to keep our treatment current. You will soon realize that there is no simple answer to the question: What is (or isn't) ethical research practice? The issues are just too complicated and the relevant principles too subject to different interpretations. But, I do promise that by the time you finish this chapter, you will be aware of the major issues in research ethics and be able to make informed, defensible decisions about the ethical conduct of social science research.

Avoid Harming Research Participants

Although this standard may seem straightforward, it can be difficult to interpret in specific cases and harder yet to define in a way agreeable to all social scientists. Does it mean that subjects should not be harmed psychologically as well as physically at all? That they should feel no anxiety or distress whatsoever during the study or only after their involvement ends? Should the possibility of any harm, no matter how remote, deter research?

Before we address these questions with respect to Milgram's experiments, a verbatim transcript of one session will give you an idea of what participants experienced (Milgram 1965:67):

150 volts delivered.	You want me to keep going?
165 volts delivered.	That guy is hollering in there. There's a lot of them here. He's liable to have a heart condition. You want me to go on?
180 volts delivered.	He can't stand it! I'm not going to kill that man in there! You hear him hollering? He's hollering. He can't stand it. . . . I mean who is going to take responsibility if anything happens to that gentleman? *[The experimenter accepts responsibility.]* All right.
195 volts delivered.	You see he's hollering. Hear that. Gee, I don't know. *[The experimenter says: "The experiment requires that you go on."]* I know it does, sir, but I mean—hugh—he don't know what he's in for. He's up to 195 volts.
210 volts delivered.	
225 volts delivered.	
240 volts delivered.	

This experimental manipulation generated "extraordinary tension" (Milgram 1963:377):

Subjects were observed to sweat, tremble, stutter, bite their lips, groan and dig their fingernails into their flesh. . . . Full-blown, uncontrollable seizures were observed for 3 subjects. [O]ne . . . seizure so violently convulsive that it was necessary to call a halt to the experiment [for that individual]. (p. 375)

An observer (behind a one-way mirror) reported (Milgram 1963), "I observed a mature and initially poised businessman enter the laboratory smiling and confident. Within 20 minutes he was reduced to a twitching, stuttering wreck, who was rapidly approaching a point of nervous collapse" (p. 377).

From critic Diana Baumrind's (1964) perspective, this emotional disturbance in subjects was "potentially harmful because it could easily effect an alteration in the subject's self-image or ability to trust adult authorities in the future" (p. 422). Stanley Milgram (1964) quickly countered that

> momentary excitement is not the same as harm. As the experiment progressed there was no indication of injurious effects in the subjects; and as the subjects themselves strongly endorsed the experiment, the judgment I made was to continue the experiment. (p. 849)

When Milgram (1964) surveyed the subjects in a follow-up, 83.7% endorsed the statement that they were "very glad" or "glad" "to have been in the experiment," 15.1% were "neither sorry nor glad," and just 1.3% were "sorry" or "very sorry" to have participated (p. 849). Interviews by a psychiatrist a year later found no evidence "of any traumatic reactions" (p. 197). Subsequently, Milgram (1977) argued that "the central moral justification for allowing my experiment is that it was judged acceptable by those who took part in it" (p. 21).

Milgram (1963) also attempted to minimize harm to subjects with postexperimental procedures "to assure that the subject would leave the laboratory in a state of well being" (p. 374). A friendly reconciliation was arranged between the subject and the victim, and an effort was made to reduce any tensions that arose as a result of the experiment.

In some cases, the "dehoaxing" (or debriefing) discussion was extensive, and all subjects were promised (and later received) a comprehensive report (Milgram 1964:849).

Baumrind (1964) was unconvinced: "It would be interesting to know what sort of procedures could dissipate the type of emotional disturbance just described [quoting Milgram]" (p. 422).

In a later article, Baumrind (1985:168) dismissed the value of the self-reported "lack of harm" of subjects who had been willing to participate in the experiment—although noting that still 16% did *not* endorse the statement that they were "glad" they had participated in the experiment. Baumrind (1985:169) also argued that research indicates most introductory psychology students (and some students in other social sciences) who have participated in a deception experiment report a decreased trust in authorities as a result—a tangible harm in itself.

Many social scientists, ethicists, and others concluded that Milgram's procedures had not harmed the subjects and so were justified for the knowledge they produced, but others sided with Baumrind's criticisms (Miller 1986:88–138). What is your opinion at this point? Does Milgram's debriefing process relieve your concerns? Are you as persuaded by the subjects' own endorsement of the procedures as was Milgram?

What about possible harm to the subjects of the famous prison simulation study at Stanford University (Haney, Banks, & Zimbardo 1973)? The study was designed to investigate the impact of social position on behavior—specifically, the impact of being either a guard or a prisoner in a prison, a "total institution." The researchers selected apparently stable and mature young male volunteers and asked them to sign a contract to work for 2 weeks as a guard or a prisoner in a simulated prison. Within the first 2 days after the prisoners were incarcerated by the "guards" in a makeshift basement prison, the prisoners began to be passive and disorganized, while the guards became "sadistic"—verbally and physically aggressive (Exhibit 3.5). Five "prisoners" were soon released for depression, uncontrollable crying, fits of rage, and, in one case, a psychosomatic rash. Instead of letting things continue for 2 weeks as planned, Zimbardo and his colleagues terminated the experiment after 6 days to avoid harming the subjects.

Through discussions in special postexperiment encounter sessions, feelings of stress among the participants who played the role of prisoner seemed to be relieved; follow-up during the next year indicated no lasting negative effects on the participants and some benefits in the form of greater insight.

Exhibit 3.5 **Chart of Guard and Prisoner Behavior**

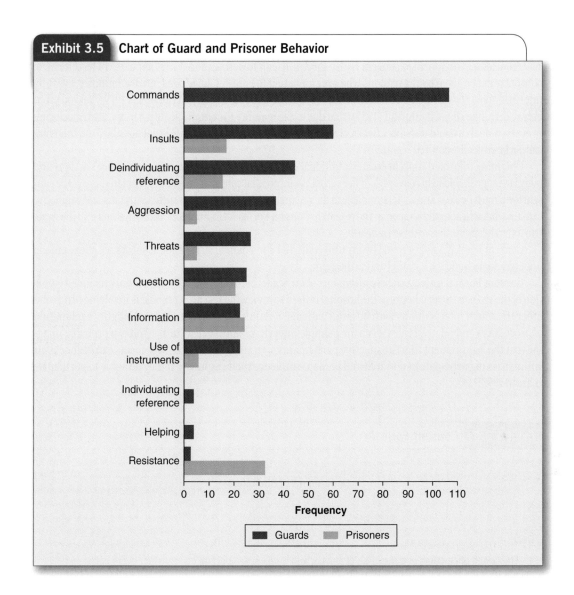

Would you ban such experiments because of the potential for harm to subjects? Does the fact that Zimbardo's and Milgram's experiments seemed to yield significant insights into the effect of a social situation on human behavior—insights that could be used to improve prisons or perhaps lessen the likelihood of another holocaust—make any difference (Reynolds 1979:133–139)? Do you believe that this benefit outweighs the foreseeable risks?

Well-intentioned researchers may also fail to foresee all the potential problems. Milgram (1974:27–31) reported that he and his colleagues were surprised by the subjects' willingness to carry out such severe shocks. In **Zimbardo's prison simulation study,** all the participants signed consent forms, but how could they have been fully informed in advance? The researchers themselves did not realize that the study participants would experience so much stress so quickly, that some prisoners would have to be released for severe negative reactions within the first few days, or that even those who were not severely stressed would soon be begging to be released from the mock prison. If this risk was not foreseeable, was it acceptable for the researchers to presume in advance that the benefits would outweigh the risks? And are you concerned, like Arthur Miller (1986), that real harm "could result from *not doing* research on destructive obedience" (p. 138) and other troubling human behaviors?

Obtain Informed Consent

The requirement of informed consent is also more difficult to define than it first appears. To be informed, consent must be given by the persons who are competent to consent, have consented voluntarily, are fully informed about the research, and have comprehended what they have been told (Reynolds 1979). Yet you probably realize, like Diana Baumrind (1985), that due to the inability to communicate perfectly, "full disclosure of everything that could possibly affect a given subject's decision to participate is not possible, and therefore cannot be ethically required" (p. 165).

Obtaining informed consent creates additional challenges for researchers. The researcher's actions and body language should help convey his or her verbal assurance that consent is voluntary. The language of the consent form must be clear and understandable to the research participants and yet sufficiently long and detailed to explain what will actually happen in the research. Consent Forms A (Exhibit 3.6) and B (Exhibit 3.7) illustrate two different approaches to these tradeoffs.

Consent Form A was approved by my university IRB for a mailed survey about substance abuse among undergraduate students. It is brief and to the point.

Consent Form B reflects the requirements of an academic hospital's IRB (I have only included a portion of the six-page form). Because the hospital is used to reviewing research proposals involving drugs and other treatment interventions with hospital patients, it requires a very detailed and lengthy explanation of procedures and related issues, even for a simple interview study such as mine. You can probably imagine that the requirement that prospective participants sign such lengthy consent forms can reduce their willingness to participate in research and perhaps influence their responses if they do agree to participate (Larson 1993:114).

Exhibit 3.6	Consent Form A

University of Massachusetts at Boston
Department of Sociology
(617) 287–6250
October 28, 1996

Dear :

  The health of students and their use of alcohol and drugs are important concerns for every college and university. The enclosed survey is about these issues at UMass/Boston. It is sponsored by University Health Services and the PRIDE Program (Prevention, Resources, Information, and Drug Education). The questionnaire was developed by graduate students in Applied Sociology, Nursing, and Gerontology.

  You were selected for the survey with a scientific, random procedure. Now it is important that you return the questionnaire so that we can obtain an unbiased description of the undergraduate student body. Health Services can then use the results to guide campus education and prevention programs.

  The survey requires only about 20 minutes to complete. Participation is completely voluntary and anonymous. No one will be able to link your survey responses to you. In any case, your standing at the University will not be affected whether or not you choose to participate. Just be sure to return the enclosed postcard after you mail the questionnaire so that we know we do not have to contact you again.

  Please return the survey by November 15th. If you have any questions or comments, call the PRIDE program at 287–5680 or Professor Schutt at 287–6250. Also call the PRIDE program if you would like a summary of our final report.

  Thank you in advance for your assistance.

Russell K. Schutt, PhD
Professor and Chair

| Exhibit 3.7 | Consent Form B |

**Research Consent Form for Social
and Behavioral Research**

Dana-Farber/Harvard Cancer Center

BIDMC/BWH/CH/DFCI/MGH/Partners Network Affiliates   OPRS 11-05

Protocol Title: <u>ASSESSING COMMUNITY HEALTH WORKERS' ATTITUDES AND KNOWLEDGE ABOUT
EDUCATING COMMUNITIES ABOUT CANCER CLINICAL TRIALS</u>

DF/HCC Principal Research Investigator / Institution: Dr. Russell Schutt, PhD / Beth Israel Deaconess
Medical Center and Univ. of Massachusetts, Boston

DF/HCC Site-Responsible Research Investigator(s) / Institution(s): Lidia Schapira, MD / Massachusetts
General Hospital

Interview Consent Form

A. INTRODUCTION

We are inviting you to take part in a research study. Research is a way of gaining new knowledge. A
person who participates in a research study is called a "subject." This research study is evaluating whether
community health workers might be willing and able to educate communities about the pros and cons of
participating in research studies.

It is expected that about 10 people will take part in this research study.

An institution that is supporting a research study either by giving money or supplying something that is
important for the research is called the "sponsor." The sponsor of this protocol is National Cancer Institute
and is providing money for the research study.

This research consent form explains why this research study is being done, what is involved in participating
in the research study, the possible risks and benefits of the research study, alternatives to participation, and
your rights as a research subject. The decision to participate is yours. If you decide to participate, please
sign and date at the end of the form. We will give you a copy so that you can refer to it while you are involved
in this research study.

If you decide to participate in this research study, certain questions will be asked of you to see if you are
eligible to be in the research study. The research study has certain requirements that must be met. If the
questions show that you can be in the research study, you will be able to answer the interview questions.

If the questions show that you cannot be in the research study, you will not be able to participate in this
research study.

Page 1 of 6

| DFCI Protocol Number: <u>06-085</u> | Date DFCI IRB Approved this Consent Form: <u>January 16, 2007</u> |
| Date Posted for Use: <u>January 16, 2007</u> | Date DFCI IRB Approval Expires: <u>August 13, 2007</u> |

(Continued)

Exhibit 3.7 (Continued)

**Research Consent Form for Social and
Behavioral Research**

Dana-Farber/Harvard Cancer Center

BIDMC/BWH/CH/DFCI/MGH/Partners Network Affiliates   OPRS 11-05

We encourage you to take some time to think this over and to discuss it with other people and to ask questions now and at any time in the future.

B. WHY IS THIS RESEARCH STUDY BEING DONE?

Deaths from cancer in general and for some specific cancers are higher for black people compared to white people, for poor persons compared to nonpoor persons, and for rural residents compared to non-rural residents. There are many reasons for higher death rates between different subpopulations. One important area for changing this is to have more persons from minority groups participate in research about cancer. The process of enrolling minority populations into clinical trials is difficult and does not generally address the needs of their communities. One potential way to increase particpation in research is to use community health workers to help educate communities about research and about how to make sure that researchers are ethical. We want to know whether community health workers think this is a good strategy and how to best carry it out.

C. WHAT OTHER OPTIONS ARE THERE?

Taking part in this research study is voluntary. Instead of being in this research study, you have the following option:

- Decide not to participate in this research study.

D. WHAT IS INVOLVED IN THE RESEARCH STUDY?

Before the research starts (screening): After signing this consent form, you will be asked to answer some questions about where you work and the type of community health work you do to find out if you can be in the research study.

If the answers show that you are eligible to participate in the research study, you will be eligible to participate in the research study. If you do not meet the eligibility criteria, you will not be able to participate in this research study.

After the screening procedures confirm that you are eligible to participate in the research study:
You will participate in an interview by answering questions from a questionnaire. The interview will take about 90 minutes. If there are questions you prefer not to answer we can skip those questions. The questions are about the type of work you do and your opinions about participating in research. If you agree, the interview will be taped and then transcribed. Your name and no other information about you will be associated with the tape or the transcript. Only the research team will be able to listen to the tapes.

Page 2 of 6

DFCI Protocol Number: <u>06-085</u>	Date DFCI IRB Approved this Consent Form: <u>January 16, 2007</u>
Date Posted for Use: <u>January 16, 2007</u>	Date DFCI IRB Approval Expires: <u>August 13, 2007</u>

**Research Consent Form for Social and
Behavioral Research**

Dana-Farber/Harvard Cancer Center

BIDMC/BWH/CH/DFCI/MGH/Partners Network Affiliates   OPRS 11-05

Immediately following the interview, you will have the opportunity to have the tape erased if you wish to withdraw your consent to taping or participation in this study. You will receive $30.00 for completing this interview.

After the interview is completed: Once you finish the interview there are no additional interventions.

...

N. DOCUMENTATION OF CONSENT

My signature below indicates my willingness to participate in this research study and my understanding that I can withdraw at any time.

_____   _____

Signature of Subject   Date
or Legally Authorized Representative

_____   _____

Person obtaining consent   Date

To be completed by person obtaining consent:

The consent discussion was initiated on _____ (date) at _____ (time.)

☐ A copy of this signed consent form was given to the subject or legally authorized representative.

For Adult Subjects

☐ The subject is an adult and provided consent to participate.

☐ The subject is an adult who lacks capacity to provide consent and his/her legally authorized representative:

  ☐ gave permission for the adult subject to participate

  ☐ did not give permission for the adult subject to participate

Page 6 of 6

DFCI Protocol Number: 06-085	Date DFCI IRB Approved this Consent Form: January 16, 2007
Date Posted for Use: January 16, 2007	Date DFCI IRB Approval Expires: August 13, 2007

Debriefing A researcher's informing subjects after an experiment about the experiment's purposes and methods and evaluating subjects' personal reactions to the experiment.

As in Milgram's study, experimental researchers whose research design requires some type of subject deception try to get around this problem by withholding some information before the experiment begins, but then debriefing subjects at the end. In a **debriefing**, the researcher explains to the subject what happened in the experiment and why, and then responds to their questions. A carefully designed debriefing procedure can help the research participants learn from the experimental research and grapple constructively with feelings elicited by the realization that they were deceived (Sieber 1992:39–41). However, even though debriefing can be viewed as a substitute, in some cases, for securing fully informed consent prior to the experiment, debriefed subjects who disclose the nature of the experiment to other participants can contaminate subsequent results (Adair, Dushenko, & Lindsay 1985). Unfortunately, if the debriefing process is delayed, the ability to lessen any harm resulting from the deception is also reduced.

For a study of the social background of men who engage in homosexual behavior in public facilities, Laud Humphreys (1970) decided that truly informed consent would be impossible to obtain. Instead, he first served as a lookout—a "watch queen"—for men who were entering a public bathroom in a city park with the intention of having sex. In a number of cases, he then left the bathroom and copied the license plate numbers of the cars driven by the men. One year later, he visited the homes of the men and interviewed them as part of a larger study of social issues. Humphreys changed his appearance so that the men did not recognize him. In *Tearoom Trade,* his book on this research, Humphreys concluded that the men who engaged in what were viewed as deviant acts were, for the most part, married, suburban men whose families were unaware of their sexual practices. But debate has continued ever since about Humphreys's failure to tell the men what he was really doing in the bathroom or why he had come to their homes for the interview. He was criticized by many, including some faculty members at the University of Washington who urged that his doctoral degree be withheld. However, many other professors and some members of the gay community praised Humphreys for helping normalize conceptions of homosexuality (Miller 1986:135).

If you were to serve on your university's IRB, would you allow this research to be conducted? Can students who are asked to participate in research by their professor be considered able to give informed consent? Do you consider *informed consent* to be meaningful if the true purpose or nature of an experimental manipulation is not revealed?

The process and even possibility of obtaining informed consent must take into account the capacity of prospective participants to give informed consent. Children cannot legally give consent to participate in research; instead, they must in most circumstances be given the opportunity to give or withhold their *assent* to participate in research, usually by a verbal response to an explanation of the research. In addition, a child's legal guardian must give written informed consent to have the child participate in research (Sieber 1992). There are also special protections for other populations that are likely to be vulnerable to coercion—prisoners, pregnant women, persons with mental disabilities, and educationally or economically disadvantaged persons. Would you allow research on prisoners, whose ability to give informed consent can be questioned? What special protections do you think would be appropriate?

Obtaining informed consent also becomes more challenging in collectivist communities in which leaders or the whole group are accustomed to making decisions for individual members. In such settings, usually in non-Western cultures, researchers may have to develop a relationship with the community before individuals can be engaged in research (Bledsoe & Hopson 2009:397-398).

Subject payments create another complication for achieving the goal of informed consent. While payments to research participants can be a reasonable way to compensate them for their time and effort, payments also serve as an inducement to participate. If the payment is a significant amount in relation to the participants' normal income, it could lead people to set aside their reservations about participating in a project—even though they may harbor those reservations (Fisher & Anushko 2008:104-105).

Avoid Deception in Research, Except in Limited Circumstances

Deception occurs when subjects are misled about research procedures to determine how they would react to the treatment if they were not research subjects. Deception is a critical component of many social psychology experiments, in part because of the difficulty of simulating real-world stresses and dilemmas in a laboratory setting. The goal is to get subjects "to accept as true what is false or to give a false impression" (Korn 1997:4). In Milgram's (1964) experiment, for example, deception seemed necessary because the subjects could not be permitted to administer real electric shocks to the "stooge," yet it would not have made sense to order the subjects to do something that they didn't find to be so troubling. Milgram (1992:187–188) insisted that the deception was absolutely essential. The results of many other social psychological experiments would be worthless if subjects understood what was really happening to them while the experiment was in progress. The real question: Is this sufficient justification to allow the use of deception?

Gary Marshall and Philip Zimbardo (1979:971–972) sought to determine the physiological basis of emotion by injecting student volunteers with adrenaline, so that their heart rate and sweating would increase, and then placing them in a room with a student "stooge" who acted silly. But, the students were told that they were being injected with a vitamin supplement to test its effect on visual acuity (Korn 1997:2–3). Piliavin and Piliavin (1972:355–356) staged fake seizures on subway trains to study helpfulness (Korn 1997:3–4). If you were a member of your university's IRB, would you vote to allow such deceptive practices in research? What about less dramatic instances of deception in laboratory experiments with students like yourself?

Do you believe that deception itself is the problem? Aronson and Mills's (1959) study of severity of initiation to groups is a good example of experimental research that does not pose greater-than-everyday risks to subjects, but still uses deception. This study was conducted at an all-women's college in the 1950s. The student volunteers who were randomly assigned to the "severe initiation" experimental condition had to read a list of embarrassing words. I think it's fair to say that even in the 1950s, reading a list of potentially embarrassing words in a laboratory setting and listening to a taped discussion were unlikely to increase the risks to which students are exposed in their everyday lives. Moreover, the researchers informed subjects that they would be expected to talk about sex and could decline to participate in the experiment if this requirement would bother them. None dropped out.

To further ensure that no psychological harm was caused, Aronson and Mills (1959) explained the true nature of the experiment to subjects after the experiment. The subjects did not seem perturbed: "None of the Ss [subjects] expressed any resentment or annoyance at having been misled. In fact, the majority were intrigued by the experiment, and several returned at the end of the academic quarter to ascertain the result" (p. 179).

Are you satisfied that this procedure caused no harm? Do you react differently to Aronson and Mills's debriefing than you did to Milgram's debriefing? The minimal deception in the Aronson and Mills experiment, coupled with the lack of any ascertainable risk to subjects and a debriefing, satisfies the ethical standards for research of most social scientists and IRBs, even today.

What scientific, educational, or applied value would make deception justifiable, even if there is some potential for harm? Who determines whether a nondeceptive intervention is "equally effective"? (Miller 1986:103). Diana Baumrind (1985:167) suggested that personal "introspection" would have been sufficient to test Milgram's hypothesis and has argued subsequently that intentional deception in research violates the ethical principles of self-determination, protection of others, and maintenance of trust between people, and so can never be justified. How much risk, discomfort, or unpleasantness might be seen as affecting willingness to participate? When should a postexperimental "attempt to correct any misconception" due to deception be deemed sufficient?

Can you see why an IRB, representing a range of perspectives, is an important tool for making reasonable, ethical research decisions when confronted with such ambiguity? Exhibit 3.8 shows a portion of the complex flowchart developed by the U.S. Department of Health and Human Services to help researchers decide what type of review will be needed for their research plans. Any research involving deception requires formal human subjects' review.

Maintain Privacy and Confidentiality

Maintaining privacy and confidentiality is another key ethical standard for protecting research participants, and the researcher's commitment to that standard should be included in the informed consent agreement (Sieber 1992). Procedures to protect each subject's privacy such as locking records and creating special identifying codes must be created to minimize the risk of access by unauthorized persons. However, statements about confidentiality should be realistic: Laws allow research records to be subpoenaed and may require reporting child abuse; a researcher may feel compelled to release information if a health- or life-threatening situation arises and participants need to be alerted. Also, the standard of confidentiality does not apply to observation in public places and information available in public records.

There is one exception to some of these constraints: The National Institutes of Health can issue a **Certificate of Confidentiality** to protect researchers from being legally required to disclose confidential information. This is intended to help researchers overcome the reluctance of individuals engaged in illegal behavior to sign a consent form or to risk exposure of their illegal activities (Sharma 2009:426). Researchers who are focusing on high-risk populations or behaviors, such as crime, substance abuse, sexual activity, or genetic information, can request such a certificate. Suspicions of child abuse or neglect must still be reported, and in some states researchers may still be required to report such crimes as elder abuse (Arwood & Panicker 2007).

Certificate of Confidentiality A certificate issued to a researcher by the National Institutes of Health that ensures the right to protect information obtained about high-risk populations or behaviors—except child abuse or neglect—from legal subpoenas.

The Health Insurance Portability and Accountability Act (HIPAA) passed by Congress in 1996 created more stringent regulations for the protection of health care data (Exhibit 3.8). As implemented by the U.S. Department of Health and Human Services in 2000 (revised in 2002), the HIPAA Final Privacy Rule applies to oral, written, and electronic information that "relates to the past, present or future physical or mental health or condition of an individual." The HIPAA Rule requires that researchers have valid authorization for any use or disclosure of "protected health information" (PHI) from a health care provider. Waivers of authorization can be granted in special circumstances (Cava, Cushman, & Goodman 2007).

The Uses of Research

Scientists must also consider the uses to which their research is put. Although many scientists believe that personal values should be left outside the laboratory, some feel that it is proper—even necessary—for scientists to concern themselves with the way their research is used.

Stanley Milgram made it clear that he was concerned about the phenomenon of obedience precisely because of its implications for peoples' welfare. As you have already learned, his first article (Milgram 1963) highlighted the atrocities committed under the Nazis by citizens and soldiers who were "just following orders." In his more comprehensive book on the obedience experiments (Milgram 1974), he also used his findings to shed light on the atrocities committed in the Vietnam War at My Lai, slavery, the destruction of

Exhibit 3.8 U.S. Department of Health and Human Services Human Subjects Decision Flowchart 4: For Tests, Surveys, Interviews, Public Behavior Observation

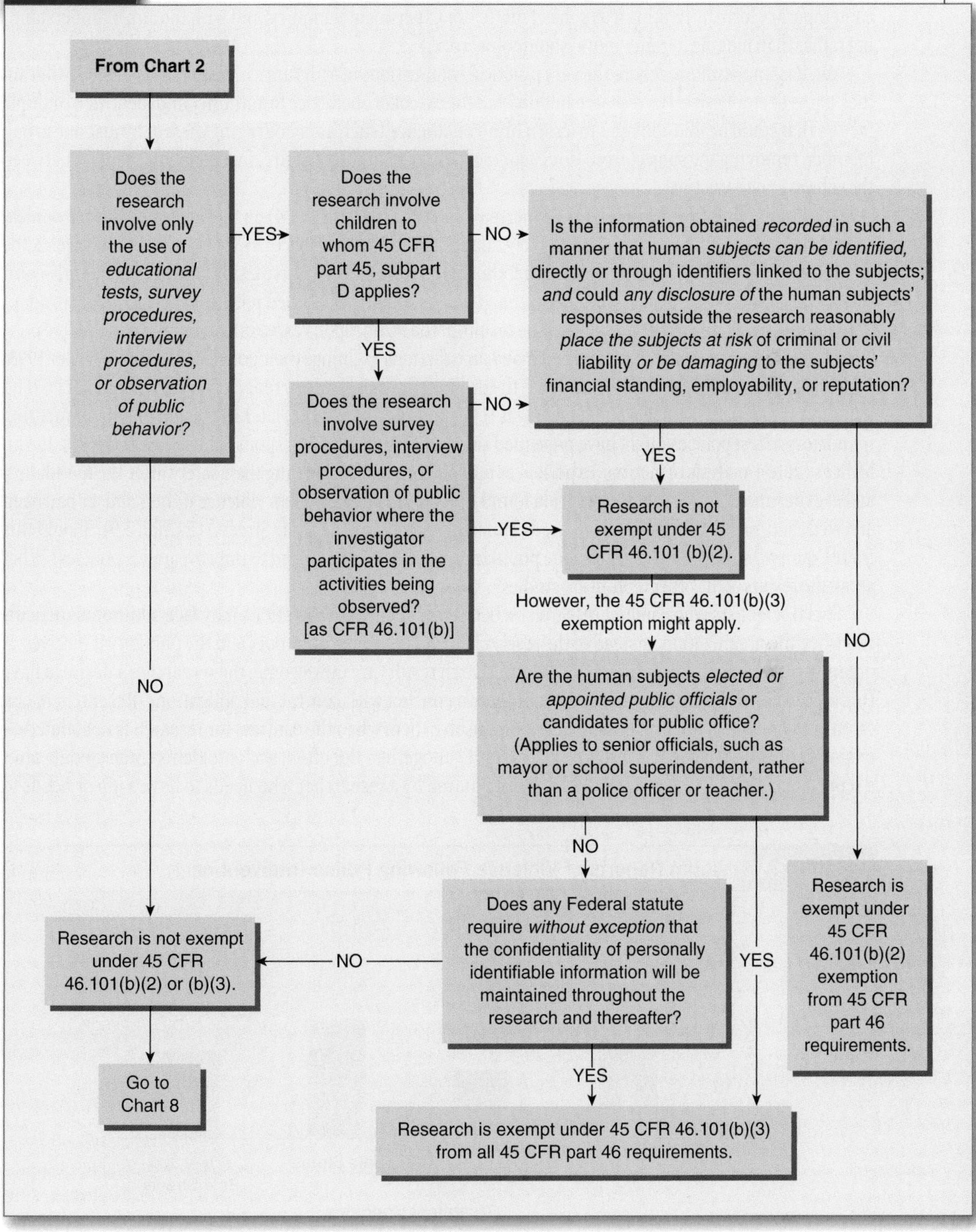

From Chart 2

Does the research involve only the use of *educational tests, survey procedures, interview procedures, or observation of public behavior?*

—YES→

Does the research involve children to whom 45 CFR part 45, subpart D applies?

— NO →

Is the information obtained *recorded* in such a manner that human *subjects can be identified,* directly or through identifiers linked to the subjects; *and* could *any disclosure of* the human subjects' responses outside the research reasonably *place the subjects at risk* of criminal or civil liability *or be damaging* to the subjects' financial standing, employability, or reputation?

YES

Does the research involve survey procedures, interview procedures, or observation of public behavior where the investigator participates in the activities being observed? [as CFR 46.101(b)]

— NO →

—YES→

Research is not exempt under 45 CFR 46.101 (b)(2).

However, the 45 CFR 46.101(b)(3) exemption might apply.

YES

Are the human subjects *elected or appointed public officials* or candidates for public office? (Applies to senior officials, such as mayor or school superindendent, rather than a police officer or teacher.)

NO

NO

Does any Federal statute require *without exception* that the confidentiality of personally identifiable information will be maintained throughout the research and thereafter?

— NO —

Research is not exempt under 45 CFR 46.101(b)(2) or (b)(3).

Go to Chart 8

YES

Research is exempt under 45 CFR 46.101(b)(2) exemption from 45 CFR part 46 requirements.

Research is exempt under 45 CFR 46.101(b)(3) from all 45 CFR part 46 requirements.

NO

the American Indian population, and the internment of Japanese Americans during World War II. Milgram makes no explicit attempt to "tell us what to do" about this problem. In fact, as a dispassionate social scientist, Milgram (1974) tells us, "What the present study [did was] to give the dilemma [of obedience to authority] contemporary form by treating it as subject matter for experimental inquiry, and with the aim of understanding rather than judging it from a moral standpoint" (p. xi).

Yet, it is impossible to ignore the very practical implications of Milgram's investigations, which Milgram took pains to emphasize. His research highlighted the extent of obedience to authority and identified multiple factors that could be manipulated to lessen blind obedience (such as encouraging dissent by just one group member, removing the subject from direct contact with the authority figure, increasing the contact between the subject and the victim).

The evaluation research by Lawrence Sherman and Richard Berk (1984) on the police response to domestic violence provides an interesting cautionary tale about the uses of science. As you recall from Chapter 2, the results of this field experiment indicated that those who were arrested were less likely to subsequently commit violent acts against their partners. Sherman (1993) explicitly cautioned police departments not to adopt mandatory arrest policies based solely on the results of the Minneapolis experiment, but the results were publicized in the mass media and encouraged many jurisdictions to change their policies (Binder & Meeker 1993; Lempert 1989). Although we now know that the original finding of a deterrent effect of arrest did not hold up in many other cities where the experiment was repeated, Sherman (1992) later suggested that implementing mandatory arrest policies might have prevented some subsequent cases of spouse abuse (pp. 150–153). JoAnn Miller's (2003) analysis of victims' experiences and perceptions concerning their safety after the mandatory arrest experiment in Dade County, Florida found that victims reported less violence if their abuser had been arrested (and/or assigned to a police-based counseling program called Safe Streets) (Exhibit 3.9). Should this Dade County finding be publicized in the popular press, so it could be used to improve police policies? What about the results of the other replication studies?

Social scientists who conduct research on behalf of specific organizations may face additional difficulties when the organization, instead of the researcher, controls the final report and the publicity it receives. If organizational leaders decide that particular research results are unwelcome, the researcher's desire to have findings used appropriately and reported fully can conflict with contractual obligations. Researchers can often anticipate such dilemmas in advance and resolve them when the contract for research is negotiated—or simply decline a particular research opportunity altogether. But often, such problems come up only after a report has been drafted, or the problems are ignored by a researcher who needs to have a job or needs to

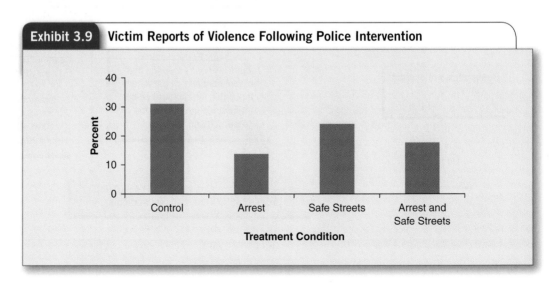

Exhibit 3.9 **Victim Reports of Violence Following Police Intervention**

maintain particular personal relationships. These possibilities cannot be avoided entirely, but because of them, it is always important to acknowledge the source of research funding in reports and to consider carefully the sources of funding for research reports written by others.

The potential of withholding a beneficial treatment from some subjects also is a cause for ethical concern. The Sherman and Berk (1984) experiment required the random assignment of subjects to treatment conditions and thus had the potential of causing harm to the victims of domestic violence whose batterers were not arrested. The justification for the study design, however, is quite persuasive: The researchers didn't know prior to the experiment which response to a domestic violence complaint would be most likely to deter future incidents (Sherman 1992). The experiment provided what seemed to be clear evidence about the value of arrest, so it can be argued that the benefits outweighed the risks.

▣ Philosophical Issues

Your general assumptions about how the social world can best be investigated—your social research philosophy—will in part shape your investigations of the social world. In this section, we focus on two general alternative research philosophies and examine some of their implications for research methods. I will review research guidelines and objectives that are consistent with both philosophies and consider examples of how the research you have learned about in Chapters 1 and 2 illustrates how to achieve these objectives and conform to these guidelines. Throughout this section, you should consider how these philosophical issues relate to the ethical issues we have just reviewed. At the end of the section, I will point out some of these relationships.

Positivism and Postpositivism

Researchers with a positivist philosophy believe that there is an objective reality that exists apart from the perceptions of those who observe it, and that the goal of science is to understand this reality better.

> Whatever nature "really" is, we assume that it presents itself in precisely the same way to the same human observer standing at different points in time and space. . . . We assume that it also presents itself in precisely the same way across different human observers standing at the same point in time and space. (Wallace 1983:461)

This is the philosophy traditionally associated with natural science, with the expectation that there are universal laws of human behavior, and with the belief that scientists must be objective and unbiased to see reality clearly (Weber 1949:72). **Positivism** asserts that a well-designed test of a specific prediction—for example, the prediction that social ties decrease among those who use the Internet more—can move us closer to understanding actual social processes.

Postpositivism is a philosophy of reality that is closely related to positivism. Postpositivists believe that there is an external, objective reality, but they are very sensitive to the complexity of this reality and to the limitations and biases of the scientists who study it (Guba & Lincoln 1994:109–111). For example, postpositivists may worry that researchers, who are heavy computer users themselves, will be biased in favor of finding positive social effects of computer use. As a result of

Positivism The belief, shared by most scientists, that there is a reality that exists quite apart from our own perception of it, that it can be understood through observation, and that it follows general laws.

Postpositivism The belief that there is an empirical reality, but that our understanding of it is limited by its complexity and by the biases and other limitations of researchers.

concerns such as this, postpositivists do not think we can ever be sure that scientific methods allow us to perceive objective reality. Instead, they believe that the goal of science is to achieve **intersubjective agreement** among scientists about the nature of reality (Wallace 1983:461). We can be more confident in the community of social researchers than in any individual social scientist (Campbell & Russo 1999:144).

> **Intersubjective agreement** An agreement by different observers on what is happening in the natural or social world.

The positivist and postpositivist philosophies consider value considerations to be beyond the scope of science: "An empirical science cannot tell anyone what he should do—but rather what he can do—and under certain circumstances—what he wishes to do" (Weber 1949:54). The idea is that developing valid knowledge about how society *is* organized, or how we live our lives, does not tell us how society *should* be organized or how we *should* live our lives. The determination of empirical facts should be a separate process from the evaluation of these facts as satisfactory or unsatisfactory (Weber 1949:11).

The idea is not to ignore value considerations, because they are viewed as a legitimate basis for selecting a research problem to investigate. In addition, many scientists also consider it acceptable to encourage government officials or private organizations to act on the basis of a study's findings, after the research is over. During a research project, however, value considerations are to be held in abeyance. The scientist's work is done when his or her research results are published or presented to other scientists.

Positivist Research Guidelines

To achieve an accurate understanding of the social world, the researcher operating within the positivist tradition must adhere to some basic guidelines about how to conduct research.

1. *Test ideas against empirical reality without becoming too personally invested in a particular outcome.* This guideline requires a commitment to "testing," as opposed to just reacting to events as they happen or looking for what we want to see (Kincaid 1996:51–54). Note how McPherson and his colleagues (2006) acknowledged the social importance of inadequate social ties but did not express personal feelings or make recommendations about that problem:

> If core discussion networks represent an important social resource, Americans are still stratified on education and race. . . . Non-whites still have smaller networks than whites. (p. 372)

2. *Plan and carry out investigations systematically.* Social researchers have little hope of conducting a careful test of their ideas if they do not think through in advance how they should go about the test and then proceed accordingly. But a systematic approach is not always easy. Here is an explanation of a portion of the systematic procedures used by McPherson et al. (2006):

> The GSS is a face-to-face survey of the noninstitutionalized U.S. adult population. The 1985 and 2004 surveys used the same questions to generate the names of confidants and identical procedures to probe for additional discussion partners. Therefore, the survey responses represent a very close replication of the same questions and procedures at two points in time, representing the same underlying population in 1985 and 2004. (pp. 356–357)

3. *Document all procedures and disclose them publicly.* Social researchers should disclose the methods on which their conclusions are based so that others can evaluate for themselves the likely soundness of these conclusions. Such disclosure is a key feature of science. It is the community of researchers, reacting to each

others' work, that provides the best guarantee against purely self-interested conclusions (Kincaid 1996). In their methodological section, McPherson and his colleagues (2006) documented the approach they used to measure social ties:

> We use the same measures of network characteristics that Marsden (1987:123–124) used in his description of the structure of 1985 American interpersonal environments. Size is the number of names mentioned in response to the "name generator" question. (p. 357)

4. *Clarify assumptions.* No investigation is complete unto itself; whatever the researcher's method, the research rests on some background assumptions. For example, Sherman and Berk (1984) identified in much research to determine whether arrest has a deterrent effect, an assumption that potential law violators think rationally and calculate potential costs and benefits prior to committing crimes. By definition, research assumptions are not tested, so we do not know for sure whether they are correct. By taking the time to think about and disclose their assumptions, researchers provide important information for those who seek to evaluate research conclusions.

5. *Specify the meaning of all the terms.* Words often have multiple or unclear meanings. *Alienation, depression, cold, crowded,* and so on can mean different things to different people. In scientific research, all terms must be defined explicitly and used consistently. For example, McPherson et al. (2006) distinguished the concept of "important matters" in their question about social ties from other possible meanings of this concept:

> While clarifying what the GSS question measures, we should also be clear about what it does not measure. Most obviously, it does not measure what people talk about in their relationships. (p. 356)

6. *Maintain a skeptical stance toward current knowledge.* The results of any particular investigation must be examined critically, although confidence about interpretations of the social or natural world increases after repeated investigations yield similar results. A general skepticism about current knowledge stimulates researchers to improve current research and expand the frontier of knowledge. Again, McPherson and his colleagues (2006) provide an example. In this next passage, they caution against a too literal interpretation of their findings:

> We would be unwise to interpret the answers to this question too literally (e.g., assuming that a specific conversation about some publicly weighty matter had occurred in the past six months). (p. 356)

7. *Replicate research and build social theory.* No one study is definitive by itself. We can't fully understand a single study's results apart from the larger body of knowledge to which it is related, and we can't place much confidence in these results until the study has been replicated. For example, Sherman and Berk (1984) needed to ensure that spouse abusers were assigned to be either arrested or not on a random basis rather than on the basis of the police officers' personal preferences. They devised a systematic procedure using randomly sequenced report sheets in different colors, but then found that police officers sometimes deviated from this procedure due to their feelings about particular cases. Subsequently, in some replications of the study, the researchers ensured compliance with their research procedures by requiring police officers to call in to a central number to receive the experimentally determined treatment. You also saw in Chapter 2 how replications of the Sherman and Berk research in other cities led to a different theory about the conditions when deterrence didn't work.

8. *Search for regularities or patterns.* Positivist and postpositivist scientists assume that the natural world has some underlying order of relationships, so that unique events and individuals can be understood at least in

part in terms of general principles (Grinnell 1992:27–29). This chapter illustrated how Stanley Milgram and others repeated his basic experiment on obedience in many settings in order to identify regularities in people's obedience to authority.

Real investigations by social scientists do not always include much attention to theory, specific definitions of all terms, and so forth. But it behooves any social researcher to study these guidelines and to consider the consequences of not following any with which they do not agree.

Interpretivism and Constructivism

Qualitative research is often guided by a different philosophy of **interpretivism**. Interpretive social scientists believe that social reality is socially constructed and that the goal of social scientists is to understand what meanings people give to reality, not to determine how reality works apart from these interpretations. In the words of Sally Lindsay and her colleagues (2007), "the researcher seeks an in-depth understanding of the experiences of the participants" (p. 101). This philosophy rejects the positivist belief that there is a concrete, objective reality that scientific methods help us understand (Lynch & Bogen 1997); instead, interpretivists believe that people construct an image of reality based on their own preferences and prejudices and their interactions with others and that this is as true of scientists as it is of everyone else in the social world. This means that we can never be sure that we have understood reality properly, that "objects and events are understood by different people differently, and those perceptions are the reality—or realities—that social science should focus on" (Rubin & Rubin 1995:35).

> **Interpretivism** The belief that reality is socially constructed and that the goal of social scientists is to understand what meanings people give to that reality. Max Weber termed the goal of interpretivist research *verstehen,* or "understanding."

> **Constructivism** A perspective that emphasizes how different stakeholders in social settings construct their beliefs.

Constructivism extends interpretivist philosophy by emphasizing the importance of exploring how different stakeholders in a social setting construct their beliefs (Guba & Lincoln 1989:44–45). It gives particular attention to the different goals of researchers and other participants in a research setting and seeks to develop a consensus among participants about how to understand the focus of inquiry (Sulkunen 2008:73). From this standpoint, "Truth is a matter of the best-informed and most sophisticated construction on which there is consensus at a given time" (Schwandt 1994:128).

In the words of Lindsay et al. (2007),

> Here we provide a descriptive account of the impact that providing home Internet access may have had on the health and social behavior of a small number of older deprived men with heart disease. The description is largely theirs and much of it is in retrospect. (pp. 99–100)

> **Hermeneutic circle** Represents the dialectical process in which the researcher obtains information from multiple stakeholders in a setting, refines his or her understanding of the setting, and then tests that understanding with successive respondents.

Constructivist inquiry uses an interactive research process, in which a researcher begins an evaluation in some social setting by identifying the different interest groups in that setting. The researcher goes on to learn what each group thinks, and then gradually tries to develop a shared perspective on the problem being evaluated (Guba & Lincoln 1989:42).

These steps are diagrammed as a circular process, called a **hermeneutic circle** (Exhibit 3.10). In this process, the researcher conducts an open-ended interview with the first respondent (R1) to learn about his or her thoughts and feelings on the subject of inquiry—the first respondent's "construction" (C1). The researcher then asks this respondent to nominate a second respondent (R2), who feels very differently. The second respondent is then interviewed in the same way, but also is asked to comment on the themes raised by the previous respondent. The process continues until all major perspectives are represented, and then may be repeated again with the same set of respondents (Guba & Lincoln 1989:180–181).

Exhibit 3.10	The Hermeneutic Circle

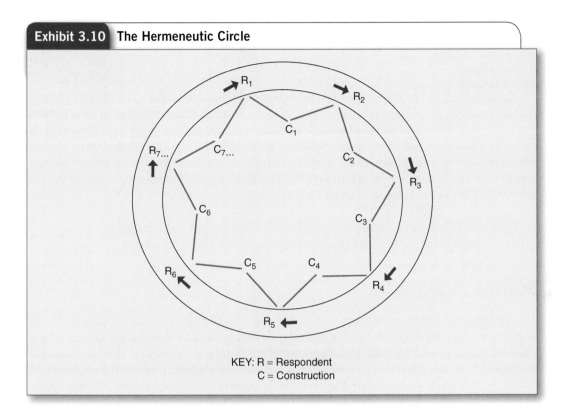

KEY: R = Respondent
C = Construction

Interpretivist/Constructivist Research Guidelines

Researchers guided by an interpretivist philosophy reject some of the positivist research guidelines. However, there are a wide variety of specific approaches that can be termed *interpretivist* and each has somewhat unique guidelines. For those working within the constructivist perspective, Guba and Lincoln (1989:42) suggest four key steps for researchers, each of which may be repeated many times in a given study:

1. Identify stakeholders and solicit their "claims, concerns, and issues."

2. Introduce the claims, concerns, and issues of each stakeholder group to the other stakeholder groups and ask for their reactions.

3. Focus further information collection on claims, concerns, and issues about which there is disagreement among stakeholder groups.

4. Negotiate with stakeholder groups about the information collected and attempt to reach consensus on the issues about which there is disagreement.

Although Lindsay and her colleagues (2007) did not follow these guidelines exactly in their interpretivist research on web-based social ties, their procedures did include a constructivist concern with eliciting feedback from their respondents:

The trustworthiness of findings was established by "peer debriefing" and "member checking." Peer debriefing took place by discussing the interpretation of the data with each of the authors. Member checking took place by discussing the key themes with two of the men who participated in the focus group to ascertain whether the results reflected their experience. (p. 101)

Feminist research Research with a focus on women's lives and often including an orientation to personal experience, subjective orientations, the researcher's standpoint, and emotions.

Feminist research is a term used to refer to research done by feminists (Reinharz 1992:6–7) and to a perspective on research that can involve many different methods (Reinharz 1992:240). The feminist perspective on research includes the interpretivist and constructivist elements of concern with personal experience and subjective feelings and with the researcher's position and standpoint (Hesse-Biber & Leavy 2007:4–5). Feminist researchers Sharlene Hesse-Biber and Patricia Lina Leavy (2007:139) emphasize the importance of viewing the social world as complex and multilayered, of sensitivity to the impact of social differences—of being an "insider" or an "outsider," and of being concerned with the researcher's position. African American feminist researcher Patricia Hill Collins (2008) suggests that researchers who are sensitive to their "outside" role within a social situation may have unique advantages:

> Outsiders within occupy a special place—they become different people and their difference sensitizes them to patterns that may be more difficult for established sociological insiders to see. (p. 317)

Scientific Paradigms

At this point, positivism may seem to represent an opposing research philosophy to interpretivism and constructivism (and, some would say, feminist methods). Researchers who think this way often refer to these philosophies as alternative **scientific paradigms**: sets of beliefs that guide scientific work in an area, including unquestioned presuppositions, accepted theories, and exemplary research findings. In his famous book on the history of science, *The Structure of Scientific Revolutions*, Thomas S. Kuhn (1970) argued that most of the time one scientific paradigm is accepted as the prevailing wisdom in a field and that scientists test ideas that make sense within that paradigm. They are conducting what Kuhn called **normal science**. It is only after a large body of contrary evidence accumulates that there may be a rapid shift to a new paradigm—that is, a **scientific revolution** (Hammersley 2008:46).

Scientific paradigm A set of beliefs that guide scientific work in an area, including unquestioned presuppositions, accepted theories, and exemplary research findings.

Normal science The gradual, incremental research conducted by scientists within the prevailing scientific paradigm.

Scientific revolution The abrupt shift from one dominant scientific paradigm to an alternative paradigm that may be developed after accumulation of a large body of evidence that contradicts the prevailing paradigm.

Paradigm wars The intense debate from the 1970s to the 1990s between social scientists over the value of positivist and interpretivist research philosophies.

If you think about positivism (or postpositivism) and interpretivism (or constructivism) as alternative paradigms, it may seem that you should choose the one philosophy that seems closest to your preferences and condemn the other as "unrealistic," "unscientific," "uncaring," perhaps even "unethical." For this reason, some social researchers refer to the intense debate over positivist and interpretivist philosophies in the 1970s and 1980s as the **paradigm war** (Alastalo 2008:34–35; Bryman 2008:15).

Ethics are important no matter which research philosophy guides a researcher. However, a researcher's approach to ethics may also vary with his or her preferred scientific paradigm. From a positivist standpoint, the researcher is "in charge" and the research participants tend to be viewed as "subjects" who the responsible researcher must protect from harm. From a constructivist standpoint, the research participants are more likely to be viewed as "people like us" with whom the researcher collaborates in an investigation and must treat as the researcher himself or herself would want to be treated.

In spite of these many differences between more positivist and more interpretivist philosophies, there are good reasons to prefer a research philosophy that integrates some of the differences between them (Smith 1991). Researchers influenced by a positivist philosophy should be careful to consider how their own social background and values shape their research approaches and interpretations—just as interpretivist researchers caution us to do (Clegg & Slife 2009:35). We also need to be sensitive to the insights that can be provided by other stakeholders in the settings we investigate. Researchers influenced more by an interpretivist philosophy should be careful to ensure that they use rigorous procedures to check the trustworthiness of their interpretations of data (Riessman 2008:185–199). If we are not willing to "ask hard questions" about our projects and the evidence we collect, we are not ready to investigate the social world (Riessman 2008:200).

You will learn in Chapter 10 how some social researchers combine methods favored by positivists and interpretivists, rejecting the idea that these are truly alternative paradigms. However, most still work primarily within one framework. In research articles published in sociology journals, C. David Gartrell and John W. Gartrell (2002) found that positivism continues to be the dominant perspective in the United States, but it has become much less common in British sociology journals.

▣ Conclusions

The extent to which ethical issues are a problem for researchers and their subjects varies dramatically with the type of research design. Survey research, in particular, creates few ethical problems. In fact, researchers from Michigan's Institute for Survey Research interviewed a representative national sample of adults and found that 68% of those who had participated in a survey were somewhat or very interested in participating in another; the more times respondents had been interviewed, the more willing they were to participate again. Presumably, they would have felt differently if they had been treated unethically (Reynolds 1979:56–57). On the other hand, some experimental studies in the social sciences that have put people in uncomfortable or embarrassing situations have generated vociferous complaints and years of debate about ethics (Reynolds 1979; Sjoberg 1967).

The evaluation of ethical issues in a research project should be based on a realistic assessment of the overall potential for harm and benefit to research subjects rather than an apparent inconsistency between any particular aspect of a research plan and a specific ethical guideline. For example, full disclosure of "what is really going on" in an experimental study is unnecessary if subjects are unlikely to be harmed. Nevertheless, researchers should make every effort to foresee all possible risks and to weigh the possible benefits of the research against these risks. They should consult with individuals with different perspectives to develop a realistic risk-benefit assessment, and they should try to maximize the benefits to, as well as minimize the risks for, subjects of the research (Sieber 1992:75–108).

Ultimately, these decisions about ethical procedures are not just up to you, as a researcher, to make. Your university's IRB sets the human subjects' protection standards for your institution and will require that researchers—even, in most cases, students—submit their research proposal to the IRB for review. So, I leave you with the instruction to review the human subjects guidelines of the ASA or other professional association in your field, consult your university's procedures for the conduct of research with human subjects, and then proceed accordingly.

You can now also understand why the debate continues between positivist and interpretivist philosophies, why researchers should think about the philosophy that guides their research, and how research can sometimes be improved by drawing on insights from both philosophies (Turner 1980:99).

Key Terms

Highlights

- Stanley Milgram's obedience experiments led to intensive debate about the extent to which deception could be tolerated in social science research and how harm to subjects should be evaluated.

- Egregious violations of human rights by researchers, including scientists in Nazi Germany and researchers in the Tuskegee syphilis study, led to the adoption of federal ethical standards for research on human subjects.

- The 1979 *Belmont Report* developed by a national commission established three basic ethical standards for the protection of human subjects: (1) respect for persons, (2) beneficence, and (3) justice.

- The Department of Health and Human Services adopted in 1991 a Federal Policy for the Protection of Human Subjects. This policy requires that every institution seeking federal funding for biomedical or behavioral research on human subjects have an institutional review board to exercise oversight.

- The ASA's standards for the protection of human subjects require avoiding harm, obtaining informed consent, avoiding deception except in limited circumstances, and maintaining privacy and confidentiality.

- Scientific research should maintain high standards for validity and be conducted and reported in an honest and open fashion.

- Effective debriefing of subjects after an experiment can help reduce the risk of harm due to the use of deception in the experiment.

- Positivism and postpositivism are research philosophies that emphasize the goal of understanding the real world; these philosophies guide most quantitative researchers. Interpretivism is a research philosophy that emphasizes an understanding of the meaning people attach to their experiences; it guides many qualitative researchers.

- The constructivist paradigm reflects an interpretivist philosophy. It emphasizes the importance of exploring and representing the ways in which different stakeholders in a social setting construct their beliefs. Constructivists interact with research subjects to develop a shared perspective on the issue being studied.

- Feminist researchers often emphasize interpretivist and constructivist perspectives in research and urge a concern with underprivileged groups.

- A scientific paradigm is a set of beliefs that guide most scientific work in an area. A scientific revolution occurs when the accumulation of contrary evidence leads to a shift from acceptance of one paradigm to another. Some researchers view positivism/postpositivism and interpretivism/constructivism as alternative paradigms.

STUDENT STUDY SITE

To assist in completing the web exercises, please access the study site at **www.sagepub.com/schuttisw7e**, where you will find the web exercises with accompanying links. You'll find other useful study materials such as self-quizzes and e-flashcards for each chapter, along with a group of carefully selected articles from research journals that illustrate the major concepts and techniques presented in the book.

Discussion Questions

1. Should social scientists be permitted to conduct replications of Milgram's obedience experiments? Zimbardo's prison simulation? Can you justify such research as permissible within the current ASA ethical standards? If not, do you believe that these standards should be altered so as to permit Milgram-type research?

2. How do you evaluate the current ASA ethical code? Is it too strict or too lenient, or just about right? Are the enforcement provisions adequate? What provisions could be strengthened?

3. Why does unethical research occur? Is it inherent in science? Does it reflect "human nature"? What makes ethical research more or less likely?

4. Does debriefing solve the problem of subject deception? How much must researchers reveal after the experiment is over as well as before it begins?

5. What policy would you recommend that researchers such as Sherman and Berk (1984) follow in reporting the results of their research? Should social scientists try to correct misinformation in the popular press about their research or should they just focus on what is published in academic journals? Should researchers speak to audiences like police conventions in order to influence policies related to their research results?

6. Do you favor the positivist/postpositivist or the interpretivist/constructivist philosophy as a guide for social research? Review the related guidelines for research and explain your position.

Practice Exercises

1. Pair up with one other student and select one of the research articles you have reviewed for other exercises. Criticize the research in terms of its adherence to each of the ethical principles for research on human subjects, as well as for the authors' apparent honesty, openness, and consideration of social consequences. Be generally negative but not unreasonable in your criticisms. The student with whom you are working should critique the article in the same way but from a generally positive standpoint, defending its adherence to the five guidelines, but without ignoring the study's weak points. Together, write a summary of the study's strong and weak points, or conduct a debate in the class.

2. Investigate the standards and operations of your university's IRB. Review the IRB website, record the composition of the IRB (if indicated), and outline the steps that faculty and students must take in order to secure IRB approval for human subjects research. In your own words, distinguish the types of research that can be exempted from review, that qualify for expedited review, and that require review by the full board. If possible, identify another student or a faculty member who has had a proposal reviewed by the IRB. Ask them to describe their experience and how they feel about it. Would you recommend any changes in IRB procedures? Researchers should consider their research philosophy as well as their theoretical stance prior to designing a research project. The "Theories and Philosophies" lesson in the Interactive Exercises link on the text's study site will help you think about the options.

3. To use these lessons, choose one of the four "Theories and Philosophies" exercises from the opening menu for the Interactive Exercises. Follow the instructions for entering your answers and responding to the program's comments.

4. Also from the book's study site, at www.sagepub.com/schuttisw7e, choose the Learning From Journal Articles option. Read one article based on research involving human subjects. What ethical issues did the research pose and how were they resolved? Does it seem that subjects were appropriately protected?

Web Exercises

1. The Collaborative Institutional Training Initiative (CITI) offers an extensive online training course in the basics of human subjects protections issues. Go to the public access CITI site at www.citiprogram.org/rcrpage.asp?affiliation=100 and complete the course in social and behavioral research. Write a short summary of what you have learned.

2. The U.S. Department of Health and Human Services maintains extensive resources concerning the protection of human subjects in research. Read several documents that you find on their website, www.hhs.gov/ohrp, and write a short report about them.

3. Read the entire ASA *Code of Ethics* at the website of the ASA Ethics Office, www.asanet.org/images/asa/docs/pdf/Codeof Ethics.pdf. Discuss the difference between the aspirational goals and the enforceable rules.

4. There are many interesting websites that discuss philosophy of science issues. Read the summaries of positivism and interpretivism at http://bjps.oxfordjournals.org/. What do these summaries add to your understanding of these philosophical alternatives?

Developing a Research Proposal

Now it's time to consider the potential ethical issues in your proposed study and the research philosophy that will guide your research. The following exercises involve very critical "Decisions in Research" (Exhibit 2.14, #6):

1. List the elements in your research plans that an IRB might consider to be relevant to the protection of human subjects. Rate each element from 1 to 5, where 1 indicates no more than a minor ethical issue and 5 indicates a major ethical problem that probably cannot be resolved.

2. Write one page for the application to the IRB that explains how you will ensure that your research adheres to each relevant ASA standard.

3. Draft a consent form to be administered to your subjects when they enroll in your research. Use underlining and marginal notes to indicate where each standard for informed consent statements is met.

4. Do you find yourself more attracted to the positivist/ postpositivist philosophy, or to an interpretivist/ constructivist or feminist philosophy? Why?

5. List the research guidelines that are consistent with the research philosophy you will adopt and suggest steps you will take to ensure adherence to each guideline.

CHAPTER 4

Conceptualization and Measurement

S ubstance abuse is a social problem of remarkable proportions, both on and off campus. In 2001 to 2002, almost 10 million adult Americans were alcohol abusers (Grant et al. 2004), and in 2002, about 400,000 years of life were lost due to alcohol-related traffic deaths (Yi, Williams, & Smothers 2004). Alcohol is involved in about half of all fatal traffic crashes, and more than 1 million arrests are made annually for driving under the influence. Workplace alcohol use occurs among 15% of the U.S. workforce, while illicit drug use occurs among 3% (Frone 2008). While in college, 4 out of 10 students binge drink (Wechsler et al. 2002), and about 1 out of 3 could be diagnosed as alcohol abusers (Knight et al. 2002). Drinking is a factor in at least half of the on-campus sexual assaults (Abbey 2002). All told, the annual costs of prevention and treatment for alcohol and drug abuse exceed $4 billion (Gruenewald et al. 1997).

Whether your goal is to learn how society works, to deliver useful services, to design effective social policies, or simply to try to protect yourself and your peers, at some point you might decide to read some of the research literature on substance abuse. Perhaps you will even attempt to design your own study of it. Every time you begin to review or design relevant research, you will have to answer two questions: (1) What is meant by *substance abuse* in this research? (the conceptualization issue) and (2) How was substance abuse measured? (the operationalization issue). Both types of questions must be answered when we evaluate prior research, and both types of questions must be kept in the forefront when we design new research. It is only when we conclude that a study used valid measures of its key concepts that we can have some hope that its conclusions are valid.

In this chapter, I first address the issue of conceptualization, using substance abuse and other concepts as examples. I then focus on measurement, reviewing first how measures of substance abuse have been constructed using such operations as available data, questions, observations, and less direct and obtrusive measures. Next I discuss the different possible levels of measurement and methods for assessing the validity and reliability of measures. The final topic is to consider the unique insights that qualitative methods can add to the measurement process. By the chapter's end, you should have a good understanding of measurement, the first of the three legs on which a research project's validity rests.

▣ Concepts

Although the drinking statistics sound scary, we need to be clear about what they mean before we march off to a Temperance Society meeting. What, after all, is binge drinking? The definition that Wechsler et al. (2002) used is "heavy episodic drinking"; more specifically, "we defined binge drinking as the consumption of at least 5 drinks in a row for men or 4 drinks in a row for women during the 2 weeks before completion of the questionnaire" (p. 205).

Is this what you call *binge drinking*? This definition is widely accepted among social researchers, so when they use the term they can understand each other. However, the National Institute on Alcoholism and Alcohol Abuse (College Alcohol Study 2008) provides a more precise definition: "A pattern of drinking alcohol that brings blood alcohol concentration to 0.08 gram percent or above." Most researchers consider the so-called 5/4 definition (5 drinks for men; 4 for women) to be a reasonable approximation to this more precise definition. We can't say that only one definition of binge drinking is "correct," or even that one is "better." What we can say is that we need to specify what we mean when we use the term. We also have to be sure that others know what definition we are using. And of course, the definition has to be useful for our purposes: A definition based solely on blood alcohol concentration will not be useful if we are not taking blood measures.

We call binge drinking a **concept**—a mental image that summarizes a set of similar observations, feelings, or ideas. To make that concept useful in research (and even in ordinary discourse), we have to define it. Many concepts are used in everyday discourse without consistent definition, sometimes definitions of concepts are themselves the object of intense debate, and the meanings of concepts may change over time. For example, when we read a *New York Times* article (Stille 2000) announcing a rise in the "social health" of the United States, after a precipitous decline in the 1970s and 1980s, we don't know whether we should feel relieved or disinterested. In fact, the authorities on the subject didn't even agree about what the term *social health* meant: lessening of social and economic inequalities (Marc Miringoff) or clear moral values (William J. Bennett). Most agreed that social health has to do with "things that are not measured in the gross national product" and that it is "a more subtle and more meaningful way of measuring what's important to [people]" (Stille 2000:A19), but the sparks flew over whose **conceptualization** of social health would prevail.

> **Concept** A mental image that summarizes a set of similar observations, feelings, or ideas, indicators, and overlapping dimensions.
>
> **Conceptualization** The process of specifying what we mean by a term. In deductive research, conceptualization helps translate portions of an abstract theory into specific variables that can be used in testable hypotheses. In inductive research, conceptualization is an important part of the process used to make sense of related observations.

Prejudice is an interesting example of a concept whose meaning has changed over time. As Harvard psychologist Gordon Allport (1954) pointed out, during the 1950s many people conceptualized prejudice as referring to "faulty generalizations" about other groups. The idea was that these cognitive "errors in reasoning" could be improved with better education. But by the end of the 1960s, this one-size-fits-all concept was replaced with more specific terms such as *racism, sexism,* and *anti-Semitism* that were conceptualized as referring to negative dispositions about specific groups that "ran too deep to be accessible to cursory introspection" (Nunberg 2002:WK3). The *isms* were conceived as both more serious and less easily acknowledged than prejudice.

Concepts such as social health, prejudice, and even binge drinking require an explicit definition before they are used in research because we cannot be certain that all readers will share a particular definition or that the current meaning of the concept is the same as it was when previous research was published. It is especially important to define clearly concepts that are abstract or unfamiliar. When we refer to concepts such as *social control, anomie,* or *social health,* we cannot count on others knowing exactly what we mean. Even experts may disagree about the meaning of frequently used concepts if they based their conceptualizations on different theories. That's okay. The point is not that there can only be one definition of a concept but that we have to specify clearly what we mean when we use a concept, and we must expect others to do the same.

Conceptualization in Practice

If we are to do an adequate job of conceptualizing, we must do more than just think up some definition, any definition, for our concepts (Goertz 2006). We have to turn to social theory and prior research to review appropriate definitions. We need to identify what we think is important about the phenomenon that interests us. We should understand how the definition we choose fits within the theoretical framework guiding the research, and what assumptions underlie this framework. We may decide the concept has several dimensions, or subconcepts, that should be distinguished.

Substance Abuse

What observations or images should we associate with the concept *substance abuse*? Someone leaning against a building with a liquor bottle, barely able to speak coherently? College students drinking heavily at a party? Someone in an Alcoholics Anonymous group drinking one beer? A 10-year-old boy drinking a small glass of wine in an alley? A 10-year-old boy drinking a small glass of wine at the dinner table in France? Do all these images share something in common that we should define as substance abuse for the purposes of a particular research study? Do only some of them share something in common? Should we take into account the cultural differences? Social situations? Physical tolerance for alcohol? Individual standards?

Many researchers now use the definition of substance abuse contained in the American Psychiatric Association's (2000) *Diagnostic and Statistical Manual of Mental Disorders, text revision (DSM-IV-TR):* "a maladaptive pattern of substance use manifested by recurrent and significant adverse consequences related to the repeated use of substances . . . must have occurred repeatedly during the same 12-month period or been persistent" (*DSM-IV-TR:* Substance Abuse Features section, p. 198). But, in spite of its popularity among professionals, we cannot judge the *DSM-IV-TR* definition of substance abuse as "correct" or "incorrect." Each researcher has the right to conceptualize as he or she sees fit. However, we can say that the *DSM-IV-TR* definition of substance abuse is useful, in part because it has been very widely adopted. It is also stated in a clear and precise language that minimizes differences in interpretation and maximizes understanding.

This clarity should not prevent us from recognizing that the definition reflects a particular theoretical orientation. *DSM-IV-TR* applies a medical *disease model* to mental illness (which is conceptualized, in *DSM-IV-TR,* to include substance abuse). This theoretical model emphasizes behavioral and biological criteria instead of the social expectations that are emphasized in a social model of substance abuse. How we conceptualize reflects how we theorize.

Just as we can connect concepts to theory, we also can connect them to other concepts. What this means is that the definition of any one concept rests on a shared understanding of the other terms used in the definition. So if our audience does not already have a shared understanding of terms such as *adequate social functioning, significant adverse consequences,* and *repeated use,* we must also define these terms before we are finished with the process of defining substance abuse.

Youth Gangs

Do you have a clear image in mind when you hear the term *youth gangs?* Although this is a very ordinary term, social scientists' attempts to define precisely the concept, youth gang, have not yet succeeded: "Neither gang researchers nor law enforcement agencies can agree on a common definition . . . and a concerted national effort . . . failed to reach a consensus" (Howell 2003:75). Exhibit 4.1 lists a few of the many alternative definitions of youth gangs.

What is the basis of this conceptual difficulty? Researcher James Howell (2003:27–28) suggests that defining the term *youth gangs* has been difficult for four reasons:

1. Youth gangs are not particularly cohesive.

2. Individual gangs change their focus over time.

3. Many have a "hodgepodge of features," with diverse members and unclear rules.

4. There are many incorrect but popular "myths" about youth gangs.

In addition, youth gangs are only one type of social group, and it is important to define youth gangs in a way that distinguishes them from these other types of groups, for example, childhood play groups, youth subculture groups, delinquent groups, and adult criminal organizations. You can think of *social group* as a broader concept that has multiple dimensions, one of which is youth gangs. In the same way, you can think of substance abuse as a concept with three dimensions: alcohol abuse, drug abuse, and polysubstance abuse. Whenever you define a concept, you need to consider whether the concept is unidimensional or multidimensional. If it is multidimensional, your job of conceptualization is not complete until you have specified the related subconcepts that belong under the umbrella of the larger concept (see Exhibit 4.2).

Exhibit 4.1 | **Alternative Definitions of Youth Gangs**

The term gang tends to designate collectivities that are marginal members of mainstream society, loosely organized, and without a clear, social purpose. (Ball & Curry 1995:227)

The gang is an interstitial group (between childhood and maturity) originally formed spontaneously, and then integrated through conflict. (Thrasher 1927:18)

[A gang is] any denotable adolescent group of youngsters who (a) are generally perceived as a distinct aggregation by others in the neighborhood, (b) recognize themselves as a denotable group (almost invariably with a group name), and (c) have been involved in a sufficient number of delinquent incidents to call forth a consistently negative response from neighborhood residents and/or law enforcement agencies. (Klein 1971:13)

A youth gang is a self-formed association of peers united by mutual interests with identifiable leadership and internal organization who act collectively or as individuals to achieve specific purposes, including the conduct of illegal activity and control of a particular territory, facility, or enterprise. (Miller 1992:21)

[A gang is] an age-graded peer group that exhibits some permanence, engages in criminal activity, and has some symbolic representation of membership. (Decker & Van Winkle 1996:31)

[A gang is] a self-identified group of kids who act corporately, at least sometimes, and violently, at least sometimes. (Kennedy, Piehl, & Braga 1996:158)

A Criminal Street Gang is any ongoing organization, association, or group of three or more persons, whether formal or informal, having as one of its primary activities the commission of criminal acts. (Street Terrorism Enforcement and Prevention Act, 1988, California Penal Code sec. 186.22[f])

Exhibit 4.2 | **Peer-Based Social Groups: A Multidimensional Concept**

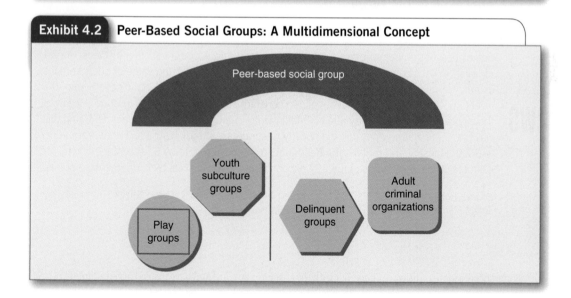

Poverty

Decisions about how to define a concept reflect the theoretical framework that guides the researchers. For example, the concept *poverty* has always been somewhat controversial, because different conceptualizations of poverty lead to different estimates of its prevalence and different social policies for responding to it.

Most of the statistics that you see in the newspapers about the poverty rate in the United States reflect a conception of poverty that was formalized by Mollie Orshansky of the Social Security Administration in 1965 and subsequently adopted by the federal government and many researchers (Putnam 1977). She defined poverty in terms of what is called an *absolute* standard, based on the amount of money required to purchase an emergency diet that is estimated to be nutritionally adequate for about 2 months. The idea is that people are truly poor if they can just barely purchase the food they need and other essential goods. This poverty standard is adjusted for household size and composition (number of children and adults), and the minimal amount of money needed for food is multiplied by three because a 1955 survey indicated that poor families spend about one-third of their incomes on food (Orshansky 1977).

Does this sound straightforward? As is often the case with important concepts, the meaning of an absolute poverty standard has been the focus of a vigorous debate (Eckholm 2006:A8). While the traditional definition of absolute poverty takes account only of a family's cash income, some argue that noncash benefits that low-income people can receive, such as food stamps, housing subsidies, and tax rebates should be added to cash income before the level of poverty is calculated. Douglas Besharov of the American Enterprise Institute terms this approach "a much needed corrective" (Eckholm 2006:A8). But some social scientists have proposed increasing the absolute standard for poverty so that it reflects what a low-income family must spend to maintain a "socially acceptable standard of living" that allows for a telephone, house repairs, and decent clothes (Uchitelle 1999). Others argue that the persistence of poverty should be taken into account, so someone who is poor for no more than a year, for example, is distinguished from someone who is poor for many years (Walker et al. 2010:367–368). Since any change in the definition of poverty will change eligibility for government benefits such as food stamps and Medicaid, the feelings about this concept run deep.

Research in the News

HOW POVERTY IS DEFINED MATTERS

Calculation of the federal poverty rate does not take into account variation in the local cost of living, expenses for health care, commuting and day care, or the value of such benefits as food stamps, housing allowances, or tax credits. According to the federal poverty rate, the level of poverty in New York City remained constant at 17.3% from 2008 to 2009. However, the city calculates the poverty rate after taking into account all these other factors. Using this way of operationalizing poverty, the poverty rate increased from 19.6% to 19.9% from 2008 to 2009. Housing and rental subsidies reduced the poverty rate by 6 percentage points, for example, while unreimbursed medical expenses increased the rate by 3.1 percentage points.

Source: Roberts, Sam. 2011. "Food Stamps and Tax Aid Kept Poverty Rate in Check." *The New York Times,* March 21:A19.

In the News

Some social scientists disagree altogether with the absolute standard and have instead urged adoption of a *relative* poverty standard (see Exhibit 4.3). They identify the poor as those in the lowest fifth or tenth of the income distribution or as those having some fraction of the average income. The idea behind this relative conception is that poverty should be defined in terms of what is normal in a given society at a particular time. "For example, while a car may be a luxury in some poor countries, in a country where most families own cars and public transportation is inadequate, a car is a basic necessity for finding and commuting to work" (Mayrl et al. 2004:10). This relative conception of poverty has largely been accepted in Europe (Walker et al. 2010:356).

Some social scientists prefer yet another conception of poverty. With the *subjective* approach, poverty is defined as what people think would be the minimal income they need to make ends meet. Of course, many have argued that this approach is influenced too much by the different standards that people use to estimate what they "need" (Ruggles 1990:20–23).

Which do you think is a more reasonable approach to defining poverty: some type of absolute standard, a relative standard, or a subjective standard? Be careful here: Conceptualization has consequences! Research using the standard absolute concept of poverty indicated that the percentage of Americans in poverty declined by 1.7% in the 1990s, but use of a relative concept of poverty led to the conclusion that poverty increased by

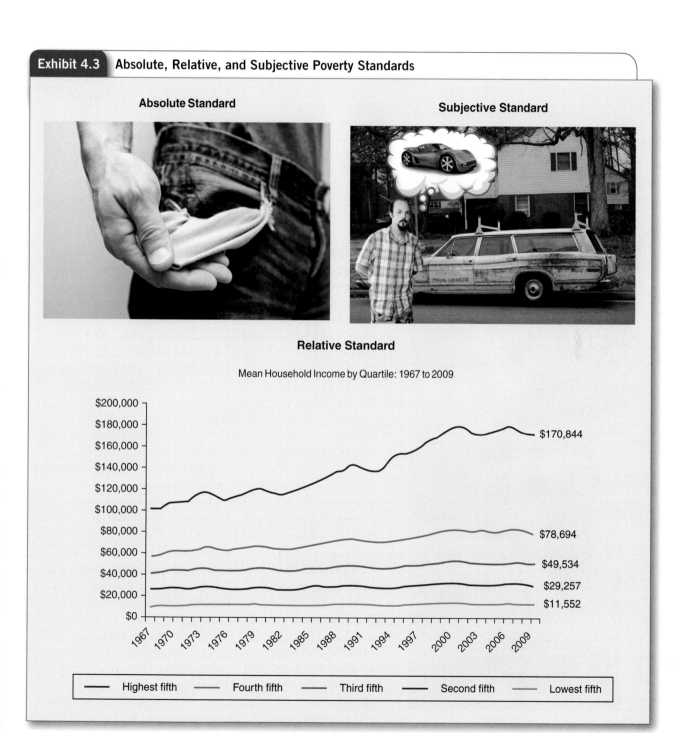

Exhibit 4.3 Absolute, Relative, and Subjective Poverty Standards

Absolute Standard

Subjective Standard

Relative Standard

Mean Household Income by Quartile: 1967 to 2009

$170,844

$78,694

$49,534

$29,257

$11,552

—— Highest fifth —— Fourth fifth —— Third fifth —— Second fifth —— Lowest fifth

2.7% (Mayrl et al. 2004:10). No matter which conceptualization we decide to adopt, our understanding of the concept of poverty will be sharpened after we consider these alternative definitions.

Trust

Take a look at Exhibit 4.4. It's a picture used by Pamela Paxton to illustrate the concept of *trust*. Do you see what it is about the picture that represents trust in other people? Have you ever thought about trust when you have left cash on a restaurant table and then walked away? Paxton (2005) defines the concept of trust with examples: "We trust others when we take a chance, yielding them some control over our money, secrets, safety, or other things we value" (p. 40).

She then distinguishes trust in people from trust in institutions. According to the survey data she reports, trust in people has been declining in America since the early 1960s, but there has not been an overall change in trust in institutions (Paxton 2005:41–44).

Operationalization

Exhibit 4.4   Picturing Trust

Identifying the concepts we will study, specifying dimensions of these concepts, and defining their meaning only begin the process of connecting our ideas to concrete observations. If we are to conduct empirical research involving a concept, we must be able to distinguish it in the world around us and determine how it may change over time or differ between persons or locations. **Operationalization** involves connecting concepts to measurement operations. You can think of it as the empirical counterpart of the process of conceptualization. When we conceptualize, we specify what we mean by a term (see Exhibit 4.5). When we operationalize, we identify specific measurements we will take to indicate that concept in empirical reality. Operationalization is the critical research step for those who identify with a positivist or postpositivist research philosophy (see Chapter 3): It is through operationalization that researchers link the abstract ideas (concepts) in our heads to concrete indicators in the real world. Researchers also find that the process of figuring out how to measure a concept helps to improve their understanding of what the concept means (Bartholomew 2010:457). Improving conceptualization and improving operationalization go hand in hand.

Exhibit 4.5 illustrates conceptualization and operationalization by using the concept of *social control,* which Donald Black (1984) defines as "all of the processes by which people define and respond to deviant behavior" (p. xi). What observations can indicate this conceptualization of social control? Billboards that condemn drunk driving? Proportion of persons arrested in a community? Average length of sentences for crimes? Types of bystander reactions to public intoxication? Gossiping among neighbors? Some combination of these? Should we distinguish formal social control such as laws and police actions from informal types of social control such as social stigma? If

Although definitions may be similar, if different operationalizations and measures are used, the result could be that survey questions will be different and, if so, prevalence estimates could not be combined across those studies. (cited in Larence 2006:53)

Although the process of achieving consistency in operationalizing intimate partner violence will not be completed quickly, the operational definition of intimate partner violence in the National Violence Against Women survey provides a good example of a useful approach:

Intimate partner violence . . . includes rape, physical assault, and stalking perpetrated by current and former dates, spouses, and cohabiting partners, with cohabiting meaning living together at least some of the time as a couple. Both same-sex and opposite-sex cohabitants are included in the definition. (Tjaden & Thoennes 2000:5)

From Observations to Concepts

Qualitative research projects usually take an inductive approach to the process of conceptualization. In an inductive approach, concepts emerge from the process of thinking about what has been observed, as compared with the deductive approach that I have just described, in which we develop concepts on the basis of theory and then decide what should be observed to indicate that concept. So instead of deciding in advance which concepts are important for a study, what these concepts mean, and how they should be measured, if you take an inductive approach, you will begin by recording verbatim what you hear in intensive interviews or see during observational sessions. You will then review this material to identify important concepts and their meaning for participants. At this point, you may also identify relevant variables and develop procedures for indicating variation between participants and settings or variation over time. As your understanding of the participants and social processes develops, you may refine your concepts and modify your indicators. The sharp boundaries in quantitative research between developing measures, collecting data with those measures, and evaluating the measures often do not exist in inductive, qualitative research.

Being "In" and "Out"

You will learn more about qualitative research in Chapter 9, but an example here will help you understand the qualitative measurement approach. For several months, Darin Weinberg (2000) observed participants in three drug abuse treatment programs in Southern California. He was puzzled by the drug abuse treatment program participants' apparently contradictory beliefs—that drug abuse is a medical disease marked by "loss of control" but that participation in a therapeutic community can be an effective treatment. He discovered that treatment participants shared an "ecology of addiction" in which they conceived of being "in" the program as a protected environment, while being in the community was considered being "out there" in a place where drug use was inevitable—in "a space one's addiction compelled one to inhabit" (Weinberg 2000:609).

I'm doin' real, real bad right now. . . . I'm havin' trouble right now staying clean for more than two days. . . . I hate myself for goin' out and I don't know if there's anything that can save me anymore. . . . I think I'm gonna die out there. (Weinberg 2000:609)

Participants contrasted their conscientiousness while in the program with the personal dissolution of those out in "the life."

So, Weinberg developed the concepts of *in* and *out* inductively, in the course of the research, and identified indicators of these concepts at the same time in the observational text. He continued to refine and evaluate the concepts throughout the research. Conceptualization, operationalization, and validation were ongoing and interrelated processes. We'll study this process in more detail in Chapter 10.

Diversity

Qualitative research techniques may also be used to explore the meaning of a concept. For example, everybody uses the term *diversity*, but Joyce M. Bell and Douglas Hartmann (2007) designed a qualitative interview project to find out "what are Americans really saying about diversity? How do they understand and experience it?" In a total of 166 interviews, they uncovered a complex pattern. Most respondents initially gave optimistic interpretations of diversity: It "makes life more fun" or "more exciting" (Bell & Hartmann 2007:899). However, many respondents went on to add a more pessimistic view: "If you have too much diversity, then you have to change the Constitution, you have to take down the Statue of Liberty, you have take down those things that set this country up as it is" (Bell & Hartmann 2007:901).

Bell and Hartmann (2007:911) concluded that the term "diversity" is often used to avoid considering the implications of social inequality—that the concept of diversity supports "happy talk" that obscures social divisions. So their understanding of the concept of diversity changed as a result of the evidence they collected. What is *your* concept of diversity?

▣ Measurement Operations

Measurement The process of linking abstract concepts to empirical indicants.

Operations A procedure for identifying or indicating the value of cases on a variable.

The deductive researcher proceeds from defining concepts in the abstract (conceptualizing) to identifying variables to measure, and finally to developing specific measurement procedures. **Measurement** is the "process of linking abstract concepts to empirical indicants" (Carmines & Zeller 1979:10). The goal is to achieve measurement validity, so the measurement **operations** must actually measure the variables they are intended to measure.

Exhibit 4.8 represents the operationalization process in three studies. The first researcher defines her concepts, binge drinking, and chooses one variable—frequency of heavy episodic drinking—to represent it. This variable is then measured with responses to a single question, or indicator: "How often within the past 2 weeks did you consume five or more drinks containing alcohol in a row?" Because "heavy" drinking is defined differently for men and women (relative to their different metabolism), the question is phrased in terms of "four or more drinks" for women. The second researcher defines her concept, poverty, as having two aspects or dimensions: subjective poverty and absolute poverty. Subjective poverty is measured with responses to a survey question: "Would you say you are poor?" Absolute poverty is measured by comparing family income to the poverty threshold. The third researcher decides that her concept, socioeconomic status, is defined by a position on three measured variables: income, education, and occupational prestige.

Social researchers have many options for operationalizing concepts. Measures can be based on activities as diverse as asking people questions, reading judicial opinions, observing social interactions, coding words in books, checking census data tapes, enumerating the contents of trash receptacles, or drawing urine and blood samples. Experimental researchers may operationalize a concept by manipulating its value. For example, to operationalize the concept of *exposure to antidrinking messages,* some subjects may listen to a talk about binge drinking while others do not. I will focus here on the operations of using published data, asking questions, observing behavior, and using unobtrusive means of measuring people's behavior and attitudes.

The variables and particular measurement operations chosen for a study should be consistent with the research question. If we ask the evaluative research question, "Are self-help groups more effective than hospital-based treatments in reducing drinking among substance abusers?" we may operationalize "form of

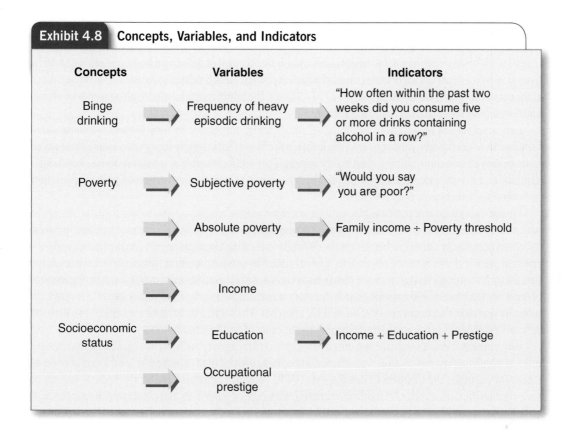

Exhibit 4.8 Concepts, Variables, and Indicators

treatment" in terms of participation in these two types of treatments. However, if we are attempting to answer the explanatory research question "What influences the success of substance abuse treatment?" we should probably consider what it is about these treatment alternatives that is associated with successful abstinence. Prior theory and research suggest that some of the important variables that differ between these treatment approaches are level of peer support, beliefs about the causes of alcoholism, and financial investment in the treatment.

Time and resource limitations must also be taken into account when we select variables and devise measurement operations. For many sociohistorical questions (e.g., "How has the poverty rate varied since 1950?"), census data or other published counts must be used. On the other hand, a historical question about the types of social bonds among combat troops in 20th-century wars probably requires retrospective interviews with surviving veterans. The validity of the data is lessened by the unavailability of many veterans from World War I and by problems of recall, but direct observation of their behavior during the war is certainly not an option.

Using Available Data

Government reports are rich and readily accessible sources of social science data. Organizations ranging from nonprofit service groups to private businesses also compile a wealth of figures that may be available to some social scientists for some purposes. In addition, the data collected in many social science surveys are archived and made available for researchers who were not involved in the original survey project.

Before we assume that available data will be useful, we must consider how appropriate they are for our concepts of interest. We may conclude that some other measure would provide a better fit with a concept or that a particular concept simply cannot be adequately operationalized with the available data. For example, law enforcement and health statistics provide several community-level indicators of substance abuse (Gruenewald et al. 1997). Statistics on arrests for the sale and possession of drugs, drunk driving arrests, and liquor law violations (such as sales to minors) can usually be obtained on an annual basis, and often quarterly, from local police departments or state crime information centers. Health-related indicators of substance abuse at the community level include single-vehicle fatal crashes, the rate of mortality due to alcohol or drug abuse, and the use of alcohol and drug treatment services.

Indicators such as these cannot be compared across communities or over time without reviewing carefully how they were constructed. The level of alcohol in the blood that is legally required to establish intoxication can vary among communities, creating the appearance of different rates of substance abuse even though drinking and driving practices may be identical. Enforcement practices can vary among police jurisdiction and over time (Gruenewald et al. 1997:14).

We also cannot assume that available data are accurate, even when they appear to measure the concept in which we are interested in a way that is consistent across communities. "Official" counts of homeless persons have been notoriously unreliable because of the difficulty in locating homeless persons on the streets, and government agencies have, at times, resorted to "guesstimates" by service providers (Rossi 1989). Even available data for such seemingly straightforward measures as counts of organizations can contain a surprising amount of error. For example, a 1990 national church directory reported 128 churches in a midwestern U.S. county; an intensive search in that county in 1992 located 172 churches (Hadaway, Marler, & Chaves 1993:744). Perhaps 30% or 40% of death certificates identify incorrectly the cause of death (Altman 1998).

Government statistics that are generated through a central agency such as the U.S. Census Bureau are often of a high quality, but caution is warranted when using official data collected by local levels of government. For example, the Uniform Crime Reports (UCR) program administered by the Federal Bureau of Investigation imposes standard classification criteria, with explicit guidelines and regular training at the local level, but data are still inconsistent for many crimes. Consider only a few of the many sources of inconsistency between jurisdictions: Variation in the classification of forcible rape cases due to differences in what is considered to be "carnal knowledge of a female"; different decisions about what is considered "more than necessary force" in the definition of "strong-arm" robberies; whether offenses in which threats were made but no physical injury occurred are classified as aggravated or simple assaults (Mosher, Miethe, & Phillips 2002:66). A new National Incident-Based Reporting System (NIBRS) corrects some of the problems with the UCR, but it requires much more training and documentation and has not yet been widely used (Mosher et al. 2002:70).

In some cases, problems with an available indicator can be lessened by selecting a more precise indicator. For example, the number of single-vehicle nighttime crashes, whether fatal or not, is a more specific indicator of the frequency of drinking and driving than just the number of single-vehicle fatal accidents (Gruenewald et al. 1997:40–41). Focusing on a different level of aggregation may also improve data quality, because procedures for data collection may differ between cities, counties, states, and so on (Gruenewald et al. 1997:40–41). It is only after factors such as legal standards, enforcement practices, and measurement procedures have been taken into account that comparisons among communities become credible.

Constructing Questions

Asking people questions is the most common and probably the most versatile operation for measuring social variables. Most concepts about individuals can be defined in such a way that measurement with one or more questions becomes an option. We associate questions with survey research, but questions are also often the basis of measures used in social experiments and in qualitative research. In this section, I introduce some options for writing single questions; in Chapter 8, I explain why single questions can be inadequate measures of some concepts, and then I examine measurement approaches that rely on multiple questions to measure a concept.

Of course, in spite of the fact that questions are, in principle, a straightforward and efficient means to measure individual characteristics, facts about events, level of knowledge, and opinions of any sort, they can easily result in misleading or inappropriate answers. Memories and perceptions of the events about which we might like to ask can be limited, and some respondents may intentionally give misleading answers. For these reasons, all questions proposed for a study must be screened carefully for their adherence to basic guidelines and then tested and revised until the researcher feels some confidence that they will be clear to the intended respondents and likely to measure the intended concept (Fowler 1995). Alternative measurement approaches will be needed when such confidence cannot be achieved.

Specific guidelines for reviewing survey questions are presented in Chapter 8; here, my focus is on the different types of questions used in social research.

Measuring variables with single questions is very popular. Public opinion polls based on answers to single questions are reported frequently in newspaper articles and TV newscasts: "Do you favor or oppose U.S. policy . . . ?" "If you had to vote today, for which candidate would you vote?" Social science surveys also rely on single questions to measure many variables: "Overall, how satisfied are you with your job?" "How would you rate your current health?"

Single questions can be designed with or without explicit response choices. The question that follows is a **closed-ended (fixed-choice) question**, because respondents are offered explicit responses from which to choose. It has been selected from the Core Alcohol and Drug Survey distributed by the Core Institute, Southern Illinois University, for the Fund for the Improvement of Postsecondary Education (FIPSE) Core Analysis Grantee Group (Presley, Meilman, & Lyerla 1994).

> **Closed-ended (fixed-choice) question** A survey question that provides preformatted response choices for the respondent to circle or check.

Compared to other campuses with which you are familiar, this campus's use of alcohol is . . . (Mark one)

_____ Greater than other campuses

_____ Less than other campuses

_____ About the same as other campuses

Most surveys of a large number of people contain primarily fixed-choice questions, which are easy to process with computers and analyze with statistics. With fixed-choice questions, respondents are also more likely to answer the questions that the researcher really wants them to answer. Including response choices reduces ambiguity and makes it easier for respondents to answer. However, fixed-response choices can obscure what people really think if the choices do not match the range of possible responses to the question; many studies show that some respondents will choose response choices that do not apply to them simply to give some sort of answer (Peterson 2000:39).

Most important, response choices should be **mutually exclusive** and exhaustive, so that every respondent can find one and only one choice that applies to him or her (unless the question is of the "Check all that apply" format). To make response choices exhaustive, researchers may need to offer at least one option with room for ambiguity. For example, a questionnaire asking college students to indicate their school status should not use freshman, sophomore, junior, senior, and graduate student as the only response choices. Most campuses also have students in a "special" category, so you might add "Other (please specify)" to the five fixed responses to this question. If respondents do not find a response option that corresponds to their answer to the question, they may skip the question entirely or choose a response option that does not indicate what they are really thinking.

> **Mutually exclusive** A variable's attributes (or values) are mutually exclusive when every case can be classified as having only one attribute (or value).

Open-ended questions, questions without explicit response choices, to which respondents write in their answers, are preferable when the range of responses cannot adequately be anticipated—namely, questions that have not previously been used in surveys and questions that are asked of new groups. Open-ended questions can also lessen confusion about the meaning of responses involving complex concepts. The next question is an open-ended version of the earlier fixed-choice question:

> **Open-ended question** A survey question to which the respondent replies in his or her own words, either by writing or by talking.

> How would you say alcohol use on this campus compares to that on other campuses?

In qualitative research, open-ended questions are often used to explore the meaning respondents give to abstract concepts. *Mental illness,* for example, is a complex concept that tends to have different meanings for different people. In a survey I conducted in homeless shelters, I asked the staff whether they believed that people at the shelter had become homeless due to mental illness (Schutt 1992). When given fixed-response choices, 47% chose "Agree" or "Strongly agree." However, when these same staff members were interviewed in depth, with open-ended questions, it became clear that the meaning of these responses varied among staff. Some believed that mental illness caused homelessness by making people vulnerable in the face of bad luck and insufficient resources:

> Mental illness [is the cause]. Just watching them, my heart goes out to them. Whatever the circumstances were that were in their lives that led them to the streets and being homeless I see it as very sad. . . . Maybe the resources weren't there for them, or maybe they didn't have the capabilities to know when the resources were there. It is misfortune. (Schutt 1992:7)

Other staff believed that mental illness caused people to reject housing opportunities:

> I believe because of their mental illness that's why they are homeless. So for them to say I would rather live on the street than live in a house and have to pay rent, I mean that to me indicates that they are mentally ill. (Schutt 1992:7)

Just like fixed-choice questions, open-ended questions should be reviewed carefully for clarity before they are used. For example, if respondents are just asked, "When did you move to Boston?" they might respond with a wide range of answers: "In 1944." "After I had my first child." "When I was 10." "Twenty years ago." Such answers would be very hard to compile. A careful review should identify potential ambiguity. To avoid it, rephrase the question to guide the answer in a certain direction, such as "In what year did you move to Boston?" or provide explicit response choices (Center for Survey Research 1987).

The decision to use closed-ended or open-ended questions can have important consequences for the information reported. Leaving an attitude or behavior off a fixed set of response choices is likely to mean that it is not reported, even if an "other" category is provided. However, any attitude or behavior is less likely to be reported if it must be volunteered in response to an open-ended question (Schwarz 2010:48).

Making Observations

Observations can be used to measure characteristics of individuals, events, and places. The observations may be the primary form of measurement in a study, or they may supplement measures obtained through questioning.

Direct observations can be used as indicators of some concepts. For example, Albert Reiss (1971a) studied police interaction with the public by riding in police squad cars, observing police-citizen interactions, and recording their characteristics on a form. Notations on the form indicated variables such as how many police-citizen contacts occurred, who initiated the contacts, how compliant citizens were with police directives, and whether police expressed hostility toward the citizens.

Using a different approach, psychologists Dore Butler and Florence Geis (1990) studied unconscious biases and stereotypes that they thought might hinder the advancement of women and minorities in work organizations. In one experiment, discussion groups of male and female students were observed from behind one-way mirrors as group leaders presented identical talks to each group. The trained observers (who were not told what the study was about) rated the number of frowns, furrowed brows, smiles, and nods of approval as the group leaders spoke. (The leaders themselves did not know what the study was about.) Group participants made disapproving expressions, such as frowns, more often when the group leader was a woman than when the leader was a man. To make matters worse, the more the women talked, the less attention they were given. Butler and Geis concluded that there was indeed a basis for unconscious discrimination in these social patterns.

Psychologists Joshua Correll, Bernadette Park, Charles M. Judd, and Bernd Wittenbrink (2002) used an even more creative approach to measure unconscious biases that could influence behavior in spite of an absence of conscious prejudice. Their approach focused on measuring reaction times to controlled observations. Correll et al. (2002) constructed a test in which individuals played a video game that required them to make a split-second decision of whether to shoot an image of a person who was holding what was a gun in some pictures and a nonlethal object such as a camera, cell phone, or bottle in others. In this ambiguous situation, white respondents were somewhat more likely to shoot a black man holding a nonlethal object than they were to shoot a white man holding a nonlethal object.

Observations may also supplement data collected in an interview study. This approach was used in a study of homeless persons participating in the Center for Mental Health Services' ACCESS (Access to Community Care and Effective Services and Supports) program. After a 47-question interview, interviewers were asked to record observations that would help indicate whether the respondent was suffering from a major mental illness. For example, the interviewers indicated, on a rating scale from 0 to 4, the degree to which the homeless participants appeared to be responding, during the interview, to voices or noises that others couldn't hear or to other private experiences (U.S. Department of Health and Human Services 1995).

Many interviews contain at least a few observational questions. Clinical studies often request a *global,* or holistic, interviewer rating of clients, based on observations and responses to questions throughout the interview. One such instrument is called the Global Assessment of Functioning Scale (American Psychiatric Association 1994).

Direct observation is often the method of choice for measuring behavior in natural settings, as long as it is possible to make the requisite observations. Direct observation avoids the problems of poor recall and self-serving distortions that can occur with answers to survey questions. It also allows measurement in a context that is more natural than an interview. But observations can be distorted, too. Observers do not see or hear everything, and what they do see is filtered by their own senses and perspectives. Disagreements about crowd size among protestors, police, and journalists are notorious, even though there is a good method of estimating crowd size based on the "carrying capacity" of public spaces (McPhail & McCarthy 2004). When the goal is to observe behavior, measurement can be distorted because the presence of an observer may cause people to act differently than the way they would otherwise (Emerson 1983). I discuss these issues in more depth in Chapter 9, but it is important to consider them whenever you read about observational measures.

Collecting Unobtrusive Measures

Unobtrusive measures allow us to collect data about individuals or groups without their direct knowledge or participation. In their classic book (now revised), Eugene Webb and his colleagues (2000) identified four types of unobtrusive measures: physical trace evidence, archives (available data), simple observation, and contrived observation (using hidden recording hardware or manipulation to elicit a response). These measures can provide valuable supplements or alternatives to more standard, survey-based measures because they lessen the possibility that subjects will make different statements to an interviewer than when they are not being studied and because they are unaffected by an interviewer's appearance or how the interviewer asks questions. We have already considered some types of archival data and observational data, so I focus here on other approaches suggested by Webb et al. (2000).

Unobtrusive measure A measurement based on physical traces or other data that are collected without the knowledge or participation of the individuals or groups that generated the data.

The physical traces of past behavior are one type of unobtrusive measure that is most useful when the behavior of interest cannot be directly observed (perhaps because it is hidden or occurred in the past) and has not been recorded in a source of available data. To measure the prevalence of drinking in college dorms or fraternity houses, we might count the number of empty bottles of alcoholic beverages in the surrounding dumpsters. Student interest in the college courses they are taking might be measured by counting the number of times that books left on reserve as optional reading are checked out or by the number of class handouts left in trash barrels outside a lecture hall. Webb and his colleagues (2000:37) suggested measuring the interest in museum exhibits by the frequency with which tiles in front of the exhibits needed to be replaced. Social variables can also be measured by observing clothing, hair length, or people's reactions to such stimuli as dropped letters or jaywalkers.

You can probably see that care must be taken to develop trace measures that are useful for comparative purposes. For instance, comparison of the number of empty bottles in dumpsters outside different dorms can be misleading; at the very least, you would need to take into account the number of residents in the dorms, the time since the last trash collection, and the accessibility of each dumpster to passersby. Counts of usage of books on reserve will only be useful if you take into account how many copies of the books are on reserve for the course, how many students are enrolled in the course, and whether reserve reading is required. Measures of tile erosion in the museum must take into account the nearness of each exhibit to doors, other popular exhibits, and so on (Webb et al. 2000:47–48).

Unobtrusive measures can also be created from diverse forms of media such as newspaper archives or magazine articles, TV or radio talk shows, legal opinions, historical documents, personal letters, or e-mail messages. Qualitative researchers may read and evaluate text, as Brian Loader and his colleagues (2002) did in their study of computer-mediated social support for people with diabetes (see Chapter 1). Quantitative researchers use content analysis to measure aspects of media such as the frequency of use of particular words or ideas or the consistency with which authors convey a particular message in their stories. An investigation of the drinking climate on campuses might include a count of the amount of space devoted to ads for alcoholic beverages in a sample of issues of the student newspaper. Campus publications also might be coded to indicate the number of times that statements discouraging substance abuse appear. With this tool, you could measure the frequency of articles reporting substance abuse–related crimes, the degree of approval of drinking expressed in TV shows or songs, or the relationship between region of the country and the amount of space devoted in the print media to drug usage.

Combining Measurement Operations

Using available data, asking questions, making observations, and using unobtrusive indicators are interrelated measurement tools, each of which may include or be supplemented by the others. From people's answers to survey questions, the U.S. Census Bureau develops widely consulted census reports containing available data on people, firms, and geographic units in the United States. Data from employee surveys may be supplemented by information available in company records. Interviewers may record observations about those whom they question. Researchers may use insights gleaned from questioning participants to make sense of the social interaction they have observed. Unobtrusive indicators can be used to evaluate the honesty of survey respondents.

The available resources and opportunities often determine the choice of a particular measurement method, but measurement is improved if this choice also takes into account the particular concept or concepts to be measured. Responses to questions such as "How socially engaged were you at the party?" or "How many days did you use sick leave last year?" are unlikely to provide information as valid as, respectively, direct observation or company records. On the other hand, observations at social gatherings may not answer our questions about why some people do not participate; we may have to ask people. Or, if no record is kept of sick leaves in a company, we may have to ask direct questions.

Questioning can be a particularly poor approach for measuring behaviors that are very socially desirable, such as voting or attending church, or that are socially stigmatized or illegal, such as abusing alcohol or drugs. The tendency of people to answer questions in socially approved ways was demonstrated in a study of church attendance in the United States (Hadaway et al. 1993). More than 40% of adult Americans say in surveys that they attend church weekly—a percentage much higher than in Canada, Australia, or Europe. However, a comparison of observed church attendance with self-reported attendance suggested that the actual rate of church attendance was much lower (see Exhibit 4.9). Always consider the possibility of measurement error when only one type of operation has been used. Of course, it is much easier to recognize this possibility than it is to determine the extent of error resulting from a particular measurement procedure. Refer to the February 1998 issue of the *American Sociological Review* for a fascinating exchange of views and evidence on the subject of measuring church attendance.

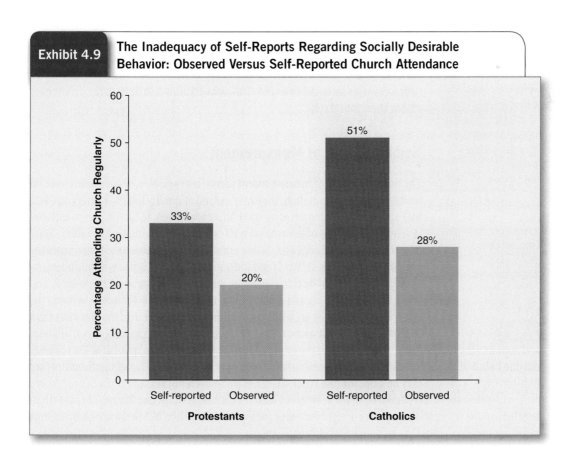

Exhibit 4.9 The Inadequacy of Self-Reports Regarding Socially Desirable Behavior: Observed Versus Self-Reported Church Attendance

Triangulation

Triangulation—the use of two or more different measures of the same variable—can strengthen measurement considerably (Brewer & Hunter 1989:17). When we achieve similar results with different measures of the same variable, particularly when they are based on such different methods as survey questions and field-based observations, we can be more confident in the validity of each measure. If results diverge with different measures, it may indicate that one or more of these measures are influenced by more measurement error than we can tolerate. Divergence between measures could also indicate that they actually operationalize different concepts. An interesting example of this interpretation of divergent results comes from research on crime. Official crime statistics only indicate those crimes that are reported to and recorded by the police; when surveys are used to measure crimes with self-reports of victims, many "personal annoyances" are included as if they were crimes (Levine 1976).

Levels of Measurement

When we know a variable's **level of measurement**, we can better understand how cases vary on that variable and so understand more fully what we have measured. Level of measurement also has important implications for the type of statistics that can be used with the variable, as you will learn in Chapter 14. There are four levels of measurement: (1) nominal, (2) ordinal, (3) interval, and (4) ratio. For most purposes, variables measured at the interval and ratio levels are treated in the same way, so I will sometimes refer to these two levels together as *interval-ratio*. Exhibit 4.10 depicts the differences among these four levels.

> **Level of measurement** The mathematical precision with which the values of a variable can be expressed. The nominal level of measurement, which is qualitative, has no mathematical interpretation; the quantitative levels of measurement—ordinal, interval, and ratio—are progressively more precise mathematically.

Nominal Level of Measurement

The **nominal level of measurement** identifies variables whose values have no mathematical interpretation; they vary in kind or quality but not in amount (they may also be called *categorical* or *qualitative variables*). In fact, it is conventional to refer to the values of nominal variables as *attributes* instead of values. *State* (referring to the United States) is one example. The variable has 50 attributes (or categories or qualities). We might indicate the specific states with numbers, so that California might be represented by the value 1, Oregon with the value 2, and so on, but these numbers do not tell us anything about the difference between the states except that they are different. California is not one unit more of *state* than Oregon, nor is it twice as much *state*. Nationality, occupation, religious affiliation, and region of the country are also measured at the nominal level. A person may be Spanish or Portuguese, but one nationality does not represent more nationality than another—just a different nationality (see Exhibit 4.10). A person may be a doctor or a truck driver, but one does not represent three units more occupation than the other. Of course, more people may identify themselves as of one nationality more than another, or one occupation may have a higher average income than another, but these are comparisons involving variables other than *nationality* or *occupation per se.*

> **Nominal level of measurement** Variables whose values have no mathematical interpretation; they vary in kind or quality, but not in amount.

Exhibit 4.10 **Levels of Measurement**

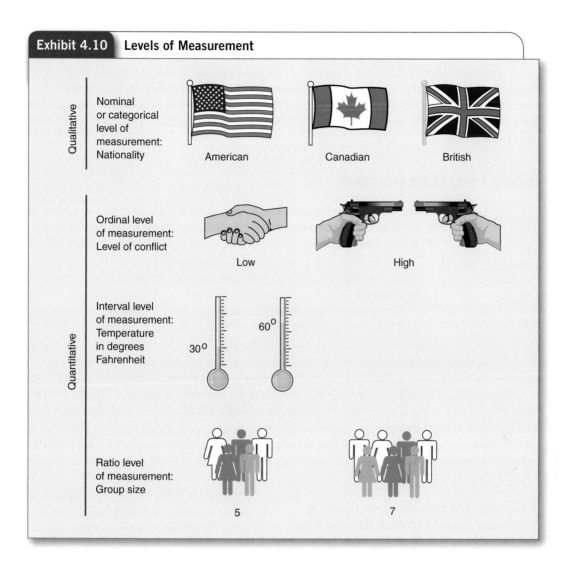

Although the attributes of categorical variables do not have a mathematical meaning, they must be assigned to cases with great care. The attributes we use to measure, or to categorize, cases must be mutually exclusive and exhaustive:

- A variable's attributes or values are mutually exclusive if every case can have only one attribute.

- A variable's attributes or values are **exhaustive** when every case can be classified into one of the categories.

> **Exhaustive** Every case can be classified as having at least one attribute (or value) for the variable.

When a variable's attributes are mutually exclusive and exhaustive, every case corresponds to one, and only one, attribute.

I know this sounds pretty straightforward, and in many cases it is. However, what we think of as mutually exclusive and exhaustive categories may really be so only because of social convention; when these conventions change, or if they differ between the societies in a multicountry study, appropriate classification at the nominal level can become much more complicated. You learned of complexities such as this in the earlier discussion of the history of measuring race.

Issues similar to these highlight the importance of informed selection of concepts, careful conceptualization of what we mean by a term, and systematic operationalization of the procedures for indicating the attributes of actual cases. The debate over the concept of race also reminds us of the value of qualitative research that seeks to learn about the meaning that people give to terms, without requiring that respondents use predetermined categories.

Ordinal Level of Measurement

> **Ordinal level of measurement**
> A measurement of a variable in which the numbers indicating a variable's values specify only the order of the cases, permitting *greater than* and *less than* distinctions.

The first of the three quantitative levels is the **ordinal level of measurement**. At this level, the numbers assigned to cases specify only the order of the cases, permitting *greater than* and *less than* distinctions. The Favorable Attitudes Toward Antisocial Behavior Scale measures attitudes toward antisocial behavior among high school students with a series of questions that permit ordinal distinctions (see Exhibit 4.11).

The response choices to each question range from "very wrong" to "not wrong at all"; there's no particular quantity of "wrongness" that these distinctions reflect, but the idea is that a student who responds that it is "not wrong at all" to a question about taking a handgun to school has a more favorable attitude toward antisocial behavior than one who says it is "a little bit wrong," which is in turn more favorable than those who respond "wrong" or "very wrong."

The properties of variables measured at the ordinal level are illustrated in Exhibit 4.10 by the contrast between the levels of conflict in two groups. The first group, symbolized by two people shaking hands, has a low level of conflict. The second group, symbolized by two persons pointing guns at each other, has a high level of conflict. To measure conflict, we would put the groups "in order" by assigning the number 1 to the low-conflict group and the number 2 to the high-conflict group. The numbers thus indicate only the relative position or order of the cases. Although low level of conflict is represented by the number 1, it is not one less unit of conflict than high level of conflict, which is represented by the number 2.

As with nominal variables, the different values of a variable measured at the ordinal level must be mutually exclusive and exhaustive. They must cover the range of observed values and allow each case to be assigned no more than one value. Often, questions that use an ordinal level of measurement simply ask respondents to rate their response to some question or statement along a continuum of, for example, strength of agreement, level of importance, or relative frequency. Like variables measured at the nominal level, variables measured at the ordinal level in this way classify cases in discrete categories and so are termed **discrete measures**. Alternatively, a series of similar questions intended to measure the same concept may be asked in a

Exhibit 4.11	**Example of Ordinal Measures: Favorable Attitudes Toward Antisocial Behavior Scale**

17. **Within the last year about how often have you used...**
(mark one for each line)

Columns: Did not use · Once/year · 6 times/year · Once/month · Twice/month · 3 times/week · 5 times/week · Once/week · Every day

a. Tobacco (smoke, chew, snuff) . . . .
b. Alcohol (beer, wine, liquor) . . . .
c. Marijuana (pot, hash, hash oil) . . . .
d. Cocaine (crack, rock, freebase) . . .
e. Amphetamines (diet pills, speed) . .
f. Sedatives (downers, ludes) . . . . . .
g. Hallucinogens (LSD, PCP) . . . .
h. Opiates (heroin, smack, horse) . . .
i. Inhalants (glue, solvents, gas) . . . .
j. Designer drugs (ecstasy, MDMA) . .
k. Steroids . . . .
l. Other illegal drugs . . . .

multi-item **index**; the numbers assigned to the ordinal responses to these questions may then be summed or averaged to create the index score. Sometimes, just numbers are offered as response choices (such as 1, 2, 3, 4, 5, 6, 7), without any verbal interpretation.

Of course, an ordinal rating scheme assumes that respondents have similar interpretations of the terms used to designate the ordered responses. This is not always the case, because rankings tend to reflect the range of alternatives with which respondents are familiar (McCarty & Shrum 2000). For example, a classic experiment shows that a square is judged larger when it is compared with many smaller squares than when it is compared with larger squares. This is particularly a problem if people in the study are from different cultures or from different backgrounds. Also, respondents with more education tend to make finer distinctions between alternatives on a rating scale.

Providing explicit "anchor points" for respondents can improve the comparability of responses to ordinal rating questions. Management faculty John McCarty and L. J. Shrum designed a procedure for rating the importance of items that is simple enough to use in mailed questionnaires. In the first step, respondents are asked to pick the most and least important (2,000) items from the entire list of items. In the second step, respondents are asked to rate the importance of each item, using the same scale. Just the act of identifying the most and least important items before rating items using phrases such as "very important" and "somewhat important" increases the extent to which respondents differentiate between responses.

> **Discrete measure** A measure that classifies cases in distinct categories.
>
> **Index** The sum or average of responses to a set of questions about a concept.

Interval Level of Measurement

The numbers indicating the values of a variable at the **interval level of measurement** represent fixed measurement units but have no absolute, or fixed, zero point. This level of measurement is represented in Exhibit 4.10 by the difference between two Fahrenheit temperatures. Although 60 degrees is 30 degrees hotter than 30 degrees, 60, in this case, is not twice as hot as 30. Why not? Because "heat" does not begin at 0 degrees on the Fahrenheit scale.

> **Interval level of measurement** A measurement of a variable in which the numbers indicating a variable's values represent fixed measurement units but have no absolute, or fixed, zero point.

An interval-level measure is created by a scale that has fixed measurement units but no absolute, or fixed, zero point. The numbers can, therefore, be added and subtracted, but ratios are not meaningful. Again, the values must be mutually exclusive and exhaustive.

There are few true interval-level measures in the social sciences, but many social scientists treat indexes created by combining responses to a series of variables measured at the ordinal level as interval-level measures. An index of this sort could be created with responses to the Core Institute's questions about friends' disapproval of substance use (see Exhibit 4.12). The survey has 13 questions on the topic, each of which has the same three response choices. If "Don't disapprove" is valued at 1, "Disapprove" is valued at 2, and "Strongly disapprove" is valued at 3, the summed index of disapproval would range from 12 to 36. A score of 20 could be treated as if it were four more units than a score of 16. Or the responses could be averaged to retain the original 1 to 3 range.

Ratio Level of Measurement

The numbers indicating the values of a variable at the **ratio level of measurement** represent fixed measuring units and an absolute zero point (zero means absolutely no amount of whatever the variable indicates). For most statistical analyses in social science research, the interval and ratio levels of measurement can be treated as equivalent. In addition to having numerical values, both the interval and ratio

> **Ratio level of measurement** A measurement of a variable in which the numbers indicating a variable's values represent fixed measuring units and an absolute zero point.

Exhibit 4.12 | **Example of Interval-Level Measures: Core Alcohol and Drug Survey**

26. **How do you think your close friends feel (or would feel) about you...** *(mark one for each line)*

Don't disapprove / Disapprove / Strongly disapprove

a. Trying marijuana once or twice .... ○ ○ ○
b. Smoking marijuana occasionally .... ○ ○ ○
c. Smoking marijuana regularly .... ○ ○ ○
d. Trying cocaine once or twice .... ○ ○ ○
e. Taking cocaine regularly .... ○ ○ ○
f. Trying LSD once or twice .... ○ ○ ○
g. Taking LSD regularly .... ○ ○ ○
h. Trying amphetamines once or twice .... ○ ○ ○
i. Taking amphetamines regularly .... ○ ○ ○
j. Taking one or two drinks of an alcoholic beverage (beer, wine, liquor) nearly every day .... ○ ○ ○
k. Taking four or five drinks nearly every day .... ○ ○ ○
l. Having five or more drinks in one sitting ...... ○ ○ ○
m. Taking steroids for body building or improved athletic performance .... ○ ○ ○

Continuous measure A measure with numbers indicating the values of variables that are points on a continuum.

levels also involve **continuous measures**: The numbers indicating the values of variables are points on a continuum, not discrete categories. But in spite of these similarities, there is an important difference between variables measured at the interval and ratio levels. On a ratio scale, 10 is 2 points higher than 8 and is also 2 *times* greater than 5—the numbers can be compared in a ratio. Ratio numbers can be added and subtracted, and because the numbers begin at an absolute zero point, they can be multiplied and divided (so ratios can be formed between the numbers). For example, people's ages can be represented by values ranging from 0 years (or some fraction of a year) to 120 or more. A person who is 30-years-old is 15 years older than someone who is 15-years-old (30 – 15 = 15) and is twice as old as that person (30/15 = 2). Of course, the numbers also are mutually exclusive and exhaustive, so that every case can be assigned one and only one value.

Exhibit 4.10 displays an example of a variable measured at the ratio level. The number of people in the first group is 5, and the number in the second group is 7. The ratio of the two groups' sizes is then 1.4, a number that mirrors the relationship between the sizes of the groups. Note that there does not actually have to be any group with a size of 0; what is important is that the numbering scheme begins at an absolute zero—in this case, the absence of any people.

It's tempting to accept the numbers that represent the values of a variable measured at the ratio level at face value, but the precision of the numbers can't make us certain about their accuracy. Income data provided in the U.S. Census is often incomplete (Scott 2001); the unemployment rate doesn't account for people who have given up looking for work (Zitner 1996), and the Consumer Price Index (CPI) does not reflect the types of goods that many groups of consumers buy (Uchitelle 1997). In each of these cases, we have to be sure that the measures that we use reflect adequately the concepts that we intend.

The Special Case of Dichotomies

Dichotomies, variables having only two values, are a special case from the standpoint of levels of measurement. The values or attributes of a variable such as gender clearly vary in kind or quality but not in amount. Thus, the variable is categorical—measured at the nominal level. Yet we can also think of the variable as indicating the presence of the attribute *female* (or *male*) or not. Viewed in this way, there is an inherent order: A female has more of the female attribute (it is present) than a male (the attribute is not present). It's also possible to think of a **dichotomy** as representing an interval level of measurement, since there is an equal interval between the two attributes. So what do you answer to the test question, "What is the level of measurement of *gender*?" "Nominal," of course, but you'll find that when a statistical procedure requires that variables be quantitative, a dichotomy can be perfectly acceptable.

Dichotomy Variable having only two values.

Comparison of Levels of Measurement

Exhibit 4.13 summarizes the types of comparisons that can be made with different levels of measurement, as well as the mathematical operations that are legitimate. All four levels of measurement allow researchers to assign different values to different cases. All three quantitative measures allow researchers to rank cases in order.

Researchers choose the levels of measurement in the process of operationalizing variables; the level of measurement is not inherent in the variable itself. Many variables can be measured at different levels, with different procedures. For example, the Core Alcohol and Drug Survey (Core Institute 1994) identifies binge drinking by asking students, "Think back over the last two weeks. How many times have you had five or more drinks at a sitting?" You might be ready to classify this as a ratio-level measure, but this would be true only if responses are recorded as the actual number of "times." Instead, the Core Survey treats this as a closed-ended question, and students are asked to indicate their answer by checking "None," "Once," "Twice," "3 to 5 times," "6 to 9 times," or "10 or more times." Use of these categories makes the level of measurement ordinal. The distance between any two cases cannot be clearly determined. A student with a response in the "6 to 9 times" category could have binged just one more time than a student who responded "3 to 5 times." You just can't tell.

It is usually a good idea to try to measure variables at the highest level of measurement possible. The more information available, the more ways we have to compare cases. We also have more possibilities for statistical analysis with quantitative than with qualitative variables. Thus, if doing so does not distort the meaning of the concept that is to be measured, measure at the highest level possible. Even if your primary concern is only to compare teenagers with young adults, measure age in years rather than in categories; you can always combine the ages later into categories corresponding to teenager and young adult.

Be aware, however, that other considerations may preclude measurement at a high level. For example, many people are very reluctant to report their exact incomes, even in anonymous questionnaires. So asking respondents to report their income in categories (such as less than $10,000, $10,000–19,999, $20,000–29,999) will result in more responses, and thus more valid data, than asking respondents for their income in dollars.

Oftentimes, researchers treat variables measured at the interval and ratio levels as comparable. They then refer to this as the **interval-ratio level of measurement**. You will learn in Chapter 14 that different statistical procedures are used for variables with fixed measurement units, but it usually doesn't matter whether there is an absolute zero point.

> **Interval-ratio level of measurement** A measurement of a variable in which the numbers indicating a variable's values represent fixed measurement units but may not have an absolute, or fixed, zero point.

| Exhibit 4.13 | Properties of Measurement Levels |

Examples of Comparison Statements	Appropriate Math Operations	Relevant Level of Measurement			
		Nominal	**Ordinal**	**Interval**	**Ratio**
A is equal to (not equal to) B	= (≠)	✓	✓	✓	✓
A is greater than (less than) B	> (<)		✓	✓	✓
A is three more than (less than) B	+ (−)			✓	✓
A is twice (half) as large as B	× (÷)				✓

Evaluating Measures

Do the operations developed to measure our variables actually do so—are they valid? If we have weighed our measurement options, carefully constructed our questions and observational procedures, and selected sensibly from the available data indicators, we should be on the right track. But we cannot have much confidence in a measure until we have empirically evaluated its validity. What good is our measure if it doesn't measure what we think it does? If our measurement procedure is invalid, we might as well go back to the starting block and try again. As a part of evaluating the validity of our measures, we must also evaluate their reliability, because reliability (consistency) is a prerequisite for measurement validity.

Measurement Validity

In Chapter 2, you learned that measurement validity refers to the extent to which measures indicate what they are intended to measure. More technically, a valid measure of a concept is one that is closely related to other apparently valid measures of the concept, and to the known or supposed correlates of that concept, but that is not related to measures of unrelated concepts, irrespective of the methods used for the other different measures (Brewer & Hunter 1989:134).

When a measure "misses the mark"—when it is not valid—our measurement procedure has been affected by measurement error. Measurement error can arise in three general ways:

1. *Idiosyncratic individual errors* are errors that affect a relatively small number of individuals in unique ways that are unlikely to be repeated in just the same way (Viswanathan 2005:289). Individuals make **idiosyncratic errors** when they don't understand a question, when some unique feelings are triggered by the wording of a question, or when they are feeling out of sorts due to some recent events.

2. *Generic individual errors* occur when the responses of groups of individuals are affected by factors that are not what the instrument is intended to measure. For example, individuals who like to please others by giving socially desirable responses may have a tendency to say that they "agree" with the statements, simply because they try to avoid saying they "disagree" with anyone. Generic individual errors may also arise when the same measure is used across cultures that differ in their understanding of the concepts underlying the measures (Church 2010:152-153).

   > **Idiosyncratic errors** Errors that affect a relatively small number of individuals in unique ways that are unlikely to be repeated in just the same way.
   >
   > **Unbalanced response choices** A fixed-choice survey question has a different number of positive and negative response choices.
   >
   > **Balanced response choices** An equal number of responses to a fixed-choice survey question express positive and negative choices in comparable language.

3. *Method factors* can also create errors in the responses of most or all respondents. Questions that are unclear may be misinterpreted by most respondents, while **unbalanced response choices** may lead most respondents to give positive rather than negative responses. For example, if respondents are asked the question with the unbalanced response choices in Exhibit 4.14, they are more likely to respond that gun ownership is wrong than if they are asked the question with the **balanced response choices** (Viswanathan 2005:142–148).

It is the job of the social scientist to try to reduce measurement errors and then to evaluate the extent to which the resulting measures are valid. The extent to which

---

Exhibit 4.14 | **Balanced and Unbalanced Response Choices**

Unbalanced response choices

How wrong do you think it is for someone who is not a hunter to own a gun?

- Very wrong
- Wrong
- A little bit wrong
- Not wrong at all

Balanced response choices

Some people think it is wrong for someone who is not a hunter to own a gun, and some people think it is a good idea to own a gun. Do you think it is very wrong, wrong, neither right nor wrong, right, or very right for someone who is not a hunter to own a gun?

- Very wrong
- Wrong
- Neither right nor wrong
- Right
- Very right

---

measurement validity has been achieved can be assessed with four different approaches: (1) face validation, (2) content validation, (3) criterion validation, and (4) construct validation. The methods of criterion and construct validation also include subtypes.

Face Validity

Researchers apply the term **face validity** to the confidence gained from careful inspection of a concept to see if it is appropriate "on its face." More precisely, we can say that a measure is face valid if it obviously pertains to the meaning of the concept being measured more than to other concepts (Brewer & Hunter 1989:131). For example, a count of the number of drinks people had consumed in the past week would be a face-valid measure of their alcohol consumption. But speaking of "face" validity, what would you think about assessing the competence of political candidates by how mature their faces look? It turns out that people are less likely to vote for candidates with more "baby-faced" features, such as rounded features and large eyes, irrespective of the candidates' records (Cook 2005). It's an unconscious bias, and, of course, it's not one that we would use as a basis for assessing competence in a social science study!

> **Face validity** The type of validity that exists when an inspection of items used to measure a concept suggests that they are appropriate "on their face."

Although every measure should be inspected in this way, face validation in itself does not provide convincing evidence of measurement validity. The question "How much beer or wine did you have to drink last week?" looks valid on its face as a measure of frequency of drinking, but people who drink heavily tend to underreport the amount they drink. So the question would be an invalid measure, at least in the study of heavy drinkers.

Content Validity

Content validity establishes that the measure covers the full range of the concept's meaning. To determine that range of meaning, the researcher may solicit the opinions of experts and review literature that identifies the different aspects, or dimensions, of the concept.

> **Content validity** The type of validity that exists when the full range of a concept's meaning is covered by the measure.

An example of a measure that covers a wide range of meaning is the Michigan Alcoholism Screening Test (MAST). The MAST includes 24 questions representing the following subscales: recognition of alcohol problems by self and others; legal, social, and work problems; help seeking; marital and family difficulties; and liver pathology (Skinner & Sheu 1982). Many experts familiar with the direct consequences of substance abuse agree that these dimensions capture the full range of possibilities. Thus, the MAST is believed to be valid from the standpoint of content validity.

Criterion Validity

When people drink an alcoholic beverage, the alcohol is absorbed into their blood and then gradually metabolized (broken down into other chemicals) in their liver (National Institute of Alcohol Abuse and Alcoholism [NIAAA] 1997). The alcohol that remains in their blood at any point, unmetabolized, impairs both thinking and behavior (NIAAA 1994). As more alcohol is ingested, cognitive and behavioral consequences multiply. The bases for these biological processes can be identified with direct measures of alcohol concentration in the blood, urine, or breath. Questions about the quantity and frequency of drinking, on the other hand, can be viewed as attempts to measure indirectly what biochemical tests measure directly.

Criterion validity is established when the scores obtained on one measure can be accurately compared with those obtained with a more direct or already validated measure of the same phenomenon (the criterion). A measure of blood-alcohol concentration or a urine test could serve as the criterion for validating a self-report measure of drinking, as long as the questions we ask about drinking refer to the same period of time. Chemical analysis of hair samples can reveal unacknowledged drug use (Mieczkowski 1997). Friends' or relatives' observations of a person's substance use also could serve, in some limited circumstances, as a criterion for validating self-report substance use measures.

> **Criterion validity** The type of validity that is established by comparing the scores obtained on the measure being validated to those obtained with a more direct or already validated measure of the same phenomenon (the criterion).

Criterion-validation studies of self-report substance abuse measures have yielded inconsistent results. Self-reports of drug use agreed with urinalysis results for about 85% of the drug users who volunteered for a health study in several cities (Weatherby et al. 1994). On the other hand, the posttreatment drinking behavior self-reported by 100 male alcoholics was substantially less than the drinking behavior observed by the alcoholics' friends or relatives (Watson et al. 1984). College students' reports of drinking are suspect too: A standard question to measure alcohol use is to ask respondents how many glasses they consume when they do drink. A criterion-validation study of this approach measured how much of the drink students poured when they had what they considered to be a "standard" drink (White et al. 2003). The students consistently overestimated how much fluid goes into a standard drink.

Inconsistent findings about the validity of a measure can occur because of differences in the adequacy of a measure across settings and populations. We cannot simply assume that a measure that was validated in one study is also valid in another setting or with a different population. The validity of even established measures has to be tested when they are used in a different context (Viswanathan 2005:297).

> **Concurrent validity** The type of validity that exists when scores on a measure are closely related to scores on a criterion measured at the same time.
>
> **Predictive validity** The type of validity that exists when a measure predicts scores on a criterion measured in the future.

The criterion that researchers select can be measured either at the same time as the variable to be validated or after that time. **Concurrent validity** exists when a measure yields scores that are closely related to scores on a criterion measured at the same time. A store might validate a question-based test of sales ability by administering it to sales personnel who are already employed and then comparing their test scores with their sales performance. Or a measure of walking speed based on mental counting might be validated concurrently with a stopwatch. **Predictive validity** is the ability of a measure to predict scores on a criterion measured in the future. For example, a store might administer a test of sales ability to new sales personnel and then validate the measure by comparing these test scores with the criterion—the subsequent sales performance of the new personnel.

An attempt at criterion validation is well worth the effort, because it greatly increases confidence that the standard is measuring what was intended. However, for many concepts of interest to social scientists, no other variable can reasonably be considered a criterion. If we are measuring feelings or beliefs or other subjective states, such as feelings of loneliness, what *direct* indicator could serve as a criterion? Even with variables for which a reasonable criterion exists, the researcher may not be able to gain access to the criterion—as would be the case with a tax return or employer document that we might wish we could use as a criterion for self-reported income.

Construct Validity

Measurement validity can also be established by showing that a measure is related to a variety of other measures as specified in a theory. This validation approach, known as **construct validity**, is commonly used in social research when no clear criterion exists for validation purposes. For example, in one study of the validity of the Addiction Severity Index (ASI), A. Thomas McLellan and his associates (1985) compared subject scores on the ASI with a number of indicators that they felt, from prior research, should be related to substance abuse: medical problems, employ-ment problems, legal problems, family problems, and psychiatric problems. They could not use a criterion validation approach because they did not have a more direct measure of abuse, such as laboratory test scores or observer reports. However, their extensive research on the subject had given them confidence that these sorts of problems were all related to substance abuse, and, indeed, they found that individuals with higher ASI rat-ings tended to have more problems in each of these areas.

> **Construct validity** The type of validity that is established by showing that a measure is related to other measures as specified in a theory.

Two other approaches to construct validation are convergent validation and discriminant validation. **Convergent validity** is achieved when one measure of a concept is associated with different types of measures of the same concept (this relies on the same type of logic as measurement triangula-tion). **Discriminant validity** is a complementary approach to construct validation. In this approach, scores on the measure to be validated are compared with scores on measures of different but related concepts. Discriminant validity is achieved if the measure to be validated is not associated strongly with the measures of different concepts. McLellan et al. (1985) found that the ASI passed the tests of convergent and discriminant validity: The ASI's measures of alcohol and drug problems were related more strongly to other measures of alcohol and drug problems than they were to mea-sures of legal problems, family problems, medical problems, and the like.

> **Convergent validity** The type of validity achieved when one measure of a concept is associated with different types of measures of the same concept.
>
> **Discriminant validity** An approach to construct validation; the scores on the measure to be validated are compared to scores on another measure of the same variable and to scores on variables that measure different but related concepts. Discriminant validity is achieved if the measure to be validated is related most strongly to its comparison measure and less so to the measures of other concepts.

The distinction between criterion validation and construct validation is not always clear. Opinions can differ about whether a particular indicator is indeed a criterion for the concept that is to be measured. For example, if you need to vali-date a question-based measure of sales ability for applicants to a sales position, few would object to using actual sales performance as a criterion. But what if you want to validate a question-based measure of the amount of social support that people receive from their friends? Should you just ask people about the social sup-port they have received? Could friends' reports of the amount of support they pro-vided serve as a criterion? Are verbal accounts of the amount of support provided adequate? What about observation of social support that people receive? Even if you could observe people in the act of counseling or otherwise supporting their friends, can an observer be sure that the interaction is indeed supportive? There isn't really a criterion here, only related concepts that could be used in a construct validation strategy. Even biochemical measures of substance abuse are questionable as criteria for validating self-reported substance use. Urine test results can be altered by ingesting certain substances, and blood tests vary in their sensitivity to the presence of drugs over a particular period.

What both construct validation and criterion validation have in common is the comparison of scores on one measure with the scores on other measures that are predicted to be related. It is not so important that researchers agree that a particular comparison measure is a criterion rather than a related construct. But it is very important to think critically about the quality of the comparison measure and whether it actually represents a different view of the same phenomenon. For example, correspondence between scores on two different self-report measures of alcohol use is a much weaker indicator of measurement validity than the correspondence of a self-report measure with an observer-based measure of substance use.

Reliability

> **Reliability** A measurement procedure yields consistent scores when the phenomenon being measured is not changing.

Reliability means that a measurement procedure yields consistent scores when the phenomenon being measured is not changing (or that the measured scores change in direct correspondence to actual changes in the phenomenon). If a measure is reliable, it is affected less by random error, or chance variation, than if it is unreliable. Reliability is a prerequisite for measurement validity: We cannot really measure a phenomenon if the measure we are using gives inconsistent results. In fact, because it is usually easier to assess reliability than validity, you are more likely to see an evaluation of measurement reliability in a research report than an evaluation of measurement validity.

There are four ways to evaluate whether a measure is reliable or unreliable: (1) test-retest, (2) interitem (internal consistency), (3) alternate forms, and (4) interobserver. For example, a test of your knowledge of research methods would be unreliable if every time you took it, you received a different score even though your knowledge of research methods had not changed in the interim, not even as a result of taking the test more than once. This is test-retest reliability. Similarly, an index composed of questions to measure knowledge of research methods would be unreliable if respondents' answers to each question were totally independent of their answers to the others. The index has interitem reliability if the component items are closely related. A measure also would be unreliable if slightly different versions of it resulted in markedly different responses (it would not achieve alternate-forms reliability). Finally, an assessment of the level of conflict in social groups would be unreliable if ratings of the level of conflict by two observers were not related to each other (it would then lack interobserver reliability).

Test-Retest Reliability

When researchers measure a phenomenon that does not change between two points separated by an interval of time, the degree to which the two measurements are related to each other is the **test-retest reliability** of the measure. If you take a test of your math ability and then retake the test 2 months later, the test is performing reliably if you receive a similar score both times—presuming that nothing happened during the 2 months to change your math ability. Of course, if events between the test and the retest have changed the variable being measured, then the difference between the test and retest scores should reflect that change.

> **Test-retest reliability** A measurement showing that measures of a phenomenon at two points in time are highly correlated, if the phenomenon has not changed, or has changed only as much as the phenomenon itself.
>
> **Intrarater (or intraobserver) reliability** Consistency of ratings by an observer of an unchanging phenomenon at two or more points in time.

When ratings by an observer, rather than ratings by the subjects themselves, are being assessed at two or more points in time, test-retest reliability is termed **intrarater (or intraobserver) reliability**.

If an observer's ratings of individuals' drinking behavior in bars are similar at two or more points in time, and the behavior has not changed, the observer's ratings of drinking behavior are reliable.

One example of how evidence about test-retest reliability may be developed is a study by Linda Sobell and her associates (1988) of alcohol abusers' past drinking behavior (using the Lifetime Drinking History Questionnaire) and life changes (using the Recent Life Changes Questionnaire). All 69 subjects in the study were patients in an addiction treatment program. They had not been drinking prior to the interview (determined by a breath test). The two questionnaires were administered by different interviewers about 2 or 3 weeks apart, both times asking the subjects to recall events 8 years prior to the interviews. Reliability was high: 92% of the subjects reported the same life events both times, and at least 81% of the subjects were classified consistently at both the interviews as having had an alcohol problem or not. When asked about their inconsistent answers, subjects reported that in the earlier interview they had simply dated an event incorrectly, misunderstood the question, evaluated the importance of an event differently, or forgotten an event. Answers to past drinking questions were less reliable when they were very specific, apparently because the questions exceeded the subjects' capacities to remember accurately.

Interitem Reliability (Internal Consistency)

When researchers use multiple items to measure a single concept, they must be concerned with **interitem reliability** (or internal consistency). For example, if we are to have confidence that a set of questions (such as those in Exhibit 4.15) reliably measures depression, the answers to these questions should be highly associated with one another. The stronger the association among the individual items and the more items included, the higher the reliability of the index. **Cronbach's alpha** is a **reliability measure** commonly used to measure interitem reliability. Of course, interitem reliability cannot be computed if only one question is used to measure a concept. For this reason, it is much better to use a multi-item index to measure an important concept (Viswanathan 2005:298–299).

Donald Hawkins, Paul Amato, and Valarie King (2007:1007) used this measure of depression in their study of adolescent well-being and obtained a high level of interitem reliability. They measured "negative outlook" with a similar set of questions (Exhibit 4.15), but the interitem reliability of this set was lower. Read through the two sets of questions. Do the sets seem to cover what you think of as being depressed and having a negative outlook? If so, they seem to be *content valid* to you.

> **Interitem reliability** An approach that calculates reliability based on the correlation among multiple items used to measure a single concept. Also known as internal consistency.
>
> **Cronbach's alpha** A statistic commonly used to measure interitem reliability.
>
> **Reliability measure** Statistics that summarize the consistency among a set of measures. Cronbach's alpha is the most common measure of the reliability of a set of items included in an index.

Alternate-Forms Reliability

Researchers test **alternate-forms reliability** by comparing the subjects' answers with slightly different versions of the survey questions (Litwin 1995:13–21). A researcher may reverse the order of the response choices in an index or modify the question wording in minor ways and then re-administer that index to the subjects. If the two sets of responses are not too different, alternate-forms reliability is established.

A related test of reliability is the **split-half reliability** approach. After a set of survey questions intended to form an index is administered, the researcher divides the questions into half by distinguishing even- and odd-numbered questions, flipping a coin, or using some other random procedure. Scores are then computed for these two sets of questions. The researchers then compare the scores for the two halves and check the relation between the subjects' scores on them. If scores on the two halves are similar and highly related to each other (so that people who score high on one half also score high on the other half, etc.), then the measure's split-half reliability is established.

> **Alternate-forms reliability** A procedure for testing the reliability of responses to survey questions in which subjects' answers are compared after the subjects have been asked slightly different versions of the questions or when randomly selected halves of the sample have been administered slightly different versions of the questions.
>
> **Split-half reliability** Reliability achieved when responses to the same questions by two randomly selected halves of a sample are about the same.

Interobserver Reliability

When researchers use more than one observer to rate the same people, events, or places, **interobserver reliability** is their goal. If observers are using the same instrument to rate the same thing, their ratings should

Exhibit 4.15	Examples of Indexes: Short Form of the Center for Epidemiologic Studies (CES-D) and "Negative Outlook" Index

At any time during the past week . . . (Circle one response on each line)	Never	Some of the Time	Most of the Time
a. Was your appetite so poor that you did not feel like eating?	1	2	3
b. Did you feel so tired and worn out that you could not enjoy anything?	1	2	3
c. Did you feel depressed?	1	2	3
d. Did you feel unhappy about the way your life is going?	1	2	3
e. Did you feel discouraged and worried about your future?	1	2	3
f. Did you feel lonely?	1	2	3

Negative outlook

How often was each of these things true during the past week? (Circle one response on each line)	A Lot, Most, or All of the Time	Sometimes	Never or Rarely
a. You felt that you were just as good as other people.	0	1	2
b. You felt hopeful about the future.	0	1	2
c. You were happy.	0	1	2
d. You enjoyed life.	0	1	2

Interobserver reliability When similar measurements are obtained by different observers rating the same persons, events, or places.

be very similar. If they are similar, we can have much more confidence that the ratings reflect the phenomenon being assessed rather than the orientations of the observers.

Assessing interobserver reliability is most important when the rating task is complex. Consider a commonly used measure of mental health, the Global Assessment of Functioning Scale (GAFS), a bit of which is shown in Exhibit 4.16. The rating task seems straightforward, with clear descriptions of the subject characteristics that are supposed to lead to high or low GAFS scores. However, the judgments that the rater must make while using this scale are very complex. They are affected by a wide range of subject characteristics, attitudes, and behaviors as well as by the rater's reactions. As a result, interobserver agreement is often low on the GAFS, unless the raters are trained carefully. Qualitative researchers confront the same issue when multiple observers attempt to observe similar phenomena in different settings or at different times. We return to this issue in the qualitative research chapters.

Ways to Improve Reliability and Validity

Whatever the concept measured or the validation method used, no measure is without some error, nor can we expect it to be valid for all times and places. For example, the reliability and validity of self-report measures

Exhibit 4.16	The Challenge of Interobserver Reliability: Excerpt From the Global Assessment of Functioning Scale (GAFS)

Consider psychological, social, and occupational functioning on a hypothetical continuum of mental health illness. Do not include impairment in functioning due to physical (or environmental) limitations.

Code (**Note**: Use intermediate codes when appropriate, e.g., 45, 68, 72.)

100 **Superior functioning in a wide range of activities, life's problems never seem to get out of hand, is sought by others because of his or her many positive qualities. No symptoms.**

90 **Absent or minimal symptoms** (e.g., mild anxiety before an exam), **good functioning in all areas, interested and involved in a wide range of activities, socially effective, generally satisfied with life, no more than everyday problems or concerns** (e.g., an occasional argument with family members).

80 **If symptoms are present, they are transient and expectable reactions to psychosocial stressors** (e.g., difficulty concentrating after family argument); **no more than slight impairment in social, occupational, or school functioning** (e.g., temporarily falling behind in schoolwork).

70 **Some mild symptoms** (e.g., depressive mood and mild insomnia) **OR some difficulty in social, occupational, or school functioning** (e.g., occasional truancy or theft within the household), **but generally functioning pretty well, has some meaningful interpersonal relationships.**

60 **Moderate symptoms** (e.g., flat affect and circumstantial speech, occasional panic attacks) **OR moderate difficulty in social, occupational, or school functioning** (e.g., few friends, conflicts with peers or co-workers).

50 **Serious symptoms** (e.g., suicidal ideation, severe obsessional rituals, frequent shoplifting) **OR any serious impairment in social, occupational, or school functioning** (e.g., no friends, unable to keep a job).

40 **Some impairment in reality testing or communication** (e.g., speech is at times illogical, obscure, or irrelevant) **OR major impairment in several areas, such as work or school, family relations, judgment, thinking, or mood** (e.g., depressed man avoids friends, neglects family, and is unable to work, child frequently beats up younger children, is defiant at home, and is failing at school).

30 **Behavior is considerably influenced by delusions or hallucinations OR serious impairment in communication or judgment** (e.g., sometimes incoherent, acts grossly inappropriately, suicidal preoccupation) **OR inability to function in almost all areas** (e.g., stays in bed all day, no job, home, or friends).

20 **Some danger of hurting self or others** (e.g., suicide attempts without clear expectation of death, frequently violent, manic excitement) **OR occasionally fails to maintain minimal personal hygiene** (e.g., smears feces) **OR gross impairment in communication** (e.g., largely incoherent or mute).

10 **Persistent danger of severely hurting self or others** (e.g., recurrent violence) **OR persistent inability to maintain minimal personal hygiene OR serious suicidal act with clear expectation of death.**

0 Inadequate information.

of substance abuse vary with factors such as whether the respondents are sober or intoxicated at the time of the interview, whether the measure refers to recent or lifetime abuse, and whether the respondents see their responses as affecting their chances of receiving housing, treatment, or some other desired outcome (Babor, Stephens, & Marlatt 1987). In addition, persons with severe mental illness are, in general, less likely to respond accurately (Corse, Hirschinger, & Zanis 1995). We should always be on the lookout for ways in which we can improve the reliability and validity of the measures we use.

Remember that a reliable measure is not necessarily a valid measure, as Exhibit 4.17 illustrates. This discrepancy is a common flaw of self-report measures of substance abuse. Most respondents answer the multiple questions in self-report indexes of substance abuse in a consistent way, so the indexes are reliable. However, a number of respondents will not admit to drinking, even though they drink a lot. Their answers to the questions are consistent, but they are consistently misleading. As a result, some indexes based on self-report are reliable but invalid. Such indexes are not useful and should be improved or discarded. Unfortunately, many measures are judged to be worthwhile on the basis of only a reliability test.

The reliability and validity of measures in any study must be tested after the fact to assess the quality of the information obtained. But then, if it turns out that a measure cannot be considered reliable and valid, little can be done to save the study. Hence, it is supremely important to select, in the first place, measures that are likely to be reliable and valid. Don't just choose the first measure you find or can think of: Consider the different strengths of different measures and their appropriateness to your study. Conduct a pretest in which you use the measure with a small sample, and check its reliability. Provide careful training to ensure a consistent approach if interviewers or observers will administer the measures. In most cases, however, the best strategy is to use measures that have been used before and whose reliability and validity have been established in other contexts. But the selection of tried and true measures still does not absolve researchers from the responsibility of testing the reliability and validity of the measure in their own studies.

When the population studied or the measurement context differs from that in previous research, instrument reliability and validity may be affected. So the researchers must take pains with the design of their study. For example, test-retest reliability has proved to be better for several standard measures used to assess

Exhibit 4.17 **The Difference Between Reliability and Validity: Drinking Behavior**

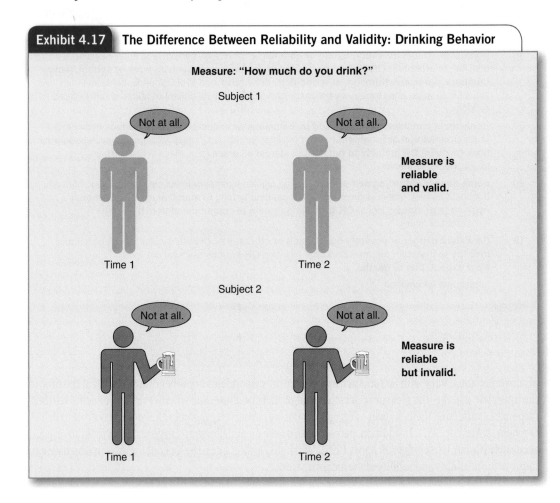

substance use among homeless persons when the interview was conducted in a protected setting and when the measures focused on factual information and referred to a recent time interval (Drake, McHugo, & Biesanz 1995). Subjects who were younger, female, recently homeless, and less severely afflicted with psychiatric problems were also more likely to give reliable answers.

It may be possible to improve the reliability and validity of measures in a study that already has been conducted if multiple measures were used. For example, in our study of housing for homeless mentally ill persons, funded by the National Institute of Mental Health, we assessed substance abuse with several different sets of direct questions as well as with reports from subjects' case managers and others (Goldfinger et al. 1996). We found that the observational reports were often inconsistent with self-reports and that different self-report measures were not always in agreement; hence, the measures were not valid. A more valid measure was initial reports of lifetime substance abuse problems, which identified all those who subsequently abused substances during the project. We concluded that the lifetime measure was a valid way to identify persons at risk for substance abuse problems. No single measure was adequate to identify substance abusers at a particular point in time during the project. Instead, we constructed a composite of observer and self-report measures that seemed to be a valid indicator of substance abuse over 6-month periods.

If the research focuses on previously unmeasured concepts, new measures will have to be devised. Researchers can use one of three strategies to improve the likelihood that new question-based measures will be reliable and valid (Fowler 1995):

Engage potential respondents in group discussions about the questions to be included in the survey. This strategy allows researchers to check for consistent understanding of terms and to hear the range of events or experiences that people will report.

Conduct cognitive interviews. Ask people a test question, then probe with follow-up questions about how they understood the question and what their answer meant.

Audiotape test interviews during the pretest phase of a survey. The researchers then review these audiotapes and systematically code them to identify problems in question wording or delivery. (pp. 104–129)

In these ways, qualitative methods help improve the validity of the fixed-response questions used in quantitative surveys.

🔲 Conclusions

Remember always that measurement validity is a necessary foundation for social research. Gathering data without careful conceptualization or conscientious efforts to operationalize key concepts often is a wasted effort.

The difficulties of achieving valid measurement vary with the concept being operationalized and the circumstances of the particular study. The examples in this chapter of difficulties in achieving valid measures of substance abuse should sensitize you to the need for caution. However, don't let these difficulties discourage you: Substance abuse is a relatively difficult concept to operationalize because it involves behavior that is socially stigmatized and often illegal. Most other concepts in social research present fewer difficulties. But even substance abuse can be measured adequately with a proper research design.

Planning ahead is the key to achieving valid measurement in your own research; careful evaluation is the key to sound decisions about the validity of measures in others' research. Statistical tests can help determine whether a given measure is valid after data have been collected, but if it appears after the fact that a measure is invalid, little can be done to correct the situation. If you cannot tell how key concepts were operationalized when you read a research report, don't trust the findings. And, if a researcher does not indicate the results of tests used to establish the reliability and validity of key measures, remain skeptical.

Key Terms

Highlights

- Conceptualization plays a critical role in research. In deductive research, conceptualization guides the operationalization of specific variables; in inductive research, it guides efforts to make sense of related observations.

- Concepts may refer to either constant or variable phenomena. Concepts that refer to variable phenomena may be very similar to the actual variables used in a study, or they may be much more abstract.

- Concepts are operationalized in research by one or more indicators, or measures, which may derive from observation, self-report, available records or statistics, books and other written documents, clinical indicators, discarded materials, or some combination of these.

- Single-question measures may be closed-ended, with fixed-response choices; open-ended; or partially closed, with fixed-response choices and an option to write another response.

- Question sets may be used to operationalize a concept.

- Indexes and scales measure a concept by combining answers to several questions and thus reducing idiosyncratic error variation. Several issues should be explored with every intended index: Does each question actually measure the same concept? Does combining items in an index obscure important relationships between individual questions and other variables? Is the index multidimensional?

- If differential weighting is used in the calculation of index scores, then we say that it is a scale.

- Level of measurement indicates the type of information obtained about a variable and the type of statistics that can be used to describe its variation. The four levels of measurement can be ordered by the complexity of the mathematical operations they permit: nominal (least complex), ordinal, interval, ratio (most complex). The measurement level of a variable is determined by how the variable is operationalized. Dichotomies, a special case, may be treated as measured at the nominal, ordinal, or interval level.

- The validity of measures should always be tested. There are four basic approaches: (1) face validation, (2) content validation, (3) criterion validation (either predictive or concurrent), and (4) construct validation. Criterion validation provides the strongest evidence of measurement validity, but there often is no criterion to use in validating social science measures.

- Measurement reliability is a prerequisite for measurement validity, although reliable measures are not necessarily valid. Reliability can be assessed through a test-retest procedure, in terms of interitem consistency, through a comparison of responses to alternate forms of the test, or in terms of consistency among observers.

STUDENT STUDY SITE

To assist in completing the web exercises, please access the study site at **www.sagepub.com/schuttisw7e**, where you will find the web exercises with accompanying links. You'll find other useful study materials such as self-quizzes and e-flashcards for each chapter, along with a group of carefully selected articles from research journals that illustrates the major concepts and techniques presented in the book.

Discussion Questions

1. What does trust mean to you? Is the picture in this chapter "worth a thousand words" about trust, or is something missing? Identify two examples of "trust in action" and explain how they represent *your* concept of trust. Now develop a short definition of trust (without checking a dictionary). Compare your definition to those of your classmates and what you find in a dictionary. Can you improve your definition based on some feedback?

2. What questions would you ask to measure level of trust among students? How about feelings of being "in" or "out" with regard to a group? Write five questions for an index and suggest response choices for each. How would you validate this measure using a construct validation approach? Can you think of a criterion validation procedure for your measure?

3. If you were given a questionnaire right now that asked you about your use of alcohol and illicit drugs in the past year, would you disclose the details fully? How do you think others would respond? What if the questionnaire was anonymous? What if there was a confidential ID number on the questionnaire so that the researcher could keep track of who responded? What criterion validation procedure would you suggest for assessing measurement validity?

4. The questions in Exhibit 4.18 on the next page are selected from my survey of shelter staff (Schutt & Fennell 1992). First, identify the level of measurement for each question. Then, rewrite each question so that it measures the same variable but at a different level. For example, you might change the question that measures seniority at the ratio level (in years, months, and days) to one that measures age at the ordinal level (in categories). Or you might change a variable measured at the ordinal level, such as highest grade in school completed, to one measured at the ratio level. For the variables measured at the nominal level, try to identify at least two underlying quantitative dimensions of variation, and write questions to measure variation along these dimensions. For example, you might change the question asking, "What is your current job title?" to two questions that ask about the pay in their current job and the extent to which their job is satisfying.

What are the advantages and disadvantages of phrasing each question at one level of measurement rather than another? Do you see any limitations on the types of questions for which levels of measurement can be changed?

Practice Exercises

1. Now, it's time to try your hand at operationalization with survey-based measures. Formulate a few fixed-choice questions to measure variables pertaining to the concepts you researched for the discussion questions, such as feelings of trust or perceptions of the level of substance abuse in your community. Arrange to interview one or two other students with the questions you have developed. Ask one fixed-choice question at a time, record your interviewee's answer, and then probe for additional comments and clarifications. Your goal is to discover how respondents understand the meaning of the concept you used in the question and what additional issues shape their response to it.

When you have finished the interviews, analyze your experience: Did the interviewees interpret the fixed-choice questions and response choices as you intended? Did you learn more about the concepts you were working on? Should your conceptual definition be refined? Should the questions be rewritten, or would more fixed-choice questions be necessary to capture adequately the variation among respondents?

Exhibit 4.18   **Selected Shelter Staff Survey Questions**

1. What is your current job title? _____

2. What is your current employment status?

   Paid, full-time _____ 1
   Paid, part-time (less than 30 hours per week) _____ 2

3. When did you start your current position? _____ / _____ / _____
   Month   Day   Year

4. In the past month, how often did you help guests deal with each of the following types of problems?
   (Circle one response on each line.)

   Very often _____ Never

Job training/placement	1	2	3	4	5	6	7
Lack of food or bed	1	2	3	4	5	6	7
Drinking problems	1	2	3	4	5	6	7

5. How likely is it that you will leave this shelter within the next year?

   Very likely _____ 1
   Moderately _____ 2
   Not very likely _____ 3
   Not likely at all _____ 4

6. What is the highest grade in school you have completed at this time?

   First through eighth grade _____ 1
   Some high school _____ 2
   High school diploma _____ 3
   Some college _____ 4
   College degree _____ 5
   Some graduate work _____ 6
   Graduate degree _____ 7

7. Are you a veteran?

   Yes _____ 1
   No _____ 2

---

2. Now, try index construction. You might begin with some of the questions you wrote for Practice Exercise 1. Try to write about four or five fixed-choice questions that each measure the same concept. Write each question so that it has the same response choices. Now, conduct a literature search to identify an index that another researcher used to measure your concept or a similar concept. Compare your index to the published index. Which seems preferable to you? Why?

3. Develop a plan for evaluating the validity of a measure. Your instructor will give you a copy of a questionnaire actually used in a study. Pick one question, and define the concept that you believe it is intended to measure. Then develop a construct validation strategy involving other measures in the questionnaire that you think should be related to the question of interest—if it measures what you think it measures.

4. What are some of the research questions you could attempt to answer with the available statistical data? Visit your library and ask for an introduction to the government documents collection. Inspect the volumes from the U.S. Census Bureau that report population characteristics by city and state. List five questions you could explore with such data. Identify six variables implied by these research questions that you could operationalize with the available data. What are the three factors that might influence variation in these measures, other than the phenomenon of interest? (Hint: Consider how the data are collected.)

5. One quick and easy way to check your understanding of the levels of measurement, reliability, and validity is with the interactive exercises on the study site. First, select one of the "Levels of Measurement" options from the Interactive Exercises link on the main menu, and then read the review information at the start of the lesson. You will then be presented with about 10 variables and response choices and asked to identify the level of measurement for each one. If you make a mistake, the program will give a brief explanation about the level of measurement. After you have reviewed one to four of these lessons, repeat the process with one or more of the "Valid and Reliable Measures" lessons.

6. Go to the book's study site and review the Methods section of two of the research articles that you find at www.sagepub.com/schuttisw7e. Write a short summary of the concepts and measures used in these studies. Which article provides clearer definitions of the major concepts? Does either article discuss possible weaknesses in measurement procedures?

Ethics Questions

1. The ethical guidelines for social research require that subjects give their *informed consent* prior to participating in an interview. How "informed" do you think subjects have to be? If you are interviewing people to learn about substance abuse and its impact on other aspects of health, is it okay to just tell respondents in advance that you are conducting a study of health issues? What if you plan to inquire about victimization experiences? Explain your reasoning.

2. Some Homeland Security practices as well as inadvertent releases of web searching records have raised new concerns about the use of unobtrusive measures of behavior and attitudes. If all identifying information is removed, do you think social scientists should be able to study the extent of prostitution in different cities by analyzing police records? How about how much alcohol different types of people use by linking credit card records to store purchases?

Web Exercises

1. How would you define *alcoholism*? Write a brief definition. Based on this conceptualization, describe a method of measurement that would be valid for a study of alcoholism (alcoholism as you define it). Now go to the American Council for Drug Education and read some their facts about alcohol at www.acde.org/common/alcohol2.pdf. Is this information consistent with your definition?

   What are the "facts" about alcoholism presented by the National Council on Alcohol and Drug Dependence (NCADD) at www.ncadd.org? How is alcoholism conceptualized? Based on this conceptualizing, give an example of one method that would be a valid measurement in a study of alcoholism.

   Now look at some of the other related links accessible from the ACDE and NCADD websites. What are some of the different conceptualizations of alcoholism that you find? How does the chosen conceptualization affect one's choice of methods of measurement?

2. What are the latest findings about student substance abuse from the Harvard School of Public Health? Check out www.hsph.harvard.edu/cas and write up a brief report.

3. A list of different measures of substance abuse is available at a site maintained by the National Institute on Alcoholism and Alcohol Abuse, http://pubs.niaaa.nih.gov/publications/Assesing%20Alcohol/. Instrument "fact sheets" can be obtained at http://pubs.niaaa.nih.gov/publications/Assesing%20Alcohol/factsheets.htm, and there is lengthy discussion of the various self-report instruments for alcohol problem screening among adults at http://pubs.niaaa.nih.gov/publications/Assesing%20Alcohol/selfreport.htm (Note the misspelling: *assesing*) (Connors & Volk 2004). Read the Connors and Volk article, and pick two of the instruments they discuss (Connors & Volk 2004:27–32). What concept of substance abuse is reflected in each measure? Is either measure multidimensional? What do you think the relative advantages of each measure might be? What evidence is provided about their reliability and validity? What other test of validity would you suggest?

SPSS Exercises

1. View the variable information for the variables *AGE, CHILDS, PARTYID3, SOCBAR, RACE,* and *INCOME06.* Click on the "variable list" icon or choose Utilities/Variables from the menu. Choose *PARTYID,* then *SOCBAR.* At which levels (nominal/categorical, ordinal, interval, ratio) are each of these variables measured? (By the way, DK means "Don't Know," NA means "No Answer," and NAP means "Not Applicable.")

2. Review the actual questions used to measure four of the variables in Question 1 or in your hypotheses in Chapter 2's SPSS exercise (Question 3). You can find most GSS questions at the following Web site: http://www.norc.org/GSS+Website/Browse+GSS+Variables. Name the variable that you believe each question measures. Discuss the face validity and content validity of each question as a measure of its corresponding variable.

Explain why you conclude that each measure is valid or not.

3. CONFED is part of an index involving the following question: How much confidence do you have in

   a. Executive branch of the federal government

   b. U.S. Supreme Court

   c. Congress

Now answer the following questions:

   a. What is the concept being measured by this index?

   b. Do you agree that each of these variables belongs in the index? Explain.

   c. What additional variables would you like to see included in this index?

Developing a Research Proposal

At this point, you can begin the process of conceptualization and operationalization. You'll need to assume that your primary research method will be conducting a survey. These next steps correspond to Exhibit 2.14, #7.

1. List at least 10 variables that will be measured in your research. No more than two of these should be sociodemographic indicators such as race or age. The inclusion of each variable should be justified in terms of theory or prior research that suggests it would be an appropriate independent or dependent variable or will have some relation to either of these.

2. Write a conceptual definition for each variable. Whenever possible, this definition should come from the existing literature—either a book you have read for

a course or the research literature that you have been searching. Ask two class members for feedback on your definitions.

3. Develop measurement procedures for each variable. Several measures should be single questions and indexes that were used in prior research (search the web and the journal literature in Sociological Abstracts or PsycINFO, the online database of Psychological Abstracts [or its full text version, PsycARTICLES]). Make up a few questions and one index yourself. Ask your classmates to answer these questions and give you feedback on their clarity.

4. Propose tests of reliability and validity for four of the measures.

CHAPTER 5

Sampling

A common technique in journalism is to put a "human face" on a story. For instance, a *Boston Globe* reporter (Johnson 2011) interviewed a participant for a story about a housing program for chronically homeless people. "Stan" had served in the Air Force and worked a series of factory jobs, but after being fired spent more than a third of his 50+ years on the street or waiting in line for a bed in a shelter. Fortunately, thanks to a new housing program developed by Boston's Pine Street Inn, he now lives in stable, permanent housing with other formerly homeless men. He feels he has developed the coping skills to live independently and so is looking forward to living alone—with a cat (Johnson 2011:B1).

It is a sad story with an all-too-uncommon happy—although uncertain—ending. Together with one other such story and comments by several service staff, the article provides a persuasive rationale for the new housing program. However, we don't know whether the two participants interviewed for the story are like most program participants, most homeless persons in Boston, or most homeless persons throughout the United States—or whether they are just two people who caught the eye of this one reporter. In other words, we don't know how generalizable their stories are, and if we don't have confidence in generalizability, then the validity of this account of how the program participants became homeless is suspect. Because we don't know whether their situation is widely shared or unique, we cannot really judge what the account tells us about the social world.

In this chapter, you will learn about sampling methods, the procedures that primarily determine the generalizability of research findings. I first review the rationale for using sampling in social research and consider two circumstances when sampling is not necessary. The chapter then turns to specific sampling methods and when they are most appropriate, using examples from research on homelessness. This section is followed by a section on sampling distributions, which introduces you to the logic of statistical inference—that is, how to determine how likely it is that our sample statistics represent the population from which the sample was drawn. By the chapter's end, you should understand which questions you need to ask to evaluate the generalizability of a study as well as what choices you need to make when designing a sampling strategy. You should also realize that it is just as important to select the "right" people or objects to study as it is to ask participants the right questions.

Sample Planning

You have encountered the problem of generalizability in each of the studies you have read about in this book. For example, Keith Hampton and Barry Wellman (1999) discussed their findings in Netville as though they could be generalized to residents of other communities; Norman Nie and Lutz Erbring (2000) generalized their Internet survey findings to the entire American adult population, and the National Geographic Society (2000) web survey findings were generalized to the entire world. Whether we are designing a sampling strategy or evaluating someone else's findings, we have to understand how and why researchers decide to sample and what the consequences of these decisions are for the generalizability of the study's findings.

Define Sample Components and the Population

> **Population** The entire set of individuals or other entities to which study findings are to be generalized.
>
> **Sample** A subset of a population that is used to study the population as a whole.
>
> **Elements** The individual members of the population whose characteristics are to be measured.

Let's say that we are designing a survey about adult homeless persons in one city. We don't have the time or resources to study the entire adult homeless **population** of the city, even though it consists of the set of individuals or other entities to which we wish to be able to generalize our findings. Even the city of Boston, which conducts an annual census of homeless persons, does not have the resources to actually survey the homeless persons they count. So instead, we resolve to study a **sample**, a subset of this population. The individual members of this sample are called **elements**, or elementary units.

In many studies, we sample directly from the elements in the population of interest. We may survey a sample of the entire population of students in a school, based on a list obtained from the registrar's office. This list, from which the elements of the population are selected, is termed the **sampling frame**. The students who are selected and interviewed from that list are the elements.

> **Sampling frame** A list of all elements or other units containing the elements in a population.

In some studies, the entities that can be reached easily are not the same as the elements from which we want information, but they include those elements. For example, we may have a list of households but not a list of the entire population of a town, even though the adults are the elements that we actually want to sample. In this situation, we could draw a sample of households so that we can identify the adult individuals in these households. The households are termed **enumeration units**, and the adults in the households are the elements (Levy & Lemeshow 1999:13–14).

Sometimes, the individuals or other entities from which we collect information are not actually the elements in our study. For example, a researcher might sample schools for a survey about educational practices and then interview a sample of teachers in each sampled school to obtain the information about educational practices. Both the schools and the teachers are termed **sampling units**, because we sample from both (Levy & Lemeshow 1999:22). The schools are selected in the first stage of the sample, so they are the *primary sampling units* (in this case, they are also the elements in the study). The teachers are *secondary sampling units* (but they are not elements, because they are used to provide information about the entire school) (see Exhibit 5.1).

It is important to know exactly what population a sample can represent when you select or evaluate sample components. In a survey of "adult Americans," the general population may reasonably be construed as all

> **Enumeration units** Units that contain one or more elements and that are listed in a sampling frame.

> **Sampling units** Units listed at each stage of a multistage sampling design.

Exhibit 5.1 | **Sample Components in a Two-Stage Study**

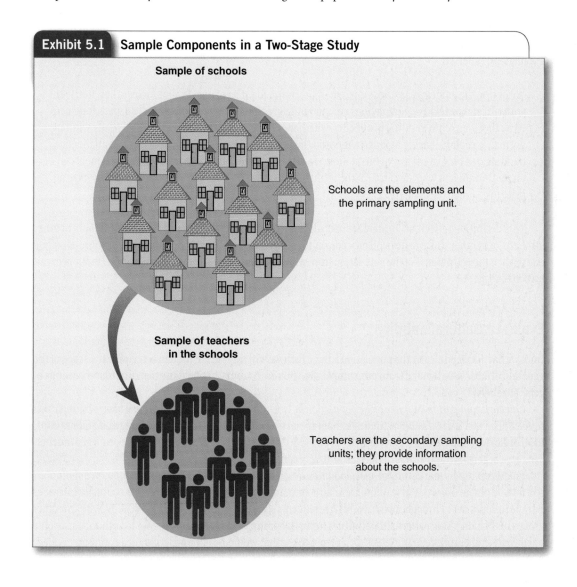

Sample of schools

Schools are the elements and the primary sampling unit.

Sample of teachers in the schools

Teachers are the secondary sampling units; they provide information about the schools.

residents of the United States who are at least 21-years-old. But always be alert to ways in which the population may have been narrowed by the sample selection procedures. For example, perhaps only English-speaking residents of the United States were surveyed. The population for a study is the aggregation of elements that we actually focus on and sample from, not some larger aggregation that we really wish we could have studied.

Some populations, such as the homeless, are not identified by a simple criterion such as a geographic boundary or an organizational membership. Clear definition of such a population is difficult but quite necessary. Anyone should be able to determine just what population was actually studied. However, studies of homeless persons in the early 1980s "did not propose definitions, did not use screening questions to be sure that the people they interviewed were indeed homeless, and did not make major efforts to cover the universe of homeless people" (Burt 1996:15). (Perhaps just homeless persons in one shelter were studied.) The result was a "collection of studies that could not be compared" (Burt 1996:15). Several studies of homeless persons in urban areas addressed the problem by employing a more explicit definition of the population: "people who had no home or permanent place to stay of their own (meaning they rented or owned it themselves) and no regular arrangement to stay at someone else's place" (Burt 1996:18).

Even this more explicit definition still leaves some questions unanswered: What is a "regular arrangement"? How permanent does a "permanent place" have to be? In a study of homeless persons in Chicago, Michael Sosin, Paul Colson, and Susan Grossman (1988) answered these questions in their definition of the population of interest:

> We define the homeless as: those current[ly] residing for at least one day but for less than fourteen with a friend or relative, not paying rent, and not sure that the length of stay will surpass fourteen days; those currently residing in a shelter, whether overnight or transitional; those currently without normal, acceptable shelter arrangements and thus sleeping on the street, in doorways, in abandoned buildings, in cars, in subway or bus stations, in alleys, and so forth; those residing in a treatment center for the indigent who have lived at the facility for less than 90 days and who claim that they have no place to go, when released. (p. 22)

This definition reflects accurately Sosin et al.'s concept of homelessness and allows researchers in other locations or at other times to develop procedures for studying a comparable population. The more complete and explicit the definition is of the population from which a sample was selected, the more precise our generalizations can be.

Evaluate Generalizability

Once we have defined clearly the population from which we will sample, we need to determine the scope of the generalizations we will make from our sample. Do you recall from Chapter 2 the two different meanings of generalizability?

Can the findings from a sample of the population be generalized to the population from which the sample was selected? Did McPherson's (2006) findings about social ties apply to the United States, Ling and Stald's (2010) to all of Norway and Denmark, or Wechsler et al.'s (2002) study of binge drinking to all U.S. college students? This type of generalizability was defined as *sample generalizability* in Chapter 2.

Can the findings from a study of one population be generalized to another, somewhat different population? Are mobile phone users in Norway and Denmark similar to those in other Scandinavian countries? In other European countries? Throughout the world? Are students similar to full-time employees, housewives, or other groups in their drinking patterns? Do findings from a laboratory study about obedience to authority at an elite northeastern U.S. college in the 1960s differ from those that would be obtained today at a commuter college in the Midwest? What is the generalizability of the results from a survey of homeless persons in one city? This type of generalizability question was defined as *cross-population generalizability* in Chapter 2.

This chapter focuses attention primarily on the problem of sample generalizability: Can findings from a sample be generalized to the population from which the sample was drawn? This is really the most basic question to ask about a sample, and social research methods provide many tools with which to address it.

Research in the News

SAMPLE POLLS INDICATE LATINO TURNOUT LIKELY TO LAG

The debate over Arizona's tough law against illegal immigrants seems to have turned off many Latinos, rather than leaving them energized. According to a nationwide phone poll of 1,357 Latinos, including 618 registered voters, only 32% of the registered voters indicated they were likely to vote, compared to 50% of all registered voters. Conducted August 17 to September 19, 2010, the survey had a margin of sampling error of plus or minus 5 percentage points for registered voters.

Source: Lacey, Marc. 2010. "Latino Turnout Likely to Lag, New Poll Finds." *The New York Times,* October 6:A1.

Sample generalizability depends on sample quality, which is determined by the amount of **sampling error**—the difference between the characteristics of a sample and the characteristics of the population from which it was selected. The larger the sampling error, the less representative the sample—and thus the less generalizable the findings. To assess sample quality when you are planning or evaluating a study, ask yourself these questions:

- From what population were the cases selected?
- What method was used to select cases from this population?
- Do the cases that were studied represent, in the aggregate, the population from which they were selected?

> **Sampling error** Any difference between the characteristics of a sample and the characteristics of a population. The larger the sampling error, the less representative the sample.
>
> **Target population** A set of elements larger than or different from the population sampled and to which the researcher would like to generalize study findings.

Cross-population generalizability involves quite different considerations. Researchers are engaging in cross-population generalizability when they project their findings onto groups or populations much larger than, or simply different from, those they have actually studied. The population to which generalizations are made in this way can be termed the **target population**—a set of elements larger than or different from the population that was sampled and to which the researcher would like to generalize any study findings. When we generalize findings to target populations, we must be somewhat speculative. We must carefully consider the validity of claims that the findings can be applied to other groups, geographic areas, cultures, or times.

Because the validity of cross-population generalizations cannot be tested empirically, except by conducting more research in other settings, I do not focus much attention on this problem here. But I return to the problem of cross-population generalizability in Chapter 7, which addresses experimental research, and in Chapter 12, which discusses methods for studying different societies.

Assess the Diversity of the Population

Sampling is unnecessary if all the units in the population are identical. Physicists don't need to select a representative sample of atomic particles to learn about basic physical processes. They can study a single atomic particle because it is identical to every other particle of its type. Similarly, biologists don't need to sample a particular type of plant to determine whether a given chemical has toxic effects on that particular type. The idea is "If you've seen one, you've seen 'em all."

What about people? Certainly, all people are not identical (nor are other animals, in many respects). Nonetheless, if we are studying physical or psychological processes that are the same among all people, sampling is not needed to achieve generalizable findings. Psychologists and social psychologists often conduct experiments on college students to learn about processes that they think are identical across individuals. They believe that most people would have the same reactions as the college students if they experienced the same experimental conditions. Field researchers who observe group processes in a small community sometimes make the same assumption.

There is a potential problem with this assumption, however: There's no way to know for sure if the processes being studied are identical across all people. In fact, experiments can give different results depending on the type of people who are studied or the conditions for the experiment. Stanley Milgram's (1965) classic experiments on obedience to authority, which you studied in Chapter 3, illustrate this point very well. You remember that the original Milgram experiments tested the willingness of male volunteers in New Haven, Connecticut, to comply with the instructions of an authority figure to give "electric shocks" to someone else, even when these shocks seemed to harm the person receiving them. In most cases, the volunteers complied. Milgram concluded that people are very obedient to authority.

Were these results generalizable to all men, to men in the United States, or to men in New Haven? The initial experiment was repeated many times to assess the generalizability of the findings. Similar results were obtained in many replications of the Milgram experiments, that is, when the experimental conditions and subjects were similar to those Milgram studied. Other studies showed that some groups were less likely to react so obediently. Given certain conditions, such as another "subject" in the room who refused to administer the shocks, subjects were likely to resist authority.

So, what do the initial experimental results tell us about how people will react to an authoritarian movement in the real world, when conditions are not so carefully controlled? In the real social world, people may be less likely to react obediently as well. Other individuals may argue against obedience to a particular leader's commands, or people may see on TV the consequences of their actions. But alternatively, people in the real world may be even more obedient to authority than were the experimental subjects, for example, when they get swept up in mobs or are captivated by ideological fervor. Milgram's initial research and the many replications of it give us great insight into human behavior, in part, because they help identify the types of people and conditions to which the initial findings (lack of resistance to authority) can be generalized. But generalizing the results of single experiments is always risky, because such research often studies a small number of people who are not selected to represent any particular population.

Representative sample A sample that "looks like" the population from which it was selected in all respects that are potentially relevant to the study. The distribution of characteristics among the elements of a representative sample is the same as the distribution of those characteristics among the total population. In an unrepresentative sample, some characteristics are overrepresented or underrepresented.

The main point is that social scientists rarely can skirt the problem of demonstrating the generalizability of their findings. If a small sample has been studied in an experiment or a field research project, the study should be replicated in different settings or, preferably, with a **representative sample** of the population to which generalizations are sought (see Exhibit 5.2). The social world and the people in it are just too diverse to be considered *identical units*. Social psychological experiments and small field studies have produced good social science, but they need to be replicated in other settings, with other subjects, to claim any generalizability. Even when we believe that we have uncovered basic social processes in a laboratory experiment or field observation, we should be very concerned with seeking confirmation in other samples and in other research.

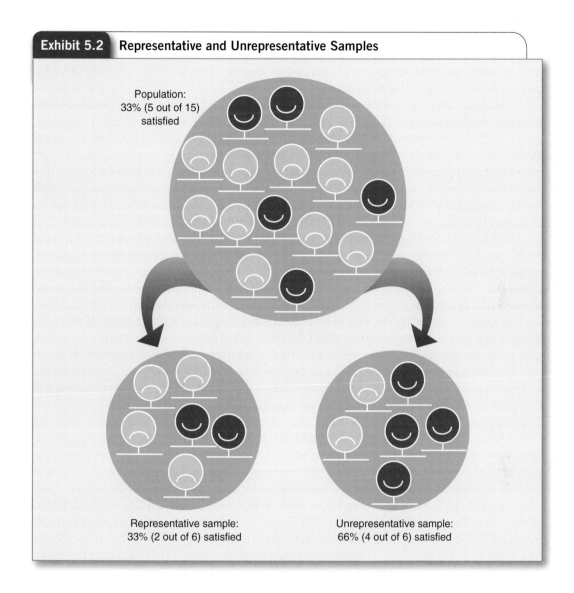

Exhibit 5.2 Representative and Unrepresentative Samples

Population:
33% (5 out of 15)
satisfied

Representative sample:
33% (2 out of 6) satisfied

Unrepresentative sample:
66% (4 out of 6) satisfied

Consider a Census

In some circumstances, it may be feasible to skirt the issue of generalizability by conducting a **census**—studying the entire population of interest—rather than drawing a sample. This is what the federal government tries to do every 10 years with the U.S. Census. Censuses also include studies of all the employees (or students) in small organizations, studies comparing all 50 states, and studies of the entire population of a particular type of organization in some area. However, in comparison with the U.S. Census and similar efforts in other countries, states, and cities, the population that is studied in these other censuses is relatively small.

The reason that social scientists don't often attempt to collect data from all the members of some large population is simply that doing so would be too expensive and time-consuming—and they can do almost as well with a sample. Some social

Census Research in which information is obtained through responses from or information about all available members of an entire population.

scientists conduct research with data from the U.S. Census, but it's the government that collects the data and it's our tax dollars that pay for the effort to get one person in about 134 million households to answer 10 questions. To conduct the 2010 Census, the U.S. Census Bureau spent more than $5.5 billion and hired 3.8 million people (U.S. Bureau of the Census 2010a, 2010b).

Even if the population of interest for a survey is a small town of 20,000 or students in a university of 10,000, researchers will have to sample. The costs of surveying "just" thousands of individuals exceed by far the budgets for most research projects. In fact, not even the U.S. Census Bureau can afford to have everyone answer all the questions that should be covered in the census. So it draws a sample. Every household must complete a short version of the census (it had 10 basic questions in 2010), but a sample of 3 million households is sent a long form (with about 60 questions) every year (U.S. Bureau of the Census 2010d). This more detailed sample survey was launched in 2005 as the American Community Survey and replaces what formerly was a long form of the census that was administered to one-sixth of the population at the same time as the regular census.

The fact that it is hard to get people to complete a survey is another reason why survey research can be costly. Even the U.S. Bureau of the Census (1999) must make multiple efforts to increase the rate of response in spite of the federal law requiring all citizens to complete their census questionnaire. Almost three-quarters (72%) of the U.S. population returned their 2010 census questionnaire through the mail (costing 42 cents per envelope) (U.S. Bureau of the Census 2010a, 2010c). However, 565,000 temporary workers and up to six follow-ups were required to contact the rest of the households that did not respond by mail, at a cost of $57 per nonrespondent (U.S. Bureau of the Census 2010a, 2010c). Even after all that, we know from the 2000 U.S. Census that some groups are still likely to be underrepresented (Armas 2002; Holmes 2001a), including minority groups (Kershaw 2000), impoverished cities (Zielbauer 2000), well-to-do individuals in gated communities and luxury buildings (Langford 2000), and even college students (Abel 2000). The number of persons missed in the 2000 census was estimated to be between 3.2 and 6.4 million (U.S. Bureau of the Census 2001).

The average survey project has far less legal and financial backing, and thus an adequate census is not likely to be possible. Consider the problems of conducting a census in Afghanistan. The first census in 23 years was conducted by its Central Statistics Office in 2003 and 2004, interrupted by snow that cut off many districts for six or seven months. Teams of census takers carried tents, sleeping bags, and satellite phones as they trekked into remote mountainous provinces. An accompanying cartographer identified the location of each village using GPS (Gall 2003:A4). Even in Russia, which spent almost $200 million to survey its population of about 145 million, resource shortages after the collapse of the Soviet Union prevented an adequate census (Myers 2002). In Vladivostok, "Many residents, angry about a recent rise in electricity prices, refused to take part. Residents on Russian Island . . . boycotted to protest dilapidated roads" (Tavernise 2002:A13). In Iraq, dominant groups may have delayed conducting a census for fear that it would document gains in population among disadvantaged groups and thereby strengthen their claims for more resources (Myers 2010:A10).

In most survey situations, it is much better to survey only a limited number from the total population so that there are more resources for follow-up procedures that can overcome reluctance or indifference about participation. (I give more attention to the problem of nonresponse in Chapter 8.)

Sampling Methods

We can now study more systematically the features of samples that make them more or less likely to represent the population from which they are selected. The most important distinction that needs to be made about the samples is whether they are based on a probability or a nonprobability sampling method. Sampling methods that allow us to know in advance how likely it is that any element of a population will be selected for the sample are termed **probability sampling methods**. Sampling methods that do not let us know in advance the likelihood of selecting each element are termed **nonprobability sampling methods**.

Probability sampling methods rely on a random, or chance, selection procedure, which is, in principle, the same as flipping a coin to decide which of two people "wins" and which one "loses." Heads and tails are equally likely to turn up in a coin toss, so both persons have an equal chance of winning. That chance, their **probability of selection**, is 1 out of 2, or .5.

Flipping a coin is a fair way to select one of two people because the selection process harbors no systematic bias. You might win or lose the coin toss, but you know that the outcome was due simply to chance, not to bias. For the same reason, a roll of a six-sided die is a fair way to choose one of six possible outcomes (the odds of selection are 1 out of 6, or .17). Dealing out a hand after shuffling a deck of cards is a fair way to allocate sets of cards in a poker game (the odds of each person getting a particular outcome, such as a full house or a flush, are the same). Similarly, state lotteries use a random process to select winning numbers. Thus, the odds of winning a lottery, the probability of selection, are known, even though they are very much smaller (perhaps 1 out of 1 million) than the odds of winning a coin toss.

There is a natural tendency to confuse the concept of **random sampling**, in which cases are selected only on the basis of chance, with a haphazard method of sampling. On first impression, "leaving things up to chance" seems to imply not exerting any control over the sampling method. But to ensure that nothing but chance influences the selection of cases, the researcher must proceed very methodically, leaving nothing to chance except the selection of the cases themselves. The researcher must follow carefully controlled procedures if a purely random process is to occur. In fact, when reading about sampling methods, do not assume that a random sample was obtained just because the researcher used a random selection method at some point in the sampling process. Look for those two particular problems: selecting elements from an incomplete list of the total population and failing to obtain an adequate response rate.

If the sampling frame is incomplete, a sample selected randomly from that list will not really be a random sample of the population. You should always consider the adequacy of the sampling frame. Even for a simple population such as a university's student body, the registrar's list is likely to be at least a bit out-of-date at any given time. For example, some students will have dropped out, but their status will not yet be officially recorded. Although you may judge the amount of error introduced in this particular situation to be negligible, the problems are greatly compounded for a larger population. The sampling frame for a city, state, or nation is always likely to be incomplete because of constant migration into and out of the area. Even unavoidable omissions from the sampling frame can bias a sample against particular groups within the population.

A very inclusive sampling frame may still yield systematic bias if many sample members cannot be contacted or refuse to participate. Nonresponse is a major hazard in survey research because **nonrespondents** are likely to differ systematically from those who take the time to participate. You should not assume that findings from a randomly selected sample will be generalizable to the population from which the sample was selected if the rate of nonresponse is considerable (certainly not if it is much above 30%).

Probability sampling method A sampling method that relies on a random, or chance, selection method so that the probability of selection of population elements is known.

Nonprobability sampling method Sampling method in which the probability of selection of population elements is unknown.

Probability of selection The likelihood that an element will be selected from the population for inclusion in the sample. In a census of all elements of a population, the probability that any particular element will be selected is 1.0. If half of the elements in the population are sampled on the basis of chance (say, by tossing a coin), the probability of selection for each element is one half, or .5. As the size of the sample as a proportion of the population decreases, so does the probability of selection.

Random sampling A method of sampling that relies on a random, or chance, selection method so that every element of the sampling frame has a known probability of being selected.

Nonrespondents People or other entities who do not participate in a study although they are selected for the sample.

Probability Sampling Methods

Probability sampling methods are those in which the probability of selection is known and is not zero (so there is some chance of selecting each element). These methods randomly select elements and therefore have no

> **Systematic bias**
> Overrepresentation or underrepresentation of some population characteristics in a sample due to the method used to select the sample. A sample shaped by systematic sampling error is a biased sample.

systematic bias; nothing but chance determines which elements are included in the sample. This feature of probability samples makes them much more desirable than nonprobability samples, when the goal is to generalize to a larger population.

Although a random sample has no systematic bias, it will certainly have some sampling error due to chance. The probability of selecting a head is .5 in a single toss of a coin and in 20, 30, or however many tosses of a coin you like. But it is perfectly possible to toss a coin twice and get a head both times. The random "sample" of the two sides of the coin is selected in an unbiased fashion, but it still is unrepresentative. Imagine selecting randomly a sample of 10 people from a population comprising 50 men and 50 women. Just by chance, can't you imagine finding that these 10 people include 7 women and only 3 men? Fortunately, we can determine mathematically the likely degree of sampling error in an estimate based on a random sample (as we'll discuss later in this chapter)—assuming that the sample's randomness has not been destroyed by a high rate of nonresponse or by poor control over the selection process.

In general, both the size of the sample and the homogeneity (sameness) of the population affect the degree of error due to chance; the proportion of the population that the sample represents does not. To elaborate,

- *The larger the sample, the more confidence we can have in the sample's representativeness.* If we randomly pick 5 people to represent the entire population of our city, our sample is unlikely to be very representative of the entire population in terms of age, gender, race, attitudes, and so on. But if we randomly pick 100 people, the odds of having a representative sample are much better; with a random sample of 1,000, the odds become very good indeed.

- *The more homogeneous the population, the more confidence we can have in the representativeness of a sample of any particular size.* Let's say we plan to draw samples of 50 from each of two communities to estimate mean family income. One community is very diverse, with family incomes varying from $12,000 to $85,000. In the other, more homogeneous community, family incomes are concentrated in a narrow range, from $41,000 to $64,000. The estimated mean family income based on the sample from the homogeneous community is more likely to be representative than is the estimate based on the sample from the more heterogeneous community. With less variation to represent, fewer cases are needed to represent the homogeneous community.

- *The fraction of the total population that a sample contains does not affect the sample's representativeness unless that fraction is large.* We can regard any sampling fraction less than 2% with about the same degree of confidence (Sudman 1976:184). In fact, sample representativeness is not likely to increase much until the sampling fraction is quite a bit higher. Other things being equal, a sample of 1,000 from a population of 1 million (with a sampling fraction of 0.001, or 0.1%) is much better than a sample of 100 from a population of 10,000 (although the sampling fraction for this smaller sample is 0.01, or 1%, which is 10 times higher). The size of the samples is what makes representativeness more likely, not the proportion of the whole that the sample represents.

Polls to predict presidential election outcomes illustrate both the value of random sampling and the problems that it cannot overcome. In most presidential elections, pollsters have predicted accurately the outcomes of the actual votes by using random sampling and, these days, phone interviewing to learn for which candidate the likely voters intend to vote. Exhibit 5.3 shows how close these sample-based predictions have been in the

Exhibit 5.3 Presidential Election Outcomes: Predicted and Actual

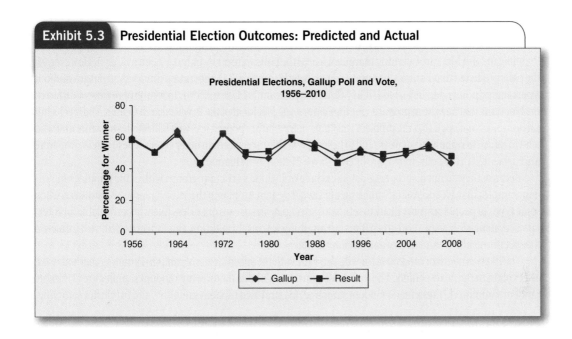

Presidential Elections, Gallup Poll and Vote, 1956–2010

last 14 contests. The exceptions were the 1980 and 1992 elections, when third-party candidates had an unpredicted effect. Otherwise, the small discrepancies between the votes predicted through random sampling and the actual votes can be attributed to random error.

The Gallup poll did quite well in predicting the result of the 2008 presidential election. The final Gallup prediction was that Barack Obama would win with 55% to John McCain's 44% (Gallup 2011). The race turned out a bit closer, with Obama winning by 53% to McCain's 46%, with other polling organizations closer to the final mark (the Rasmussen and Pew polls were exactly on target) (Panagopoulos 2008). In 2004, the final Gallup prediction of 49% for Bush was within 2 percentage points of his winning total of 51% (actually, 50.77%); the "error" is partially due to the 1% of votes cast for third-party candidate Ralph Nader. The results of different polls can vary slightly due to differences in how the pollsters estimate who will actually vote, but the overall rate of accuracy is consistently impressive.

Nevertheless, election polls have produced some major errors in prediction. The reasons for these errors illustrate some of the ways in which unintentional systematic bias can influence sample results. In 1936, a *Literary Digest* poll predicted that Alfred M. Landon would defeat President Franklin Delano Roosevelt in a landslide, but instead Roosevelt took 63% of the popular vote. The problem? The *Digest* mailed out 10 million mock ballots to people listed in telephone directories, automobile registration records, voter lists, and so on. But in 1936, during the Great Depression, only relatively wealthy people had phones and cars, and they were more likely to be Republican. Furthermore, only 2,376,523 completed ballots were returned, and a response rate of only 24% leaves much room for error. Of course, this poll was not designed as a random sample, so the appearance of systematic bias is not surprising. Gallup predicted the 1936 election results accurately with a more systematically selected sample of just 3,000 that avoided so much bias (although they did not yet use random sampling) (Bainbridge 1989:43–44).

In 1948, pollsters mistakenly predicted that Thomas E. Dewey would beat Harry S. Truman, based on the sampling method that George Gallup had used successfully since 1934. The problem was that pollsters stopped collecting data several weeks before the election, and in those weeks, many people changed their minds (Kenney 1987). The sample was systematically biased by underrepresenting shifts in voter sentiment just before the election. It was this experience that convinced Gallup to use only random sampling methods (as well as to continue polling until the election).

The fast-paced 2008 presidential primary elections were also challenging for the pollsters, primarily among Democratic Party voters. In the early New Hampshire primary, polls successfully predicted Republican John McCain's winning margin of 5.5% (the polls were off by only 0.2%, on average). However, all the polls predicted that Barack Obama would win New Hampshire's Democratic primary by a margin of about 8 percentage points, but he lost to Hillary Clinton by 12 points (47% to 35%). In a careful review of different explanations that have been proposed for that failure, the president of the Pew Research Center, Andrew Kohut (2008:A27), concluded that the problem was that voters who are poorer, less well educated, and white and who tend to refuse to respond to surveys tend to be less favorable to blacks than other voters. These voters, who were unrepresented in the polls, were more likely to favor Clinton over Obama.

Because they do not disproportionately exclude or include particular groups within the population, random samples that are successfully implemented avoid systematic bias in the selection process. However, when some types of people are more likely to refuse to participate in surveys or are less likely to be available for interviews, systematic bias can still creep into the sampling process. In addition, random error will still influence the specific results obtained from any random sample.

The likely amount of random error will also vary with the specific type of random sampling method used, as I explain in the next sections. The four most common methods for drawing random samples are (1) simple random sampling, (2) systematic random sampling, (3) stratified random sampling, and (4) cluster sampling.

Simple Random Sampling

Simple random sampling requires some procedure that generates numbers or otherwise identifies cases strictly on the basis of chance. As you know, flipping a coin or rolling a die can be used to identify cases strictly on the basis of chance, but these procedures are not very efficient tools for drawing a sample. A **random number table**, such as the one in Appendix C, simplifies the process considerably. The researcher numbers all the elements in the sampling frame and then uses a systematic procedure for picking corresponding numbers from the random number table. (Practice Exercise 1 at the end of this chapter explains the process step-by-step.) Alternatively, a researcher may use a lottery procedure. Each case number is written on a small card, and then the cards are mixed up and the sample is selected from the cards.

> **Simple random sampling** A method of sampling in which every sample element is selected only on the basis of chance, through a random process.
>
> **Random number table** A table containing lists of numbers that are ordered solely on the basis of chance; it is used for drawing a random sample.

When a large sample must be generated, these procedures are very cumbersome. Fortunately, a computer program can easily generate a random sample of any size. The researcher must first number all the elements to be sampled (the sampling frame) and then run the computer program to generate a random selection of the numbers within the desired range. The elements represented by these numbers are the sample.

Organizations that conduct phone surveys often draw random samples using another automated procedure, called **random digit dialing.** A machine dials random numbers within the phone prefixes corresponding to the area in which the survey is to be conducted. Random digit dialing is particularly useful when a sampling frame is not available. The researcher simply replaces any inappropriate number (e.g., those that are no longer in service or that are for businesses) with the next randomly generated phone number.

> **Random digit dialing** The random dialing by a machine of numbers within designated phone prefixes, which creates a random sample for phone surveys.

As the fraction of the population that has only cell phones has increased (23% in 2009), it has become essential to explicitly sample cell phone numbers as well as landline phone numbers. Those who use cell phones only tend to be younger, more male, more single, more likely to be black or Hispanic and are less likely to vote compared to those who have a landline phone. As a result, failing to include cell phone numbers in a phone survey can introduce bias (Christian et al. 2010). In fact, in a 2008 presidential election survey, those who use only cell phones were less likely to be registered voters than landline users but were considerably more favorable to Obama than landline users (Keeter 2008). (Exhibit 5.4)

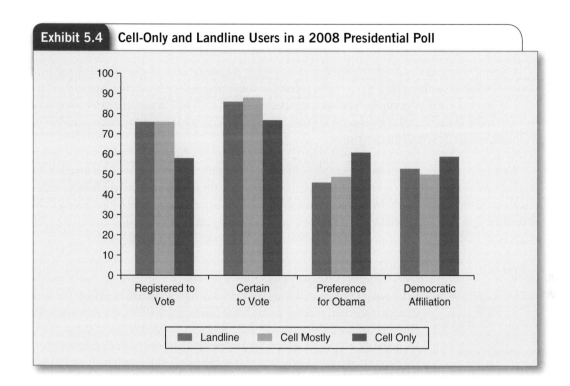

Exhibit 5.4   Cell-Only and Landline Users in a 2008 Presidential Poll

The probability of selection in a true simple random sample is equal for each element. If a sample of 500 is selected from a population of 17,000 (i.e., a sampling frame of 17,000), then the probability of selection for each element is 500 to 17,000, or .03. Every element has an equal chance of being selected, just like the odds in a toss of a coin (1 to 2) or a roll of a die (1 to 6). Thus, simple random sampling is an *equal probability of selection method,* or EPSEM.

Simple random sampling can be done either with or without replacement sampling. In **replacement sampling**, each element is returned to the sampling frame after it is selected so that it may be sampled again. In sampling without replacement, each element selected for the sample is then excluded from the sampling frame. In practice, it makes no difference whether sampled elements are replaced after selection as long as the population is large and the sample is to contain only a small fraction of the population. Random sampling with replacement is, in fact, rarely used.

> **Replacement sampling** A method of sampling in which sample elements are returned to the sampling frame after being selected, so they may be sampled again. Random samples may be selected with or without replacement.

In a study involving simple random sampling, Bruce Link and his associates (1996) used random digit dialing to contact adult household members in the continental United States for an investigation of public attitudes and beliefs about homeless people. Of the potential interviewees, 63% responded. The sample actually obtained was not exactly comparable with the population sampled: Compared with U.S. Census figures, the sample overrepresented women, people age 25 to 54, married people, and those with more than a high school education; it underrepresented Latinos.

How does this sample strike you? Let's assess sample quality using the questions posed earlier in the chapter:

- *From what population were the cases selected?* There is a clearly defined population: the adult residents of the continental United States (who live in households with phones).

- *What method was used to select cases from this population?* The case selection method is a random selection procedure, and there are no systematic biases in the sampling.

- *Do the cases that were studied represent, in the aggregate, the population from which they were selected?* The findings will very likely represent the population sampled because there were no biases in the sampling and a very large number of cases were selected. However, 37% of those selected for interviews could not be contacted or chose not to respond. This rate of nonresponse seems to create a small bias in the sample for several characteristics.

We must also consider the issue of cross-population generalizability: Do findings from this sample have implications for any larger group beyond the population from which the sample was selected? Because a representative sample of the entire U.S. adult population was drawn, this question has to do with cross-national generalizations. Link and his colleagues (1996) don't make any such generalizations. There's no telling what might occur in other countries with different histories of homelessness and different social policies.

Systematic Random Sampling

> **Systematic random sampling**
> A method of sampling in which sample elements are selected from a list or from sequential files.

Systematic random sampling is a variant of simple random sampling. The first element is selected randomly from a list or from sequential files, and then every *n*th element is selected. This is a convenient method for drawing a random sample when the population elements are arranged sequentially. It is particularly efficient when the elements are not actually printed (i.e., there is no sampling frame) but instead are represented by folders in filing cabinets.

Systematic random sampling requires the following three steps:

1. The total number of cases in the population is divided by the number of cases required for the sample. This division yields the sampling interval, the number of cases from one sampled case to another. If 50 cases are to be selected out of 1,000, the **sampling interval** is 20; every 20th case is selected.

2. A number from 1 to 20 (or whatever the sampling interval is) is selected randomly. This number identifies the first case to be sampled, counting from the first case on the list or in the files.

3. After the first case is selected, every *n*th case is selected for the sample, where *n* is the sampling interval. If the sampling interval is not a whole number, the size of the sampling interval is varied systematically to yield the proper number of cases for the sample. For example, if the sampling interval is 30.5, the sampling interval alternates between 30 and 31. In almost all sampling situations, systematic random sampling yields what is essentially a simple random sample. The exception is a situation in which the sequence of elements is affected by **periodicity**—that is, the sequence varies in some regular, periodic pattern. For example, the houses in a new development with the same number of houses in each block (e.g., 8) may be listed by block, starting with the house in the northwest corner of each block and continuing clockwise. If the sampling interval is 8, the same as the periodic pattern, all the cases selected will be in the same position (see Exhibit 5.5). But in reality, periodicity and the sampling interval are rarely the same.

> **Sampling interval** The number of cases from one sampled case to another in a systematic random sample.
>
> **Periodicity** A sequence of elements (in a list to be sampled) that varies in some regular, periodic pattern.

Stratified Random Sampling

Although all probability sampling methods use random sampling, some add steps to the sampling process to make sampling more efficient or easier. **Stratified random sampling** uses information known

Exhibit 5.5 The Effect of Periodicity on Systematic Random Sampling

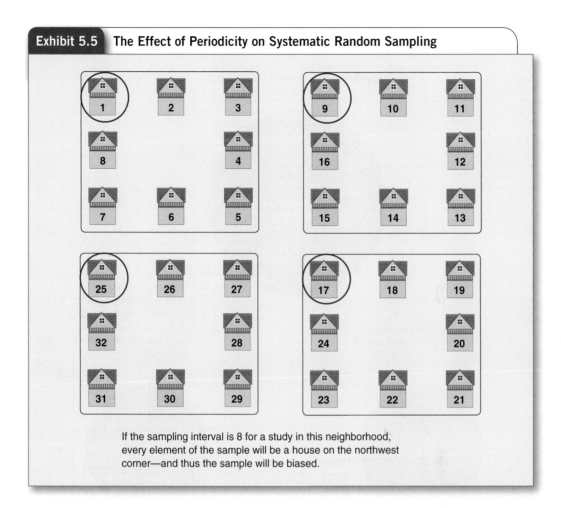

If the sampling interval is 8 for a study in this neighborhood, every element of the sample will be a house on the northwest corner—and thus the sample will be biased.

about the total population prior to sampling to make the sampling process more efficient. First, all elements in the population (i.e., in the sampling frame) are distinguished according to their value on some relevant characteristic. That characteristic forms the sampling strata. Next, elements are sampled randomly from within these strata. For example, race may be the basis for distinguishing individuals in some population of interest. Within each racial category, individuals are then sampled randomly. Of course, using this method requires more information prior to sampling than is the case with simple random sampling. It must be possible to categorize each element in one and only one stratum, and the size of each stratum in the population must be known.

This method is more efficient than drawing a simple random sample because it ensures appropriate representation of elements across strata. Imagine that you plan to draw a sample of 500 from an ethnically diverse neighborhood. The neighborhood population is 15% black, 10% Hispanic, 5% Asian, and 70% white. If you drew a simple random sample, you might end up with somewhat disproportionate numbers of each group. But if you created sampling strata based on race and ethnicity, you could randomly select cases from each stratum: 75 blacks (15% of the sample), 50 Hispanics (10%), 25 Asians (5%), and 350 whites (70%). By using **proportionate stratified sampling**, you would eliminate any possibility of sampling error in the sample's distribution of ethnicity. Each stratum would be represented exactly in proportion to its size in the population from which the sample was drawn (see Exhibit 5.6).

Stratified random sampling A method of sampling in which sample elements are selected separately from population strata that are identified in advance by the researcher.

Proportionate stratified sampling Sampling method in which elements are selected from strata in exact proportion to their representation in the population.

Exhibit 5.6 **Stratified Random Sampling**

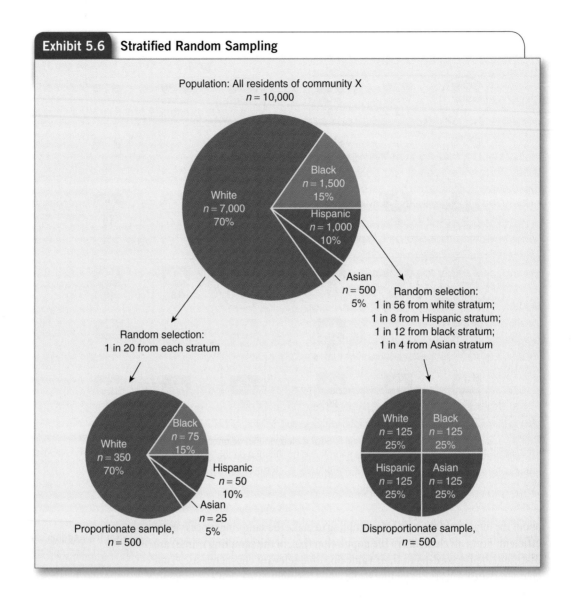

This is the strategy used by Brenda Booth et al. (2002) in a study of homeless adults in two Los Angeles county sites with large homeless populations. Specifically, Booth et al. (2002:432) selected subjects at random from homeless shelters, meal facilities, and from literally homeless populations on the streets. Respondents were sampled proportionately to their numbers in the downtown and Westside areas, as determined by a one-night enumeration. They were also sampled proportionately to their distribution across three nested sampling strata: the population using shelter beds, the population using meal facilities, and the unsheltered population using neither.

In **disproportionate stratified sampling**, the proportion of each stratum that is included in the sample is intentionally varied from what it is in the population. In the case of the sample stratified by ethnicity, you might select equal numbers of cases from each racial or ethnic group: 125 blacks (25% of the sample), 125 Hispanics (25%), 125 Asians (25%), and 125 whites (25%). In this type of sample, the probability of selection of every case is known but unequal between strata. You know what the proportions are in the population, and so you can easily adjust your combined sample statistics to reflect these true proportions. For instance, if you want to combine the ethnic groups and estimate the average income

Disproportionate stratified sampling Sampling in which elements are selected from strata in different proportions from those that appear in the population.

employees who have time to talk when they pick up their paycheck at a personnel office, or approaching particular individuals at opportune times while observing activities in a social setting. You may find yourself interviewing available students at campus hangouts as part of a course assignment. To study sexual risk-taking among homeless youth in Minneapolis, Linda Halcón and Alan Lifson (2004:73) hired very experienced street youth outreach workers who approached youth known or suspected to be homeless and asked if they would be willing to take part in a 20- to 30-minute interview. The interviewers then conducted the 44-question interview, after which they gave respondents some risk reduction and referral information and a $20 voucher.

A participant observation study of a group may require no more sophisticated approach. When Philippe Bourgois, Mark Lettiere, and James Quesada (1997) studied homeless heroin addicts in San Francisco, they immersed themselves in a community of addicts living in a public park. These addicts became the availability sample.

An availability sample is often appropriate in social research—for example, when a field researcher is exploring a new setting and trying to get some sense of the prevailing attitudes or when a survey researcher conducts a preliminary test of a new set of questions.

Now, use the sample evaluation questions (p. 139) to evaluate person-in-the-street interviews of the homeless. If your answers are something like "The population was unknown," "The method for selecting cases was haphazard," and "The cases studied do not represent the population," you're right! There is no clearly definable population from which the respondents were drawn, and no systematic technique was used to select the respondents. There certainly is not much likelihood that the interviewees represent the distribution of sentiment among homeless persons in the Boston area, or of welfare mothers, of impoverished rural migrants, or whatever we imagine the relevant population is in a particular study.

In a similar vein, perhaps person-in-the-street comments to news reporters suggest something about what homeless persons think, or maybe they don't; we can't really be sure. But let's give reporters their due: If they just want to have a few quotes to make their story more appealing, nothing is wrong with their sampling method. However, their approach gives us no basis for thinking that we have an overview of community sentiment. The people who happen to be available in any situation are unlikely to be just like those who are unavailable. We can't be at all certain that what we learn can be generalized with any confidence to a larger population of concern.

Availability sampling often masquerades as a more rigorous form of research. Popular magazines periodically survey their readers by printing a questionnaire for readers to fill out and mail in. A follow-up article then appears in the magazine under a title such as "What You Think About Intimacy in Marriage." If the magazine's circulation is large, a large sample can be achieved in this way. The problem is that usually only a tiny fraction of readers return the questionnaire, and these respondents are probably unlike other readers who did not have the interest or time to participate. So the survey is based on an availability sample. Even though the follow-up article may be interesting, we have no basis for thinking that the results describe the readership as a whole—much less the population at large.

Do you see now why availability sampling differs so much from random sampling methods, which require that "nothing but chance" affects the actual selection of cases? What makes availability sampling "haphazard" is precisely that a great many things other than chance can affect the selection of cases, ranging from the prejudices of the research staff to the work schedules of potential respondents. To truly leave the selection of cases up to chance, we have to design the selection process very carefully so that other factors are not influential. There's nothing haphazard about selecting cases randomly.

Quota Sampling

Quota sampling is intended to overcome the most obvious flaw of availability sampling—that the sample will just consist of whoever or whatever is available, without any concern for its similarity to the population of interest. The distinguishing feature of a quota sample is that quotas are set to ensure that the sample represents certain characteristics in proportion to their prevalence in the population.

Quota sampling A nonprobability sampling method in which elements are selected to ensure that the sample represents certain characteristics in proportion to their prevalence in the population.

Suppose that you wish to sample adult residents of a town in a study of support for a tax increase to improve the town's schools. You know from the town's annual report what the proportions of town residents are in terms of gender, race, age, and number of children. You think that each of these characteristics might influence support for new school taxes, so you want to be sure that the sample includes men, women, whites, blacks, Hispanics, Asians, older people, younger people, big families, small families, and childless families in proportion to their numbers in the town population.

This is where quotas come in. Let's say that 48% of the town's adult residents are men and 52% are women, and that 60% are employed, 5% are unemployed, and 35% are out of the labor force. These percentages and the percentages corresponding to the other characteristics become the quotas for the sample. If you plan to include a total of 500 residents in your sample, 240 must be men (48% of 500), 260 must be women, 300 must be employed, and so on. You may even set more refined quotas, such as certain numbers of employed women, employed men, unemployed men, and so on. With the quota list in hand, you (or your research staff) can now go out into the community looking for the right number of people in each quota category. You may go door to door, bar to bar, or just stand on a street corner until you have surveyed 240 men, 260 women, and so on.

The problem is that even when we know that a quota sample is representative of the particular characteristics for which quotas have been set, we have no way of knowing if the sample is representative in terms of any other characteristics. In Exhibit 5.11, for example, quotas have been set for gender only. Under these circumstances, it's no surprise that the sample is representative of the population only in terms of gender, not in terms of race. Interviewers are only human; they may avoid potential respondents with menacing dogs in the front yard, or they could seek out respondents who are physically attractive or who look like they'd be easy to interview. Realistically, researchers can set quotas for only a small fraction of the characteristics relevant to a study, so a quota sample is really not much better than an availability sample (although following careful, consistent procedures for selecting cases within the quota limits always helps).

This last point leads me to another limitation of quota sampling: You must know the characteristics of the entire population to set the right quotas. In most cases, researchers know what the population looks like in terms of no more than a few of the characteristics relevant to their concerns—and in some cases, they have no such information on the entire population.

If you're now feeling skeptical of quota sampling, you've gotten the drift of my remarks. Nonetheless, in some situations, establishing quotas can add rigor to sampling procedures. It's almost always better to maximize possibilities for comparison in research, and quota sampling techniques can help qualitative researchers do this. For instance, Doug Timmer, Stanley Eitzen, and Kathryn Talley (1993:7) interviewed homeless persons in several cities and other locations for their book on the sources of homelessness. Persons who were available were interviewed, but the researchers paid some attention to generating a diverse sample. They interviewed 20 homeless men who lived on the streets without shelter and 20 mothers who were found in family shelters. About half of those whom the researchers selected in the street sample

Exhibit 5.11   Quota Sampling

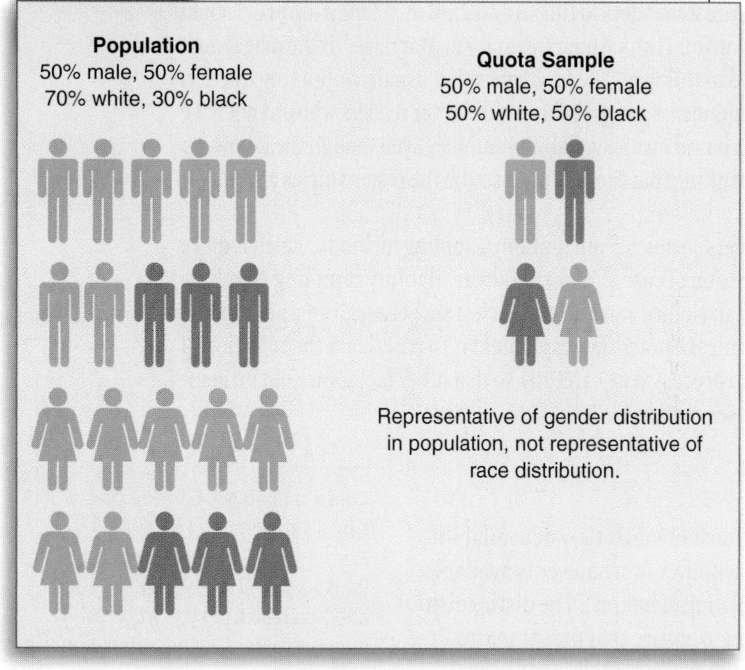

Population
50% male, 50% female
70% white, 30% black

Quota Sample
50% male, 50% female
50% white, 50% black

Representative of gender distribution in population, not representative of race distribution.

were black, and about half were white. Although the researchers did not use quotas to try to match the distribution of characteristics among the total homeless population, their informal quotas helped ensure some diversity in key characteristics.

Does quota sampling remind you of stratified sampling? It's easy to understand why, since they both select sample members, in part, on the basis of one or more key characteristics. Exhibit 5.12 summarizes the differences between quota sampling and stratified random sampling. The key difference, of course, is quota sampling's lack of random selection.

| Exhibit 5.12 | Comparison of Stratified and Quota Sampling Methods |

Feature	Stratified	Quota
Unbiased (random) selection of cases	Yes	No
Sampling frame required	Yes	No
Ensures representation of key strata	Yes	Yes

Purposive Sampling

In **purposive sampling**, each sample element is selected for a purpose, usually because of the unique position of the sample elements. Purposive sampling may involve studying the entire population of some limited group (directors of shelters for homeless adults) or a subset of a population (mid-level managers with a reputation for efficiency). Or a purposive sample may be a *key informant survey,* which targets individuals who are particularly knowledgeable about the issues under investigation.

Purposive sampling A nonprobability sampling method in which elements are selected for a purpose, usually because of their unique position.

Herbert Rubin and Irene Rubin (1995) suggest three guidelines for selecting informants when designing any purposive sampling strategy. Informants should be

- Knowledgeable about the cultural arena or situation or experience being studied
- Willing to talk
- Represent[ative of] the range of points of view (p. 66)

In addition, Rubin and Rubin (1995) suggest continuing to select interviewees until you can pass two tests:

Completeness: What you hear provides an overall sense of the meaning of a concept, theme, or process. (p. 72)

Saturation: You gain confidence that you are learning little that is new from subsequent interview[s]. (p. 73)

Adhering to these guidelines will help ensure that a purposive sample adequately represents the setting or issues studied.

Of course, purposive sampling does not produce a sample that represents some larger population, but it can be exactly what is needed in a case study of an organization, community, or some other clearly defined and relatively limited group. In an intensive organizational case study, a purposive sample of organizational leaders might be complemented with a probability sample of organizational members. Before designing her probability samples of hospital patients and homeless persons, Dee Roth (1990:146–147) interviewed a purposive sample of 164 key informants from organizations that had contact with homeless people in each of the counties she studied.

Snowball Sampling

Snowball sampling is useful for hard-to-reach or hard-to-identify populations for which there is no sampling frame, but the members of which are somewhat interconnected (at least some members of the population know each other). It can be used to sample members of groups such as drug dealers, prostitutes, practicing criminals, participants in Alcoholics Anonymous groups, gang leaders, informal organizational leaders, and homeless persons. It may also be used for charting the relationships among members of some group (a sociometric study), for exploring the population of interest prior to developing a formal sampling plan, and for developing what becomes a census of informal leaders of small organizations or communities. However, researchers using snowball sampling normally cannot be confident that their sample represents the total population of interest, so generalizations must be tentative.

Snowball sampling A method of sampling in which sample elements are selected as they are identified by successive informants or interviewees.

Rob Rosenthal (1994) used snowball sampling to study homeless persons living in Santa Barbara, California:

I began this process by attending a meeting of homeless people I had heard about through my housing advocate contacts. . . . One homeless woman . . . invited me to . . . where she promised to introduce me around. Thus a process of snowballing began. I gained entree to a group through people I knew, came to know others, and through them gained entree to new circles. (pp. 178, 180)

One problem with this technique is that the initial contacts may shape the entire sample and foreclose access to some members of the population of interest:

Sat around with [my contact] at the Tree. Other people come by, are friendly, but some regulars, especially the tougher men, don't sit with her. Am I making a mistake by tying myself too closely to her? She lectures them a lot. (Rosenthal 1994:181)

More systematic versions of snowball sampling can reduce the potential for bias. For example, *respondent-driven sampling* gives financial incentives to respondents to recruit diverse peers (Heckathorn 1997). Limitations on the number of incentives that any one respondent can receive increase the sample's diversity. Targeted incentives can steer the sample to include specific subgroups. When the sampling is repeated through several waves, with new respondents bringing in more peers, the composition of the sample converges on a more representative mix of characteristics than would occur with uncontrolled snowball sampling. Exhibit 5.13 shows how the sample spreads out through successive recruitment waves to an increasingly diverse pool (Heckathorn 1997:178). Exhibit 5.14 shows that even if the starting point were all white persons, respondent-driven sampling would result in an appropriate ethnic mix from an ethnically diverse population (Heckathorn 2002:17).

Lessons About Sample Quality

Some lessons are implicit in my evaluations of the samples in this chapter:

- We can't evaluate the quality of a sample if we don't know what population it is supposed to represent. If the population is unspecified because the researchers were never clear about the population they were trying to sample, then we can safely conclude that the sample itself is no good.

- We can't evaluate the quality of a sample if we don't know how cases in the sample were selected from the population. If the method was specified, we then need to know whether cases were selected in a systematic fashion and on the basis of chance. In any case, we know that a haphazard method of sampling (as in person-on-the-street interviews) undermines generalizability.

- Sample quality is determined by the sample actually obtained, not just by the sampling method itself. If many of the people selected for our sample are nonrespondents or people (or other entities) who do

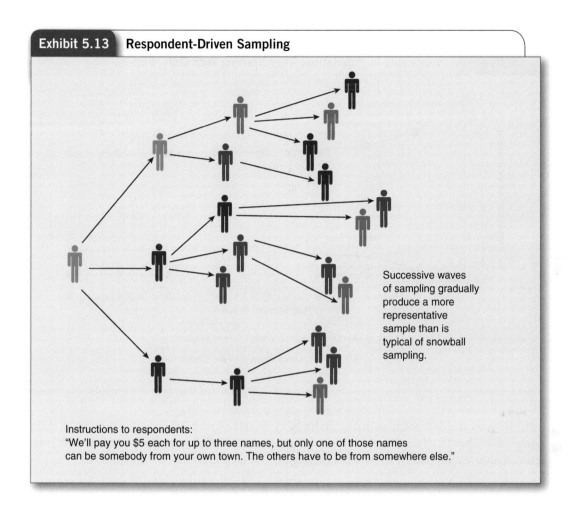

Exhibit 5.13 Respondent-Driven Sampling

Successive waves of sampling gradually produce a more representative sample than is typical of snowball sampling.

Instructions to respondents:
"We'll pay you $5 each for up to three names, but only one of those names can be somebody from your own town. The others have to be from somewhere else."

not participate in the study although they have been selected for the sample, the quality of our sample is undermined—even if we chose the sample in the best possible way.

- We need to be aware that even researchers who obtain very good samples may talk about the implications of their findings for some group that is larger than, or just different from, the population they actually sampled. For example, findings from a representative sample of students in one university often are discussed as if they tell us about university students in general. And maybe they do; we just don't know for sure.

- A sample that allows for comparisons involving theoretically important variables is better than one that does not allow such comparisons. Even when we study people or social processes in depth, it is best to select individuals or settings with an eye to how useful they will be for examining relationships. Limiting an investigation to just one setting or just one type of person will inevitably leave us wondering what it is that makes a difference.

Generalizability in Qualitative Research

Qualitative research often focuses on populations that are hard to locate or very limited in size. In consequence, nonprobability sampling methods such as availability sampling and snowball sampling are often

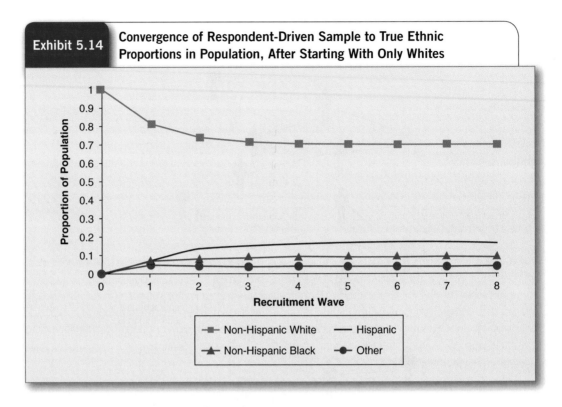

Exhibit 5.14 Convergence of Respondent-Driven Sample to True Ethnic Proportions in Population, After Starting With Only Whites

used. However, this does not mean that generalizability should be ignored in qualitative research, or that a sample should be studied simply because it is convenient (Gobo 2008:206). Janet Wards Schofield (2002) suggests two different ways of increasing the generalizability of the samples obtained in such situations:

> *Studying the Typical.* Choosing sites on the basis of their fit with a typical situation is far preferable to choosing on the basis of convenience. (p. 181)

> *Performing Multisite Studies.* A finding emerging repeatedly in the study of numerous sites would appear to be more likely to be a good working hypothesis about some as yet unstudied site than a finding emerging from just one or two sites. . . . Generally speaking, a finding emerging from the study of several very heterogenous sites would be more . . . likely to be useful in understanding various other sites than one emerging from the study of several very similar sites. (p. 184)

Giampietro Gobo (2008:204–205) highlights another approach to improving generalizability in qualitative research. A case may be selected for in-depth study because it is atypical, or deviant. Investigating social processes in a situation that differs from the norm will improve understanding of how social processes work in typical situations: "the exception that proves the rule."

Some qualitative researchers do question the value of generalizability, as most researchers understand it. The argument is that understanding the particulars of a situation in depth is an important object of inquiry in itself. In the words of sociologist Norman Denzin,

> The interpretivist rejects generalization as a goal and never aims to draw randomly selected samples of human experience. . . . Every instance of social interaction . . . represents a slice from the life world that is the proper subject matter for interpretive inquiry. (Denzin cited in Schofield 2002:173)

▣ Sampling Distributions

A well-designed probability sample is one that is likely to be representative of the population from which it was selected. But as you've seen, random samples still are subject to sampling error owing just to chance. To deal with that problem, social researchers take into account the properties of a sampling distribution, a hypothetical distribution of a statistic across all the random samples that could be drawn from a population. Any single random sample can be thought of as just one of an infinite number of random samples that, in theory, could have been selected from the population. If we had the finances of Gatsby and the patience of Job and were able to draw an infinite number of samples, and we calculated the same type of statistic for each of these samples, we would then have a sampling distribution. Understanding sampling distributions is the foundation for understanding how statisticians can estimate sampling error.

What does a sampling distribution look like? Because a sampling distribution is based on some statistic calculated for different samples, we need to choose a statistic. Let's focus on the arithmetic average, or mean. I will explain the calculation of the mean in Chapter 14, but you may already be familiar with it: You add up the values of all the cases and divide by the total number of cases. Let's say you draw a random sample of 500 families and find that their average (mean) family income is $58,239. Imagine that you then draw another random sample. That sample's mean family income might be $60,302. Imagine marking these two means on graph paper and then drawing more random samples and marking their means on the graph. The resulting graph would be a sampling distribution of the mean.

Exhibit 5.15 demonstrates what happened when I did something very similar to what I have just described—not with an infinite number of samples and not from a large population, but through the same process using the 2006 General Social Survey (GSS) sample as if it were a population. First, I drew 49 different random samples, each consisting of 30 cases, from the 2006 GSS. (The standard notation for the number of cases in each sample is $n = 30$.) Then I calculated for each random sample the approximate mean family income (approximate because the GSS does not record actual income in dollars). I then graphed the means of the 49 samples. Each bar in Exhibit 5.15 shows how many samples had a particular family income. The mean for the population (the total GSS sample) is $59,213, and you can see that many of the samples in the sampling distribution are close to this value. However, although many of the sample means are close to the population mean, some are quite far from it. If you had calculated the mean from only one sample, it could have been anywhere in this sampling distribution, but it is unlikely to have been far from the population mean—that is, unlikely to have been close to either end (or "tail") of the distribution.

Estimating Sampling Error

We don't actually observe sampling distributions in real research; researchers just draw the best sample they can and then are stuck with the results—one sample, not a distribution of samples. A sampling distribution is a theoretical distribution. However, we can use the properties of sampling distributions to calculate the amount of sampling error that was likely with the random sample used in a study. The tool for calculating sampling error is called **inferential statistics**.

Sampling distributions for many statistics, including the mean, have a "normal" shape. A graph of a normal distribution looks like a bell, with one "hump" in the middle, centered on the population mean, and the number of cases tapering off

> **Inferential statistics** A mathematical tool for estimating how likely it is that a statistical result based on data from a random sample is representative of the population from which the sample is assumed to have been selected.

on both sides of the mean. Note that a normal distribution is symmetric: If you folded it in half at its center (at the population mean), the two halves would match perfectly. This shape is produced by **random sampling error**—variation owing purely to chance. The value of the statistic varies from sample to sample because of chance, so higher and lower values are equally likely.

The partial sampling distribution in Exhibit 5.15 does not have a completely normal shape because it involves only a small number of samples (49), each of which has only 30 cases. Exhibit 5.16 shows what the

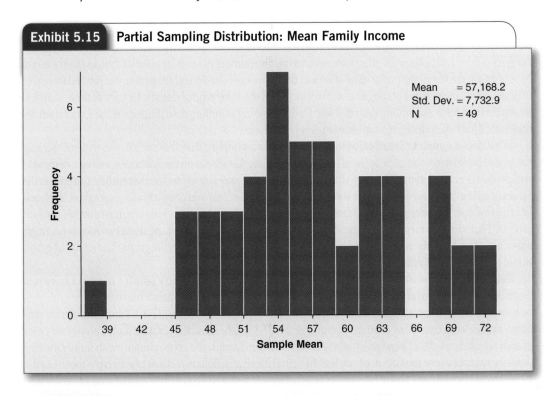

Exhibit 5.15 **Partial Sampling Distribution: Mean Family Income**

Mean = 57,168.2
Std. Dev. = 7,732.9
N = 49

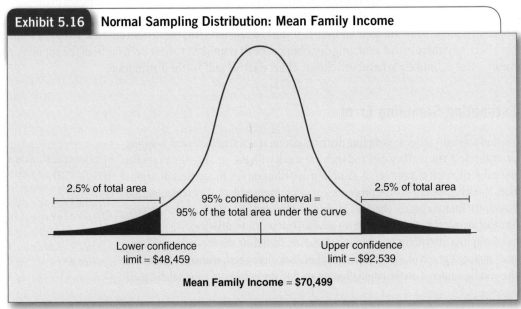

Exhibit 5.16 **Normal Sampling Distribution: Mean Family Income**

2.5% of total area

2.5% of total area

95% confidence interval = 95% of the total area under the curve

Lower confidence limit = $48,459

Upper confidence limit = $92,539

Mean Family Income = $70,499

sampling distribution of family incomes would look like if it formed a perfectly normal distribution—if, rather than 49 random samples, I had selected thousands of random samples.

The properties of a sampling distribution facilitate the process of statistical inference. In the sampling distribution, the most frequent value of the **sample statistic**—the statistic (such as the mean) computed from sample data—is identical to the **population parameter**—the statistic computed for the entire population. In other words, we can have a lot of confidence that the value at the peak of the bell curve represents the norm for the entire population. A population parameter also may be termed the *true value* for the statistic in that population. A sample statistic is an estimate of a population parameter.

In a normal distribution, a predictable proportion of cases fall within certain ranges. Inferential statistics takes advantage of this feature and allows researchers to estimate how likely it is that, given a particular sample, the true population value will be within some range of the statistic. For example, a statistician might conclude from a sample of 30 families that "we can be 95% confident that the true mean family income in the total population is between $33,813 and $53,754." The interval from $33,813 to $53,754 would then be called the *95% confidence interval for the mean*. The lower ($33,813) and upper ($53,754) bounds of this interval are termed the *confidence limits*. Exhibit 5.16 marks such confidence limits, indicating the range that encompasses 95% of the area under the normal curve; 95% of all sample means would fall within this range, as does the mean of our hypothetical sample of 30 cases.

Although all normal distributions have these same basic features, they differ from one another in the extent to which they cluster around the mean. A sampling distribution is more compact when it is based on larger samples. Stated another way, we can be more confident in estimates based on larger random samples because we know that a larger sample creates a more compact sampling distribution. Compare the two sampling distributions of mean family income shown in Exhibit 5.17. Both depict the results for about 50 samples. However, in one study, each sample consisted of 100 families, and in the other study each sample consisted of only 5 families. Clearly, the larger samples result in a sampling distribution that is much more tightly clustered around the mean (range of 34 to 44) than is the case with the smaller samples (range of 17 to 57). The 95% confidence interval for mean family income for the entire 2006 GSS sample of 3,873 cases (the ones that had valid values of family income) was $57,416 to $61,009—an interval only $3,593 wide. But the 95% confidence interval for the mean family income in one GSS subsample of 100 cases was much wider, with limits of $48,891 and $72,501. And for a subsample of only 5 cases, the 95% confidence interval was very broad indeed: from $16,566 to $161,310. As you can see, such small samples result in statistics that actually give us very little useful information about the population.

Other confidence intervals, such as the 99% confidence interval, can be reported. As a matter of convention, statisticians use only the 95%, 99%, and 99.9% confidence limits to estimate the range of values that are likely to contain the true value. These conventional limits reflect the conservatism inherent in classical statistical inference: Don't make an inferential statement unless you are very confident (at least 95% confident) that it is correct.

The less precise an estimate of a particular statistic from a particular sample is, the more confident we can be—and the wider the confidence interval. As I mentioned previously, the 95% confidence interval for the entire 2006 GSS sample is $57,416 to $61,009 (a width of $3,593); the 99% confidence interval is $56,848 to $61,578 (a width of $4,730).

I will explain how to calculate confidence intervals in Chapter 14 and how to express the variability in a sample estimate with a statistic called the standard error. The basic statistics that I introduce in that chapter will make it easier to understand these other statistics. If you have already completed a statistics course, you

Random sampling error (chance sampling error) Differences between the population and the sample that are due only to chance factors (random error), not to systematic sampling error. Random sampling error may or may not result in an unrepresentative sample. The magnitude of sampling error due to chance factors can be estimated statistically.

Sample statistic The value of a statistic, such as a mean, computed from sample data.

Population parameter The value of a statistic, such as a mean, computed using the data for the entire population; a sample statistic is an estimate of a population parameter.

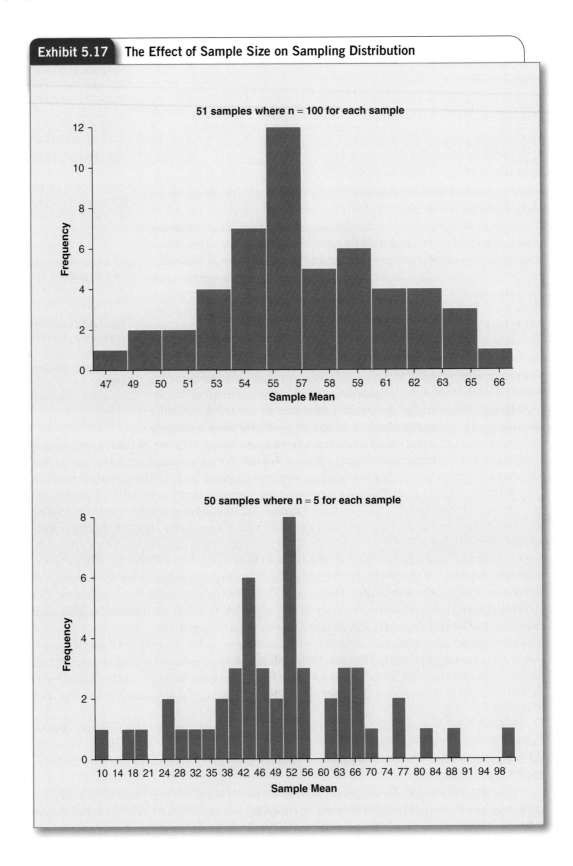

Exhibit 5.17 The Effect of Sample Size on Sampling Distribution

might want to turn now to Chapter 14's confidence interval section for a quick review. In any case, you should now have a sense of how researchers make inferences from a random sample of a population.

Sample Size Considerations

You have learned now that more confidence can be placed in the generalizability of statistics from larger samples, so that you may be eager to work with random samples that are as large as possible. Unfortunately, researchers often cannot afford to sample a very large number of cases. Therefore, they try to determine during the design phase of their study how large a sample they must have to achieve their purposes. They have to consider the degree of confidence desired, the homogeneity of the population, the complexity of the analysis they plan, and the expected strength of the relationships they will measure.

- The less sampling error desired, the larger the sample size must be.

- Samples of more homogeneous populations can be smaller than samples of more diverse populations. Stratified sampling uses prior information on the population to create more homogeneous population strata from which the sample can be selected, so stratified samples can be smaller than simple random samples.

- If the only analysis planned for a survey sample is to describe the population in terms of a few variables, a smaller sample is required than if a more complex analysis involving sample subgroups is planned. If much of the analysis will focus on estimating the characteristics of subgroups within the sample, it is the size of the subgroups that must be considered, not the size of the total sample (Levy & Lemeshow 1999:74).

- When the researchers expect to find very strong relationships among the variables when they test hypotheses, they will need a smaller sample to detect these relationships than if they expect weaker relationships.

Researchers can make more precise estimates of the sample size required through a method called *statistical power analysis* (Kraemer & Thiemann 1987). Statistical power analysis requires a good advance estimate of the strength of the hypothesized relationship in the population. In addition, the math is complicated, so it helps to have some background in mathematics or to be able to consult a statistician. For these reasons, many researchers do not conduct formal power analyses when deciding how many cases to sample.

Exhibit 5.18 shows the results of a power analysis conducted to determine the sample size required to estimate a proportion in the population, with the null hypothesis that the proportion is .50. For the sake of simplicity, it is assumed that the researcher wants to be 95% confident that the actual proportion differs from the null hypothesis of .50; in other words, the researcher wants a sample size that will identify a difference from .5 that is significant at the .05 level. You can see that if the true proportion in the population is actually .55, a sample larger than 800 will be needed to detect this difference at the .05 level of significance. However, if the true proportion in the population is .60, then a random sample of only 200 cases is necessary. The required sample size falls off very gradually beyond this point, as the actual proportion in the population rises beyond .60.

It should be clear from Exhibit 5.18 that you must have a good estimate of the true population value of the statistic you are going to calculate. You also have to decide what significance level (such as .05) you want to achieve in your statistical test. Both of these factors can have a major impact on the number of cases you need to obtain.

You can obtain some general guidance about sample sizes from the current practices of social scientists. For professional studies of the national population in which only a simple description is desired, professional social science studies typically have used a sample size of between 1,000 and 1,500 people, with up to 2,500

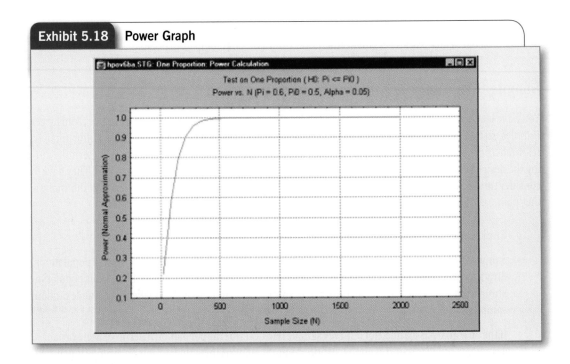

Exhibit 5.18   Power Graph

being included if detailed analyses are planned. Studies of local or regional populations often sample only a few hundred people, in part because these studies lack sufficient funding to draw larger samples. Of course, the sampling error in these smaller studies is considerably larger than in a typical national study (Sudman 1976:87).

Conclusions

Sampling is a powerful tool for social science research. Probability sampling methods allow a researcher to use the laws of chance, or probability, to draw samples from which population parameters can be estimated with a high degree of confidence. A sample of just 1,000 or 1,500 individuals can be used to estimate reliably the characteristics of the population of a nation comprising millions of individuals.

But researchers do not come by representative samples easily. Well-designed samples require careful planning, some advance knowledge about the population to be sampled, and adherence to systematic selection procedures—all so that the selection procedures are not biased. And even after the sample data are collected, the researcher's ability to generalize from the sample findings to the population is not completely certain. The best that he or she can do is to perform additional calculations that state the degree of confidence that can be placed in the sample statistic.

The alternatives to random, or probability-based, sampling methods are almost always much less palatable for quantitative studies, even though they are typically much cheaper. Without a method of selecting cases likely to represent the population in which the researcher is interested, research findings will have to be carefully qualified. Qualitative researchers whose goal is to understand a small group or setting in depth may necessarily have to use unrepresentative samples, but they must keep in mind that the generalizability of their findings will not be known. Additional procedures for sampling in qualitative studies are introduced in Chapter 9.

Social scientists often seek to generalize their conclusions from the population that they studied to some larger target population. The validity of generalizations of this type is necessarily uncertain, because having a representative sample of a particular population does not at all ensure that what we find will hold true in other populations. Nonetheless, as you will see in Chapter 15, the cumulation of findings from studies based on local or otherwise unrepresentative populations can provide important information about broader populations.

Key Terms

Availability sampling   154
Census   141
Cluster   152
Cluster sampling   152
Disproportionate stratified sampling   150
Elements   136
Enumeration units   137
Inferential statistics   161
Nonprobability sampling method   143
Nonrespondents   143
Periodicity   148
Population   136

Population parameter   163
Probability of selection   143
Probability sampling method   143
Proportionate stratified sampling   149
Purposive sampling   157
Quota sampling   155
Random digit dialing   146
Random number table   146
Random sampling   143
Random sampling error   162
Replacement sampling   147
Representative sample   140

Sample   136
Sample statistic   163
Sampling error   139
Sampling frame   136
Sampling interval   148
Sampling units   137
Simple random sampling   146
Snowball sampling   157
Stratified random sampling   148
Systematic bias   144
Systematic random sampling   148
Target population   139

Highlights

- Sampling theory focuses on the generalizability of descriptive findings to the population from which the sample was drawn. It also considers whether statements can be generalized from one population to another.

- Sampling is unnecessary when the elements that would be sampled are identical, but the complexity of the social world makes it difficult to argue very often that different elements are identical. Conducting a complete census of a population also eliminates the need for sampling, but the resources required for a complete census of a large population are usually prohibitive.

- Nonresponse undermines sample quality: It is the obtained sample, not the desired sample, that determines sample quality.

- Probability sampling methods rely on a random selection procedure to ensure no systematic bias in the selection of elements. In a probability sample, the odds of selecting elements are known, and the method of selection is carefully controlled.

- A sampling frame (a list of elements in the population) is required in most probability sampling methods. The adequacy of the sampling frame is an important determinant of sample quality.

- Simple random sampling and systematic random sampling are equivalent probability sampling methods in most of the situations. However, systematic random sampling is inappropriate for sampling from lists of elements that have a regular, periodic structure.

- Stratified random sampling uses prior information about a population to make sampling more efficient. Stratified sampling may be either proportionate or disproportionate. Disproportionate stratified sampling is useful when a research question focuses on a stratum or on strata that make up a small proportion of the population.

- Cluster sampling is less efficient than simple random sampling, but it is useful when a sampling frame is unavailable. It is also useful for large populations spread out across a wide area or among many organizations.

- Nonprobability sampling methods can be useful when random sampling is not possible, when a research question does not concern a larger population, and when a preliminary exploratory study is appropriate. However, the representativeness of nonprobability samples cannot be determined.

- The likely degree of error in an estimate of a population characteristic based on a probability sample decreases when the size of the sample and the homogeneity of the population from which the sample was selected increases. The proportion of the population that is sampled does not affect sampling error, except when that proportion is large. The degree of sampling error affecting a sample statistic can be estimated from the characteristics of the sample and knowledge of the properties of sampling distributions.

STUDENT STUDY SITE

To assist in completing the web exercises, please access the study site at **www.sagepub.com/schuttisw7e**, where you will find the web exercises with accompanying links. You'll find other useful study materials such as self-quizzes and e-flashcards for each chapter, along with a group of carefully selected articles from research journals that illustrate the major concepts and techniques presented in the book.

Discussion Questions

1. When (if ever) is it reasonable to assume that a sample is not needed because "everyone is the same"—that is, the population is homogeneous? Does this apply to research such as that of Stanley Milgram's on obedience to authority? What about investigations of student substance abuse? How about investigations of how people (or their bodies) react to alcohol? What about research on likelihood of voting (the focus of Chapter 14)?

2. All adult U.S. citizens are required to participate in the decennial census, but some do not. Some social scientists have argued for putting more resources into a large representative sample, so that more resources are available to secure higher rates of response from hard-to-include groups. Do you think that the U.S. Census should shift to a probability-based sampling design? Why or why not?

3. What increases sampling error in probability-based sampling designs? Stratified rather than simple random sampling? Disproportionate (rather than proportionate) stratified random sampling? Stratified rather than cluster random sampling? Why do researchers select *disproportionate* (rather than proportionate) stratified samples? Why do they select cluster rather than simple random samples?

4. What are the advantages and disadvantages of probability-based sampling designs compared with nonprobability-based designs? Could any of the research described in this chapter with a nonprobability-based design have been conducted instead with a probability-based design? What are the difficulties that might have been encountered in an attempt to use random selection? How would you discuss the degree of confidence you can place in the results obtained from research using a nonprobability-based sampling design?

Practice Exercises

1. Select a random sample using the table of random numbers in Appendix C. Compute a statistic based on your sample, and compare it with the corresponding figure for the entire population. Here's how to proceed:

   a. First, select a very small population for which you have a reasonably complete sampling frame. One possibility would be the list of asking prices for houses advertised in your local paper. Another would be the listing of some characteristic of states in a U.S. Census Bureau publication, such as average income or population size.

   b. The next step is to create your sampling frame, a numbered list of all the elements in the population. If you are using a complete listing of all elements, as from a U.S. Census Bureau publication, the sampling frame is the same as the list. Just number the elements (states). If your population is composed of housing ads in the local paper, your sampling frame will be those ads that contain a housing price. Identify these ads, and then number them sequentially, starting with 1.

c. Decide on a method of picking numbers out of the random number table in Appendix C, such as taking every number in each row, row by row (or you may move down or diagonally across the columns). Use only the first (or last) digit in each number if you need to select 1 to 9 cases, or only the first (or last) two digits if you want fewer than 100 cases.

d. Pick a starting location in the random number table. It's important to pick a starting point in an unbiased way, perhaps by closing your eyes and then pointing to some part of the page.

e. Record the numbers you encounter as you move from the starting location in the direction you decided on in advance, until you have recorded as many random numbers as the number of cases you need in the sample. If you are selecting states, 10 might be a good number. Ignore numbers that are too large (or too small) for the range of numbers used to identify the elements in the population. Discard duplicate numbers.

f. Calculate the average value in your sample for some variable that was measured—for example, population size in

a sample of states or housing price for the housing ads. Calculate the average by adding up the values of all the elements in the sample and dividing by the number of elements in the sample.

g. Go back to the sampling frame and calculate this same average for all elements in the list. How close is the sample average to the population average?

h. Estimate the range of sample averages that would be likely to include 90% of the possible samples.

2. Draw a snowball sample of people who are involved in bungee jumping or some other uncommon sport that does not involve teams. Ask friends and relatives to locate a first contact, and then call or visit this person and ask for names of others. Stop when you have identified a sample of 10. Review the problems you encountered, and consider how you would proceed if you had to draw a larger sample.

3. Two lesson sets from the Interactive Exercises link on the study site will help you review terminology, "Identifying Sampling Techniques" and the logic of "Assessing Generalizability."

4. Identify one article at the book's study site that used a survey research design. Describe the sampling procedure. What type was it? Why did the author(s) use this particular type of sample?

Ethics Questions

1. How much pressure is too much pressure to participate in a probability-based sample survey? Is it okay for the U.S. government to mandate legally that all citizens participate in the decennial census? Should companies be able to require employees to participate in survey research about work-related issues? Should students be required to participate in surveys about teacher performance? Should parents be required to consent to the participation of their high school–age students in a survey about substance abuse and health issues? Is it okay to give monetary incentives for participation in a survey of homeless shelter clients? Can monetary incentives be coercive? Explain your decisions.

2. Federal regulations require special safeguards for research on persons with impaired cognitive capacity. Special safeguards are also required for research on prisoners and on children. Do you think special safeguards are necessary? Why or why not? Do you think it is possible for individuals in any of these groups to give "voluntary consent" to research participation? What procedures might help make consent to research truly voluntary in these situations? How could these procedures influence sampling plans and results?

Web Exercises

1. Research on homelessness has been rising in recent years as housing affordability has declined. Search the web for sites

that include the word *homelessness* and see what you find. You might try limiting your search to those that also contain the word *census*. Pick a site and write a paragraph about what you learned from it.

2. Check out the "people and households" section of the U.S. Census Bureau website: www.census.gov. Based on some of the data you find there, write a brief summary of some aspects of the current characteristics of the American population.

SPSS Exercises

1. Take a look again at the distribution of support for capital punishment (CAPPUN), this time with what is called a *frequency distribution*.

   a. Click Analyze/Descriptive Statistics/Frequencies.

   b. Highlight CAPPUN and click on the arrow that sends it over to the Variables window, then click OK.

   Examine the percentages in the Valid percent column. What percentage of the American population in 2010 favored capital punishment?

2. Now select random samples of the GSS2010x respondents:

   a. Go to the Data Editor window, and select a random sample containing 40 of the respondents. From the menu:

   b. Click Data/Select cases/All Cases/OK.

   c. Click Select cases/Random sample of cases/Sample.

   d. Select exactly 40 cases from the first 100 cases.

   e. Click Continue/OK. (Before you click OK, be sure that the "Filter out unselected cases" box is checked.)

   f. Determine the percentage of the subsample that favored capital punishment by repeating the steps in SPSS Exercise 1. Record the subsample characteristics and its percentage.

   g. Now, repeat Steps 2a and 2b 10 times. Each time, add 100 to the "first 100 cases" request (so that on the last step you will be requesting "Exactly 40 cases from the first 1,000 cases").

   h. Select a random sample containing five of the respondents. Now repeat Steps 2a through 2c (10 times), this time for samples of 5.

3. How does the distribution of CAPPUN in these subsamples compares with that for the total GSS sample?

   a. Plot the results of Steps 2c and 2d on separate sheets of graph paper. Each graph's horizontal axis will represent the possible range of percentages (from 0 to 100, perhaps in increments of 5); the vertical axis will represent the number of samples in each range of percentages (perhaps ranging from 0 to 10). Make an X to indicate this percentage for each sample. If two samples have the same percentage, place the corresponding Xs on top of each other. The X for each sample should be one unit high on the vertical axis.

   b. Draw a vertical line corresponding to the point on the horizontal axis that indicates the percentage of the total GSS sample that favors capital punishment.

   c. Describe the shape of both the graphs. These are the sampling distributions for the two sets of samples. Compare them with each other. Do the percentages from the larger samples tend to be closer to the mean of the entire sample (as obtained in SPSS Exercise 1)? What does this tell you about the relationship between sample size and sampling error?

Developing a Research Proposal

Consider the possibilities for sampling (Exhibit 2.14, #8).

1. Propose a sampling design that would be appropriate if you were to survey students on your campus only. Define the population, identify the sampling frame(s), and specify the elements and any other units at different stages. Indicate the exact procedure for selecting people to be included in the sample.

2. Propose a different sampling design for conducting your survey in a larger population, such as your city, state, or the entire nation.

Research Design and Causation

I dentifying causes—figuring out why things happen—is the goal of most social science research. Unfortunately, valid explanations of the causes of social phenomena do not come easily. Why did the rates of serious violent crime rise in the early 1990s and then begin a sustained drop that has continued into the 2000s, even into the recession (Bureau of Justice Statistics 2011a; Savage 2010)? (Exhibit 6.1) Arizona State University criminologist Scott Decker points to the low levels of crime committed by illegal immigrants to explain the falling crime rate in his state (Archibold 2010) and sociologist Robert J. Sampson (2008:29) draws attention to the rising level of immigration through the 1990s to help explain the national decline in the crime rate. Criminal justice advocates in Texas point to the state's investment in community treatment and diversion programs (Grissom 2011). Police officials in New York City point to the effectiveness of Compstat, the city's computer program that indicates to the police where crimes are clustering (Dewan 2004a:A25; Dewan 2004b:A1; Kaplan 2002:A3), while other New Yorkers credit the increase in the ranks of

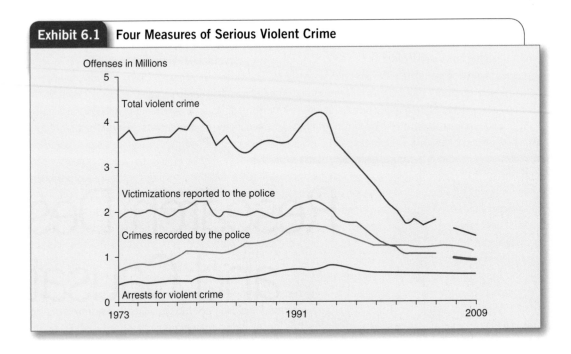

Exhibit 6.1 **Four Measures of Serious Violent Crime**

Offenses in Millions

Total violent crime

Victimizations reported to the police

Crimes recorded by the police

Arrests for violent crime

1973   1991   2009

New York's police officers due to its Safe Streets, Safe Cities program (Rashbaum 2002). Yet another possible explanation in New York City is the declining level of crack cocaine use (Dewan 2004b:C16). But then should we worry about the increasing number of drug arrests nationally (Bureau of Justice Statistics 2011b) and a rise in abuse of prescription drugs (Goodnough 2010)? Should we look for cautionary lessons to Japan, where the crime rate has risen sharply after being historically very low (Onishi 2003)? To explain changes in the rate of serious crime, we must design our research strategies carefully.

▣ Research Design Alternatives

I begin this chapter by discussing three key elements of research design: the design's units of analysis, its use of cross-sectional or longitudinal data, and whether its methods are primarily quantitative or qualitative. Whenever we design research, we must decide whether to use individuals or groups as our units of analysis and whether to collect data at one or several points in time. The decisions that we make about these design elements will affect our ability to draw causal conclusions in our analysis. Whether the design is primarily quantitative or qualitative in its methods also affects the type of causal explanation that can be developed: Quantitative projects lead to nomothetic causal explanations, while qualitative projects that have a causal focus can lead to idiographic explanations. After reviewing these three key design elements, I will also review the criteria for achieving explanations that are causally valid from a nomothetic perspective. By the end of the chapter, you should have a good grasp of the different meanings of causation and be able to ask the right questions to determine whether causal inferences are likely to be valid. You also may have a better answer about the causes of crime and violence.

Units of Analysis

In nonexperimental research designs, we can be misled about the existence of an association between two variables when we do not know to what **units of analysis** the measures in our study refer—that is, the level of social life on which the research question is focused, such as individuals, groups, towns, or nations. I first discuss this important concept before explaining how it can affect causal conclusions.

> **Units of analysis** The level of social life on which a research question is focused, such as individuals, groups, towns, or nations.

Individual and Group

In most sociological and psychological studies, the units of analysis are individuals. The researcher may collect survey data from individuals, analyze the data, and then report on, say, how many individuals felt socially isolated and whether substance abuse by individuals was related to their feelings of social isolation.

The units of analysis may instead be groups of some sort, such as families, schools, work organizations, towns, states, or countries. For example, a researcher may collect data from town and police records on the number of accidents in which a driver was intoxicated and the presence or absence of a server liability law in the town. (These laws make those who serve liquor liable for accidents caused by those to whom they served liquor.) The researcher can then analyze the relationship between server liability laws and the frequency of accidents due to drunk driving (perhaps also taking into account the town's population). Because the data describe the town, towns are the units of analysis.

In some studies, groups are the units of analysis, but data are collected from individuals. For example, in their study of influences on violent crime in Chicago neighborhoods, Robert Sampson, Stephen Raudenbush, and Felton Earls (1997:919) hypothesized that efficacy would influence neighborhood crime rates. *Collective efficacy* was defined conceptually as a characteristic of the neighborhood: the extent to which residents were likely to help other residents and were trusted by other residents. However, they measured this variable in a survey of individuals. The responses of individual residents about their perceptions of their neighbors' helpfulness and trustworthiness were averaged together to create a collective efficacy score for each neighborhood. It was this neighborhood measure of collective efficacy that was used to explain variation in the rate of violent crime between neighborhoods. The data were collected from individuals and were about individuals, but they were combined (aggregated) so as to describe neighborhoods. The units of analysis were thus groups (neighborhoods).

In a study such as that of Sampson et al.'s (1997), we can distinguish the concept of units of analysis from the **units of observation**. Data were collected from individuals, the units of observation in this study, and then the data were aggregated and analyzed at the group level. In most studies, the units of observation and the units of analysis are the same. For example, Yili Xu, Mora L. Fiedler, and Karl H. Flaming (2005), in collaboration with the Colorado Springs Police Department, surveyed a stratified random sample of 904 residents to test whether their sense of collective efficacy and other characteristics would predict their perceptions of crime, fear of crime, and satisfaction with police. Their data were collected from individuals and analyzed at the individual level. They concluded that collective efficacy was not as important as in Sampson et al.'s (1997) study.

> **Units of observation** The cases about which measures actually are obtained in a sample.

The important point is to know what the units of observation are, what the level of analysis is, and then to evaluate whether the conclusions are appropriate to these study features. A conclusion that "crime increases with joblessness" could imply either that individuals who lose their jobs are more likely to commit a crime or that a community with a high unemployment rate is likely to have a high crime rate—or both. Whether we are drawing conclusions from data we collected or interpreting others' conclusions, it is important to be clear about which relationship is being referred to.

We also have to know what the units of analysis are to interpret statistics appropriately. Measures of association tend to be stronger for group-level than for individual-level data because measurement errors at the individual level tend to cancel out at the group level (Bridges & Weis 1989:29–31).

In the News

Research in the News

GROUP INTELLIGENCE, NOT INDIVIDUALS' INTELLIGENCE, IMPROVES GROUP PERFORMANCE

Do groups do better solving problems when they are composed of more intelligent individuals or have a more intelligent leader? Research led by an MIT management professor finds that what is important for group performance is collective intelligence, not the intelligence of individuals. Social sensitivity in the group and an egalitarian pattern of involvement seem to be the keys to collective intelligence.

Source: Johnson, Carolyn Y. 2010. "Group IQ: What Makes One Team of People Smarter than Another? A New Field of Research Finds Surprising Answers." *Boston Sunday Globe,* December 19:K1, K2.

The Ecological Fallacy and Reductionism

Researchers should make sure that their causal conclusions reflect the units of analysis in their study. Conclusions about processes at the individual level should be based on individual-level data; conclusions about group-level processes should be based on data collected about groups. In most cases, when this rule is violated, we can be misled about the existence of an association between two variables.

A researcher who draws conclusions about individual-level processes from group-level data could be making what is termed an **ecological fallacy** (see Exhibit 6.2). The conclusions may or may not be correct, but we must recognize the fact that group-level data do not necessarily reflect solely individual-level processes. For example, a researcher may examine factory records and find that the higher the percentage of unskilled workers in factories, the higher the rate of employee sabotage in those factories. But the researcher would commit an ecological fallacy if he or she then concluded that individual unskilled factory workers are more likely to engage in sabotage. This conclusion is about an individual-level causal process (the relationship between the occupation and criminal propensities of individuals), even though the data describe groups (factories). It could actually be that white-collar workers are the ones more likely to commit sabotage in factories with more unskilled workers, perhaps because the white-collar workers feel they won't be suspected in these settings.

> **Ecological fallacy** An error in reasoning in which incorrect conclusions about individual-level processes are drawn from group-level data.

On the other hand, when data about individuals are used to make inferences about group-level processes, a problem occurs that can be thought of as the mirror image of the ecological fallacy: the **reductionist fallacy**, also known as *reductionism,* or the *individualist fallacy* (see Exhibit 6.2). For example, William Julius Wilson (1987:58) noted that we can be misled into concluding from individual-level data that race has a causal effect on violence. His reasoning went like this: There is an association at the individual level between race and the likelihood of arrest for violent crime, but community-level data reveal that almost 40% of poor blacks live in extremely poor areas, compared with only 7% of poor whites (in 1980). The concentration of African Americans in poverty areas, not the race or other characteristics of the individuals in these areas, may be the cause of higher rates of violence. Explaining violence in this case requires community-level data.

> **Reductionist fallacy (reductionism)** An error in reasoning that occurs when incorrect conclusions about group-level processes are based on individual-level data. Also known as an individualist fallacy.

The fact that errors in causal reasoning can be made in this way should not deter you from conducting research with group-level data nor make you unduly critical of researchers who make inferences about individuals on the basis of group-level data. When considered broadly, many research questions point to relationships that

Exhibit 6.2 Errors in Causal Conclusions

You make conclusions about

		Groups	Individuals
You collect data from	**Groups**	More homogeneous groups tend to have stronger social bonds.	Groups with a higher average age are more conservative, so older people are more conservative. *Possible Ecological Fallacy*
	Individuals	Students who socialize more have lower grades, so schools with more social engagement will have poorer student performance. *Possible Reductionist Fallacy*	Older people tend to be more conservative.

could be manifested in many ways and on many levels. Sampson's (1987) study of urban violence is a case in point. His analysis involved only aggregate data about cities, and he explained his research approach as, in part, a response to the failure of other researchers to examine this problem at the structural, aggregate level. Moreover, Sampson argued that the rates of joblessness and family disruption in communities influence community social processes, not just individuals who are unemployed or who grew up without two parents. Yet Sampson suggested that the experience of joblessness and poverty is what tends to reduce the propensity of individual men to marry and that the experience of growing up in a home without two parents, in turn, increases the propensity of individual juveniles to commit crimes. These conclusions about individual behavior seem consistent with the patterns Sampson found in his aggregate, city-level data, so it seems unlikely that he committed an ecological fallacy when he proposed them.

The solution is to know what the units of analysis and units of observation were in a study and to take these into account in weighing the credibility of the researcher's conclusions. The goal is not to reject out of hand conclusions that refer to a level of analysis different from what was actually studied. Instead, the goal is to consider the likelihood that an ecological fallacy or a reductionist fallacy has been made when estimating the causal validity of the conclusions.

Cross-Sectional and Longitudinal Designs

Research designs can be either cross-sectional or longitudinal. In **cross-sectional research designs**, all data are collected at one point in time. Identifying the **time order** of effects—what happened first, and so on—is critical for developing a causal analysis, but can be an insurmountable problem with a cross-sectional design. In **longitudinal research designs**, data are collected at two or more points in time, and so identification of the time order of effects can be quite straightforward.

Cross-Sectional Designs

Much of the research you have encountered so far in this text—the observations of computer use in Chapter 1, the surveys of binge drinking in Chapter 4 and of homeless persons in Chapter 5—has been cross-sectional. Although each of these studies took some time to carry out, they measured the actions, attitudes, and characteristics of respondents at only one point in time.

Cross-sectional research design A study in which data are collected at only one point in time.

Time order A criterion for establishing a causal relation between two variables. The variation in the presumed cause (the independent variable) must occur before the variation in the presumed effect (the dependent variable).

Longitudinal research design A study in which data are collected that can be ordered in time; also defined as research in which data are collected at two or more points in time.

Sampson and Raudenbush (1999) used a very ambitious cross-sectional design to study the effect of visible public social and physical disorder on the crime rate in Chicago neighborhoods. Their theoretical framework focused on the concept of informal social control: the ability of residents to regulate social activity in their neighborhoods through their collective efforts according to desired principles. They believed that informal social control would vary between neighborhoods, and they hypothesized that it was the strength of informal social control that would explain variation in crime rates rather than just the visible sign of disorder. They contrasted this prediction to the "broken windows" theory: the belief that signs of disorder themselves cause crime. In the theory Sampson and Raudenbush proposed, both visible disorder and crime were consequences of low levels of informal social control (measured with an index of collective efficacy). One did not cause the other (Exhibit 6.3).

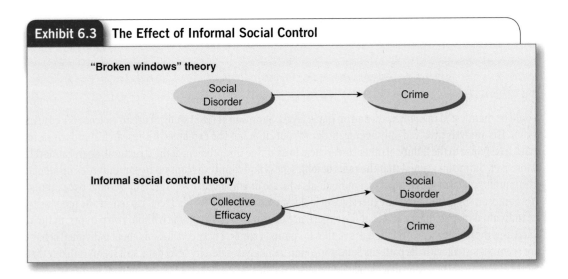

Exhibit 6.3   The Effect of Informal Social Control

Sampson and Raudenbush (1999) measured visible disorder through direct observation: Trained observers rode slowly around every street in 196 Chicago census tracts. They also conducted a survey of residents and examined police records. Both survey responses and police records were used to measure crime levels. The level of neighborhood informal social control and other variables were measured with the average resident responses to several survey questions. Both the crime rate and the level of social and physical disorder varied between neighborhoods in relation to the level of informal social control. Informal social control (collective efficacy) was a much more important factor in the neighborhood crime rate than was visible social and physical disorder, measured at the same time (Exhibit 6.4).

There are four special circumstances in which we can be more confident in drawing

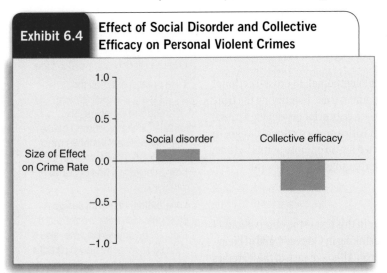

Exhibit 6.4   Effect of Social Disorder and Collective Efficacy on Personal Violent Crimes

conclusions about time order on the basis of cross-sectional data. Because in these special circumstances the data can be ordered in time, they might even be thought of as longitudinal designs (Campbell 1992). These four special circumstances are as follows:

1. *The independent variable is fixed at some point prior to the variation in the dependent variable.* So-called demographic variables that are determined at birth—such as sex, race, and age—are fixed in this way. So are variables such as education and marital status, if we know when the value of cases on these variables was established and if we know that the value of cases on the dependent variable was set some time afterward. For example, say we hypothesize that education influences the type of job individuals have. If we know that respondents completed their education before taking their current jobs, we would satisfy the time order requirement even if we were to measure education at the same time we measure type of job. However, if some respondents possibly went back to school as a benefit of their current job, the time order requirement would not be satisfied.

2. *We believe that respondents can give us reliable reports of what happened to them or what they thought at some earlier point in time.* Julie Horney, D. Wayne Osgood, and Ineke Haen Marshall (1995) provide an interesting example of the use of such retrospective data. The researchers wanted to identify how criminal activity varies in response to changes in life circumstances. They interviewed 658 newly convicted male offenders sentenced to a Nebraska state prison. In a 45- to 90-minute interview, they recorded each inmate's report of his life circumstances and of his criminal activities for the preceding 2 to 3 years. They then found that criminal involvement was related strongly to adverse changes in life circumstances, such as marital separation or drug use. Retrospective data are often inadequate for measuring variation in past feelings, events, or behaviors, however, because we may have difficulty recalling what we have felt or what has happened in the past and what we do recall is likely to be influenced by what we feel in the present (Elliott et al. 2008:229). For example, retrospective reports by both adult alcoholics and their parents appear to overestimate greatly the frequency of childhood problems (Vaillant 1995). People cannot report reliably the frequency and timing of many past events, from hospitalization to hours worked. However, retrospective data tend to be reliable when it concerns major, persistent experiences in the past, such as what type of school someone went to or how a person's family was structured (Campbell 1992).

3. *Our measures are based on the records that contain information on cases in earlier periods.* Government, agency, and organizational records are an excellent source of time-ordered data after the fact. However, sloppy record keeping and changes in data collection policies can lead to inconsistencies, which must be taken into account. Another weakness of such archival data is that they usually contain measures of only a fraction of the variables that we think are important.

4. *We know that the value of the dependent variable was similar for all cases prior to the treatment.* For example, we may hypothesize that a training program (independent variable) improves the English-speaking abilities (dependent variable) of a group of recent immigrants. If we know that none of the immigrants could speak English prior to enrolling in the training program, we can be confident that any subsequent variation in their ability to speak English did not precede exposure to the training program. This is one way that traditional experiments establish time order: Two or more equivalent groups are formed prior to exposing one of them to some treatment.

Longitudinal Designs

In longitudinal research, data are collected that can be ordered in time. By measuring the value of cases on an independent variable and a dependent variable at different times, the researcher can determine whether variation in the independent variable precedes variation in the dependent variable.

In some longitudinal designs, the same sample (or panel) is followed over time; in other designs, sample members are rotated or completely replaced. The population from which the sample is selected may be defined broadly, as when a longitudinal survey of the general population is conducted. Or the population may be defined narrowly, as when the members of a specific age group are sampled at multiple points in time. The frequency of follow-up measurement can vary, ranging from a before-and-after design with just the one follow-up to studies in which various indicators are measured every month for many years.

Certainly, it is more difficult to collect data at two or more points in time than at one time. Quite frequently researchers simply cannot, or are unwilling to, delay completion of a study for even 1 year to collect follow-up data. But think of the many research questions that really should involve a much longer follow-up period: What is the impact of job training on subsequent employment? How effective is a school-based program in improving parenting skills? Under what conditions do traumatic experiences in childhood result in mental illness? It is safe to say that we will never have enough longitudinal data to answer many important research questions. Nonetheless, the value of longitudinal data is so great that every effort should be made to develop longitudinal research designs when they are appropriate for the research question asked. The following discussion of the three major types of longitudinal designs will give you a sense of the possibilities (see Exhibit 6.5).

Repeated cross-sectional design (trend study) A type of longitudinal study in which data are collected at two or more points in time from different samples of the same population.

Repeated cross-sectional designs. Studies that use a repeated cross-sectional design, also known as trend studies, have become fixtures of the political arena around election time. Particularly in presidential election years, we have all become accustomed to reading weekly, even daily, reports on the percentage of the population that supports each candidate. Similar polls are conducted to track sentiment on many other social issues. For example, a 1993 poll reported that 52% of adult Americans supported a ban on the possession of handguns, compared

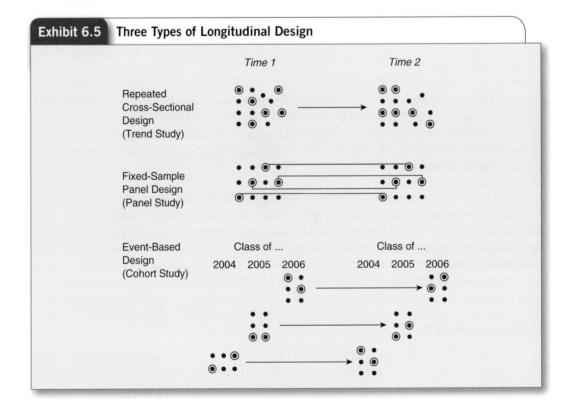

Exhibit 6.5 | **Three Types of Longitudinal Design**

with 41% in a similar poll conducted in 1991. According to pollster Louis Harris, this increase indicated a "sea change" in public attitudes (cited in Barringer 1993). Another researcher said, "It shows that people are responding to their experience [of an increase in handgun-related killings]" (cited in Barringer 1993:1).

Repeated cross-sectional surveys are conducted as follows:

1. A sample is drawn from a population at Time 1, and data are collected from the sample.

2. As time passes, some people leave the population and others enter it.

3. At Time 2, a different sample is drawn from this population.

These features make the repeated cross-sectional design appropriate when the goal is to determine whether a population has changed over time. Has racial tolerance increased among Americans in the past 20 years? Are employers more likely to pay maternity benefits today than they were in the 1950s? These questions concern the changes in the population as a whole, not just the changes in individuals within the population. We want to know whether racial tolerance increased in society, not whether this change was due to migration that brought more racially tolerant people into the country or to individual U.S. citizens becoming more tolerant. We are asking whether employers overall are more likely to pay maternity benefits today than they were yesterday, not whether any such increase was due to recalcitrant employers going out of business or to individual employers changing their maternity benefits. When we do need to know whether individuals in the population changed, we must turn to a panel design.

Fixed-sample panel designs. Panel designs *allow* us to identify changes in individuals, groups, or whatever we are studying. This is the process for conducting **fixed-sample panel designs**:

1. A sample (called a panel) is drawn from a population at Time 1, and data are collected from the sample.

2. As time passes, some panel members become unavailable for follow-up, and the population changes.

3. At Time 2, data are collected from the same people as at Time 1 (the panel)—except for those people who cannot be located.

> **Fixed-sample panel design (panel study)** A type of longitudinal study in which data are collected from the same individuals—the panel—at two or more points in time. In another type of panel design, panel members who leave are replaced with new members.

Because a panel design follows the same individuals, it is better than a repeated cross-sectional design for testing causal hypotheses. For example, Robert Sampson and John Laub (1990) used a fixed-sample panel design to investigate the effect of childhood deviance on adult crime. They studied a sample of white males in Boston when the subjects were between 10 and 17 years old and then followed up when the subjects were in their adult years. Data were collected from multiple sources, including the subjects themselves and criminal justice records. Sampson and Laub (1990:614) found that children who had been committed to a correctional school for persistent delinquency were much more likely than other children in the study to commit crimes as adults: 61% were arrested between the ages of 25 and 32, compared with 14% of those who had not been in correctional schools as juveniles. In this study, juvenile delinquency unquestionably occurred before adult criminality. If the researchers had used a cross-sectional design to study the past of adults, the juvenile delinquency measure might have been biased by memory lapses, by self-serving recollections about behavior as juveniles, or by loss of agency records.

Christopher Schreck, Eric Steward, and Bonnie Fischer (2006) wanted to identify predictors of adolescent victimization and wondered if the cross-sectional studies that had been conducted about victimization might have provided misleading results. Specifically, they suspected that adolescents with lower levels of self-control might be more prone to victimization and so needed to collect or find longitudinal data in which self-control was measured before experiences of victimization. The theoretical model they proposed to test included several other concepts that criminologists have identified as related to delinquency and that also might be influenced by levels of self-control: having delinquent peers, engaging in more delinquency, and being less attached to parents and school (Exhibit 6.6). They analyzed data available from a panel study of delinquency and found that low self-control at an earlier time made it more likely that adolescents would subsequently experience victimization, even taking account of other influences. Their use of a panel design allowed them to be more confident that the self-control—victimization relationship was causal than if they had used a cross-sectional design.

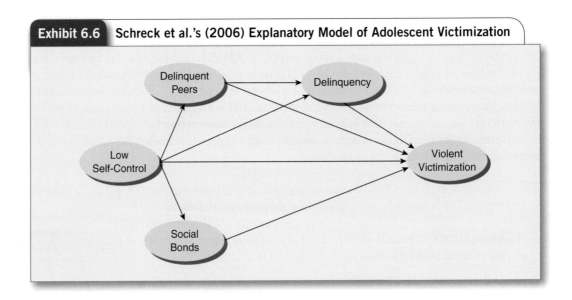

Exhibit 6.6 **Schreck et al.'s (2006) Explanatory Model of Adolescent Victimization**

In spite of their value in establishing time order of effects, panel designs are a challenge to implement successfully, so they often are not even attempted. There are two major difficulties:

1. *Expense and attrition.* It can be difficult, and very expensive, to keep track of individuals over a long period, and inevitably the proportion of panel members who can be located for follow-up will decline over time. Panel studies often lose more than one-quarter of their members through attrition (Miller 1991:170), and those who are lost are often not necessarily like those who remain in the panel. As a result, a high rate of subject attrition may mean that the follow-up sample will no longer be representative of the population from which it was drawn and may no longer provide a sound basis for estimating change. Subjects who were lost to follow-up may have been those who changed the most, or the least, over time. For example, between 5% and 66% of subjects are lost in substance abuse prevention studies, and the dropouts typically had begun the study with higher rates of tobacco and marijuana use (Snow, Tebes, & Arthur 1992:804).

It does help to compare the baseline characteristics of those who are interviewed at follow-up with characteristics of those lost to follow-up. If these two groups of panel members were not very different at baseline, it is less likely that changes had anything to do with characteristics of the missing panel members. Even better, subject attrition can be reduced substantially if sufficient staff can be used to keep track of panel members. In their panel study, Sampson and Laub (1990) lost only 12% of the juveniles in the original sample (8% if you do not count those who had died).

that the researcher may have made. The first three of the criteria are generally considered the most important bases for identifying a nomothetic causal effect: (1) empirical association, (2) appropriate time order, and (3) nonspuriousness (Hammersley 2008:43). The features of experimental research designs are particularly well suited to meeting these criteria and for testing nomothetic causal explanations. However, we must also consider the degree to which these criteria are met when evaluating nonexperimental research that is designed to test causal hypotheses.

The other two criteria that I introduce in this chapter are a bit different. They are not necessary for establishing that a causal connection exists, but they help us understand it better. Identifying a causal mechanism, the fourth criterion, helps us understand *why* a causal connection exists. The fifth, specifying the context in which a causal effect occurs, helps us understand *when* or *under what conditions* it occurs. Answers to both questions can considerably strengthen causal explanations (Hammersley 2008:44-45).

In the following subsections, I will indicate how researchers attempt to meet the five criteria with both experimental and nonexperimental designs. Illustrations of experimental design features will use a 2002 study by M. Lyn Exum on the effect of intoxication and anger on aggressive intentions. Most illustrations of nonexperimental design features will be based on the study by Sampson and Raudenbush (1999) of neighborhood social control, which I have already introduced.

Exum (2002) and her assistants recruited 84 male students of legal drinking age at a mid-Atlantic university, using classroom announcements and fliers (women were not included because it was not possible for them to be screened for pregnancy prior to participation, as required by federal guidelines). Students who were interested in participating were given some background information that included the explanation that the study was about alcohol and cognitive skills. All participants were scheduled for individual appointments. When they arrived for the experiment, they completed a mood questionnaire and engaged in a meaningless video game. Those who were randomly assigned to the Alcohol condition were then given 1.5 ounces of 50% ethanol (vodka) per 40 pounds of body weight in orange juice (those in the No Alcohol condition just drank orange juice).

The other part of the experimental manipulation involved inducing anger among a randomly selected half of the participants. This was accomplished by the experimenter, who falsely accused the selected students of having come to the experiment 30 minutes late, informing them that as a consequence of their tardiness, they would not be paid as much for their time, and then loudly saying "bullshit" when the students protested. After these manipulations, the students read a fictional scenario involving the student and another man in a conflict about a girlfriend. The students were asked to rate how likely they would be to physically assault the other man, and what percentage of other male students they believed would do so (see Exhibit 6.8).

The students in the Alcohol condition (who were intoxicated) and had also been angered predicted that more students would react with physical aggression to the events depicted in the scenario (see Exhibit 6.9). The students in the four experimental conditions did not differ in their reports of their own likely aggressiveness, but Exum suggested this could be a result of the well-established phenomenon of *self-enhancement bias*—the tendency to evaluate oneself more positively than others. She concluded that she found mixed support for her hypothesis that alcohol increases violent decision making among persons who are angry.

Was this causal conclusion justified? How confident can we be in its internal validity? Do you think that college students' reactions in a controlled setting with a fixed amount of alcohol are likely to be generalized to other people and settings? Does it help to know that Exum carefully confirmed that the students in the Alcohol condition were indeed intoxicated and that those in the Anger condition were indeed angry when they read the scenario? What about the causal conclusion by Sampson and Raudenbush (1999) that social and physical disorder does not directly cause neighborhood crime? Were you convinced? In the next sections, I will show how well the features of the research designs used by Exum and by Sampson and Raudenbush meet the criteria for nomothetic causal explanation, and thus determine the confidence we can place in their causal conclusions. I will also identify those features of a *true experiment* that make this research design particularly well suited to testing nomothetic causal hypotheses.

Exhibit 6.8 An Experiment to Test the Effect of Intoxication and Anger on Intention to Aggress

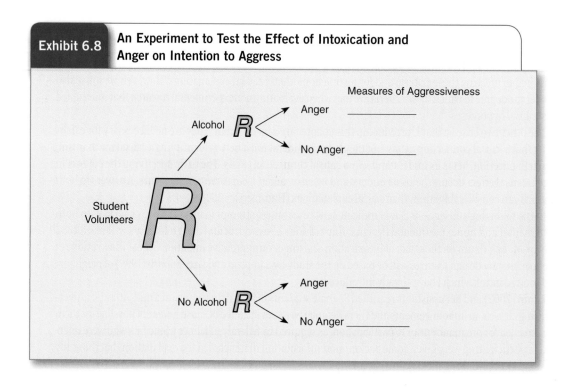

Exhibit 6.9 Measures of Aggression by Experimental Condition

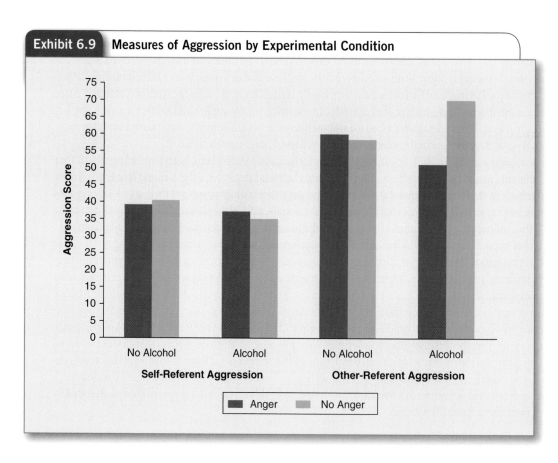

Association

We say that there was an **association** between aggressive intentions and intoxication (for angry students) in Exum's (2002) experiment because the level of aggressive intentions varied according to whether students were intoxicated. An empirical (or observed) association between the independent and dependent variables is the first criterion for identifying a nomothetic causal effect.

> **Association** A criterion for establishing a nomothetic causal relationship between two variables: Variation in one variable is related to variation in another variable.

We can determine whether an association exists between the independent and dependent variables in a true experiment, because there are two or more groups that differ in terms of their value on the independent variable. One group receives some "treatment," such as reading a cathartic message, that manipulates the value of the independent variable. This group is termed the *experimental group*. In a simple experiment, there may be one other group that does not receive the treatment; it is termed the *control group*. The Exum study compared four groups created with two independent variables; other experiments may compare only two groups that differ in terms of one independent variable, or more groups that represent multiple values of the independent variable or combinations of the values of more than two independent variables.

In nonexperimental research, the test for an association between the independent and dependent variables is like that used in experimental research—seeing whether values of cases that differ on the independent variable tend to differ in terms of the dependent variable. The difference with nonexperimental research designs is that the independent variable is not a treatment to which the researcher assigns some individuals. In their nonexperimental study of neighborhood crime, Sampson and Raudenbush (1999) studied the association between the independent variable (level of social and physical disorder) and the crime rate, but they did not assign individuals to live in neighborhoods with low or high levels of disorder.

Time Order

Association is a necessary criterion for establishing a causal effect, but it is not sufficient. We must also ensure that the variation in the dependent variable occurred after the variation in the independent variable. This is the criterion of time order. Our research design determines our ability to determine time order.

Experimental Designs

In a true experiment, the researcher determines the time order. Exum (2002) first had some students drink alcohol and some experience the anger-producing manipulation and then measured their level of aggressive intentions. If we find an association between intoxication or anger and aggressiveness outside of an experimental situation, the criterion of time order may not be met. People who are more inclined to interpersonal aggression may be more likely than others to drink to the point of intoxication or to be angered by others in the first place. This would result in an association between intoxication and aggressive intentions, but the association would reflect the influence of being an aggressive person on drinking behavior rather than the other way around.

Nonexperimental Designs

You have already learned that nonexperimental research designs can be either cross-sectional or longitudinal. Because cross-sectional designs do not establish the time order of effects, their conclusions about causation must be more tentative. For example, although Sampson and Raudenbush (1999) found that lower rates of crime were associated with more informal social control (collective efficacy), their cross-sectional design could not establish directly that the variation in the crime rate occurred after variation in informal social control. Maybe it was a high crime rate that led residents to stop trying to exert much control over deviant activities in the neighborhood. It is difficult to discount such a possibility when only cross-sectional data are available, even though we can diagram hypothetical relations between the variables as if they are ordered in time (see Exhibit 6.10, panel 1).

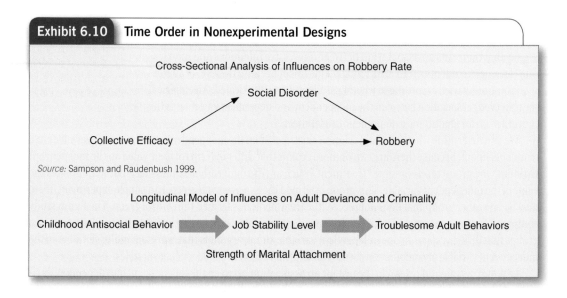

Exhibit 6.10   Time Order in Nonexperimental Designs

Cross-Sectional Analysis of Influences on Robbery Rate

Social Disorder

Collective Efficacy → Robbery

Source: Sampson and Raudenbush 1999.

Longitudinal Model of Influences on Adult Deviance and Criminality

Childhood Antisocial Behavior → Job Stability Level → Troublesome Adult Behaviors

Strength of Marital Attachment

In contrast, Sampson and Laub's (1990) longitudinal study of the effects of childhood deviance (antisocial behavior) on adult crime provided strong evidence of appropriate time order. Data on juvenile delinquency were collected when subjects were between 10- and 17-years-old, so there's no question that the delinquency occurred before the job and marital experiences and then the adult criminality and other troublesome behaviors with which it was associated (see Exhibit 6.10, panel 2).

Nonspuriousness

Nonspuriousness A criterion for establishing a causal relation between two variables; when a relationship between two variables is not due to variation in a third variable.

Spurious relationship A relationship between two variables that is due to variation in a third variable.

Nonspuriousness is another essential criterion for establishing the existence of a causal effect of an independent variable on a dependent variable; in some respects, it is the most important criterion. We say that a relationship between two variables is not spurious when it is not due to variation in a third variable. Have you heard the old adage "Correlation does not prove causation"? It is meant to remind us that an association between two variables might be caused by something other than an effect of the presumed independent variable on the dependent variable—that is, it might be a **spurious relationship**. If we measure children's shoe sizes and their academic knowledge, for example, we will find a positive association. However, the association results from the fact that older children have larger feet as well as more academic knowledge. Shoe size does not cause knowledge, or vice versa.

Do storks bring babies? If you believe that correlation proves causation, then you might think so. The more storks that appear in certain districts in Holland, the more babies are born. But the association in Holland between number of storks and number of babies is spurious. In fact, both the number of storks and the birthrate are higher in rural districts than in urban districts. The rural or urban character of the districts (the **extraneous variable**) causes variation in the other two variables.

If you think this point is obvious, consider a social science example. Do schools with more resources produce better student outcomes? Before you answer the question, consider the fact that parents with more education and higher income

Extraneous variable A variable that influences both the independent and dependent variables so as to create a spurious association between them that disappears when the extraneous variable is controlled.

tend to live in neighborhoods that spend more on their schools. These parents also are more likely to have books in the home and provide other advantages for their children. Do the parents cause variation in both school resources and student performance? If so, there would be an association between school resources and student performance that was at least partially spurious (Exhibit 6.11).

Randomization

A true experiment like Bushman, Baumeister, and Stacks's (1999) study of catharsis uses a technique called **randomization** to reduce the risk of spuriousness. Students in Bushman's experiment were asked to select a message to read by drawing a random number out of a bag. That is, the students were assigned randomly to a treatment condition. If students were assigned to only two groups, a coin toss could have been used (see Exhibit 6.12). **Random assignment** ensures that neither the students' aggressiveness nor any of their

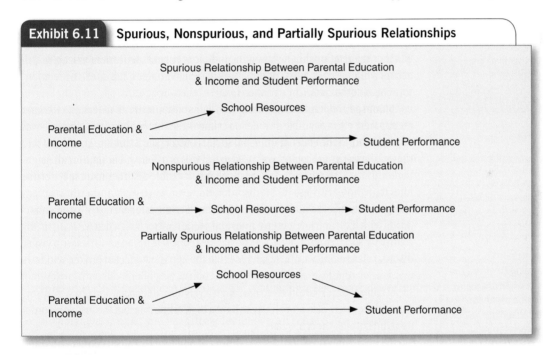

Exhibit 6.11 Spurious, Nonspurious, and Partially Spurious Relationships

Spurious Relationship Between Parental Education
& Income and Student Performance

Nonspurious Relationship Between Parental Education
& Income and Student Performance

Partially Spurious Relationship Between Parental Education
& Income and Student Performance

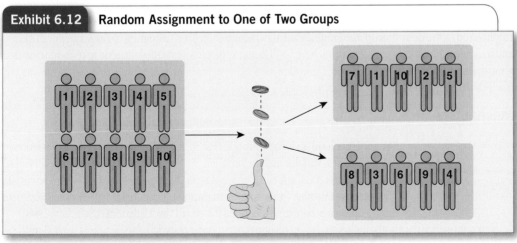

Exhibit 6.12 Random Assignment to One of Two Groups

other characteristics or attitudes could influence which of the messages they read. As a result, the different groups are likely to be equivalent in all respects at the outset of the experiment. The greater the number of cases assigned randomly to the groups, the more likely that the groups will be equivalent in all respects. Whatever the preexisting sources of variation among the students, these could not explain why the group that read the procatharsis message became more aggressive, whereas the others didn't.

> **Randomization** The random assignment of cases, as by the toss of a coin.
>
> **Random assignment** A procedure by which each experimental subject is placed in a group randomly, a distinction from random sampling.

Statistical Control

A nonexperimental study like Sampson and Raudenbush's (1999) cannot use random assignment to comparison groups to minimize the risk of spurious effects. Even if we wanted to, we couldn't randomly assign people to live in neighborhoods with different levels of informal social control. Instead, nonexperimental researchers commonly use an alternative approach to try to achieve the criterion of nonspuriousness. The technique of **statistical control** allows researchers to determine whether the relationship between the independent and dependent variables still occurs while we hold constant the values of other variables. If it does, the relationship could not be caused by variation in these other variables.

> **Statistical control** A method in which one variable is held constant so that the relationship between two (or more) other variables can be assessed without the influence of variation in the control variable.
>
> *Example:* In a different study, Sampson (1987) found a relationship between rates of family disruption and violent crime. He then classified cities by their level of joblessness (the control variable) and found that same relationship between the rates of family disruption and violent crime among cities with different levels of joblessness. Thus, the rate of joblessness could not have caused the association between family disruption and violent crime.

Sampson and Raudenbush designed their study, in part, to determine whether the apparent effect of visible disorder on crime—the "broken windows" thesis—was spurious due to the effect of informal social control (see Exhibit 6.3). Exhibit 6.13 shows how statistical control was used to test this possibility. The data for all neighborhoods show that neighborhoods with much visible disorder had higher crime rates than those with less visible disorder. However, when we examine the relationship between visible disorder and neighborhood crime rate separately for neighborhoods with high and low levels of informal social control (i.e., when we statistically control for social control level), we see that the crime rate no longer varies with visible disorder. Therefore, we must conclude that the apparent effect of broken windows was spurious due to level of informal social control. Neighborhoods with low levels of social control were more likely to have high levels of visible social and physical disorder, and they were also more likely to have a high crime rate, but the visible disorder itself did not alter the crime rate.

We can strengthen our understanding of nomothetic causal connections, and increase the likelihood of drawing causally valid conclusions, by considering two additional criteria: causal mechanism and causal context. These two criteria are emphasized in the definition of idiographic causal explanation, with its attention to the sequence of events and the context in which they happen, but here I will limit my discussion of these criteria to research oriented toward nomothetic causal explanations.

Mechanism

> **Mechanism** A discernible process that creates a causal connection between two variables.

A causal **mechanism** is some process that creates the connection between variation in an independent variable and the variation in the dependent variable it is hypothesized to cause (Cook & Campbell 1979:35; Marini & Singer 1988). Many social scientists (and scientists in other fields) argue that no nomothetic causal explanation is adequate until a causal mechanism is identified (Costner 1989; Hedström & Swedberg 1998).

Our confidence in causal conclusions based on nonexperimental research also increases with identification of a causal mechanism (Shrout 2011:15-16). Such mechanisms help us understand how variation in the

Exhibit 6.13 The Use of Statistical Control to Reduce Spuriousness

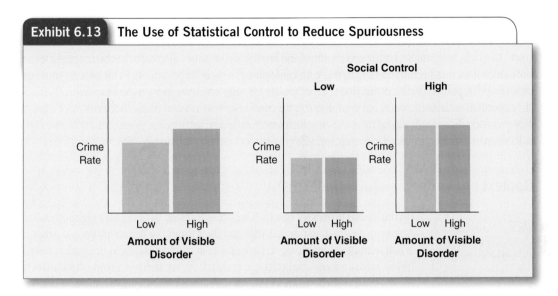

independent variable results in variation in the dependent variable. For example, in a study that re-analyzed data from Sheldon Glueck and Elenor Glueck's (1950) pathbreaking study of juvenile delinquency, Sampson and Laub (1994) found that children who grew up with structural disadvantages such as family poverty and geographic mobility were more likely to become juvenile delinquents. Why did this occur? Their analysis indicated that these structural disadvantages led to lower levels of informal social control in the family (less parent-child attachment, less maternal supervision, and more erratic or harsh discipline). In turn, lower levels of informal social control resulted in a higher probability of delinquency (Exhibit 6.14). Informal social control thus intervened in the relationship between structural disadvantage and juvenile delinquency.

In their study of deterrence of spouse abuse (introduced in Chapter 2), Lawrence Sherman and Richard Berk (1984) designed follow-up experiments to test or control several causal mechanisms that they wondered about after their first experiment: Did recidivism decrease for those who were arrested for spouse abuse *because* of the exemplary work of the arresting officers? Did recidivism increase for arrestees *because* they experienced more stressors with their spouses as time passed? Investigating these and other possible causal mechanisms enriched Sherman and Berk's eventual explanation of how arrest influences recidivism.

Exhibit 6.14 Intervening Variables in Nonexperimental Research: Structural Disadvantage and Juvenile Delinquency

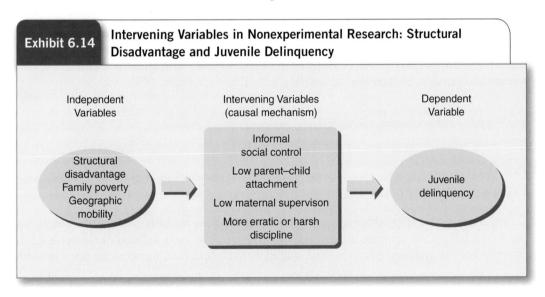

Of course, you might ask why structural disadvantage tends to result in lower levels of family social control or how family social control influences delinquency. You could then conduct research to identify the mechanisms that link, for example, family social control and juvenile delinquency. (Perhaps the children feel they're not cared for, so they become less concerned with conforming to social expectations.) This process could go on and on. The point is that identification of a mechanism through which the independent variable influences the dependent variable increases our confidence in the conclusion that a causal connection does indeed exist. However, identification of a causal mechanism in turn requires concern for the same causal criteria that we consider when testing the original relationship, including time order and nonspuriousness (Shrout 2011: 15–21).

Context

Contextual effects Relationships among variables that vary among geographic units or other social settings.

Context A focus of idiographic causal explanation; a particular outcome is understood as part of a larger set of interrelated circumstances.

Do the causal processes in which we are interested vary across neighborhoods? Among organizations? Across regions? Over time? For different types of people? When relationships among variables differ across geographic units such as counties or across other social settings, researchers say there is a **contextual effect**. Identification of the **context** in which a causal relationship occurs can help us understand that relationship. The changes in the crime rate with which we began this chapter differed for blacks and whites, for youth and adults, and in urban and rural areas (Ousey & Lee 2004:359–360). These contextual effects suggest that single-factor explanations about these changes are incorrect (Rosenfeld 2004:89).

Sampson and Laub (1993) found support for a contextual effect in their study of 538,000 juvenile justice cases in 322 U.S. counties: In counties having a relatively large underclass and poverty concentrated among minorities, juvenile cases were more likely to be treated harshly. These relationships occurred for both African American and white juveniles, but were particularly strong for African Americans. The results of this research suggest the importance of taking social context into account when examining criminal justice processes (see also Dannefer & Schutt 1982; Schutt & Dannefer 1988).

Exum (2002) tested the effect of intoxication itself on aggressive intentions, but also in relation to whether students were angry or not. She found that it was only in the context of being angry that alcohol led the students to express more aggressive intentions (in terms of how they said other students would act (see p. 186)).

Context was also important in Sherman and Berk's (1984) research on domestic violence. Arrest was less effective in reducing subsequent domestic violence in cities with high levels of unemployment than in cities with low levels of unemployment. This seemed to be more evidence of the importance of individuals having a "stake in conformity" (Berk et al. 1992). Awareness of contextual differences helps us make sense of the discrepant findings from local studies. Always remember that the particular cause on which we focus in a given research design may be only one among a set of interrelated factors required for the effect; when we take context into account, we specify these other factors (Hage & Meeker 1988; Papineau 1978).

▣ Conclusions

Causation and the means for achieving causally valid conclusions in research is the last of the three legs on which the validity of research rests. In this chapter, you have learned about two alternative meanings of causation (nomothetic and idiographic). You have studied the five criteria used to evaluate the extent to which particular research designs may achieve causally valid findings. You have learned how our ability to meet these criteria is shaped by research design features such as units of analysis, use of a cross-sectional or longitudinal design, and use of randomization or statistical control to deal with the problem of spuriousness. You have also seen why the distinction between experimental and nonexperimental designs has so many consequences for how, and how well, we are able to meet nomothetic criteria for causation.

I should reemphasize that the results of any particular study are part of an always-changing body of empirical knowledge about social reality. Thus, our understandings of causal relationships are always partial. Researchers always wonder whether they have omitted some relevant variables from their controls, whether their experimental results would differ if the experiment were conducted in another setting, or whether they have overlooked a critical historical event. But by using consistent definitions of terms and maintaining clear standards for establishing the validity of research results—and by expecting the same of others who do research—social researchers can contribute to a growing body of knowledge that can reliably guide social policy and social understanding.

When you read the results of a social scientific study, you should now be able to evaluate critically the validity of the study's findings. If you plan to engage in social research, you should now be able to plan an approach that will lead to valid findings. And with a good understanding of three dimensions of validity (measurement validity, generalizability, and causal validity) under your belt, and with sensitivity also to the goal of *authenticity,* you are ready to focus on the major methods of data collection used by social scientists.

Key Terms

Association   187
Causal effect (idiographic perspective)   184
Causal effect (nomothetic perspective)   182
Ceteris paribus   182
Cohort   181
Context   192
Contextual effects   192
Counterfactual   182
Cross-sectional research design   175

Ecological fallacy   174
Event-based design (cohort study)   181
Extraneous variable   188
Fixed-sample panel design (panel study)   179
Idiographic causal explanation   183
Longitudinal research design   175
Mechanism   190
Nomothetic causal explanation   182
Nonspuriousness   188
Random assignment   189

Randomization   189
Reductionist fallacy (reductionism)   174
Repeated cross-sectional design (trend study)   178
Spurious relationship   188
Statistical control   190
Subject fatigue   181
Time order   175
Units of analysis   173
Units of observation   173

Highlights

- We do not fully understand the variables in a study until we know to which units of analysis—what level of social life—they refer.
- Invalid conclusions about causality may occur when relationships between variables measured at the group level are assumed to apply at the individual level (the ecological fallacy) and when relationships between variables measured at the level of individuals are assumed to apply at the group level (the reductionist fallacy). Nonetheless, many research questions

point to relationships at multiple levels and so may profitably be investigated at multiple units of analysis.

- Longitudinal designs are usually preferable to cross-sectional designs for establishing the time order of effects. Longitudinal designs vary in terms of whether the same people are measured at different times, how the population of interest is defined, and how frequently follow-up measurements are taken. Fixed-sample panel designs provide the strongest test for the time order of effects, but they can be difficult to carry out successfully because of their expense as well as subject attrition and fatigue.

- Causation can be defined in either nomothetic or idiographic terms. Nomothetic causal explanations deal with effects on average. Idiographic causal explanations deal with the sequence of events that led to a particular outcome.

- The concept of nomothetic causal explanation relies on a comparison. The value of cases on the dependent variable is measured after they have been exposed to variation in an independent variable. This measurement is compared with what the value of cases on the dependent variable would have been if they had not been exposed to the variation in the independent variable (the counterfactual). The validity of nomothetic causal conclusions rests on how closely the comparison group comes to the ideal counterfactual.

- From a nomothetic perspective, three criteria are generally viewed as necessary for identifying a causal relationship: (1) association between the variables, (2) proper time order, and (3) nonspuriousness of the association. In addition, the basis for concluding that a causal relationship exists is strengthened by the identification of a causal mechanism and the context for the relationship.

- Association between two variables is in itself insufficient evidence of a causal relationship. This point is commonly made with the expression, "Correlation does not prove causation."

- Experiments use random assignment to make comparison groups as similar as possible at the outset of an experiment in order to reduce the risk of spurious effects due to extraneous variables.

- Nonexperimental designs use statistical controls to reduce the risk of spuriousness. A variable is controlled when it is held constant so that the association between the independent and dependent variables can be assessed without being influenced by the control variable.

- Ethical and practical constraints often preclude the use of experimental designs.

- Idiographic causal explanations can be difficult to identify, because the starting and ending points of particular events and the determination of which events act as causes in particular sequences may be ambiguous.

STUDENT STUDY SITE

To assist in completing the web exercises, please access the study site at **www.sagepub.com/schuttisw7e**, where you will find the web exercises with accompanying links. You'll find other useful study materials such as self-quizzes and e-flashcards for each chapter, along with a group of carefully selected articles from research journals that illustrate the major concepts and techniques presented in the book.

Discussion Questions

1. There's a lot of "sound and fury" in the social science literature about units of analysis and levels of explanation. Some social researchers may call another a *reductionist* if they explain a problem such as substance abuse as due to "lack of self-control." The idea is that the behavior requires consideration of social structure—a group level of analysis rather than an individual level of analysis. Another researcher may be said to commit an *ecological fallacy* if she assumes that group-level characteristics explain behavior at the individual level (such as saying that "immigrants are more likely to commit crime" because the *neighborhoods* with higher proportions of immigrants have higher crime rates). Do you favor causal explanations at the individual or the group (or social structural) level? If you were forced to mark on a scale

from 0 to 100, the percentage of crime that is due to problems with individuals rather than to problems with the settings in which they live, where would you make your mark? Explain your decision.

2. Researchers often try to figure out how people have changed over time by conducting a cross-sectional survey of people of different ages. The idea is that if people who are in their 60s tend to be happier than people who are in their 20s, it is because people tend to "become happier" as they age. But maybe people who are in their 60s now were just as happy when they were in their 20s, and people in their 20s now will be just as unhappy when they are in their 60s. (That's called a *cohort effect*.) We can't be sure unless we conduct a panel

study (survey the same people at different ages). What, in your experience, are the major differences between the generations today in social attitudes and behaviors? Which would you attribute to changes as people age, and which to differences between cohorts in what they have experienced (such as common orientations among baby boomers)? Explain your reasoning.

3. The chapter begins with some alternative explanations for recent changes in the crime rate. Which of the explanations make the most sense to you? Why? How could you learn more about the effect on crime of one of the "causes" you have identified in a laboratory experiment? What type

of study could you conduct in the community to assess its causal impact?

4. This chapter discusses both experimental and nonexperimental approaches to identifying causes. What are the advantages and disadvantages of both approaches for achieving each of the five criteria identified for causal explanations?

5. Construct an idiographic causal explanation for a recent historical or a personal event. For example, what was the sequence of events that led to the outcome of the 2008 U.S. presidential election? What was the sequence of events that led to the invasion of Iraq?

Practice Exercises

1. The study site contains lessons on units of analysis and the related problems of ecological fallacy and reductionism in the Interactive Exercises. Choose the "Units of Analysis" lesson from the main menu. It describes several research projects and asks you to identify the units of analysis in each. Then it presents several conclusions for particular studies, and asks you to determine whether an error has been made.

2. Thomas Rotolo and Charles R. Tittle (2006) were puzzled by a contradictory finding about the relationship between city population size and crime rates: The results of most cross-sectional studies differ from those typically obtained in longitudinal studies. In order to test different causal hypotheses about this relationship, they obtained data about 348 cities that had at least 25,000 residents and adequate data about crime at four time points from 1960 to 1990. Let's review different elements of their arguments and use them to review criteria for causality.

   a. Cross-sectional studies tend to find that cities with more people have a higher crime rate. What criterion for causality does this establish? Review each of the other criteria for causality and explain what noncausal bases they suggest could account for this relationship.

   b. Some have argued that larger cities have higher rates of crime because large size leads to less social integration, which in turn leads to more crime. Which causal criterion does this explanation involve? How much more confident would you be that there is a causal effect of size on crime if you knew that this other relationship occurred also? Explain your reasoning.

   c. Evidence from longitudinal studies has been more mixed, but most do not find a relationship between city size and

the crime rate. What do you think could explain the lack of a longitudinal relationship in spite of the cross-sectional relationship? Explain.

   d. Some have proposed that the presence of transients in large cities is what leads to higher crime rates, because transients (those who are not permanent residents) are more likely to commit crimes. What causal criterion does this involve? Draw a diagram that shows your reasoning.

   e. In their analysis, Rotolo and Tittle (2006) control for region because they suggest that in regions that are traditionally very urban, people may be accustomed to rapid patterns of change, while in newly urbanizing regions this may not be the case. What type of causal criterion would region be? What other factors like this do you think the analysis should take into account? Explain your reasoning.

   f. Now you can examine the Rotolo and Tittle (2006) article online (if your library subscribes to the *Journal of Quantitative Criminology*) and read the details.

3. Search Sociological Abstracts or another index to the social science literature for several articles on studies using any type of longitudinal design. You will be searching for article titles that use words such as *longitudinal, panel, trend,* or *over time.* How successful were the researchers in carrying out the design? What steps did the researchers who used a panel design take to minimize panel attrition? How convinced are you by those using repeated cross-sectional designs that they have identified a process of change in individuals? Did any researchers use retrospective questions? How did they defend the validity of these measures?

Ethics Questions

1. Randomization is a key feature of experimental designs that are often used to investigate the efficacy of new treatments for serious and often incurable, terminal diseases. What ethical issues do these techniques raise in studies of experimental treatments for incurable, terminal diseases? Would you make an ethical argument that there are situations when it is *more* ethical to use random assignment than usual procedures for deciding whether patients receive a new treatment?

2. You learned in this chapter that Sampson and Raudenbush (1999) had observers drive down neighborhood streets in Chicago and record the level of disorder they observed. What should have been the observers' response if they observed a crime in progress? What if they just suspected that a crime was going to occur? What if the crime was a drug dealer interacting with a driver at the curb? What if it was a prostitute soliciting a customer? What, if any, ethical obligation does a researcher studying a neighborhood have to residents in that neighborhood? Should research results be shared at a neighborhood forum?

3. Exum's (2002) experimental manipulation included having some students drink to the point of intoxication. This was done in a carefully controlled setting, with a measured amount of alcohol, and the students who were intoxicated were kept in a room after the experiment was finished until they were sober. Exum also explained the experiment to all the students when they finished the experiment. If you were a student member of your university's Institutional Review Board, would you vote to approve a study with these features? Why or why not? Would you ban an experiment like this involving alcohol altogether, or would you set even more stringent criteria? If the latter, what would those criteria be? Do you think Exum should have been required to screen prospective female students for pregnancy so that some women could have been included in the study (who were not pregnant)? Can you think of any circumstances in which you would allow an experiment involving the administration of illegal drugs?

Web Exercises

1. Go to the Disaster Center website, www.disastercenter.com/crime. Review the crime rate nationally, and, by picking out links to state reports, compare the recent crime rates in two states. Report on the prevalence of the crimes you have examined. Propose a causal explanation for variation in crime between states over time, or both. What research design would you propose to test this explanation? Explain.

2. Go to the Crime Stoppers USA (CSUSA) website at www.crimestoppersusa.com/. Check out "About Us" and then "What Is Crime Stoppers." How is CSUSA "fighting crime"? What does CSUSA's approach assume about the cause of crime? Do you think CSUSA's approach to fighting crime is based on valid conclusions about causality? Explain.

3. What are the latest trends in crime? Write a short statement after inspecting the FBI's Uniform Crime Reports at www.fbi.gov (go to the "Crime Statistics/UCR" section under "Stats & Services").

SPSS Exercises

We can use the GSS2010x data to learn how causal hypotheses can be evaluated with nonexperimental data.

1. Specify four hypotheses in which CAPPUN is the dependent variable and the independent variable is also measured with a question in the 2010 GSS. The independent variables should have no more than 10 valid values (check the variable list).

   a. Inspect the frequency distributions of each independent variable in your hypotheses. If it appears that any has little valid data or was coded with more than 10 categories, substitute another independent variable.

   b. Generate cross-tabulations that show the association between CAPPUN and each of the independent variables. Make sure that CAPPUN is the row variable and that you select "Column Percents."

c. Does support for capital punishment vary across the categories of any of the independent variables? By how much? Would you conclude that there is an association, as hypothesized, for any pairs of variables?

d. Might one of the associations you have just identified be spurious due to the effect of a third variable? What might such an extraneous variable be? Look through the variable list and find a variable that might play this role. If you can't think of any possible extraneous variables, or if you didn't find an association in support of any of your hypotheses, try this: Examine the association between CAPPUN and WRKSTAT2. In the next step, control for sex (gender).

The idea is that there is an association between work status and support for capital punishment that might be spurious due to the effect of sex (gender). Proceed with the following steps:

i. Select Analyze/Descriptive statistical/Crosstabs.

ii. In the Crosstabs window, highlight CAPPUN and then click the right arrow to move it into Rows, Move WRKSTAT2 into Columns and SEX into Layer 1 of 1.

iii. Select Cells/Percentages Column/Continue/OK.

Is the association between employment status and support for capital punishment affected by gender? Do you conclude that the association between CAPPUN and WRKSTAT2 seems to be spurious due to the effect of SEX?

2. Does the association between support for capital punishment and any of your independent variables vary with social context? Marian Borg (1997) concluded that it did. Test this by reviewing the association between attitude toward African Americans (HELPBLK) and CAPPUN. Follow the procedures in SPSS Exercise 1d, but click HELPBLK into columns and REGION4 into Layer 1 of 1. (You must first return the variables used previously to the variables list.) Take a while to study this complex three-variable table. Does the association between CAPPUN and HELPBLK vary with region? How would you interpret this finding?

3. Now, how about the influence of an astrological sign on support for capital punishment? Create a cross-tabulation in which ZODIAC is the independent (column) variable and CAPPUN is the dependent (row) variable (with column percents). What do you make of the results?

Developing a Research Proposal

How will you try to establish the causal effects you hypothesize (Exhibit 2.14, #9, #10, #11, #16)?

1. Identify at least one hypothesis involving what you expect is a causal relationship. Be sure to specify whether your units of analysis will be individuals or groups.

2. Identify the key variables that should be controlled in your survey design to increase your ability to avoid arriving at a spurious conclusion about the hypothesized causal effect. Draw on relevant research literature and social theory to identify these variables.

3. Add a longitudinal component to your research design. Explain why you decided to use this particular longitudinal design.

4. Review the criteria for establishing a nomothetic causal effect and discuss your ability to satisfy each one. If you have decided to adopt an idiographic causal approach, explain your rationale.

Experiments

I thought some of your answers seemed a little goofy. A traveler spends time in airports? A librarian has a lot of books? Couldn't you think of anything better than that?

This is the type of feedback that some students received from another student with whom they were exchanging written comments about a picture of other people. Would you have been offended? What if you had commented on a picture about a student of another race and then been told that your comment was "a little offensive" and involved "stereotypes"? Would this type of confrontational response have made you less likely to say anything that might sound like stereotyping? This is the primary question that Alexander M. Czopp, Margo J. Monteith, and Aimee Y. Mark (2006) attempted to answer in their study about reducing bias through interpersonal confrontation.

Czopp and his colleagues did not study pairs of students who just happened to be looking at pictures together. Rather, these interpersonal confrontations were part of a carefully controlled experiment in which particular types of feedback were provided to students according to the experimental condition to which they were randomly assigned. I will use this research and others to illustrate the features of a true experimental design. As you learned in Chapter 6, experiments provide the strongest research design for testing nomothetic causal hypotheses. They are preferred when the research question is about the effect of a treatment or some other variable whose values can be manipulated by the researcher. M. Lyn Exum's (2002) research about the impact of intoxication and anger on aggressive intentions was one example (Chapter 6); Lawrence Sherman and Richard Berk's (1984) research on the effect of arrest on recidivism in cases of domestic violence (Chapter 2) was another.

This chapter examines experimental methodology in greater detail. First, you will read about different types of experimental design (both true experiments and quasi-experiments) and how they compare with nonexperimental designs. Next, you will learn about the ability of particular designs to establish causally valid conclusions and to achieve generalizable results. Finally, you will consider ethical issues that should be given particular attention when considering experimental research.

▣ True Experiments

True experiments must have at least three features:

1. Two groups (in the simplest case, an experimental and a control group)

2. Variation in the independent variable before assessment of change in the dependent variable

3. Random assignment to the two (or more) comparison groups

> **True experiment** Experiment in which subjects are assigned randomly to an experimental group that receives a treatment or other manipulation of the independent variable, and a comparison group that does not receive the treatment or receives some other manipulation. Outcomes are measured in a posttest.

The combination of these features permits us to have much greater confidence in the validity of causal conclusions than is possible in other research designs. As you learned in Chapter 6, two more things further enhance our confidence in the validity of an experiment's findings:

1. Identification of the causal mechanism

2. Control over the context of an experiment

You will learn more about each of these key features of experimental design as you review three different experimental studies about social processes. I use simple diagrams to help describe and compare the experiments' designs. These diagrams also show at a glance how well suited any experiment is to identifying causal relationships, by indicating whether the experiment has a comparison group, a pretest and a posttest, and randomization.

Experimental and Comparison Groups

Experimental group In an experiment, the group of subjects that receives the treatment or experimental manipulation.

Comparison group In an experiment, a group that has been exposed to a different treatment (or value of the independent variable) than the experimental group.

Control group A comparison group that receives no treatment.

True experiments must have at least one **experimental group** (subjects who receive some treatment) and at least one **comparison group** (subjects to whom the experimental group can be compared). The comparison group differs from the experimental group in terms of one or more independent variables, whose effects are being tested. In other words, variation in the independent variable determines the difference between the experimental and comparison groups.

In many experiments, the independent variable indicates the presence or absence of something, such as receiving a treatment program or not receiving it. In these experiments, the comparison group, consisting of the subjects who do not receive the treatment, is termed a **control group**. You learned in Chapter 6 that an experiment can have more than two groups. There can be several treatment groups, corresponding to different values of the independent variable, and several comparison groups, including a control group that receives no treatment.

Czopp and his colleagues (2006) used a control group in one of the experiments they conducted about interpersonal confrontation. For this experiment, Czopp et al. recruited 111 white students from introductory psychology classes and assigned them randomly to one of the four groups. Earlier in the semester, before the experiment began, all the students had completed a long survey that included an "Attitude Toward Blacks" (ATB) scale. At the time of the experiment, individual students came to a laboratory, sat before a computer, and were told that they would be working with a student in another room to complete a task. All interaction with this "other subject" would be through the computer. Unknown to the student recruits, the other student was actually the experimenter. Also without their knowledge, the student recruits were assigned randomly to one of the four conditions in the experiment.

The students first answered some "getting to know you" questions from the "other student." The "other student" did not mention his or her gender, but did mention that he was white. After this brief Q&A, the student subjects were shown images of various people, of different races, and asked to write a comment about them to the "other student." The "other student" did the same thing, according to a standard script, and also had an opportunity to give some feedback about these comments. Here is the type of feedback that students in the Racial Confrontation condition soon received (the particular words varied according to what the student subject had actually written):

> i thought some of your answers seemed a little offensive. the Black guy wandering the streets could be a lost tourist and the Black woman could work for the government. people shouldn't use stereotypes, you know? (Czopp et al. 2006:795)

Students in the Nonracial Confrontation condition received the type of response with which I began this chapter. Students in the No-Confrontation Control condition were told only, "I thought you typed fast. good job." Shortly after this, the subjects completed the ATB again as well as other measures of their attitudes.

The study results indicated that the interpersonal confrontations were effective in curbing stereotypic responding, whether the confronter was white or black (see Exhibit 7.1). Students who were "confronted" in this situation also indicated a more negative conception of themselves after the confrontation.

Pretest and Posttest Measures

Posttest In experimental research, the measurement of an outcome (dependent) variable after an experimental intervention or after a presumed independent variable has changed for some other reason.

All true experiments have a **posttest**—that is, measurement of the outcome in both groups after the experimental group has received the treatment. Many true

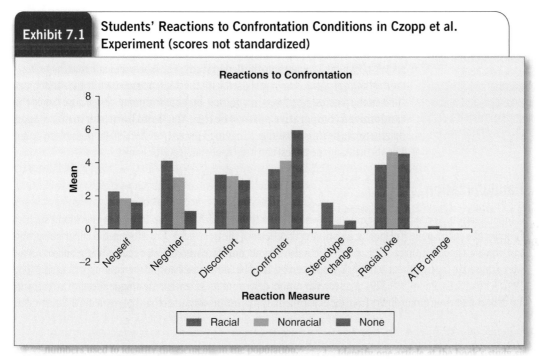

Exhibit 7.1 Students' Reactions to Confrontation Conditions in Czopp et al. Experiment (scores not standardized)

Note: Negself = negative attitude toward self; Negother = negative attitude toward others; ATB = attitude toward blacks.

experiments also have **pretests** that measure the dependent variable prior to the experimental intervention. A pretest is exactly the same as a posttest, just administered at a different time. Pretest scores permit a direct measure of how much the experimental and comparison groups changed over time, such as the change in stereotypic responding among the students that Czopp et al. studied.

Pretest scores also allow the researcher to verify that randomization was successful (that chance factors did not lead to an initial difference between the groups). In addition, by identifying subjects' initial scores on the dependent variable, a pretest provides a more complete picture of the conditions in which the intervention had (or didn't have) an effect (Mohr 1992:46–48). A randomized experimental design with a pretest and posttest is termed a **randomized comparative change design** or a **pretest-posttest control group design**.

An experiment may have multiple posttests and perhaps even multiple pretests. Multiple posttests can identify just when the treatment has its effect and for how long. This is particularly important when treatments are delivered over a period of time (Rossi & Freeman 1989:289–290).

However, strictly speaking, a true experiment does not require a pretest. When researchers use random assignment to the experimental and comparison groups, the groups' initial scores on the dependent variable and on all other variables are very likely to be similar. Any difference in outcome between the experimental and comparison groups is therefore likely to be due to the intervention (or to other processes occurring during the experiment), and the likelihood of a difference just on the basis of chance can be calculated. This is fortunate, because the dependent variable in some experiments cannot be measured in a pretest. For example, Czopp et al. (2006:797) measured the attitudes of their student subjects toward the "other student" who had confronted them. They weren't able

Pretest In experimental research, the measurement of an outcome (dependent) variable prior to an experimental intervention or change in a presumed independent variable for some other reason. The pretest is exactly the same "test" as the posttest, but it is administered at a different time.

Randomized comparative change design The classic true experimental design in which subjects are assigned randomly to two groups; both these groups receive a pretest, and then one group receives the experimental intervention, and then both groups receive a posttest. Also known as a *pretest-posttest control group design*.

Randomized comparative posttest design A true experimental design in which subjects are assigned randomly to two groups—one group then receives the experimental intervention and both groups receive a posttest; there is no pretest. Also known as a *posttest-only control group design.*

to measure this attitude until after the interaction in which they manipulated the confrontation. Thus, Exhibit 7.1 includes some measures that represent change from the pretest to the posttest (ATB scores) and some that represent only scores at the posttest. Exhibit 7.2 diagrams the Czopp study. The labels indicate that the *pretest-posttest control group design* was used with the ATB measure, while the other measures in the posttest were used in a *posttest-only control group design,* also called the **randomized comparative posttest design.** You'll also learn later in this chapter that there can be a disadvantage to having a pretest, even when it is possible to do so: The act of taking the pretest can itself cause subjects to change.

Randomization

Randomization, or random assignment, is what makes the comparison group in a true experiment such as Czopp et al.'s such a powerful tool for identifying the effects of the treatment. A randomized comparison group can provide a good estimate of the counterfactual—the outcome that would have occurred if the subjects who were exposed to the treatment actually had not been exposed but otherwise had had the same experiences (Mohr 1992:3; Rossi & Freeman 1989:229). A researcher cannot determine for sure what the unique effects of a treatment are if the comparison group differs from the experimental group in any way other than not receiving the treatment.

Research in the News

NUDGE THE VOTE

Randomized experiments about voter turnout have been changing the tactics used by many political campaigns. Direct voter-contact mail, get-out-the-vote robocalls, lawn signs, visits to editorial boards, newspaper ads? All useless. What works? Calls that ask potential voters how they will get to the polls (increases turnout among those who live alone). Mailings that list likely voters' voting histories, including the elections they missed, as well as those of their neighbors, with a promise to send updated information after the election. Plain envelopes rather than glossy four-color brochures.

Source: Issenberg, Sasha. 2010. "Nudge the Vote: What's Even Better Than Getting Lady Gaga to Play Your Election Day Rally? Sending Out a Mailing that Applies a Subtle Dose of Peer Pressure. How Behavioral Science is Remaking Politics." *The New York Times Magazine,* October 31:28–31.

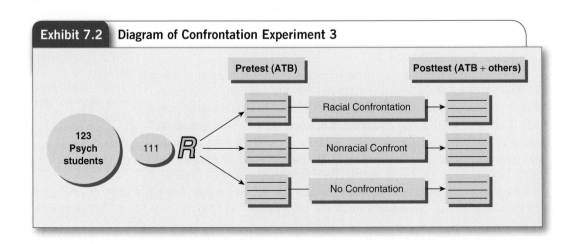

Exhibit 7.2 **Diagram of Confrontation Experiment 3**

Pretest (ATB) Posttest (ATB + others)

123 Psych students → 111 → R → Racial Confrontation →
Nonracial Confront →
No Confrontation →

Assigning subjects randomly to the experimental and comparison groups ensures that systematic bias does not affect the assignment of subjects to groups. Of course, random assignment cannot guarantee that the groups are perfectly identical at the start of the experiment. Randomization removes bias from the assignment process, but only by relying on chance, which itself can result in some intergroup differences (Bloom 2008:116). Fortunately, researchers can use statistical methods to determine the odds of ending up with groups that differ very much on the basis of change, and these odds are low even for groups of moderate size. The larger the group, the less likely it is that even modest differences will occur on the basis of chance and the more possible it becomes to draw conclusions about causal effects from relatively small differences in the outcome.

Note that the random assignment of subjects to experimental and comparison groups is not the same as random sampling of individuals from some larger population (see Exhibit 7.3). In fact, random assignment (randomization) does not help at all to ensure that the research subjects are representative of some larger

---

Exhibit 7.3 | **Random Sampling Versus Random Assignment**

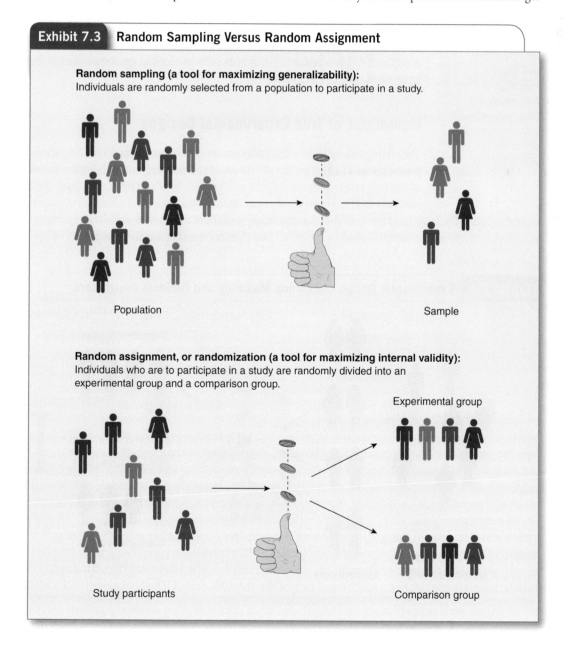

Random sampling (a tool for maximizing generalizability):
Individuals are randomly selected from a population to participate in a study.

Population

Sample

Random assignment, or randomization (a tool for maximizing internal validity):
Individuals who are to participate in a study are randomly divided into an experimental group and a comparison group.

Experimental group

Study participants

Comparison group

population; instead, representativeness is the goal of random sampling. What random assignment does—create two (or more) equivalent groups—is useful for maximizing the likelihood of internal validity, not generalizability (Bloom 2008:116).

Matching is another procedure used to equate experimental and comparison groups, but by itself it is a poor substitute for randomization. Matching of individuals in a treatment group with those in a comparison group might involve pairing persons on the basis of similarity of gender, age, year in school, or some other characteristic. The basic problem is that, as a practical matter, individuals can be matched on only a few characteristics; as a result, unmatched differences between the experimental and comparison groups may still influence outcomes. When matching is used as a substitute for random assignment, the research becomes quasi-experimental instead of being a true experiment. However, matching combined with randomization, also called *blocking*, can reduce the possibility of differences due to chance (Bloom 2008:124). For example, if individuals are matched in terms of gender and age, and then the members of each matched pair are assigned randomly to the experimental and comparison groups, the possibility of outcome differences due to differences in the gender and age composition of the groups is eliminated (see Exhibit 7.4).

> **Matching** A procedure for equating the characteristics of individuals in different comparison groups in an experiment. Matching can be done on either an individual or an aggregate basis. For individual matching, individuals who are similar in terms of key characteristics are paired prior to assignment, and then the two members of each pair are assigned to the two groups. For aggregate matching, also termed *blocking*, groups that are chosen for comparison are similar in terms of the distribution of key characteristics.

Limitations of True Experimental Designs

The distinguishing features of true experiments—experimental and comparison group, pretests (which are not always used) and posttests, and randomization—do not help researchers identify the mechanisms by which treatments have their effects. In fact, this question of causal mechanisms often is not addressed in experimental research. The hypothesis test itself does not require any analysis of mechanism, and if the experiment was conducted under carefully controlled conditions during a limited span of time, the causal effect (if any)

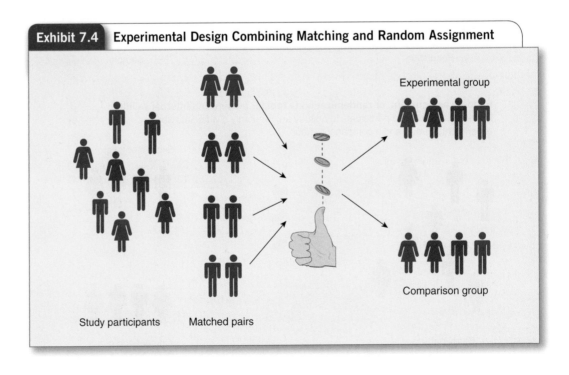

Exhibit 7.4   Experimental Design Combining Matching and Random Assignment

Study participants   Matched pairs   Experimental group   Comparison group

social relationships, and social beings—of people and the groups to which they belong—is so great that it often defies reduction to the simplicity of a laboratory or restriction to the requirements of experimental design. Yet the virtues of experimental designs mean that they should always be considered when explanatory research is planned.

Key Terms

Assignment variable   211
Before-and-after design   206
Comparison group   200
Compensatory rivalry
    (John Henry effect)   217
Contamination   217
Control group   200
Demoralization   217
Differential attrition (mortality)   215
Distribution of benefits   223
Double-blind procedure   218
Endogenous change   216
Ex post facto control group design   212
Expectancies of experimental staff   218
Experimental group   200
External events   217

Factorial survey   219
Field experiment   205
Hawthorne effect   218
History effect   217
Intent-to-treat
    analysis   216
Matching   204
Multiple group before-and-after
    design   209
Nonequivalent control group
    design   206
Placebo effect   218
Posttest   200
Pretest   201
Quasi-experimental
    design   206

Randomized comparative change design
    (pretest-posttest control group
    design)   201
Randomized comparative posttest
    design (posttest-only control
    group design)   202
Regression-discontinuity design   211
Regression effect   216
Repeated measures panel design   210
Selection bias   215
Selective distribution of benefits
    (see distribution of benefits)   223
Solomon four-group design   221
Time series design   210
Treatment misidentification   215
True experiment   199

Highlights

- The independent variable in an experiment is represented by a treatment or other intervention. Some subjects receive one type of treatment; others may receive a different treatment or no treatment. In true experiments, subjects are assigned randomly to comparison groups.

- Experimental research designs have three essential components: (1) use of at least two groups of subjects for comparison, (2) measurement of the change that occurs as a result of the experimental treatment, and (3) use of random assignment. In addition, experiments may include identification of a causal mechanism and control over experimental conditions.

- Random assignment of subjects to experimental and comparison groups eliminates systematic bias in group assignment. The odds of a difference between the experimental and comparison groups because of chance can be calculated. They become very small for experiments with at least 30 subjects per group.

- Both random assignment and random sampling rely on a chance selection procedure, but their purposes differ. Random

assignment involves placing predesignated subjects into two or more groups on the basis of chance; random sampling involves selecting subjects out of a larger population on the basis of chance. Matching of cases in the experimental and comparison groups is a poor substitute for randomization, because identifying in advance all important variables on which to make the match is not possible. However, matching can improve the comparability of groups, when it is used to supplement randomization.

- Quasi-experiments include features that maximize the comparability of the control and experimental groups and make it unlikely that self-selection determines group membership. Nonexperiments rely only on naturally occurring groups without any particular criteria to reduce the risk of selection bias.

- Causal conclusions derived from experiments can be invalid because of selection bias, endogenous change, effects of external events, cross-group contamination, or treatment misidentification. In true experiments, randomization should eliminate selection bias and bias due to endogenous change. External events, cross-group contamination, and treatment misidentification can threaten the validity of causal conclusions in both true experiments and quasi-experiments.

- The generalizability of experimental results declines if the study conditions are artificial and the experimental subjects are unique. Field experiments are likely to produce more generalizable results than experiments conducted in the laboratory.

- The external validity of causal conclusions is determined by the extent to which they apply to different types of individuals and settings. When causal conclusions do not apply to all the subgroups in a study, they are not generalizable to corresponding subgroups in the population—and so they are not externally valid with respect to those subgroups. Causal conclusions can also be considered externally invalid when they occur only under the experimental conditions.

- Subject deception is common in laboratory experiments and poses unique ethical issues. Researchers must weigh the potential harm to subjects and debrief subjects who have been deceived. In field experiments, a common ethical problem is selective distribution of benefits. Random assignment may be the fairest way of allocating treatment when treatment openings are insufficient for all eligible individuals and when the efficacy of the treatment is unknown.

STUDENT STUDY SITE

To assist in completing the web exercises, please access the study site at **www.sagepub.com/schuttisw7e**, where you will find the web exercises with accompanying links. You'll find other useful study materials such as self-quizzes and e-flashcards for each chapter, along with a group of carefully selected articles from research journals that illustrate the major concepts and techniques presented in the book.

Discussion Questions

1. Read the original article reporting one of the experiments described in this chapter. Critique the article using as your guide the article review questions presented in Appendix A. Focus on the extent to which experimental conditions were controlled and the causal mechanism was identified. Did inadequate control over conditions or inadequate identification of the causal mechanism make you feel uncertain about the causal conclusions?

2. Select a true experiment, perhaps from the *Journal of Experimental and Social Psychology, Journal of Personality and Social Psychology,* or from sources suggested in class. Diagram the experiment using the exhibits in this chapter as a model. Discuss the extent to which experimental conditions were controlled and the causal mechanism was identified. How confident can you be in the causal conclusions from the study, based on review of the threats to internal validity discussed in this chapter: selection bias, endogenous change, external events, contamination, and treatment misidentification? How generalizable do you think the study's results are to the population from which cases were selected? To specific subgroups in the study? How thoroughly do the researchers discuss these issues?

3. Repeat Discussion Question 2 with a quasi-experiment.

Practice Exercises

1. Arrange with an instructor in a large class to conduct a multiple pretest-posttest study of the impact of watching a regularly scheduled class movie. Design a 10-question questionnaire to measure knowledge about the topics in the film.

Administer this questionnaire shortly before and shortly after the film is shown and then again 1 week afterward. After scoring the knowledge tests, describe the immediate and long-term impact of the movie.

2. Volunteer for an experiment! Contact the psychology department, and ask about opportunities for participating in laboratory experiments. Discuss the experience with your classmates.

3. Take a few minutes to review the "Sources of Internal Invalidity" lesson from the Interactive Exercises link on the study site. It will be time well spent.

4. Select an article that used an experimental design from the book's study site, at www.sagepub.com/schuttisw7e. Diagram the design and identify sources of internal and external invalidity that are not controlled by the experimental design.

Ethics Questions

1. What specific rules do you think should guide researchers' decisions about subject deception and the selective distribution of benefits? How much deception should be allowed, and under what circumstances? Was the deception in the Milgram study acceptable? What about deception in "Psych 101" lab experiments such as Bushman's experiments on aggression (Chapter 6)? Do you think it would be acceptable to distribute free bicycles to a random selection of commuters for 2 years to see if they change their habits more than those in a matched control group? What about the complaints of a member of the control group (after all, she pays her taxes too!)?

2. Under what conditions do you think that the randomized assignment of subjects to a specific treatment is ethical in social science research? Was it ethical for Sherman and Berk (1984) and the researchers who conducted the replication studies to randomly assign individuals accused of domestic violence to an arrest or nonarrest treatment? What about randomly assigning some welfare recipients to receive higher payments than others? And what about randomly assigning some students to receive a different instructional method than others?

Web Exercises

1. Go to Sociosite at www.sociosite.net/index.php. Choose "Subject Areas." Choose a sociological subject area you are interested in. How would you conduct a study on your chosen subject using experimental methods? Choose at least five of the key terms listed at the end of this chapter that are relevant to and incorporated in the research experiment you have located on the web. Explain how each of the five key terms you have chosen plays a role in the research example you have found on the web.

2. Try out the process of randomization. Go to the website www.randomizer.org and click on "Randomize Now" at the bottom of the page. Type numbers into the randomizer for an

experiment with 2 groups and 20 individuals per group. Repeat the process for an experiment with 4 groups and 10 individuals per group. Plot the numbers corresponding to each individual in each group. Does the distribution of numbers within each group truly seem to be random?

3. Participate in a social psychology experiment on the web. Go to www.socialpsychology.org/expts.htm. Pick an experiment in which to participate and follow the instructions. After you finish, write up a description of the experiment and evaluate it using the criteria discussed in the chapter.

SPSS Exercises

Because the GSS2010x doesn't provide experimental data to work with, we'll pause in our study of support for capital punishment and examine some relationships involving workplace variables such as some of those that were the focus of research reviewed in this chapter.

Do the features of work influence attitudes about the work experience? We can test some hypothetical answers to this question with the GSS2010x data set (although not within the context of an experimental design).

1. Describe the feelings of working Americans about their jobs and economic rewards, based on their responses to questions about balancing work and family demands, their satisfaction with their finances, and their job satisfaction. Generate the frequencies as follows:

    a. Click Analyze/Descriptive statistics/Frequencies.

    b. Select SATFIN, SATJOB.

How satisfied are working people with their jobs and their pay?

2. Do these feelings vary with work features?

    a. Pose at least three hypotheses in which either SATFIN or SATJOB is the dependent variable and one of the following two variables is the independent variable: earnings or work status. Now test these hypotheses by comparing average scores on the attitudinal variables between categories of the independent variables:

        i. Click Analyze/Compare Means/Means

        ii. Select Dependent List: SATFIN, SATJOB

        iii. Independent List: RINCOM4, WRKSTAT2

    b. Which hypotheses appear to be supported? (Remember to review the distributions of the dependent variables [SATFIN, SATJOB] to remind yourself what a higher average score indicates on each variable.)

Developing a Research Proposal

Your work in this section should build on your answers to the proposal development questions in the last chapter, assuming that you will use an experimental design (Exhibit 2.14, #13, #14, #17).

1. Design a laboratory experiment to test one of your hypotheses or a related hypothesis. Describe the experimental design, commenting on each component of a true experiment. Specify clearly how the independent variable will be manipulated and how the dependent variable will be measured.

2. Assume that your experiment will be conducted on campus. Formulate recruitment and randomization procedures.

3. Discuss the extent to which each source of internal invalidity is a problem in the study. Propose procedures to cope with these sources of invalidity.

4. How generalizable would you expect the study's findings to be? What can be done to increase generalizability?

5. Develop appropriate procedures for the protection of human subjects in your experiment. Include among these procedures a consent form. Give particular attention to any aspects of the study that are likely to raise ethical concerns.

CHAPTER 8

Survey Research

"Education forms a unique dimension of social status, with qualities that make it especially important to health." John Mirowsky and Catherine E. Ross (2003:1) make this claim at the start of *Education, Social Status, and Health* and then present evidence to support it throughout the book. Most of their evidence comes from two surveys. In this chapter, we will focus on one of them, the Aging, Status, and the Sense of Control (ASOC) survey that was funded by the National Institute on Aging.

I begin this chapter with a brief review of the reasons for using survey methods, but I will then focus attention on the Mirowsky and Ross ASOC survey and use it to illustrate some key features of survey research. I will explain the major steps in questionnaire design and then discuss the features of four types of surveys, highlighting the unique problems attending each one and suggesting some possible solutions. I will give particular attention to the ways in which new means of communication such as cell phones and the Internet have been changing survey research since the first ASOC survey in 1995 (there have been two more, in 1998 and 2001). I discuss ethics issues in the final section. By the chapter's end, you should be well on your way to becoming an informed consumer of survey reports and a knowledgeable developer of survey designs. As you read the chapter, I also hope that you will occasionally reflect on how education influences social status and health.

Survey Research in the Social Sciences

Survey research Research in which information is obtained from a sample of individuals through their responses to questions about themselves or others.

Survey research involves the collection of information from a sample of individuals through their responses to questions. Mirowsky and Ross (2003) turned to survey research for their study of education, social status, and health because it is an efficient method for systematically collecting data from a broad spectrum of individuals and social settings. As you probably have observed, a great many social scientists—as well as newspaper editors, political pundits, government agencies, and marketing gurus—make the same methodological choice. In fact, surveys have become a multibillion-dollar industry in the United States that shapes what we read in the newspapers, see on TV, and find in government reports (Converse 1984; Tourangeau 2004:776).

Attractions of Survey Research

Survey research owes its popularity to three features: versatility, efficiency, and generalizability. Each of these features is changing as a result of new technologies.

Versatility

First and foremost, survey methods are versatile. Although a survey is not the ideal method for testing all hypotheses or learning about every social process, a well-designed survey can enhance our understanding of just about any social issue. Mirowsky and Ross's (2003) survey covered a range of topics about work and health, and there is hardly any other topic of interest to social scientists that has not been studied at some time with survey methods. Politicians campaigning for election use surveys, as do businesses marketing a product, governments assessing community needs, agencies monitoring program effectiveness, and lawyers seeking to buttress claims of discrimination or select favorable juries.

Computer technology has made surveys even more versatile. Computers can be programmed so that different types of respondents are asked different questions. Short videos or pictures can be presented to respondents on a computer screen. An interviewer may give respondents a laptop on which to record their answers to sensitive personal questions, such as about illegal activities, so that not even the interviewer will know what they said (Tourangeau 2004:788–794).

Efficiency

Surveys also are popular because data can be collected from many people at relatively low cost and, depending on the survey design, relatively quickly. John Mirowsky and Catherine Ross (2003:207) contracted with the Survey Research Laboratory (SRL) of the University of Illinois for their 25-minute 2003 telephone survey of 2,495 adult Americans. SRL estimated that the survey would incur direct costs of $183,000—that's $73.35 per respondent—and take up to 1 year to complete. Both this cost and the length of time required were relatively high, because SRL made special efforts to track down respondents from the first wave of interviews in 1995. One-shot telephone interviews can cost as little as $30 per subject (Ross 1990). Large mailed surveys cost even less, about $10 to $15 per potential respondent, although the costs can increase greatly when intensive follow-up efforts are made. Surveys of the general population using personal interviews are much more expensive, with costs ranging from about $100 per potential respondent, for studies in a limited geographical area, to $300 or more when lengthy travel or repeat visits are needed to connect with respondents (F. Fowler, personal communication, January 7, 1998; see also Dillman 1982; Groves & Kahn 1979). Surveys through the web have become the quickest way to gather survey data, but there are problems with this method that I will soon discuss.

Surveys are efficient because many variables can be measured without substantially increasing the time or cost. Mailed questionnaires can include up to 10 pages of questions before respondents begin to balk. In-person interviews can be much longer. For example, the 2006 General Social Survey (GSS) had seven different versions that measured about 1,200 variables. The upper limit for phone surveys seems to be about 45 minutes.

Of course, these efficiencies can be attained only in a place with a reliable communications infrastructure (Labaw 1980:xiii–xiv). A reliable postal service, which is required for mail surveys, generally has been available in the United States—although residents of the Bronx, New York, have complained that delivery of local first-class mail often takes 2 weeks or more, almost ruling out mail surveys (Purdy 1994). The British postal service, the Royal Mail, has been accused of even worse performance: a "total shambles," with mail abandoned in some cases and purposely misdelivered in other cases (Lyall 2004:A4). Phone surveys have been very effective in countries such as the United States, where 96% of households have phones (Tourangeau 2004:777). Also important to efficiency are the many survey research organizations—about 120 academic and nonprofit organizations in the United States—that provide trained staff and proper equipment (Survey Research Laboratory 2008).

Modern information technology has been a mixed blessing for survey efficiency. The Internet makes it easier to survey some populations, but it leaves out important segments. Caller ID and answering machines make it easy to screen out unwanted calls, but these tools also make it harder to reach people in phone surveys. In addition, as discussed in Chapter 5, a growing number of people use only cell phones. As a result, the percentage of U.S. households without landline telephones climbed to 23% by 2009, after a long decline to below 5% in 2001 (Christian et al. 2010) (see Exhibit 8.1). As a result of these trends, survey researchers must spend more time and money to reach potential respondents (Tourangeau 2004:781–782).

Generalizability

Survey methods lend themselves to probability sampling from large populations. Thus, survey research is very appealing when sample generalizability is a central research goal. In fact, survey research is often the only means available for developing a representative picture of the attitudes and characteristics of a large population.

Surveys are also the method of choice when cross-population generalizability is a key concern, because they allow a range of social contexts and subgroups to be sampled. The consistency of relationships can then be

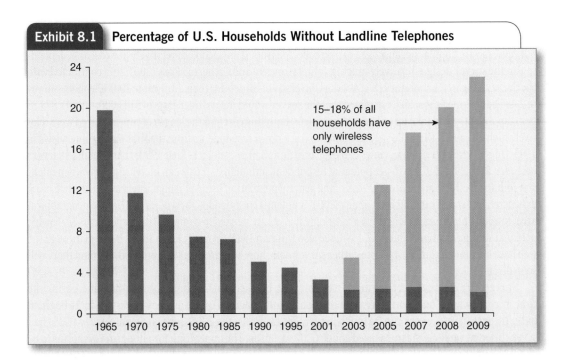

Exhibit 8.1 Percentage of U.S. Households Without Landline Telephones

15–18% of all households have only wireless telephones

examined across the various subgroups. An ambitious Internet-based international survey sponsored by the National Geographic Society (2000) was completed by 80,012 individuals from 178 countries and territories.

Unfortunately (for survey researchers), the new technologies that are lowering the overall rate of response to phone surveys are also making it more difficult to obtain generalizable samples. In the United States, 30% of households in 2008 didn't have access to the Internet at home or work, and in these households persons tend to be older and poorer than those who are "connected" (de Leeuw 2008:321; Tourangeau 2004:792). Those who rely exclusively on cell phones tend to be younger, on average, than those who also have a phone installed in their home. In addition, the growing size of the foreign-born population in the United States, almost 12% in 2002, requires either foreign-language versions of survey forms or acknowledging an inability to generalize survey results to the entire population (Tourangeau 2004:783).

The Omnibus Survey

An omnibus survey shows just how versatile, efficient, and generalizable a survey can be. An **omnibus survey** covers a range of topics of interest to different social scientists, in contrast to the typical survey that is directed at a specific research question. It has multiple sponsors or is designed to generate data useful to a broad segment of the social science community rather than to answer a particular research question. It is usually directed to a sample of some general population, so the questions, about a range of different issues, are appropriate to at least some sample members.

Omnibus survey A survey that covers a range of topics of interest to different social scientists.

One of sociology's most successful omnibus surveys is the GSS of the National Opinion Research Center at the University of Chicago. It is an extensive interview administered biennially to a probability sample of at least 3,000 Americans (4,510 in 2006), with a wide range of questions and topic areas chosen by a board of overseers. Some questions are asked of only a randomly selected subset of respondents. This **split-ballot design** allows more questions without increasing the survey's cost. It also facilitates experiments on the effect of question wording: Different forms of the same question are included in the split-ballot subsets. The GSS is widely available to universities, instructors, and students (Davis & Smith 1992; National Opinion Research Center 2011), as are many other survey data sets archived by the Inter-University Consortium for Political and Social Research (ICPSR) (more details about the ICPSR are in Chapter 13). John Mirowsky and Catherine Ross contributed their survey data set to the ICPSR.

Errors in Survey Research

It might be said that surveys are too easy to conduct. Organizations and individuals often decide that a survey will help solve some important problem because it seems so easy to write up some questions and distribute them. But without careful attention to sampling, measurement, and overall survey design, the effort is likely to be a flop.

> **Split-ballot design** Unique questions or other modifications in a survey administered to randomly selected subsets of the total survey sample, so that more questions can be included in the entire survey or so that responses to different question versions can be compared.

Such flops are too common for comfort, so the responsible survey researcher must take the time to design surveys properly and to convince sponsoring organizations that this time is worth the effort (Turner & Martin 1984:68).

For a survey to succeed, it must minimize four types of error (Groves 1989:vi, 10–12): (1) poor measurement, (2) nonresponse, (3) inadequate coverage of the population, and (4) sampling error.

Poor measurement. Measurement error was a key concern in Chapter 4, but there is much more to be learned about how to minimize these errors of observation in the survey process. The theory of *satisficing* can help us understand the problem. It takes effort to answer survey questions carefully: Respondents have to figure out what each question means, then recall relevant information and finally decide which answer is most appropriate. Survey respondents *satisfice* when they reduce the effort required to answer a question by interpreting questions superficially and giving what they think will be an acceptable answer (Krosnick 1999:547–548). Presenting clear and interesting questions in a well-organized questionnaire will help reduce measurement error by encouraging respondents to answer questions carefully and to take seriously the request to participate in the survey. Tailoring questions to the specific population surveyed is also important. In particular, persons with less education are more likely to satisfice in response to more challenging questions (Holbrook et al. 2003; Narayan & Krosnick 1996). The focus of the next section is on writing questions for a survey, and questionnaire design is discussed later in the chapter.

Nonresponse. Nonresponse is a major and growing problem in survey research, although it is a problem that varies between particular survey designs. Social exchange theory can help us understand why nonresponse rates have been growing in the United States and Western Europe since the early 1950s (Dillman 2000:14–15; Groves & Couper 1998:155–189; Tourangeau 2004:782). According to social exchange theory, a well-designed survey effort will maximize the social rewards for survey participation and minimize its costs, as well as establish trust that the rewards will outweigh the costs (Blau 1964). The perceived benefits of survey participation have declined with decreasing levels of civic engagement and with longer work hours (Groves, Singer, & Corning 2000; Krosnick 1999:539–540). Perceived costs have increased with the widespread use of telemarketing and the ability of many people to screen out calls from unknown parties with answering machines and caller ID. In addition, recipients pay for time on cell phone calls, so the ratio of costs to benefits worsens for surveys attempting to reach persons using cell phones (Nagourney 2002). We will review more specifics about nonresponse in this chapter's sections on particular survey methods.

Inadequate coverage of the population. A poor sampling frame can invalidate the results of an otherwise well-designed survey. We considered the importance of a good sampling frame in Chapter 5; in this chapter, I will discuss special coverage problems related to each of the particular survey methods.

Sampling error. The process of random sampling can result in differences between the characteristics of the sample members and the population simply on the basis of chance. I introduced this as a topic in Chapter 5. You will learn how to calculate sampling error in Chapter 14.

▣ Writing Survey Questions

Questions are the centerpiece of survey research. Because the way they are worded can have a great effect on the way they are answered, selecting good questions is the single most important concern for survey researchers. All hope for achieving measurement validity is lost unless the questions in a survey are clear and convey the intended meaning to respondents.

You may be thinking that you ask people questions all the time and have no trouble understanding the answers you receive, but can't you also think of times when you've been confused in casual conversation by misleading or misunderstood questions? Now, consider just a few of the differences between everyday conversations and standardized surveys that make writing survey questions much more difficult:

- Survey questions must be asked of many people, not just one.

- The same survey question must be used with each person, not tailored to the specifics of a given conversation.

- Survey questions must be understood in the same way by people who differ in many ways.

- You will not be able to rephrase a survey question if someone doesn't understand it because that would result in a different question for that person.

- Survey respondents don't know you and so can't be expected to share the nuances of expression that help you and your friends and family to communicate.

Writing questions for a particular survey might begin with a brainstorming session or a review of previous surveys. Then, whatever questions are being considered must be systematically evaluated and refined. Although most professionally prepared surveys contain previously used questions as well as some new ones, every question that is considered for inclusion must be reviewed carefully for its clarity and ability to convey the intended meaning. Questions that were clear and meaningful to one population may not be so to another. Nor can you simply assume that a question used in a previously published study was carefully evaluated.

Adherence to a few basic principles will go a long way toward ensuring clear and meaningful questions. Each of these principles summarizes a great deal of research, although none of them should be viewed as an inflexible mandate (Alwin & Krosnick 1991). As you will learn in the next section, every question must be considered in terms of its relationship to the other questions in a survey. Moreover, every survey has its own unique requirements and constraints; sometimes violating one principle is necessary to achieve others.

Avoid Confusing Phrasing

What's a confusing question? Try this one that I received years ago from the Planetary Society in their National Priorities Survey for United States Space Program:

> The Moon may be a place for an eventual scientific base, and even for engineering resources. Setting up a base or mining experiment will cost tens of billions of dollars in the next century. Should the United States pursue further manned and unmanned scientific research projects on the surface of the Moon?

> Yes ☐   No ☐   No opinion ☐

Does a "yes" response mean that you favor spending tens of billions of dollars for a base or mining experiment? Does "the next century" refer to the 21st century or to the 100 years after the survey (which was distributed in the 1980s)? Could you favor further research projects on the Moon but oppose funding a scientific base or engineering resources? Are engineering resources supposed to have something to do with a mining experiment? Does a mining experiment occur "on the surface of the Moon"? How do you answer if you favor unmanned scientific research projects on the moon but not manned projects?

There are several ways to avoid such confusing phrasing. In most cases, a simple direct approach to asking a question minimizes confusion. Use shorter rather than longer words and sentences: "brave" rather than "courageous"; "job concerns" rather than "work-related employment issues" (Dillman 2000:52). Try to keep the total number of words to 20 or fewer and the number of commas to 3 or fewer (Peterson 2000:50). On the other hand, questions shouldn't be abbreviated in a way that results in confusion: To ask, "In what city or town do you live?" is to focus attention clearly on a specific geographic unit, a specific time, and a specific person (you); the simple format,

Residential location: _____,

does not do this.

Sometimes, when sensitive issues or past behaviors are the topic, longer questions can provide cues that make the respondent feel comfortable or aid memory (Peterson 2000:51).

Breaking up complex issues into simple parts also reduces confusion. In a survey about health services (such as the one by Mirowsky and Ross), you might be tempted to ask a complex question like this (Schaeffer & Presser 2003):

During the past 12 months since July 1st, 1987, how many times have you seen or talked with a doctor or a medical assistant about your health? Do not count any times you might have seen a doctor while you were a patient in a hospital, but count all the other times you actually saw or talked to a medical doctor of any kind about your health. (pp. 70–71)

This question can be simplified, thereby reducing confusion, by breaking it up into several shorter questions:

Have you been a patient in the hospital overnight in the past 12 months since July 1st, 1987?

(Not counting when you were in a hospital overnight) During the past 12 months since July 1st, 1987, how many times did you actually see any medical doctor about your own health?

During the past 12 months since July 1st, 1987, were there any times when you didn't actually see the doctor but saw a nurse or other medical assistant working for the doctor?

During the past 12 months since July 1st, 1987, did you get any medical advice, prescriptions, or results of tests over the telephone from a medical doctor, nurse, or medical assistant working for a doctor? (Cannell et al. 1989, Appendix A, p. 1)

A sure way to muddy the meaning of a question is to use **double negatives**: "Do you *disagree* that there should *not* be a tax increase?" Respondents have a hard time figuring out which response matches their sentiments. Such errors can easily be avoided with minor wording changes, but even experienced survey researchers can make this mistake unintentionally, perhaps while trying to avoid some other wording problem. For instance, in a survey commissioned by the American Jewish Committee, the Roper polling organization wrote a question about the Holocaust that was carefully worded to be neutral and

> **Double negative** A question or statement that contains two negatives, which can muddy the meaning of the question.

value free: "Does it seem possible or does it seem impossible to you that the Nazi extermination of the Jews never happened?" Among a representative sample of adult Americans, 22% answered that it was possible the extermination never happened (Kifner 1994:A12). Many Jewish leaders and politicians were stunned, wondering how one in five Americans could be so misinformed. But a careful reading of the question reveals how confusing it is: Choosing "possible," the seemingly positive response, means that you don't believe the Holocaust happened. In fact, the Gallup organization then rephrased the question to avoid the double negative, giving a brief definition of the Holocaust and then asking, "Do you doubt that the Holocaust actually happened or not?" Only 9% responded that they doubted it happened. When a wider range of response choices was given, only 2.9% said that the Holocaust "definitely" or "probably" did not happen. To be safe, it's best just to avoid using negative words such as "don't" and "not" in questions.

So-called **double-barreled questions** are also guaranteed to produce uninterpretable results because they actually ask two questions but allow only one answer. For example, during the Watergate scandal, Gallup poll results indicated that when the question was "Do you think President Nixon should be impeached and compelled to leave the presidency, or not?" only about a third of Americans supported impeaching President Richard M. Nixon. But when the Gallup organization changed the question to ask respondents if they "think there is enough evidence of possible wrongdoing in the case of President Nixon to bring him to trial before the Senate, or not," over half answered yes. Apparently, the first, double-barreled version of the question confused support for impeaching Nixon—putting him on trial before the Senate—with concluding that he was guilty before he had had a chance to defend himself (Kagay & Elder 1992:E5).

> **Double-barreled question** A single survey question that actually asks two questions but allows only one answer.

It is also important to identify clearly what kind of information each question is to obtain. Some questions focus on attitudes, or what people say they want or how they feel. Some questions focus on beliefs, or what people think is true. Some questions focus on behavior, or what people do. And some questions focus on attributes, or what people are like or have experienced (Dillman 1978:79–118; Gordon 1992). Rarely can a single question effectively address more than one of these dimensions at a time.

Whichever type of information a question is designed to obtain, be sure it is asked of only the respondents who may have that information. If you include a question about job satisfaction in a survey of the general population, first ask respondents whether they have a job. You will only annoy respondents if you ask a question that does not apply to them (Schaeffer & Presser 2003:74). These **filter questions** create **skip patterns**. For example, respondents who answer *no* to one question are directed to skip ahead to another question, but respondents who answer *yes* go on to the **contingent question**. Skip patterns should be indicated clearly with an arrow or other mark in the questionnaire as demonstrated in Exhibit 8.2.

> **Filter question** A survey question used to identify a subset of respondents who then are asked other questions.
>
> **Skip pattern** The unique combination of questions created in a survey by filter questions and contingent questions.
>
> **Contingent question** A question that is asked of only a subset of survey respondents.

Minimize the Risk of Bias

Specific words in survey questions should not trigger biases, unless that is the researcher's conscious intent. Biased or loaded words and phrases tend to produce misleading answers. For example, a 1974 survey found that 18% of respondents supported sending U.S. troops "if a situation like Vietnam were to develop in another part of the world." But when the question was reworded to mention sending troops to "stop a communist takeover"—"communist takeover" being a loaded phrase—favorable responses rose to 33% (Schuman & Presser 1981:285).

Answers can also be biased by more subtle problems in phrasing that make certain responses more or less attractive to particular groups. To minimize biased responses, researchers have to test reactions to the phrasing of a question. For example, Mirowsky and Ross (personal email, 2009) wanted to ask people, "Do you feel

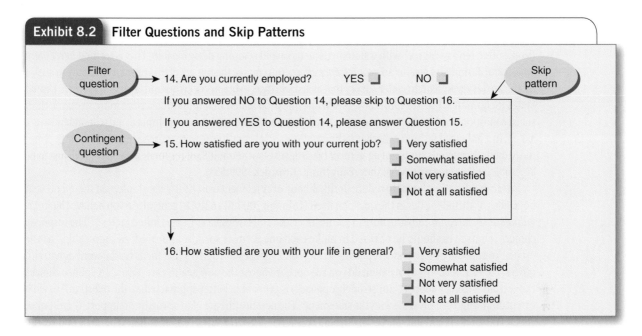

Exhibit 8.2 Filter Questions and Skip Patterns

Filter question → 14. Are you currently employed?   YES ☐   NO ☐

If you answered NO to Question 14, please skip to Question 16.

If you answered YES to Question 14, please answer Question 15.

Contingent question → 15. How satisfied are you with your current job?   ☐ Very satisfied
☐ Somewhat satisfied
☐ Not very satisfied
☐ Not at all satisfied

Skip pattern

16. How satisfied are you with your life in general?   ☐ Very satisfied
☐ Somewhat satisfied
☐ Not very satisfied
☐ Not at all satisfied

fit?" However, when the Survey Research Lab tried out this question with a small sample, people did not seem to understand that they were being asked about their level of energy and general feelings of fitness; they just focused on whether they had some type of health problem. It seemed that people had a biased concept of health as involving only problems rather than including the concept of positive health. As a result, Mirowsky and Ross rephrased the question to be more explicit: "Do you feel physically fit?"

Responses can also be biased when response alternatives do not reflect the full range of possible sentiment on an issue. When people pick a response choice, they seem to be influenced by where they are placing themselves relative to the other response choices. For example, the Detroit Area Study (Turner & Martin 1984:252) asked the following question: "People feel differently about making changes in the way our country is run. In order to keep America great, which of these statements do you think is best?" When the only response choices were "We should be very cautious of making changes" and "We should be free to make changes," only 37% said that we should be free to make changes. However, when a response choice was added that suggested we should "constantly" make changes, 24% picked that response and another 32% chose the "free to make changes" response, for a total of 56% who seemed open to making changes in the way our country is run (Turner & Martin 1984:252). Including the more extreme positive alternative ("constantly" make changes) made the less extreme positive alternative more attractive.

If the response alternatives for a question fall on a continuum from positive to negative, the number of positive and negative categories should be balanced so that one end of the continuum doesn't seem more attractive than the other (Dillman 2000:57–58). If you ask respondents, "How satisfied are you with the intramural sports program here?" and include "completely satisfied" as the most positive possible response, then "completely dissatisfied" should be included as the most negative possible response. This is called a *bipolar* scale.

Of course, the advice to minimize the risk of bias means nothing to those who conduct surveys to elicit bias. This is the goal of *push polling*, a technique that has been used in some political campaigns. In a push poll, the pollsters for a candidate call potential voters and ask them a series of questions that convey negative information about the opposing candidate. It's really not a survey at all—just a propaganda effort—but it casts reputable survey research (and ethical political polling firms) in a bad light (Connolly & Manning 2001).

Avoid Making Either Disagreement or Agreement Disagreeable

People often tend to "agree" with a statement just to avoid seeming disagreeable. This is termed *agreement bias*, *social desirability bias*, or an *acquiescence effect*. You can see the impact of this human tendency in a 1974 Michigan Survey Research Center survey that asked who was to blame for crime and lawlessness in the United States (Schuman & Presser 1981:208). When one question stated that individuals were more to blame than social conditions, 60% of the respondents agreed. But when the question was rephrased and respondents were asked, in a balanced fashion, whether individuals or social conditions were more to blame, only 46% chose individuals. Numerous studies of agreement bias suggest that about 10% of respondents will "agree" just to be agreeable, without regard to what they really think (Krosnick 1999:553).

You can take several steps to reduce the likelihood of agreement bias. As a general rule, you should present both sides of attitude scales in the question itself (Dillman 2000:61–62): "In general, do you believe that *individuals* or *social conditions* are more to blame for crime and lawlessness in the United States?" The response choices themselves should be phrased to make each one seem as socially approved, as "agreeable," as the others. You should also consider replacing a range of response alternatives that focus on the word *agree* with others. For example, "To what extent do you support or oppose the new health care plan?" (response choices range from "strongly support" to "strongly oppose") is probably a better approach than the question "To what extent do you agree or disagree with the statement: 'The new health care plan is worthy of support'?" (response choices range from "strongly agree" to "strongly disagree"). For the same reason, simple true/false and yes/no response choices should be avoided (Schaeffer & Presser 2003:80–81).

You may also gain a more realistic assessment of respondents' sentiment by adding to a question a counterargument in favor of one side to balance an argument in favor of the other side. Thus, don't just ask in an employee survey whether employees should be required to join the union; instead, ask whether employees should be required to join the union or be able to make their own decision about joining. In one survey, 10% more respondents said they favored mandatory union membership when the counterargument was left out than when it was included. It is reassuring to know, however, that this approach does not change the distribution of answers to questions about which people have very strong beliefs (Schuman & Presser 1981:186).

When an illegal or socially disapproved behavior or attitude is the focus, we have to be concerned that some respondents will be reluctant to agree that they have ever done or thought such a thing. In this situation, the goal is to write a question and response choices that make agreement seem more acceptable. For example, Dillman (2000:75) suggests that we ask, "Have you ever taken anything from a store without paying for it?" rather than "Have you ever shoplifted something from a store?" Asking about a variety of behaviors or attitudes that range from socially acceptable to socially unacceptable will also soften the impact of agreeing with those that are socially unacceptable.

Minimize Fence-Sitting and Floating

Two related problems in writing survey questions also stem from people's desire to choose an acceptable answer. There is no uniformly correct solution to these problems; researchers have to weigh the alternatives in light of the concept to be measured and whatever they know about the respondents.

Fence-sitters Survey respondents who see themselves as being neutral on an issue and choose a middle (neutral) response that is offered.

Fence-sitters, people who see themselves as being neutral, may skew the results if you force them to choose between opposites. In most cases, about 10% to 20% of such respondents—those who do not have strong feelings on an issue—will choose an explicit middle, neutral alternative (Schuman & Presser 1981:161–178).

Having an explicit neutral response option is generally a good idea: It identifies fence-sitters and tends to increase measurement reliability (Schaeffer & Presser 2003:78).

Even more people can be termed **floaters**: respondents who choose a substantive answer when they really don't know or have no opinion. A third of the public will provide an opinion on a proposed law that they know nothing about if they are asked for their opinion in a closed-ended survey question that does not include "Don't know" as an explicit response choice. However, 90% of these persons will select the "Don't know" response if they are explicitly given that option. On average, offering an explicit response option increases the "Don't know" responses by about a fifth (Schuman & Presser 1981:113–160).

> **Floaters** Survey respondents who provide an opinion on a topic in response to a closed-ended question that does not include a "Don't know" option, but who will choose "Don't know" if it is available.

Exhibit 8.3 depicts the results of one study that tested the effect of giving respondents an explicit "No opinion" option to the question "Are government leaders smart?" Notice how many more people chose "No opinion" when they were given that choice than when their only explicit options were "Smart" and "Not smart."

In spite of the prevalence of floating, people often have an opinion but are reluctant to express it. In fact, most political pollsters use **forced-choice questions** without a "Don't know" option. Just after President Clinton's victory, Frank Newport, editor in chief of the Gallup poll, defended pollsters' efforts to get all prospective voters to declare a preferred candidate:

> **Forced-choice questions** Closed-ended survey questions that do not include "Don't know" as an explicit response choice.

> It would not be very instructive for pollsters . . . to allow large numbers of voters to claim they are undecided all through the election season. We would miss the dynamics of change, we would be unable to tell how well candidates were doing in response to events, and publicly released polls would be out of synchronization with private, campaign polls. (Newport 1992:A28)

Because there are so many floaters in the typical survey sample, the decision to include an explicit "Don't know" option for a question is important. Unfortunately, the inclusion of an explicit "Don't know" response choice leads some people who do have a preference to take the easy way out—to satisfice—and choose "Don't know." This is particularly true in surveys of less-educated populations—except for questions that are really impossible to decipher, to which more educated persons are likely to say they "don't know" (Schuman & Presser 1981:113–146). As a result, survey experts now recommend that questions not include "Don't know" or "No opinion" options (Krosnick 1999:558; Schaeffer & Presser 2003:80). Adding an open-ended question in which respondents are asked to discuss their opinions can help identify respondents who are floaters (Smith, 1984).

Researchers who use in-person or telephone interviews (rather than self-administered questionnaires) may get around the dilemma somewhat by

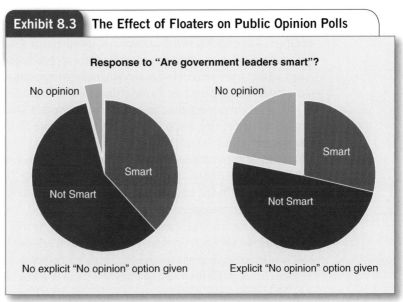

Exhibit 8.3 The Effect of Floaters on Public Opinion Polls

Response to "Are government leaders smart"?

No explicit "No opinion" option given

Explicit "No opinion" option given

reading the response choices without a middle or "Don't know" alternative but recording a noncommittal response if it is offered. Mirowsky and Ross's (2001) questionnaire for their phone survey about education and health included the following example (responses in ALL CAPS were not read):

My misfortunes are the result of mistakes I have made.

(Do you . . .)

1. <1> Strongly agree,
2. <2> Agree,
3. <3> Disagree, or
4. <4> Strongly disagree?
5. <7> NO CODED RESPONSE APPLICABLE
6. <8> DON'T KNOW
7. <9> REFUSED

Maximize the Utility of Response Categories

Questions with fixed response choices must provide one and only one possible response for everyone who is asked the question—that is, the response choices must be exhaustive and mutually exclusive. Ranges of ages, incomes, years of schooling, and so forth should not overlap and should provide a response option for all respondents.

There are two exceptions to this principle: (1) Filter questions may tell some respondents to skip over a question (the response choices do not have to be exhaustive), and (2) respondents may be asked to "check all that apply" (the response choices are not mutually exclusive). Even these exceptions should be kept to a minimum. Respondents to a self-administered questionnaire should not have to do a lot of skipping around, or they may lose interest in answering carefully all the applicable questions. Some survey respondents react to a "check all that apply" request by just checking enough responses so that they feel they have "done enough" for that question and then ignoring the rest of the choices (Dillman 2000:63).

Vagueness in the response choices is also to be avoided. Questions about thoughts and feelings will be more reliable if they refer to specific times or events (Turner & Martin 1984:300). Usually a question like "On how many days did you read the newspaper in the last week?" produces more reliable answers than one like "How often do you read the newspaper? (frequently, sometimes, never)." In their survey, Mirowsky and Ross (2001:2) sensibly asked the question "Do you currently smoke 7 or more cigarettes a week?" rather than the vaguer question "Do you smoke?" Of course, being specific doesn't help if you end up making unreasonable demands of your respondents' memories. One survey asked, "During the past 12 months, about how many times did you see or talk to a medical doctor?" According to their written health records, respondents forgot 60% of their doctor visits (Goleman 1993b:C11). So unless your focus is on major events that are unlikely to have been forgotten, limit questions about specific past experiences to the past month.

Sometimes, problems with response choices can be corrected by adding questions. For example, if you ask, "How many years of schooling have you completed?" someone who dropped out of high school but completed the requirements for a General Equivalency Diploma (GED) might not be sure how to respond.

By asking a second question, "What is the highest degree you have received?" you can provide the correct alternative for those with a GED as well as for those who graduated from high school.

Adding questions may also improve memory about specific past events. Imagine the problem you might have answering the question, "How often did you receive help from classmates while preparing for exams or completing assignments during the last month? (very often, somewhat often, occasionally, rarely, or never)." Now, imagine a series of questions that asks you to identify the exams and assignments you had in the past month and, for each one, inquires whether you received each of several types of help from classmates: study suggestions, study sessions, related examples, general encouragement, and so on. The more specific focus on particular exams and assignments should result in more complete recall (Dykema & Schaeffer 2000).

How many response categories are desirable? Five categories work well for unipolar ratings, while seven will capture most variation on bipolar ratings (Krosnick 2006; Schaeffer & Presser 2003:78–79). Responses are more reliable when these categories are labeled rather than identified only by numbers (Krosnick 1999:544; Schaeffer & Presser 2003:78). Exhibit 8.4 shows these alternatives, based on a question and response choices in the Mirowsky and Ross (2001) questionnaire.

Exhibit 8.4   **Labeled Unipolar, Unlabeled Unipolar, and Bipolar Response Options**

Original

>Q72a< How free do you feel to disagree with the person who
supervises your work? Are you . . .

<1> Not at all free,
<2> Somewhat free,
<3> Largely but not completely free, or
<4> Completely free to disagree?
<7> NO CODED RESPONSE APPLICABLE
<8> DON'T KNOW
<9> REFUSED

Labeled unipolar version

How comfortable do you feel disagreeing with the person who supervises your work? (Please circle one number to indicate your response.)

1. Extremely comfortable
2. Very comfortable
3. Quite comfortable
4. Somewhat comfortable
5. Not at all comfortable

Labeled bipolar version

Do you feel comfortable or uncomfortable disagreeing with the person who supervises your work? (Please circle one number to indicate your response.)

1. Very comfortable
2. Mostly comfortable
3. Slightly comfortable
4. Feel neither comfortable nor uncomfortable

(Continued)

Exhibit 8.4 **(Continued)**

5. Slightly uncomfortable
6. Mostly uncomfortable
7. Very uncomfortable

Unlabeled unipolar version

Please circle a number from 1 to 10 to indicate how comfortable you feel disagreeing with the person who supervises your work. 1 means "not at all comfortable" and 10 means "extremely comfortable."

How comfortable do you feel disagreeing with the person who supervises your work?

Not at all									Extremely
1	2	3	4	5	6	7	8	9	10

Combining Questions in Indexes

Idiosyncratic variation Variation in responses to questions that is caused by individuals' reactions to particular words or ideas in the question instead of by variation in the concept that the question is intended to measure.

Writing single questions that yield usable answers is always a challenge. Simple though they may seem, single questions are prone to error due to **idiosyncratic variation**, which occurs when individuals' responses vary because of their reactions to particular words or ideas in the question. Differences in respondents' backgrounds, knowledge, and beliefs almost guarantee that some will understand the same question differently.

In some cases, the effect of idiosyncratic variation can be dramatic. For example, when people were asked in a survey whether they would "forbid" public speeches against democracy, 54% agreed. When the question was whether they would "not allow" public speeches against democracy, 75% agreed (Turner & Martin 1984:chap. 5). Respondents are less likely to respond affirmatively to the question, "Did you see a broken headlight?" than they are to the question, "Did you see *the* broken headlight?" (Turner & Martin 1984:chap. 9).

The guidelines in this chapter for writing clear questions should help reduce idiosyncratic variation caused by different interpretations of questions. But, the best option is often to develop multiple questions about a concept and then to average the responses to those questions in a composite measure termed an *index* or *scale*.

The idea is that idiosyncratic variation in response to particular questions will average out, so that the main influence on the combined measure will be the concept upon which all the questions focus. The index can be considered a more complete measure of the concept than can any one of the component questions.

Creating an index is not just a matter of writing a few questions that seem to focus on a concept. Questions that seem to you to measure a common concept might seem to respondents to concern several different issues. The only way to know that a given set of questions does, in fact, form an index is to administer the questions to people like those you plan to study. If a common concept is being measured, people's responses to the different questions should display some consistency. In other words, responses to the different questions should be correlated. Exhibit 8.5 illustrates an index in which responses to the items are correlated; the substantial area of overlap indicates that the questions are measuring a common concept.

thoughtful—and potentially less valid. Without a consistent approach, information obtained from different respondents will not be comparable—less reliable and less valid.

Balancing Rapport and Control

Adherence to some basic guidelines for interacting with respondents can help interviewers maintain an appropriate balance between personalization and standardization:

- Project a professional image in the interview: that of someone who is sympathetic to the respondent but nonetheless has a job to do.

- Establish rapport at the outset by explaining what the interview is about and how it will work and by reading the consent form. Ask the respondent if he or she has any questions or concerns, and respond to these honestly and fully. Emphasize that everything the respondent says is confidential.

- During the interview, ask questions from a distance that is close but not intimate. Stay focused on the respondent and make sure that your posture conveys interest. Maintain eye contact, respond with appropriate facial expressions, and speak in a conversational tone of voice.

- Be sure to maintain a consistent approach; deliver each question as written and in the same tone of voice. Listen empathetically, but avoid self-expression or loaded reactions.

- Repeat questions if the respondent is confused. Use nondirective probes—such as "Can you tell me more about that?"—for open-ended questions.

As with phone interviewing, computers can be used to increase control of the in-person interview. In a **computer-assisted personal interviewing (CAPI)** project, interviewers carry a laptop computer that is programmed to display the interview questions and to process the responses that the interviewer types in, as well as to check that these responses fall within allowed ranges (Tourangeau 2004:790–791). Interviewers seem to like CAPI, and the data obtained are comparable in quality to data obtained in a noncomputerized interview (Shepherd et al. 1996). A CAPI approach also makes it easier for the researcher to develop skip patterns and experiment with different types of questions for different respondents without increasing the risk of interviewer mistakes (Couper et al. 1998).

> **Computer-assisted personal interview (CAPI)** A personal interview in which the laptop computer is used to display interview questions and to process responses that the interviewer types in, as well as to check that these responses fall within allowed ranges.

The presence of an interviewer may make it more difficult for respondents to give honest answers to questions about socially undesirable behaviors such as drug use, sexual activity, and not voting (Schaeffer & Presser 2003:75). CAPI is valued for this reason, since respondents can enter their answers directly in the laptop without the interviewer knowing what their response is. Alternatively, interviewers can simply hand respondents a separate self-administered questionnaire containing the more sensitive questions. After answering these questions, the respondent seals the separate questionnaire in an envelope so that the interviewer does not know the answers. When this approach was used for the GSS questions about sexual activity, about 21% of men and 13% of women who were married or had been married admitted to having cheated on a spouse ("Survey on Adultery" 1993:A20). The degree of rapport becomes a special challenge when survey questions concern issues related to such demographic characteristics as race or gender (Groves 1989). If the interviewer and respondent are similar on the characteristics at issue, the responses to these questions may differ from those that would be given if the interviewer and respondent differ on these characteristics. For example, a white respondent may not disclose feelings of racial prejudice to a black interviewer that he would admit to a white interviewer.

Although in-person interview procedures are typically designed with the expectation that the interview will involve only the interviewer and the respondent, one or more other household members are often within earshot. In a mental health survey in Los Angeles, for example, almost half the interviews were conducted in the presence of another person (Pollner & Adams 1994). It is reasonable to worry that this third-party presence will influence responses about sensitive subjects—even more so because the likelihood of a third party being present may correspond with other subject characteristics. For example, in the Los Angeles survey, another person was present in 36% of the interviews with Anglos, in 47% of the interviews with African Americans, and in 59% of the interviews with Hispanics. However, there is no consistent evidence that respondents change their answers because of the presence of another person. Analysis of this problem with the Los Angeles study found very little difference in reports of mental illness symptoms between respondents who were alone and those who were in the presence of others.

Maximizing Response to Interviews

Even if the right balance has been struck between maintaining control over interviews and achieving good rapport with respondents, in-person interviews still can be problematic. Because of the difficulty of finding all the members of a sample, response rates may suffer. Exhibit 8.15 displays the breakdown of nonrespondents to the 1990 GSS. Of the total original sample of 2,165, only 86% (1,857) were determined to be valid selections of dwelling units with potentially eligible respondents. Among these potentially eligible respondents, the response rate was 74%. The GSS is a well-designed survey using carefully trained and supervised interviewers, so this response rate indicates the difficulty of securing respondents from a sample of the general population even when everything is done "by the book."

Several factors affect the response rate in interview studies. Contact rates tend to be lower in central cities, in part, because of difficulties in finding people at home and gaining access to high-rise apartments and, in

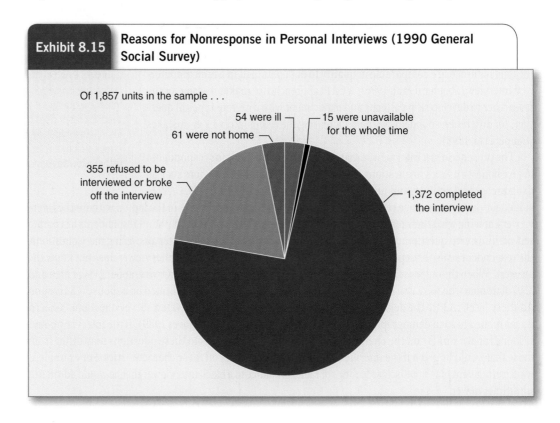

Exhibit 8.15 **Reasons for Nonresponse in Personal Interviews (1990 General Social Survey)**

Of 1,857 units in the sample . . .

54 were ill — 15 were unavailable for the whole time

61 were not home

355 refused to be interviewed or broke off the interview

1,372 completed the interview

part, because of interviewer's reluctance to visit some areas at night, when people are more likely to be home (Fowler 1988:45–60). Single-person households also are more difficult to reach, whereas households with young children or elderly adults tend to be easier to contact (Groves & Couper 1998:119–154).

Refusal rates vary with some respondents' characteristics. People with less education participate somewhat less in surveys of political issues (perhaps because they are less aware of current political issues). Less education is also associated with higher rates of "Don't know" responses (Groves 1989). High-income persons tend to participate less in surveys about income and economic behavior (perhaps because they are suspicious about why others want to know about their situation). Unusual strains and disillusionment in a society can also undermine the general credibility of research efforts and the ability of interviewers to achieve an acceptable response rate. These problems can be lessened with an advance letter introducing the survey project and by multiple contact attempts throughout the day and evening, but they cannot entirely be avoided (Fowler 1988:52–53; Groves & Couper 1998). Encouraging interviewers to tailor their response when potential respondents express reservations about participating during the initial conversation can also lead to lower rates of refusal: Making small talk to increase rapport and delaying asking a potential respondent to participate may reduce the likelihood of a refusal after someone first expresses uncertainty about participating (Maynard, Freese & Schaeffer 2010:810).

Web Surveys

Web-based surveys have become an increasingly useful survey method for two reasons: growth in the fraction of the population using the Internet and technological advances that make web-based survey design relatively easy. Many specific populations have very high rates of Internet use, so a web-based survey can be a good option for groups such as professionals, middle-class communities, members of organizations and, of course, college students. Due to the Internet's global reach, web-based surveys also make it possible to conduct large, international surveys. However, coverage remains a problem with many populations. About 30% of American households are not connected to the Internet (U.S. Bureau of the Census 2010e), so it is not yet possible to survey directly a representative sample of the U.S. population on the web—and given a plateau in the rate of Internet connections, this coverage problem may persist for the near future (Couper & Miller 2008:832). Rates of Internet usage are much lower in other parts of the world, with a worldwide average of 28.7% and rates as low as 10.9% in Africa and 21.5% in all of Asia (Internetworldstats.com 2011). Since households without Internet access also tend to be older, poorer, and less educated than those who are connected, web-based surveys of the general population can result in seriously biased estimates (Tourangeau 2004:792–793).

> **Web survey** A survey that is accessed and responded to on the World Wide Web.

There are several different approaches to conducting web-based surveys, each with unique advantages and disadvantages and somewhat different effects on the coverage problem. Many web-based surveys begin with an e-mail message to potential respondents that contains a direct "hotlink" to the survey website (Gaiser & Schreiner 2009:70). This approach is particularly useful when a defined population with known e-mail addresses is to be surveyed. The researcher can then send e-mail invitations to a representative sample without difficulty. To ensure that the appropriate people respond to a web-based survey, researchers may require that respondents enter a PIN (personal identification number) to gain access to the web survey (Dillman 2000:378). However, lists of unique e-mail addresses for the members of defined populations generally do not exist outside of organizational settings. Many people have more than one e-mail address and often there is no apparent link between an e-mail address and the name or location of the person to whom it is assigned. As a result, there is no available method for drawing a random sample of e-mail addresses for people from any general population, even if the focus is only on those with Internet access (Dillman 2007:449).

because they actually have different opinions. For example, when equivalent samples were asked by phone or mail, "Is the gasoline shortage real or artificial?" many more phone respondents than mail respondents answered that it was "very real" (Peterson 2000:24). Respondents to phone survey questions tend to endorse more extreme responses to scalar questions (which range from more to less) than respondents to mail or web surveys (Dillman 2007:456–457). Responses may also differ between questions—one-third of the questions in one survey—when asked in web-based and phone survey modes, even with comparable samples (Rookey et al. 2008:974). When responses differ by survey mode, there is often no way to know which responses are more accurate, although it appears that web-based surveys are likely to result in more admissions of socially undesirable experiences (Kreuter et al. 2008; Peterson 2000:24). Use of the same question structures, response choices, and skip instructions across modes substantially reduces the likelihood of mode effects, as does using a small number of response choices for each question (Dillman 2000:232–240; Dillman & Christian 2005), but web-based survey researchers are only beginning to identify the effect of visual appearance on the response to questions (Dillman 2007:472–487).

A Comparison of Survey Designs

Which survey design should be used when? Group-administered surveys are similar, in most respects, to mailed surveys, except that they require the unusual circumstance of having access to the sample in a group setting. We therefore don't need to consider this survey design by itself; what applies to mailed surveys applies to group-administered survey designs, with the exception of sampling issues. The features of mixed-mode surveys depend on the survey types that are being combined. Thus, we can focus our comparison on the four survey designs that involve the use of a questionnaire with individuals sampled from a larger population: (1) mailed surveys, (2) phone surveys, (3) in-person surveys, and (4) electronic surveys. Exhibit 8.17 summarizes their strong and weak points.

The most important consideration in comparing the advantages and disadvantages of the four methods is the likely response rate they will generate. Mailed surveys must be considered the least preferred survey design from a sampling standpoint, although declining rates of response to phone surveys are changing this comparison.

Contracting with an established survey research organization for a phone survey is often the best alternative to a mailed survey. The persistent follow-up attempts that are necessary to secure an adequate response rate are much easier over the phone than in person. But, as explained earlier, the process requires an increasing number of callbacks to many households and rates of response have been declining. Current federal law prohibits automated dialing of cell phone numbers, so it is very costly to include the increasing growing number of cell phone-only individuals in a phone survey.

In-person surveys are preferable in terms of the possible length and complexity of the questionnaire itself, as well as with respect to the researcher's ability to monitor conditions while the questionnaire is completed. Mailed surveys often are preferable for asking sensitive questions, although this problem can be lessened in an interview by giving respondents a separate sheet to fill out or a laptop in which to enter their answers. Although interviewers may themselves distort results, either by changing the wording of questions or by failing to record answers properly, survey research organizations can reduce this risk through careful interviewer training and monitoring. Some survey supervisors will have interviews tape-recorded so that they can review the dialogue between interviewer and respondents and provide feedback to the interviewers to help improve their performance. Some survey organizations have also switched to having in-person interviews completed entirely by the respondent on a laptop as they listen to prerecorded questions.

Exhibit 8.17	Advantages and Disadvantages of the Four Survey Designs

Characteristics of Design	Mail Survey	Phone Survey	In-Person Survey	Web Survey
Representative sample				
Opportunity for inclusion is known				
For completely listed populations	High	High	High	Medium
For incompletely listed populations	Medium	Medium	High	Low
Selection within sampling units is controlled (e.g., specific family members must respond)	Medium	High	High	Low
Respondents are likely to be located				
If samples are heterogeneous	Medium	Medium	High	Low
If samples are homogeneous and specialized	High	High	High	High
Questionnaire construction and question design				
Allowable length of questionnaire	Medium	Medium	High	Medium
Ability to include				
Complex questions	Medium	Low	High	High
Open questions	Low	High	High	Medium
Screening questions	Low	High	High	High
Tedious, boring questions	Low	High	High	Low
Ability to control question sequence	Low	High	High	High
Ability to ensure questionnaire completion	Medium	High	High	Low
Distortion of answers				
Odds of avoiding social desirability bias	High	Medium	Low	High
Odds of avoiding interviewer distortion	High	Medium	Low	High
Odds of avoiding contamination by others	Medium	High	Medium	Medium
Administrative goals				
Odds of meeting personnel requirements	High	High	Low	Medium
Odds of implementing quickly	Low	High	Low	High
Odds of keeping costs low	High	Medium	Low	High

A phone survey limits the length and complexity of the questionnaire but offers the possibility of very carefully monitoring interviewers (Dillman 1978; Fowler 1988:61–73):

> Supervisors in [one organization's] Telephone Centers work closely with the interviewers, monitor their work, and maintain records of their performance in relation to the time schedule, the quality of their work, and help detect and correct any mistakes in completed interviews prior to data reduction and processing. (J. E. Blair, personal communication to C. E. Ross, April 10, 1989)

However, people interviewed by phone tend to get less interested in the survey than those interviewed in person, they tend to satisfice more—apparently in a desire to complete the survey more quickly, and they tend to be less trusting of the survey motives (Holbrook et al. 2003).

The advantages and disadvantages of electronic surveys must be weighed in light of the population that is to be surveyed and capabilities at the time that the survey is to be conducted. At this time, too many people lack Internet connections for survey researchers to use the Internet to survey the general population.

These various points about the different survey designs lead to two general conclusions. First, in-person interviews are the strongest design and generally preferable when sufficient resources and a trained interview staff are available; telephone surveys have many of the advantages of in-person interviews at much less cost, but response rates are an increasing problem. Second, the "best" survey design for any particular study will be determined by the study's unique features and goals rather than by any absolute standard of what the best survey design is.

▣ Ethical Issues in Survey Research

Survey research usually poses fewer ethical dilemmas than do experimental or field research designs. Potential respondents to a survey can easily decline to participate, and a cover letter or introductory statement that identifies the sponsors of, and motivations for, the survey gives them the information required to make this decision. The methods of data collection are quite obvious in a survey, so little is concealed from the respondents. Only in group-administered surveys might the respondents be, in effect, a captive audience (probably of students or employees), and so these designs require special attention to ensure that participation is truly voluntary. (Those who do not wish to participate may be told they can just hand in a blank form.)

Current federal regulations to protect human subjects allow survey research to be exempted from formal review unless respondents can be identified and disclosure of their responses could place them at risk. Specifically, the Code of Federal Regulations (2009) states,

> (2) Research involving the use of educational tests (cognitive, diagnostic, aptitude, achievement), survey procedures, interview procedures or observation of public behavior, unless:

> (i) information obtained is recorded in such a manner that human subjects can be identified, directly or through identifiers linked to the subjects; and (ii) any disclosure of the human subjects' responses outside the research could reasonably place the subjects at risk of criminal or civil liability or be damaging to the subjects' financial standing, employability, or reputation. (p. 46.101(b)2)

Confidentiality is most often the primary focus of ethical concern in survey research. Many surveys include some essential questions that might, in some way, prove damaging to the subjects if their answers were disclosed. To prevent any possibility of harm to subjects due to disclosure of such information, the researcher must preserve subject confidentiality. Nobody but research personnel should have access to information that could be used to link respondents to their responses, and even that access should be limited to what is necessary for specific research purposes. Only numbers should be used to identify respondents on their questionnaires, and the researcher should keep the names that correspond to these numbers in a safe, private location, unavailable to staff and others who might otherwise come across them. Follow-up mailings or contact attempts that require linking the ID numbers with names and addresses should be carried out by trustworthy assistants under close supervision. For electronic surveys, encryption technology should be used to make information provided over the Internet secure from unauthorized persons.

> **Confidentiality** Provided by research in which identifying information that could be used to link respondents to their responses is available only to designated research personnel for specific research needs.

Mirowsky and Ross (1999) focused special attention on the maintenance of respondent confidentiality because of the longitudinal nature of the ASOC survey. To recontact people for interviews after the first wave in 1995, they had to keep files with identifying information. Here is how John Mirowsky (1999) described their procedures in their application for funding of the third wave (the 2001 survey):

> In order to follow respondents, their phone numbers and first names were recorded. They also were asked to give the phone number and first name of someone outside the household who would know how to contact them if they moved. Multiple attempts were made to contact each respondent at 18 months after the initial interview, as described in the section on tracking in the progress report. Those contacted were asked again about someone who will know how to contact them if they move. If a respondent is not reached at the old number, the outside contact is called and asked to help locate the person.
>
> The privacy of respondents is maintained by having separate groups of people collect data and analyze it. The staff of the Survey Research Laboratory of the University of Illinois at Chicago collects the data. Computer-aided interviewing software gave each respondent a sequence number in two separate data sets. One data set consists solely of the first names, phone numbers, and contacts that could identify a respondent. That data set is secured by the Survey Research Laboratory. It is only available to laboratory staff when it is needed for follow-up. It is not available at any time to the principal investigator, his associates, or anyone else not on the survey laboratory interviewing staff. The second data set contains disposition codes and responses to questions. It does not contain any information that could identify individual respondents. The second data set is available to the principal investigator and his associates for analysis. (p. 41)

Not many surveys can provide true **anonymity**, so that no identifying information is ever recorded to link respondents with their responses. The main problem with anonymous surveys is that they preclude follow-up attempts to encourage participation by initial nonrespondents, and they prevent panel designs, which measure change through repeated surveys of the same individuals. In-person surveys rarely can be anonymous because an interviewer must, in almost all cases, know the name and address of the interviewee. However, phone surveys that are meant only to sample opinion at one point in time, as in political polls, can safely be completely anonymous. When no future follow-up is desired, group-administered surveys also can be anonymous. To provide anonymity in a mail survey, the researcher should omit identifying codes from the questionnaire but could include a self-addressed, stamped postcard so the respondent can notify the researcher that the questionnaire has been returned without creating any linkage to the questionnaire itself (Mangione 1995:69).

> **Anonymity** Provided by research in which no identifying information is recorded that could be used to link respondents to their responses.

🔲 Conclusions

Survey research is an exceptionally efficient and productive method for investigating a wide array of social research questions. Mirowsky and Ross (2003) and Mirowsky (1999) were able to survey representative samples of Americans and older Americans and follow them for 6 years. These data allowed Mirowsky and Ross to investigate the relationships among education, social status, and health and how these relationships are changing.

In addition to the potential benefits for social science, considerations of time and expense frequently make a survey the preferred data collection method. One or more of the six survey designs reviewed in this chapter (including mixed mode) can be applied to almost any research question. It is no wonder that surveys have become the most popular research method in sociology and that they frequently inform discussion and planning about important social and political questions. As use of the Internet increases, survey research should become even more efficient and popular.

The relative ease of conducting at least some types of survey research leads many people to imagine that no particular training or systematic procedures are required. Nothing could be further from the truth. But as a result of this widespread misconception, you will encounter a great many nearly worthless survey results. You must be prepared to examine carefully the procedures used in any survey before accepting its findings as credible. And if you decide to conduct a survey, you must be prepared to invest the time and effort that proper procedures require.

Key Terms

Anonymity   273
Behavior coding   247
Cognitive interview   246
Computer-assisted personal interview
   (CAPI)   265
Computer-assisted telephone interview
   (CATI)   263
Confidentiality   272
Context effects   250
Contingent question   236
Cover letter   255
Double-barreled question   236

Double negative   235
Electronic survey   254
Fence-sitters   238
Filter question   236
Floaters   239
Forced-choice questions   239
Group-administered survey   258
Idiosyncratic variation   242
In-person interview   264
Interactive voice response (IVR)   264
Interpretive questions   247
Interview schedule   244

Mailed survey   254
Mixed-mode survey   269
Omnibus survey   232
Part-whole
   question effects   250
Phone survey   258
Questionnaire   244
Skip pattern   236
Split-ballot design   232
Survey pretest   246
Survey research   230
Web survey   267

Highlights

- Surveys are the most popular form of social research because of their versatility, efficiency, and generalizability. Many survey data sets, such as the GSS, are available for social scientists to use in teaching and research.

- Omnibus surveys cover a range of topics of interest and generate data useful to multiple sponsors.

- Survey designs must minimize the risk of errors of observation (measurement error) and errors of nonobservation (errors due to inadequate coverage, sampling error, and nonresponse). The likelihood of both types of error varies with the survey goals. For example, political polling can produce inconsistent results because of rapid changes in popular sentiment.

- Social exchange theory asserts that behavior is motivated by the return expected to the individual for the behavior. Survey designs must maximize the social rewards, minimize the costs of participating, and establish trust that the rewards will outweigh the costs.

- A survey questionnaire or interview schedule should be designed as an integrated whole, with each question and section serving some clear purpose and complementing the others.

- Questions must be worded carefully to avoid confusing respondents, encouraging a less-than-honest response, or triggering biases. Inclusion of "Don't know" choices and neutral responses may help, but the presence of such options also affects the distribution of answers. Open-ended questions can be used to determine the meaning that respondents attach to their answers. Answers to any survey questions may be affected by the questions that precede them in a questionnaire or interview schedule.

- Questions can be tested and improved through review by experts, focus group discussions, cognitive interviews, behavior coding, and pilot testing. Every questionnaire and interview schedule should be pretested on a small sample that is like the sample to be surveyed.

- Interpretive questions should be used in questionnaires to help clarify the meaning of responses to critical questions.

- The cover letter for a mailed questionnaire should be credible, personalized, interesting, and responsible.

- Response rates in mailed surveys are typically well below 70% unless multiple mailings are made to nonrespondents and the questionnaire and cover letter are attractive, interesting, and carefully planned. Response rates for group-administered surveys are usually much higher.

- Phone interviews using random digit dialing allow fast turnaround and efficient sampling. Multiple callbacks are often required, and the rate of nonresponse to phone interviews is rising. Phone interviews should be limited in length to about 30 to 45 minutes.

- In-person interviews have several advantages over other types of surveys: They allow longer and more complex interview schedules, monitoring of the conditions when the questions are answered, probing for respondents' understanding of the questions, and high response rates. However, the interviewer must balance the need to establish rapport with the respondent with the importance of maintaining control over the delivery of the interview questions.

- Electronic surveys may be e-mailed or posted on the web. Interactive voice response systems using the telephone are another option. At this time, use of the Internet is not sufficiently widespread to allow web surveys of the general population, but these approaches can be fast and efficient for populations with high rates of computer use.

- Mixed-mode surveys allow the strengths of one survey design to compensate for the weaknesses of another. However, questions and procedures must be designed carefully to reduce the possibility that responses to the same question will vary as a result of the mode of delivery.

- In deciding which survey design to use, researchers must take into account the unique features and goals of the study. In general, in-person interviews are the strongest, but most expensive, survey design.

- Most survey research poses few ethical problems because respondents are able to decline to participate—an option that should be stated clearly in the cover letter or introductory statement. Special care must be taken when questionnaires are administered in group settings (to "captive audiences") and when sensitive personal questions are to be asked; subject confidentiality should always be preserved.

STUDENT STUDY SITE

To assist in completing the web exercises, please access the study site at **www.sagepub.com/schuttisw7e**, where you will find the web exercises with accompanying links. You'll find other useful study materials such as self-quizzes and e-flashcards for each chapter, along with a group of carefully selected articles from research journals that illustrate the major concepts and techniques presented in the book.

Discussion Questions

1. Response rates to phone surveys are declining, even as phone usage increases. Part of the problem is that lists of cell phone numbers are not available and wireless service providers do not allow outside access to their networks. Cell phone users may also have to pay for incoming calls. Do you think regulations should be passed to increase the ability of survey researchers to include cell phones in their random digit dialing surveys? How would you feel about receiving survey calls on your cell phone? What problems might result from "improving" phone survey capabilities in this way?

2. In-person interviews have for many years been the "gold standard" in survey research, because the presence of an interviewer increases the response rate, allows better rapport with the interviewee, facilitates clarification of questions and instructions, and provides feedback about the interviewee's situation. However, researchers who design in-person interviewing projects are now making increasing use of technology to ensure consistent questioning of respondents and to provide greater privacy while answering questions. But, having a respondent answer questions on a laptop while the interviewer waits is a very different social process than actually asking the questions verbally. Which approach would you favor in survey research? What tradeoffs can you suggest there might be in terms of quality of information collected, rapport building, and interviewee satisfaction?

3. Each of the following questions was used in a survey that I received at some time in the past. Evaluate each question and its response choices using the guidelines for writing survey questions presented in this chapter. What errors do you find? Try to rewrite each question to avoid such errors and improve question wording.

   a. The first question is an *Info World* (computer publication) product evaluation survey:

      How interested are you in PostScript Level 2 printers?

      ____Very ____Somewhat ____Not at all

   b. From the Greenpeace National Marine Mammal Survey:

      Do you support Greenpeace's nonviolent, direct action to intercept whaling ships, tuna fleets, and other commercial fishermen to stop their wanton destruction of thousands of magnificent marine mammals?

      ____Yes ____No ____Undecided

   c. From a U.S. Department of Education survey of college faculty:

      How satisfied or dissatisfied are you with each of the following aspects of your instructional duties at this institution?

	Very Dissat.	Somewhat Dissat.	Somewhat Satisf.	Very Satisf.
a. The authority I have to make decisions about what courses I teach	1	2	3	4
b. Time available for working with students as advisor, mentor	1	2	3	4

   d. From a survey about affordable housing in a Massachusetts community:

      Higher than single-family density is acceptable to make housing affordable.

Strongly Agree	Undecided	Disagree	Strongly Agree	Disagree
1	2	3	4	5

   e. From a survey of faculty experience with ethical problems in research:

      Are you reasonably familiar with the codes of ethics of any of the following professional associations?

	Very Familiar	Familiar	Not Too Familiar
American Sociological Association	1	2	0
Society for the Study of Social Problems	1	2	0
American Society of Criminology	1	2	0

   If you are familiar with any of the above codes of ethics, to what extent do you agree with them?

   Strongly Agree   Agree   No opinion   Disagree   Strongly Disagree

   Some researchers have avoided using a *professional code of ethics* as a guide for the following reason. Which responses, if any, best describe your reasons for not using all or any of parts of the codes?

	Yes	No
1. Vagueness	1	0
2. Political pressures	1	0
3. Codes protect only individuals, not groups	1	0

f. From a survey of faculty perceptions:

Of the students you have observed while teaching college courses, please indicate the percentage who significantly improve their performance in the following areas.

Reading ___%

Organization ___%

Abstraction ___%

g. From a University of Massachusetts, Boston, student survey:

A person has a responsibility to stop a friend or relative from driving when drunk.

Strongly Agree_____ Agree_____ Disagree_____ Strongly Disagree_____

Even if I wanted to, I would probably not be able to stop most people from driving drunk.

Strongly Agree_____ Agree_____ Disagree_____ Strongly Disagree_____

Practice Exercises

1. Consider how you could design a split-ballot experiment to determine the effect of phrasing a question or its response choices in different ways. Check recent issues of the local newspaper for a question used in a survey of attitudes about some social policy or political position. Propose some hypothesis about how the wording of the question or its response choices might have influenced the answers people gave, and devise an alternative that differs only in this respect. Distribute these questionnaires to a large class (after your instructor makes the necessary arrangements) to test your hypothesis.

2. I received in my university mailbox some years ago a two-page questionnaire that began with the following cover letter at the top of the first page:

Critique the cover letter found at the end of the list, and then draft a more persuasive one.

3. Test your understanding of survey research terminology by completing one set of Interactive Exercises from the study site on survey design. Be sure to review the text on the pages indicated in relation to any answers you missed.

4. Review one of the survey research studies available on the book's study site, www.sagepub.com/schuttisw7e. Describe the sampling and measurement methods used and identify both strong and weak points of the survey design. Would a different type of survey design (in-person, phone, mailed, web-based) have had any advantages? Explain you answer.

Faculty Questionnaire

This survey seeks information on faculty perception of the learning process and student performance in their undergraduate careers. Surveys have been distributed in universities in the Northeast, through random deposit in mailboxes of selected departments. This survey is being conducted by graduate students affiliated with the School of Education and the Sociology Department. We greatly appreciate your time and effort in helping us with our study.

Ethics Questions

1. Group-administered surveys are easier to conduct than other types of surveys, but they always raise an ethical dilemma. If a teacher allows a social research survey to be distributed in their class, or if an employer allows employees to complete a survey on company time, is the survey truly voluntary? Is it sufficient to read a statement to the group stating that their

participation is entirely up to them? How would you react to a survey in your class? What general guidelines should be followed in such situations?

2. Tjaden and Thoennes (2000) sampled adults with random digit dialing to study violent victimization from a nationally representative sample of adults. What ethical dilemmas do you see in reporting victimizations that are identified in a survey? What about when the survey respondents are under the age of 18? What about children under the age of 12?

Web Exercises

1. Who does survey research and how do they do it? These questions can be answered through careful inspection of ongoing surveys and the organizations that administer them at www.ciser.cornell.edu/info/polls.shtml. Spend some time reading about the different survey research organizations, and write a brief summary of the types of research they conduct, the projects in which they are involved, and the resources they offer on their web sites. What are the distinctive features of different survey research organizations?

2. Go to the Research Triangle Institute site at www.rti.org. Click on "Survey Research & Services" then "Innovations."

Read about their methods for computer-assisted interviewing and their cognitive laboratory methods for refining questions. What does this add to my treatment of these topics in this chapter?

3. Go to the Survey Resources Network at http://surveynet.ac.uk/sqb/. Click on the "Surveys" link on the left-side menu, and then click on one of the listed surveys or survey sections that interests you. Review 10 questions used in the survey, and critique them in terms of the principles for question writing that you have learned. Do you find any question features that might be attributed to the use of British English?

SPSS Exercises

What can we learn from the GSS data about the orientations of people who support capital punishment? Is it related to religion? Reflective of attitudes toward race? What about political views? Is it a guy thing? Do attitudes and behavior concerning guns have some relation to support for capital punishment?

1. To answer these questions, we will use some version of each of the following variables in our analysis: PARTYID3, GUNLAW, HELPBLK, RACDIF1, FUND, OWNGUN, and CAPPUN. Check the wording of each of these questions at the University of Michigan's GSS website (click on "Browse GSS Variables" and use the mnemonic listing of variables to find those in the list above): www.norc.org/GSS+Website

   How well does each of these questions meet the guidelines for writing survey questions? What improvements would you suggest?

2. Now generate cross-tabulations to show the relationship between each of these variables, treated as independent variables, and support for capital punishment. A cross-tabulation can be used to display the distribution of responses on the dependent variable for each category of the independent variable. For this purpose, you should substitute several slightly different versions of the variables you just reviewed. From the menu, select Analyze/Descriptive Statistics/Crosstabs:

   Rows: CAPPUN

   Columns: SEX, PARTYID3, GUNLAW, HELPBLK, RACDIF1, FUND, OWNGUNR

   Cells: column percentages

   (If you have had a statistics course, you will also want to request the chi-square statistic for each of the above tables.)

   Describe the relationship you have found in the tables, noting the difference in the distribution of the dependent (row) variable—support for capital punishment—between the categories of each of the independent (column) variables.

3. Summarize your findings. What attitudes and characteristics are associated strongly with support for the death penalty?

4. What other hypotheses would you like to test? What else do you think needs to be taken into account to understand

the relationships you have identified? For example, should you take into account the race of the respondents? Why or why not?

5. Let's take a minute to learn about recoding variables. If you generate the frequencies for POLVIEWS and for POLVIEWS3, you'll see how I recoded POLVIEWS3. Why? Because I wanted to use a simple categorization by political party views in the cross-tabulation. You can try to replicate my recoding in SPSS. From the menu, click Transform/Recode/Into different variables. Identify the old variable name and type in the new one. Type in the appropriate sets of old values and the corresponding new values. You may need to check the numerical codes corresponding to the old values with the variable list pull-down menu (the ladder icon with a question mark).

Developing a Research Proposal

These steps focus again on the "Research Design" decisions, but this time assuming that you will use a survey design (Exhibit 2.14, #13 to #17).

1. Write 10 questions for a one-page questionnaire that concerns your proposed research question. Your questions should operationalize at least three of the variables on which you have focused, including at least one independent and one dependent variable (you may have multiple questions to measure some variables). Make all but one of your questions closed-ended. If you completed the "Developing a Research Proposal" exercises in Chapter 4, you can select your questions from the ones you developed for those exercises.

2. Conduct a preliminary pretest of the questionnaire by conducting cognitive interviews with two students or other persons like those to whom the survey is directed. Follow up the closed-ended questions with open-ended probes that ask the students what they meant by each response or what came to mind when they were asked each question. Take account of the feedback you receive when you revise your questions.

3. Polish up the organization and layout of the questionnaire, following the guidelines in this chapter. Prepare a rationale for the order of questions in your questionnaire. Write a cover letter directed to the appropriate population that contains appropriate statements about research ethics (human subjects' issues).

CHAPTER 9

Qualitative Methods

Observing, Participating, Listening

H urricane Katrina roared ashore in Louisiana on August 29, 2005 at 7:10 a.m. Eastern Standard Time (NOAA 2007). The combined force of winds up to 125 miles per hour and heavy rain soon breached several levees that protect New Orleans from surrounding lakes; by August 31, 80% of New Orleans was under up to 20 feet of flood water (Exhibit 9.1). With more than 1,000 deaths,

1 million displaced and total costs in excess of $100 billion, Katrina was one of the most devastating natural disasters in U.S. history. The resulting disruptions in individual lives and social patterns were equally profound. In the words of one 30-year-old New Orleans resident,

> I have stopped planning ahead. All I thought I had has been taken away. I lost my job, my home. When I was evacuated, I didn't know where I would go. You had to live one day at a time. (Davis & Land 2007:76)

Within days, graduate students and then faculty researchers from the University of Delaware's Disaster Research Center began to arrive in affected communities to study the storm's impact and the response to it (Thomas 2005). The research they designed used intensive interviews and participant observation as well as analysis of documents (Rodríguez, Trainor, & Quarantelli 2006). Other social researchers interviewed other survivors and organized focus group discussions about the experience (Davis & Land 2007; Elder et al. 2007). In this chapter, I use this research on Hurricane Katrina and social science studies of other disasters to illustrate how sociologists learn by observing as they participate in a natural setting and related qualitative methods. These examples will help you understand how some of our greatest insights into social processes can result from what appear to be very ordinary activities: observing, participating, listening, and talking.

Exhibit 9.1 **New Orleans After Hurricane Katrina**

But you will also learn that qualitative research is much more than just doing what comes naturally in social situations. Qualitative researchers must observe keenly, take notes systematically, question respondents strategically, and prepare to spend more time and invest more of their whole selves than often occurs with experiments or surveys. Moreover, if we are to have any confidence in a qualitative study's conclusions, each element of its design must be reviewed as carefully as we would review the elements of an experiment or survey. The result of careful use of these methods can also be insights into the features of the social world that are ill suited to investigation with experiments or surveys, and to social processes that defy quantification.

The chapter begins with an overview of the major features of qualitative research, as reflected in the research on Hurricane Katrina and several other disasters. The next section discusses the various approaches to participant observation research, which is the most distinctive qualitative method, and reviews the steps involved in participant observation. I then discuss the method of systematic observation, which is actually a quantitative approach to recording observational data; the contrast with participant observation will be instructive. In the following section, I review, in some detail, the issues involved in intensive interviewing before briefly explaining focus groups, an increasingly popular qualitative method. The last section covers ethical issues that are of concern in any type of qualitative research project. By the chapter's end, you should appreciate the hard work required to translate "doing what comes naturally" into systematic research, be able to recognize strong and weak points in qualitative studies, and be ready to do some of it yourself.

▣ Fundamentals of Qualitative Methods

Qualitative methods refer to several distinctive research techniques, including **participant observation**, **intensive (depth) interviewing**, and **focus groups**. Participant observation and intensive interviewing are often used in the same project, while focus groups combine some elements of these two approaches into a unique data-collection strategy. These techniques often can be used to enrich experiments and surveys. Qualitative methods can also be used in the study of textual or other documents as well as in historical and comparative research, but we will leave these research techniques for other chapters.

Participant observation A qualitative method for gathering data that involves developing a sustained relationship with people while they go about their normal activities.

Intensive (depth) interviewing A qualitative method that involves open-ended, relatively unstructured questioning in which the interviewer seeks in-depth information on the interviewer's feeling, experiences, and perceptions (Lofland & Lofland 1984:12).

Focus groups A qualitative method that involves unstructured group interviews in which the focus group leader actively encourages discussion among participants on the topics of interest.

Features of Qualitative Research

Although these three qualitative designs differ in many respects, they share several features that distinguish them from experimental and survey research designs (Denzin & Lincoln 1994; Maxwell 1996; Wolcott 1995):

Collection primarily of qualitative rather than quantitative data. Any research design may collect both qualitative and quantitative data, but qualitative methods emphasize observations about natural behavior and artifacts that capture social life as participants experience it, rather than in categories the researcher predetermines. For example, the Disaster Research Center (DRC) researchers observed the response to the unprecedented threat posed by Katrina in the major New Orleans hotels and concluded that the hotels went through three stages of "improvisation": (1) the hotels encouraged all guests who were not stranded to leave, (2) the hotel chains sent in food and other provisions for guests and staff, and (3) the hotels reorganized so that they could provide semipermanent lodging for federal disaster employees and evacuees (Rodríguez et al. 2006:87–89). The researchers described the different activities in the hotels at each of these "stages."

Exploratory research questions, with a commitment to inductive reasoning. Qualitative researchers typically begin their projects seeking not to test preformulated hypotheses but to discover what people think, how they act, and why, in some social setting. Only after many observations do qualitative researchers try to develop general principles to account for their observations. The DRC researchers asked, "How did people, groups, and organizations in Louisiana react to the impact of Hurricane Katrina in September 2005?" (Rodríguez et al. 2006:83).

A focus on previously unstudied processes and unanticipated phenomena. Previously unstudied attitudes and actions can't adequately be understood with a structured set of questions or within a highly controlled experiment. So, qualitative methods have their greatest appeal when we need to explore new issues, investigate hard-to-study groups, or determine the meaning people give to their lives and actions. Disasters such as Hurricane Katrina certainly meet the criteria of unanticipated and hard-to-study. Dag Nordanger (2007:174) used qualitative methods to study loss and bereavement in war-torn Tigray, Ethiopia, because preliminary information indicated that people in this culture adjusted to loss in a very different way than people in Western societies.

An orientation to social context, to the interconnections between social phenomena rather than to their discrete features. The context of concern may be a program or organization, a community, or a broader social context. This feature of qualitative research is evident in Elif Kale-Lostuvali's (2007) description of Gölcük, a Turkish town, after the İzmit earthquake:

> For the first few months, the majority of the population lived in tents located either in tent cities or near partially damaged homes. Around mid-December, eligible survivors began to move into prefabricated houses built on empty hills around the center of the town. Many survivors had lost their jobs because of the earthquake. . . . Hence, daily life revolved around finding out about forms of provision. (p. 752)

A focus on human subjectivity, on the meanings that participants attach to events and that people give to their lives. "Through life stories, people 'account for their lives.' . . . The themes people create are the means by which they interpret and evaluate their life experiences and attempt to integrate these experiences to form a self-concept" (Kaufman 1986:24–25). You can see this emphasis in an excerpt from an interview Dag Nordanger (2007) conducted with a Tigrayan woman who had lost her property and all her nine children in the preceding decades of war in Ethiopia:

> My name is the same. I was Mrs. NN, and now I am Mrs. NN. But I am not like the former. The former Mrs. NN had everything at hand, and was highly respected by others. People came to me for advice and help. But the recent Mrs. NN is considered like a half person. Though she does not go out for begging, she has the lifestyle of a beggar, so she is considered to be a beggar. (p. 179)

Use of idiographic rather than nomothetic causal explanation. With its focus on particular actors and situations and the processes that connect them, qualitative research tends to identify causes as particular events embedded within an unfolding, interconnected action sequence (Maxwell 1996:20–21). The language of variables and hypotheses appears only rarely in the qualitative literature. Havidán Rodríguez, Joseph Trainor, and Enrico L. Quarantelli (2006) include in their analysis of "emergent and prosocial behavior following Hurricane Katrina" the following sequence of events in New Orleans hospitals:

> The floodwaters from the levee breaks created a new kind of crisis. Basements with stored food, water, and fuel, as well as morgues, were inundated. . . . As emergency generators ran out of fuel, the water, sewage, and air-conditioning systems failed. Patients who died in the hospitals had to be temporarily stored in stairwells. Eventually, waste of all kinds was strewn almost everywhere. The rising temperatures made most diagnostic equipment inoperable. . . . Regular hospital procedures simply stopped, but personnel improvised to try to provide at least minimum health care. For instance, physicians, nurses, and volunteers fanned patients to keep them cool, sometimes using manually operated devices to keep them breathing. (pp. 89–90)

Reflexive research design, in which the design develops as the research progresses.

> Each component of the design may need to be reconsidered or modified in response to new developments or to changes in some other component. . . . The activities of collecting and analyzing data, developing and modifying theory, elaborating or refocusing the research questions, and identifying and eliminating validity threats are usually all going on more or less simultaneously, each influencing all of the others. (Maxwell 1996:2–3)

You can see this reflexive quality in Elif Kale-Lostuvali's (2007) description of his qualitative research work as he studied state-citizen encounters in the aftermath of the İzmit earthquake:

> I made my initial trip to Gölcük at the beginning of October 1999, six weeks after the earthquake. From then until the end of July 2000, I made two to three day-long trips per month, spending a total of 25 days in Gölcük. During these trips, I spent time mainly in the Gölcük Crisis Center, the administrative

offices of two major tent cities, and two major prefab areas observing interactions. . . . As I got to know some of the state agents and survivors better, I began to hear their responses after specific interactions and their views of the provision and distribution process in general. In addition, I often walked around and spoke with many people in tent cities, in prefab areas, and in the center of the town. Sometimes, people I met in this way invited me to their homes and offices. (p. 752)

Ultimately, Kale-Lostuvali (2007:752) reported conversations with approximately 100 people, in-depth interviews with 30 carefully selected people, and many observational notes.

Sensitivity to the subjective role of the researcher. Qualitative researchers recognize that their perspective on social phenomena will reflect in part their own background and current situation. Who the researcher is and "where he or she is coming from" can affect what the research "finds." Some qualitative researchers believe that the goal of developing a purely "objective" view of the social world is impossible, but they discuss in their publications their own feelings about what they have studied so that others can consider how these feelings affected their findings. You can imagine how anthropology graduate student Hannah Gill (2004) had to consider her feelings when she encountered crime in the community she was studying in the Dominican Republic:

> On the second day I found myself flattened under a car to avoid getting shot by a woman seeking revenge for her husband's murder in the town market, and on the third I was sprinting away from a knife fight at a local hang-out. (p. 2)

Rather than leaving, Hannah Gill assessed the danger, realized that although she stood out in the community as a "young, white American graduate student," she "was not a target and with necessary precautions, I would be relatively safe." She decided to use her experiences as "a constructive experience" that would give her insight into how people respond to risk.

Sociologist Barrie Thorne (1993) shared with readers her subjective reactions as she observed girls on a school playground:

> I felt closer to the girls not only through memories of my own past, but also because I knew more about their gender-typed interactions. I had once played games like jump rope and statue buyer, but I had never ridden a skateboard and had barely tried sports like basketball and soccer. . . . Were my moments of remembering, the times when I felt like a ten-year-old girl, a source of distortion or insight? (p. 26)

William Miller and Benjamin Crabtree (1999a) captured the entire process of qualitative research in a simple diagram (Exhibit 9.2). In this diagram, qualitative research begins with the qualitative researcher reflecting on the setting and his or her relation to it and interpretations of it. The researcher then describes the goals and means for the research. This description is followed by sampling and collecting data, describing the data, and organizing those data. Thus, the *gathering process* and the *analysis process* proceed together, with repeated description and analysis of data as they are collected and reflexive attention to the researcher's engagement in the process. As the data are organized, connections are identified between different data segments, and efforts are made to corroborate the credibility of these connections. This *interpretive process* begins to emerge in a written account that represents what has been done and how the data have been interpreted. Each of these steps in the research process informs the others and is repeated throughout the research process.

History of Qualitative Research

Anthropologists and sociologists laid the foundation for modern qualitative methods while doing field research in the early decades of the 20th century. Dissatisfied with studies of native peoples who relied on

Exhibit 9.2 | **Qualitative Research Process**

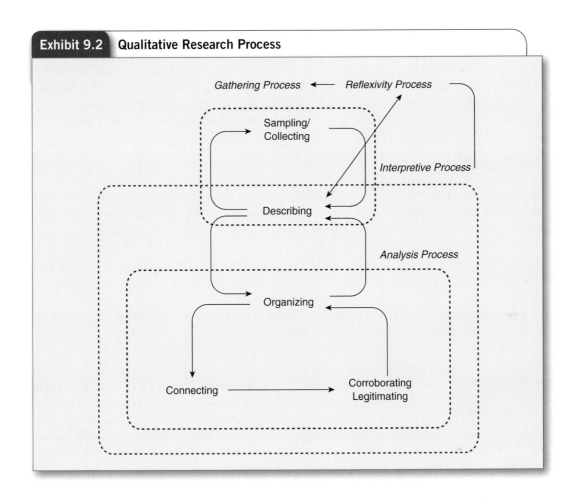

secondhand accounts and inspection of artifacts, anthropologists Franz Boas and Bronislaw Malinowski went to live in or near the communities they studied. Boas visited Native American villages in the American Northwest; Malinowski lived among New Guinea natives. Neither truly participated in the ongoing social life of those they studied (Boas collected artifacts and original texts, and Malinowski reputedly lived as something of a noble among the natives he studied), but both helped establish the value of intimate familiarity with the community of interest and thus laid the basis for modern anthropology (Emerson 1983:2–5).

Many of sociology's **field research** pioneers were former social workers and reformers. Some brought their missionary concern with the spread of civic virtue among new immigrants to the Department of Sociology and Anthropology at the University of Chicago. Their successors continued to focus on the sources of community cohesion and urban strain but came to view the city as a social science "laboratory" rather than as a focus for reform. They adapted the fieldwork methods of anthropology to studying the "natural areas" of the city and the social life of small towns (Vidich & Lyman 2004). By the 1930s, 1940s, and 1950s, qualitative researchers were emphasizing the value of direct participation in community life and sharing in subjects' perceptions and interpretations of events (Emerson 1983:6–13). This naturalistic focus continued to dominate qualitative research into the 1960s, and qualitative researchers refined and formalized methods to develop the most realistic understanding of the natural social world (Denzin & Lincoln 2000:14–15).

> **Field research** Research in which natural social processes are studied as they happen and left relatively undisturbed.

The next two decades saw increasing emphasis on the way in which participants in the social world construct the reality that they experience and increasing disbelief that researchers could be disinterested observers of social reality or develop generalizable knowledge about it. Adherents of the *postmodern* perspective that you learned about in Chapter 1 urged qualitative researchers to describe particular events, rituals, and customs and to recognize that the interpretations they produced were no more "privileged" than others' interpretations (Denzin & Lincoln 2000:15). "The making of every aspect of human existence is culturally created and determined in particular, localized circumstances about which no generalizations can be made. Even particularized meaning, however, is . . . relative and temporary" (Spretnak 1991:13–14).

The Case Study

Qualitative research projects often have the goal of developing an understanding of an entire slice of the social world, not just discrete parts of it. What was the larger social context in New Orleans after Hurricane Katrina (Rodríguez et al. 2006:87)? What was Chicago like during the 1995 Heat Wave, when thousands were hospitalized and more than 700 died of heat-related causes (Klinenberg 2002:1–9)? Sociologist Kai Erikson sent me the following verbal "picture" of New Orleans, as he observed it during a research trip a few days after Katrina:

> The carnage stretches out almost endlessly: more than a hundred thousand [crumpled] homes, at least fifty thousand [flattened] automobiles, the whole mass being covered by a crust of grey mud, dried as hard as fired clay by the sun. It was the silence of it, the emptiness of it; that is the story.

Case study A setting or group that the analyst treats as an integrated social unit that must be studied holistically and in its particularity.

Questions and images such as these reflect a concern with developing a **case study**. Case study is not so much a single method as it is a way of thinking about what qualitative data analysis can, or perhaps should, focus on. The case may be an organization, community, social group, family, or even an individual; as far as the qualitative researcher is concerned, it must be understood in its entirety. The idea is that the social world really functions as an integrated whole; social researchers therefore need to develop "deep understanding of particular instances of phenomena" (Mabry 2008:214). By contrast, from this perspective, the quantitative research focus on variables and hypotheses mistakenly "slices and dices" reality in a way that obscures how the social world actually functions.

Educational researcher Robert Stake (1995) presents the logic of the case study approach thus:

> Case study is the study of the particularity and complexity of a single case, coming to understand its activity within important circumstances. . . . The qualitative researcher emphasizes episodes of nuance, the sequentiality of happenings in context, the wholeness of the individual. (pp. xi–xii)

Central to much qualitative case study research is the goal of creating a **thick description** of the setting studied—a description that provides a sense of what it is like to experience that setting from the standpoint of the natural actors in that setting (Geertz 1973). Stake's (1995) description of "a case within a case," a student in a school he studied, illustrates how a thick description gives a feel of the place and persons within it:

Thick description A rich description that conveys a sense of what it is like from the standpoint of the natural actors in that setting.

> At 8:30 a.m. on Thursday morning. Adam shows up at the cafeteria door. Breakfast is being served but Adam doesn't go in. The woman giving out meal chits has her hands on him, seems to be sparring with him, verbally. And then he disappears. Adam is one of five siblings, all arrive at school in the morning with less than usual parent attention. Short, with a beautifully sculpted head . . . Adam is a person of notice.

At 8:55 he climbs the stairs to the third floor with other upper graders, turning to block the girls behind them and thus a string of others. Adam manages to keep the girls off balance until Ms Crain . . . spots him and gets traffic moving again. Mr. Garson . . . notices Adam, has a few quiet words with him before a paternal shove toward the room. (p. 150)

You will learn in the next sections how qualitative methodologists design research that can generate such thick descriptions of particular cases.

▣ Participant Observation

Participant observation, termed *fieldwork* in anthropology, was used by Havidán Rodríguez and his colleagues (2006) to study the aftermath of Hurricane Katrina, by Dag Nordanger (2007) to study the effects of trauma in Ethiopia, and by Elif Kale-Lostuvali (2007) to study the aftermath of the İzmit earthquake. Participant observation is a qualitative method in which natural social processes are studied as they happen (in "the field" rather than in the laboratory) and left relatively undisturbed. It is the classic field research method—a means for seeing the social world as the research subjects see it, in its totality, and for understanding subjects' interpretations of that world (Wolcott 1995:66). By observing people and interacting with them in the course of their normal activities, participant observers seek to avoid the artificiality of experimental design and the unnatural structured questioning of survey research (Koegel 1987:8). This method encourages consideration of the context in which social interaction occurs, of the complex and interconnected nature of social relations, and of the sequencing of events (Bogdewic 1999:49).

The term *participant observer* actually refers to several different specific roles that a qualitative researcher can adopt (see Exhibit 9.3). As a **covert observer**, a researcher observes others without participating in social interaction and does not self-identify as a researcher.

This role is often adopted for studies in public places where there is nothing unusual about someone sitting and observing others. However, in many settings, a qualitative researcher will function as a **complete observer**, who does not participate in group activities and is publicly defined as a researcher. These two relatively passive roles contrast with the roles of covert and overt participation. A qualitative researcher is a **covert participant** when she acts just like other group members and does not disclose her research role. If she publicly acknowledges being a researcher but nonetheless participates in group activities, she can be termed an *overt participant*, or true **participant observer**.

Choosing a Role

The first concern of every participant observer is to decide what balance to strike between observing and participating and whether to reveal one's role as a researcher. These decisions must take into account the specifics of the social

Covert observer A role in participant observation in which the researcher does not participate in group activities and is not publicly defined as a researcher.

Complete observer A role in participant observation in which the researcher does not participate in group activities and is publicly defined as a researcher.

Covert participant A role in field research in which the researcher does not reveal his or her identity as a researcher to those who are observed, while participating.

Participant observer A researcher who gathers data through participating and observing in a setting where he or she develops a sustained relationship with people while they go about their normal activities. The term participant observer is often used to refer to a continuum of possible roles, from complete observation, in which the researcher does not participate along with others in group activities, to complete participation, in which the researcher participates without publicly acknowledging being an observer.

Exhibit 9.3 The Participant Observation Continuum

To study an activist group, you could take the role of covert observer.

> I can see who is protesting. I wonder whether their behavior will suggest why?

You could take the role of overt observer:

> Hello, I am a researcher. Tell me, why do you participate in these activities?

You could take the role of participant and observer:

> Hello, I am a researcher and an activist. Tell me, why do you participate in these activities?

You could take the role of covert participant:

situation being studied, the researcher's own background and personality, the larger socio-political context, and ethical concerns. Which balance of participating and observing is most appropriate also changes during most projects, often many times. And the researcher's ability to maintain either a covert or an overt role will many times be challenged.

Covert Observation

In both observational roles, researchers try to see things as they happen, without actively participating in these events. Although there is no fixed formula to guide the observational process, observers try to identify the who, what, when, where, why, and how of the activities in the setting. Their observations will usually become more focused over time, as the observer develops a sense of the important categories of people and activities and gradually develops a theory that accounts for what is observed (Bogdewic 1999:54–56).

In social settings involving many people, in which observing while standing or sitting does not attract attention, covert observation is possible and is unlikely to have much effect on social processes. You may not even want to call this "covert" observation, because your activities as an observer may be no different from those of others who are simply observing others to pass the time. However, when you take notes, when you systematically check out the different areas of a public space or different people in a crowd, when you arrive and leave at particular times to do your observing, you are acting differently in important respects from others in the setting. Moreover, when you write up what you have observed and, possibly, publish it, you have taken something unique from the people in that setting. If you adopt the role of covert observer, you should always remember to evaluate how your actions in the setting and your purposes for being there may affect the actions of others and your own interpretations.

Overt Observation

When a researcher announces her role as a research observer, her presence is much more likely to alter the social situation being observed. This is the problem of **reactive effects**. It is not "natural" in most social situations for someone to be present who will record his or her observations for research and publication purposes, and so individuals may alter their behavior. The overt, or complete, observer is even more likely to have an impact when the social setting involves few people or if observing is unlike the usual activities in the setting. Observable differences between the observer and those being observed also increase the likelihood of reactive effects. For example, some children observed in the research by Barrie Thorne (1993:16–17) treated her as a teacher when she was observing them in a school playground and so asked her to resolve disputes. No matter how much she tried to remain aloof, she still appeared to children as an adult authority figure and so experienced pressure to participate (Thorne 1993:20). However, in most situations, even overt observers find that their presence seems to be ignored by participants after a while and to have no discernible impact on social processes.

> **Reactive effects** The changes in individual or group behavior that are due to being observed or otherwise studied.

Overt Participation (Participant Observer)

Most **field researchers** adopt a role that involves some active participation in the setting. Usually, they inform at least some group members of their research interests, but then they participate in enough group activities to develop rapport with members and to gain a direct sense of what group members experience. This is not an easy balancing act.

> **Field researcher** A researcher who uses qualitative methods to conduct research in the field.

> The key to participant observation as a fieldwork strategy is to take seriously the challenge it poses to participate more, and to play the role of the aloof observer less. Do not think of yourself as someone who needs to wear a white lab coat and carry a clipboard to learn about how humans go about their everyday lives. (Wolcott 1995:100)

Dag Nordanger (2007) described how, accompanied by a knowledgeable Tigrayan research assistant, he developed rapport with community members in Tigray, Ethiopia:

> Much time was spent at places where people gathered, such as markets, "sewa houses" (houses for homebrewed millet beer: sewa), cafés, and bars, and for the entire study period most invitations for a drink or to go to people's homes for *injerra* (the sour pancake that is their staple food) and coffee ceremonies were welcomed. The fact that the research topic garnered interest and engagement made access to relevant information easy. Numerous informal interviews derived from these settings, where the researcher discussed his interests with people. (p. 176)

Participating and observing have two clear ethical advantages as well. Because group members know the researcher's real role in the group, they can choose to keep some information or attitudes hidden. By the same token, the researcher can decline to participate in unethical or dangerous activities without fear of exposing his or her identity.

Most field researchers who opt for disclosure get the feeling that, after they have become known and at least somewhat trusted figures in the group, their presence does not have any palpable effect on members' actions. The major influences on individual actions and attitudes are past experiences, personality, group structure, and so on, so the argument goes, and these continue to exert their influence even when an outside observer is present. The participant observer can then be ethical about identity disclosure and still observe the natural social world. In practice, however, it can be difficult to maintain a fully open research role in a setting in which new people come and go, often without providing appropriate occasions during which the researcher can disclose his or her identity.

Of course, the argument that the researcher's role can be disclosed without affecting the social process under investigation is less persuasive when the behavior to be observed is illegal or stigmatized, so that participants have reasons to fear the consequences of disclosure to any outsider. Konstantin Belousov and his colleagues (2007) provide a dramatic example of this problem from their fieldwork on regulatory enforcement in the Russian shipping industry. In a setting normally closed to outsiders and linked to organized crime, the permission of a port official was required. However, this official was murdered shortly after the research began. After that "our presence was now barely tolerated, and to be avoided at all costs. . . . explanations became short and respondents clearly wished to get rid of us as soon as possible" (pp. 164–165).

Even when researchers maintain a public identity as researchers, ethical dilemmas arising from participation in the group activities do not go away. In fact, researchers may have to "prove themselves" to the group members by joining in some of their questionable activities. For example, police officers gave John Van Maanen (1982) a nonstandard and technically prohibited pistol to carry on police patrols. Harold Pepinsky (1980) witnessed police harassment of a citizen but did not intervene when the citizen was arrested. Trying to strengthen his ties with a local political figure in his study of a poor Boston community he called Cornerville, William Foote Whyte (1955) illegally voted multiple times in a local election.

Experienced participant observers try to lessen some of the problems of identity disclosure by evaluating both their effect on others in the setting and the effect of others on the observers writing about these effects throughout the time they are in the field and while they analyze their data. They also are sure, while in the field, to preserve some physical space and regular time when they can concentrate on their research and schedule occasional meetings with other researchers to review the fieldwork. Participant observers modify their role as circumstances seem to require, perhaps not always disclosing their research role at casual social gatherings or group outings, but being sure to inform new members of it.

Research in the News

TAPING AND ANALYZING FAMILY LIFE

In the News

Researchers at the University of California, Los Angeles, recruited 32 local families and then videotaped them almost continuously while they were awake and at-home during one week. A researcher roamed the houses with a handheld computer, recording at 10-minute intervals each family member's location and activities; family members also participated in in-depth interviews and completed questionnaires. Couples reported less stress if they had a more rigid division of labor, whether equal or not. Half the fathers spent at least as much time as their wives along with their children when they were at home, and were more likely to engage in physical activities, while their wives were more likely to watch TV with their children. No one spent time in the yard.

Source: Carey, Benedict. 2010. "Families' Every Hug and Fuss, Taped, Analyzed and Archived." *The New York Times,* May 23:A1.

Covert Participation

To lessen the potential for reactive effects and to gain entry to otherwise inaccessible settings, some field researchers have adopted the role of covert participants, keeping their research secret and trying their best to act similar to other participants in a social setting or group. Laud Humphreys (1970) took the role of a covert participant when he served as a "watch queen" so that he could learn about the men engaging in homosexual acts in a public restroom. Randall Alfred (1976) joined a group of Satanists to investigate the group members and their interaction. Erving Goffman (1961) worked as a state hospital assistant while studying the treatment of psychiatric patients.

Although the role of covert participant lessens some of the reactive effects encountered by the complete observer, covert participants confront other problems:

- *Covert participants cannot take notes openly or use any obvious recording devices.* They must write up notes based solely on their memory and must do so at times when it is natural for them to be away from the group members.

- *Covert participants cannot ask questions that will arouse suspicion.* Thus, they often have trouble clarifying the meaning of other participants' attitudes or actions.

- *The role of a covert participant is difficult to play successfully.* Covert participants will not know how the regular participants would act in every situation in which the researchers find themselves. Regular participants have entered the situation from different social backgrounds and with goals different from that of the researchers. Researchers' spontaneous reactions to every event are unlikely to be consistent with those of the regular participants (Mitchell 1993). Suspicion that researchers are not "one of us" may then have reactive effects, obviating the value of complete participation (Erikson 1967). In his study of the Satanists, for example, Alfred (1976) pretended to be a regular group participant until he completed his research, at which time he informed the group leader of his covert role. Rather than act surprised, the leader told Alfred that he had long considered Alfred to be "strange," not similar to the other people—and we will never know for sure how Alfred's observations were affected.

- *Covert participants need to keep up the act at all times while in the setting under study.* Researchers may experience enormous psychological strain, particularly in situations where they are expected to choose sides in intragroup conflict, to participate in criminal or other acts. Of course, some covert observers may become so wrapped up in the role they are playing that they adopt not only just the mannerisms but also the perspectives and goals of the regular participants—that is, they "go native." At this point, they abandon research goals and cease to evaluate critically what they are observing.

Ethical issues have been at the forefront of debate over the strategy of covert participation. Kai Erikson (1967) argued that covert participation is, by its very nature, unethical and should not be allowed except in public settings. Erikson points out that covert researchers cannot anticipate the unintended consequences of their actions for research subjects. If other people suspect the identity of the researcher or if the researcher contributes to or impedes group action, the consequences can be adverse. In addition, other social scientists are harmed either when covert research is disclosed during the research or on its publication, because distrust of social scientists increases and access to research opportunities may decrease.

However, a total ban on covert participation would "kill many a project stone dead" (Punch 1994:90). Studies of unusual religious or sexual practices and of institutional malpractice would rarely be possible. "The crux of the matter is that some deception, passive or active, enables you to get at data not obtainable by other means" (Punch 1994:91). Richard G. Mitchell Jr. (1993) presents the argument of some researchers that the social world "is presumed to be shot through with misinformation, evasion, lies, and fronts at every level, and research in kind—secret, covert, concealed, and disguised—is necessary and appropriate" (p. 30). Therefore, some field researchers argue that covert participation is legitimate in some of the settings. If the researcher maintains the confidentiality of others, keeps commitments to others, and does not directly lie to others, some degree of deception may be justified in exchange for the knowledge gained (Punch 1994:90).

Entering the Field

Entering the field, the setting under investigation, is a critical stage in a participant observation project because it can shape many subsequent experiences. Some background work is necessary before entering the field—at least enough to develop a clear understanding of what the research questions are likely to be and to review one's personal stance toward the people and problems that are likely to be encountered. With participant observation, researchers must also learn in advance how participants dress and what their typical activities are so as to avoid being caught completely unaware. Finding a participant who can make introductions is often critical (Rossman & Rallis 1998:102–103), and formal permission may be needed in an organization's setting (Bogdewic 1999:51–53). It may take weeks or even months until entry is possible.

Timothy Diamond (1992) applied to work as an assistant to conduct research as a participant observer in a nursing home. His first effort failed miserably:

> My first job interview. . . . The administrator of the home had agreed to see me on [the recommendation of two current assistants]. The administrator . . . probed suspiciously, "Now why would a white guy want to work for these kinds of wages?" . . . He continued without pause, "Besides, I couldn't hire you if I wanted to. You're not certified." That, he quickly concluded, was the end of our interview, and he showed me to the door. (pp. 8–9)

After taking a course and receiving his certificate, Diamond was able to enter the role of nursing assistant as others did.

Many field researchers avoid systematic study and extensive reading about a setting for fear that it will bias their first impressions, but entering without any sense of the social norms can lead to a disaster. Whyte came close to such a disaster when he despaired of making any social contacts in Cornerville and decided to try an unconventional entry approach (i.e., unconventional for a field researcher). In *Street Corner Society,* Whyte (1995) describes what happened when he went to a hotel bar in search of women to talk to:

> I looked around me again and now noticed a threesome: one man and two women. It occurred to me that here was a maldistribution of females which I might be able to rectify. I approached the group and opened with something like this: "Pardon me. Would you mind if I joined you?" There was a moment of silence while the man stared at me. He then offered to throw me downstairs. I assured him that this would not be necessary and demonstrated as much by walking right out of there without any assistance. (p. 289)

Whyte needed a **gatekeeper** who could grant him access to the setting; he finally found one in "Doc" (Rossman & Rallis 1998:108–111). A helpful social worker at the local settlement house introduced Whyte to this respected leader, who agreed to help:

> **Gatekeeper** A person in a field setting who can grant researchers access to the setting.

> Well, any nights you want to see anything, I'll take you around. I can take you to the joints—gambling joints—I can take you around to the street corners. Just remember that you're my friend. That's all they need to know [so they won't bother you]. (Whyte 1955:291)

You have already learned that Dag Nordanger (2007:176) relied on a gatekeeper to help him gain access to local people in Tigray, Ethiopia.

When participant observing involves public figures who are used to reporters and researchers, a more direct approach may secure entry into the field. Richard Fenno (1978:257) used this direct approach in his study of members of the U.S. Congress: He simply wrote and asked permission to observe selected members of the Congress at work. He received only two refusals, attributing this high rate of subject cooperation to such reasons as interest in a change in the daily routine, commitment to making themselves available, a desire for more publicity, the flattery of scholarly attention, and interest in helping to teach others about politics. Other groups have other motivations, but in every case, some consideration of these potential motives in advance should help smooth entry into the field.

In short, field researchers must be very sensitive to the impression they make and to the ties they establish when entering the field. This stage lays the groundwork for collecting data from people who have different perspectives and for developing relationships that the researcher can use to surmount the problems in data collection that inevitably arise in the field. The researcher should be ready with a rationale for his or her participation and some sense of the potential benefits to participants. Discussion about these issues with key participants or gatekeepers should be honest and identify what the participants can expect from the research, without necessarily going into detail about the researcher's hypotheses or research questions (Rossman & Rallis 1998:51–53, 105–108).

Developing and Maintaining Relationships

Researchers must be careful to manage their relationships in the research setting so that they can continue to observe and interview diverse members of the social setting throughout the long period typical of participant observation (Maxwell 1996:66). Every action the researcher takes can develop or undermine this relationship. Interaction early in the research process is particularly sensitive, because participants don't know the researcher and the researcher doesn't know the routines. Barrie Thorne (1993) felt she had gained access to kids' more private world "when kids violated rules in my presence, such as swearing or openly blowing bubble gum where these acts were forbidden, or swapping stories about recent acts of shoplifting" (pp. 18–19). On the other hand, Van Maanen (1982) found his relationship with police officers undermined by one incident:

> Following a family beef call in what was tagged the Little Africa section of town, I once got into what I regarded as a soft but nonetheless heated debate with the officer I was working with that evening on the merits of residential desegregation. My more or less liberal leanings on the matter were bothersome to this officer, who later reported my disturbing thoughts to his friends in the squad. Before long, I was an anathema to this friendship clique and labeled by them undesirable. Members of this group refused to work with me again. (p. 110)

So Van Maanen failed to maintain a research (or personal) relationship with this group. Do you think he should have kept his opinions about residential desegregation to himself? How honest should field researchers be about their feelings? Should they "go along to get along"?

William Foote Whyte used what, in retrospect, was a sophisticated two-part strategy to develop and maintain a relationship with the Cornerville street-corner men. The first part of Whyte's strategy was to maintain good relations with Doc and, through Doc, to stay on good terms with the others. Doc became a **key informant** in the research setting—a knowledgeable insider who knew the group's culture and was willing to share access and insights with the researcher (Gilchrist & Williams 1999). The less obvious part of Whyte's strategy was a consequence of his decision to move into Cornerville, a move he decided was necessary to really understand and be accepted in the community. The room he rented in a local family's home became his base of operations. In some respects, this family became an important dimension of Whyte's immersion in the community: He tried to learn Italian by speaking with the family members, and they conversed late at night as if Whyte were a real family member. But Whyte recognized that he needed a place to unwind after his days of constant alertness in the field, so he made a conscious decision not to include the family as an object of study. Living in this family's home became a means for Whyte to maintain standing as a community insider without becoming totally immersed in the demands of research (Whyte 1955:294–297).

> **Key informant** An insider who is willing and able to provide a field researcher with superior access and information, including answers to questions that arise in the course of the research.

Experienced participant observers have developed some sound advice for others seeking to maintain relationships in the field (Bogdewic 1999:53–54; Rossman & Rallis 1998: 105–108; Whyte 1955:300–306; Wolcott 1995:91–95):

- Develop a plausible (and honest) explanation for yourself and your study.

- Maintain the support of key individuals in groups or organizations under study.

- Be unobtrusive and unassuming. Don't "show off" your expertise.

- Don't be too aggressive in questioning others (e.g., don't violate implicit norms that preclude discussion of illegal activity with outsiders). Being a researcher requires that you do not simultaneously try to be the guardian of law and order. Instead, be a reflective listener.

- Ask very sensitive questions only of informants with whom your relationship is good.

- Be self-revealing, but only up to a point. Let participants learn about you as a person, but without making too much of yourself.

- Don't fake your social similarity with your subjects. Taking a friendly interest in them should be an adequate basis for developing trust.

- Avoid giving or receiving monetary or other tangible gifts but without violating norms of reciprocity. Living with other people, taking others' time for conversations, and going out for a social evening all create expectations and incur social obligations, and you can't be an active participant without occasionally helping others. But you will lose your ability to function as a researcher if you come to be seen as someone who gives away money or other favors. Such small forms of assistance as an occasional ride to the store or advice on applying to college may strike the right balance.

- Be prepared for special difficulties and tensions if multiple groups are involved. It is hard to avoid taking sides or being used in situations of intergroup conflict.

Sampling People and Events

In qualitative research, the need to intensively study the people, places, or phenomena of interest guide sampling decisions. In fact, most qualitative researchers limit their focus to just one or a few sites or programs, so that they can focus all their attention on the social dynamics of those settings. This focus on a limited number of cases does not mean that sampling is unimportant. The researcher must be reasonably confident about gaining access and that the site can provide relevant information. The sample must be appropriate and adequate for the study, even if it is not representative. The qualitative researcher may select a *critical case* that is unusually rich in information pertaining to the research question, a *typical case* precisely because it is judged to be typical, and/or a *deviant case* that provides a useful contrast (Kuzel 1999). Within a research site, plans may be made to sample different settings, people, events, and artifacts (see Exhibit 9.4).

Studying more than one case or setting almost always strengthens the causal conclusions and makes the findings more generalizable (King et al. 1994). The DRC researchers (Rodríguez et al. 2006:87) studied emergent behavior in five social "groupings": hotels, hospitals, neighborhood groups, rescue teams, and the Joint Field Office (JFO). To make his conclusions more generalizable, Timothy Diamond (1992:5) worked in three different Chicago nursing homes "in widely different neighborhoods" that had very different proportions of residents supported by Medicaid. He then "visited many homes across the United States to validate my observations" (p. 5). Klinenberg (2002:79–128) contrasted the social relations in two Chicago neighborhoods.

| Exhibit 9.4 | Sampling Plan for a Participant Observation Project in Schools |

| Information Source* | Type of Information to Be Obtained | | | | |
	Collegiality	Goals and Community	Action Expectations	Knowledge Orientation	Base
Settings					
Public places (halls, main offices)					
Teacher's lounge	X	X		X	X
Classrooms		X	X	X	X
Meeting rooms	X		X	X	
Gymnasium or locker room		X			
Events					
Faculty meetings	X		X		X
Lunch hour	X*				X
Teaching		X	X	X	X
People					
Principal		X	X	X	X
Teachers	X*	X	X	X	X
Students		X	X	X	
Artifacts					
Newspapers		X	X		X
Decorations		X*			
*Selected examples in each category					

Thorne (1993:6–7) observed in a public elementary school in California for 8 months and then, 4 years later, in a public elementary school in Michigan for 3 months.

Other approaches to sampling in field research are more systematic. You have already learned in Chapter 5 about some of the nonprobability sampling methods that are used in field research. For instance, purposive sampling can be used to identify the opinion of the leaders and representatives of different roles. With snowball sampling, field researchers learn from participants about who represents different subgroups in a setting. Quota sampling also may be employed to ensure the representation of particular categories of participants. Using some type of intentional sampling strategy within a particular setting can allow tests of some hypotheses, which would otherwise have to wait until comparative data could be collected from several other settings (King et al. 1994).

Theoretical sampling A sampling method recommended for field researchers by Glaser and Strauss (1967). A theoretical sample is drawn in a sequential fashion, with settings or individuals selected for study as earlier observations or interviews indicate that these settings or individuals are influential.

Experience sampling method (ESM) A technique for drawing a representative sample of everyday activities, thoughts, and experiences. Participants carry a pager and are beeped at random times over several days or weeks; on hearing the beep, participants complete a report designed by the researcher.

Theoretical sampling is a systematic approach to sampling in participant observation studies (Glaser & Strauss 1967). When field researchers discover in an investigation that particular processes seem to be important, inferring that certain comparisons should be made or that similar instances should be checked, the researchers then choose new settings or individuals that permit these comparisons or checks (Ragin 1994:98–101) (see Exhibit 9.5). Spencer Moore and colleague's (2004) strategy for selecting key informants in their research on North Carolina's Hurricane Floyd experience exemplifies this type of approach:

> Sixteen key informant interviews were conducted with volunteers for local nonprofit organizations, community and religious leaders, and local government officials in all five HWC-project counties. These representatives were chosen on the basis of their county or city administrative position (e.g., emergency management or assistant county managers), as well as on the basis of leadership in flood-related relief activities, as identified by local officials or as reported in local newspapers. (p. 209)

When field studies do not require ongoing, intensive involvement by researchers in the setting, the **experience sampling method (ESM)** can be used. The experiences, thoughts, and feelings of a number of people are sampled randomly as they go about their daily activities. Participants in an ESM study carry an electronic pager and fill out reports when they are beeped. For example, 107 adults carried pagers in Robert Kubey's (1990) ESM study of television habits and family quality of life. Participants' reports indicated that heavy TV viewers were less active during non-TV family activities, although heavy TV viewers also spent more time with their families and felt as positively toward other family members as did those who watched less TV.

Although ESM is a powerful tool for field research, it is still limited by the need to recruit people to carry pagers. Ultimately, the generalizability of ESM findings relies on the representativeness, and reliability, of the persons who cooperate in the research.

Taking Notes

Written notes are the primary means of recording participant observation data (Emerson, Fretz, & Shaw 1995). Of course, "written" no longer means handwritten; many field researchers jot down partial notes while observing and then retreat to their computer to write up more complete notes on a daily basis. The computerized text can then be inspected and organized after it is printed out, or it can be marked up and organized for analysis using one of several computer programs designed especially for the task.

Jottings Brief notes written in the field about highlights of an observation period.

Field notes Notes that describe what has been observed, heard, or otherwise experienced in a participant observation study. These notes usually are written after the observational session.

It is almost always a mistake to try to take comprehensive notes while engaged in the field—the process of writing extensively is just too disruptive. The usual procedure is to jot down brief notes about highlights of the observation period. These brief notes, called **jottings**, can then serve as memory joggers when writing the actual **field notes** at a later session. It will also help maintain a daily log in which each day's activities are recorded (Bogdewic 1999:58–67). With the aid of the jottings and some practice, researchers usually remember a great deal of what happened—as long as the comprehensive field notes are written immediately afterward, or at least within the next 24 hours, and before they have discussed them with anyone else.

Exhibit 9.5 **Theoretical Sampling**

Original cases interviewed in a study of cocaine users:

Realization: Some cocaine users are businesspeople.
Add businesspeople to sample:

Realization: Sample is low on women.
Add women to sample:

Realization: Some female cocaine users are mothers of young children.
Add mothers to sample:

The following excerpts shed light on the note-taking processes that Tim Diamond and Barrie Thorne used while in the field. Taking notes was more of a challenge for Diamond (1992), because many people in the setting did not know that he was a researcher:

While I was getting to know nursing assistants and residents and experiencing aspects of their daily routines, I would surreptitiously take notes on scraps of paper, in the bathroom or otherwise out of sight, jotting down what someone had said or done. (pp. 6–7)

Thorne (1993) was able to take notes openly:

I went through the school days with a small spiral notebook in hand, jotting descriptions that I later expanded into field notes. When I was at the margins of a scene, I took notes on the spot. When I was more fully involved, sitting and talking with kids at a cafeteria table or playing a game of jump rope, I held observations in my memory and recorded them later. (p. 17)

Usually, writing up notes takes much longer—at least three times longer—than the observing. Field notes must be as complete, detailed, and true as possible to what was observed and heard. Direct quotes should be distinguished clearly from paraphrased quotes, and both should be set off from the researcher's observation and reflections. Pauses and interruptions should be indicated. The surrounding context should receive as much attention as possible, and a map of the setting should always be included with indications of where the individuals were at different times. The following excerpt from field notes collected by the DRC team members in the government's Joint Field Office show how notes can preserve a picture of the context:

> In the course of several weeks the [JFO] building has been wired to accommodate the increased electrical needs and the computer needs of the personnel. People were sleeping in bunk beds on site, in closets, and in corners of any room. The operation runs 24–7. Maps are hung on almost every wall with every type of imaginable data, from flooded areas to surge areas; total population and population density; number of housing, buildings, and people impacted by Katrina. . . . There is a logistics supply store that is full of materials and supplies, with a sign reminding people to take only what they need and this was a "no looting zone." The DRC team also observed flyers focusing on "stress management," as well as "how to cope with over-stressed workers." (Rodríguez et al. 2006:96)

Careful note taking yields a big payoff. On page after page, field notes will suggest new concepts, causal connections, and theoretical propositions. Social processes and settings can be described in rich detail, with ample illustrations. Exhibit 9.6, for example, contains field notes recorded by an anthropologist studying living arrangements for homeless mentally ill persons in the housing study for which I was a coinvestigator (Schutt 2011). The notes contain observations of the setting, the questions the anthropologist asked, the answers she received, and her analytic thoughts about one of the residents. What can be learned from just this one page of field notes? The mood of the house at this time is evident, with joking, casual conversation, and close friendships. "Dick" remarks on problems with household financial management, and, at the same time, we learn a bit about his own activities and personality (a regular worker who appears to like systematic plans). We see how a few questions and a private conversation elicit information about the transition from the shelter to the house, as well as about household operations. The field notes also provide the foundation for a more complete picture of one resident, describing "Jim's" relationships with others, his personal history, his interests and personality, and his orientation to the future. We can also see analytic concepts emerge in the notes, such as the concept of pulling himself together, and of some house members working as a team. You can imagine how researchers can go on to develop a theoretical framework for understanding the setting and a set of concepts and questions to inform subsequent observations.

Complete field notes must provide even more than a record of what was observed or heard. Notes also should include descriptions of the methodology: where researchers were standing or sitting while they observed, how they chose people for conversation or observation, what counts of people or events they made and why. Sprinkled throughout the notes also should be a record of the researchers' feelings and thoughts while observing: when they were disgusted by some statement or act, when they felt threatened or intimidated, why their attention shifted from one group to another, and what ethical concerns arose. Notes such as these provide a foundation for later review of the likelihood of bias or of inattention to some salient features of the situation.

Notes may, in some situations, be supplemented by still pictures, videotapes, and printed material circulated or posted in the research setting. Such visual material can bring an entirely different qualitative dimension into the analysis and call attention to some features of the social situation and actors within it that were missed in the notes (Grady 1996). Commentary on this material can be integrated with the written notes (Bogdewic 1999:67–68).

> **Exhibit 9.6** Field Notes From an Evolving Consumer Household (ECH)
>
> I arrive around 4:30 p.m. and walk into a conversation between Jim and somebody else as to what color jeans he should buy. There is quite a lot of joking going on between Jim and Susan. I go out to the kitchen and find Dick about to take his dinner out to the picnic table to eat (his idea?) so I go ask if I can join him. He says yes. In the course of the conversation, I find out that he works 3 days a week in the "prevoc" program at the local day program, Food Services branch, for which he gets $10 per week. Does he think the living situation will work out? Yes. All they need is a plan for things like when somebody buys something and then everybody else uses it. Like he bought a gallon of milk and it was gone in two days, because everyone was using it for their coffee. I ask if he's gone back to the shelter to visit and he says, "No. I was glad to get out of there." He came to [the ECH] from [a shelter] through homeless outreach [a Department of Mental Health program]. Had been at [the shelter] since January. Affirms that [the ECH] is a better place to live than the shelter. Why? Because you have your own room and privacy and stuff. How have people been getting along with each other? He says, "Fine."
>
> I return to the living room and sit down on the couch with Jim and Susan. Susan teases Jim and he jokes back. Susan is eating a T.V. dinner with M & M's for dessert. There is joking about working off the calories from the M & M's by doing sit-ups, which she proceeds to demonstrate. This leads to a conversation about exercise during which Jim declares his intention to get back into exercise by doing sports, like basketball.
>
> Jim seems to have his mind on pulling himself together, which he characterizes as "getting my old self back." When I ask him what he's been doing since I saw him last, he says, "Working on my appearance." And in fact, he has had a haircut, a shave, and washed his clothes. When I ask him what his old self was like, he says, "You mean before I lost everything?" I learn that he used to work two jobs, had "a family" and was into "religion." This seems to have been when he was quite young, around eighteen. He tells me he was on the street for 7–8 years, from 1978–1985, drinking the whole time. I ask him whether he thinks living at [the ECH] will help him to get his "old self back" and he says that it will "help motivate me." I observe that he seems pretty motivated already. He says yes, "but this will motivate me more."
>
> Jim has a warm personality, likes to joke and laugh. He also speaks up—in meetings he is among the first to say what he thinks and he talks among the most. His "team" relationship with Bill is also important to him—"me and Bill, we work together."

Managing the Personal Dimensions

Our overview of participant observation would not be complete without considering its personal dimensions. Because field researchers become a part of the social situation they are studying, they cannot help but be affected on a personal, emotional level. At the same time, those being studied react to researchers not just as researchers but as personal acquaintances—often as friends, sometimes as personal rivals. Managing and learning from this personal side of field research is an important part of any project.

The impact of personal issues varies with the depth of researchers' involvement in the setting. The more involved researchers are in the multiple aspects of the ongoing social situation, the more important personal issues become and the greater the risk of "going native." Even when researchers acknowledge their role, "increased contact brings sympathy, and sympathy in its turn dulls the edge of criticism" (Fenno 1978:277). Fenno minimized this problem by returning frequently to the university and by avoiding involvement in the personal lives of the congressional representatives he was studying. To study the social life of "corner boys," however, Whyte could not stay so disengaged. He moved into an apartment with a Cornerville family and lived for about 4 years in the community he was investigating:

> The researcher, like his informants, is a social animal. He has a role to play, and he has his own personality needs that must be met in some degree if he is to function successfully. Where the researcher operates out of a university, just going into the field for a few hours at a time, he can keep his personal

social life separate from field activity. His problem of role is not quite so complicated. If, on the other hand, the researcher is living for an extended period in the community he is studying, his personal life is inextricably mixed with his research. (Whyte 1955:279)

Hannah Gill (2004) tried not to get too inured to crime as she studied communities in the Dominican Republic: "After several weeks my initial fear faded and I found it necessary to make periodic 'reality checks' with supervisors and colleagues abroad, which remind a researcher desensitized to violence and crime not to get too comfortable" (p. 3).

Thorne (1993) wondered whether "my moments of remembering, the times when I felt like a ten-year-old girl, [were] a source of distortion or insight?" She concluded that they were both: "Memory, like observing, is a way of knowing and can be a rich resource," but "when my own responses . . . were driven by emotions like envy or aversion, they clearly obscured my ability to grasp the full social situation" (p. 26). Deborah Ceglowski (2002) found that

the feelings well up in my throat when Brian [a child in the Head Start program she studied] asks me to hold his hand. It's the gut reaction to hearing Ruth [a staff member] tell about Brian's expression when they pull into his yard and his mother isn't there. It is the caring connection of sitting next to Steven [another child] and hearing him say, "I miss my mom." (p. 15)

There is no formula for successfully managing the personal dimension of a field research project. It is much more an art than a science and flows more from the researcher's own personality and natural approach to other people than from formal training. Sharing similarities such as age, race, or gender with those who are studied may help to create mutual feelings of comfort, but such social similarities may mask more important differences in perspective due to education, social class, and having the role of researcher (Doucet & Mauthner 2008:334). Furthermore, novice field researchers often neglect to consider how they will manage personal relationships when they plan and carry out their projects. Then, suddenly, they find themselves doing something they don't believe they should just to stay in the good graces of research subjects or juggling the emotions resulting from conflict within the group. As Whyte (1955) noted,

The field worker cannot afford to think only of learning to live with others in the field. He has to continue living with himself. If the participant observer finds himself engaging in behavior that he has learned to think of as immoral, then he is likely to begin to wonder what sort of a person he is after all. Unless the field worker can carry with him a reasonably consistent picture of himself, he is likely to run into difficulties. (p. 317)

If you plan a field research project, follow these guidelines:

- Take the time to consider how you want to relate to your potential subjects as people.

- Speculate about what personal problems might arise and how you will respond to them.

- Keep in touch with other researchers and personal friends outside the research setting.

- Maintain standards of conduct that make you comfortable as a person and that respect the integrity of your subjects. (Whyte 1955:300–317)

When you evaluate participant observers' reports, pay attention to how they defined their role in the setting and dealt with personal problems. Don't place too much confidence in such research unless the report provides this information. The primary strengths of participant observation—learning about the social world from the participants' perspectives, as they experience it, and minimizing the distortion of these perspectives by the methods used to measure them—should not blind us to its primary weaknesses—the lack of consistency in the data collected, particularly when different observers are used, and the many opportunities for direct influence of the researchers' perspective on what is observed. Whenever we consider using the method of participant observation, we also must realize that the need to focus so much attention on each setting studied will severely restrict the possible number of settings or people we can study.

🖳 Systematic Observation

Observations can be made in a more systematic, quantitative design that allows systematic comparisons and more confident generalizations. A researcher using systematic observation develops a standard form on which to record variation within the observed setting in terms of variables of interest. Such variables might include the frequency of some behavior(s), the particular people observed, the weather or other environmental conditions, and the number and state of repair of physical structures. In some systematic observation studies, records will be obtained from a random sample of places or times.

> **Systematic observation** A strategy that increases the reliability of observational data by using explicit rules that standardize coding practices across observers.

Robert Sampson and Stephen Raudenbush's (1999) study of disorder and crime in urban neighborhoods provides an excellent example of systematic observation methods. Although you learned about some features of this pathbreaking research in Chapter 6, in this section I elaborate on their use of systematic social observation to learn about these neighborhoods. A systematic observational strategy increases the reliability of observational data by using explicit rules that standardize coding practices across observers (Reiss 1971b). It is a method particularly well suited to overcome one of the limitations of survey research on crime and disorder: Residents who are fearful of crime perceive more neighborhood disorder than do residents who are less fearful, even though both are observing the same neighborhood (Sampson & Raudenbush 1999:606).

This ambitious multiple-methods investigation combined observational research, survey research, and archival research. The observational component involved a stratified probability (random) sample of 196 Chicago census tracts. A specially equipped sport-utility vehicle was driven down each street in these tracts at the rate of 5 miles per hour. Two video recorders taped the blocks on both sides of the street, while two observers peered out of the vehicle's windows and recorded their observations in the logs. The result was an observational record of 23,816 face blocks (the block on one side of the street is a face block). The observers recorded in their logs codes that indicated land use, traffic, physical conditions, and evidence of physical disorder (see Exhibit 9.7). The videotapes were sampled and then coded for 126 variables, including housing characteristics, businesses, and social interactions. Physical disorder was measured by counting such features as cigarettes or cigars in the street, garbage, empty beer bottles, graffiti, condoms, and syringes. Indicators of social disorder included adults loitering, drinking alcohol in public, fighting, and selling drugs. To check for reliability, a different set of coders recoded the videos for 10% of the blocks. The repeat codes achieved 98% agreement with the original codes.

Sampson and Raudenbush also measured crime levels with data from police records, census tract socioeconomic characteristics with census data, and resident attitudes and behavior with a survey. As you learned in Chapter 6, the combination of data from these sources allowed a test of the relative impact on the crime rate of informal social control efforts by residents and of the appearance of social and physical disorder.

Peter St. Jean (2007) extended the research of Sampson and Raudenbush with a mixed-methods study of high crime areas that used resident surveys, participant observation, in-depth interviews with

| Exhibit 9.7 | Neighborhood Disorder Indicators Used in Systematic Observation Log |

Variable	Category	Frequency
Physical Disorder		
Cigarettes, cigars on street or gutter	no	6,815
	yes	16,758
Garbage, litter on street or sidewalk	no	11,680
	yes	11,925
Empty beer bottles visible in street	no	17,653
	yes	5,870
Tagging graffiti	no	12,859
	yes	2,252
Graffiti painted over	no	13,390
	yes	1,721
Gang graffiti	no	14,138
	yes	973
Abandoned cars	no	22,782
	yes	806
Condoms on sidewalk	no	23,331
	yes	231
Needles/syringes on sidewalk	no	23,392
	yes	173
Political message graffiti	no	15,097
	yes	14
Social Disorder		
Adults loitering or congregating	no	14,250
	yes	861
People drinking alcohol	no	15,075
	yes	36
Peer group, gang indicators present	no	15,091
	yes	20
People intoxicated	no	15,093
	yes	18
Adults fighting or hostilely arguing	no	15,099
	yes	12
Prostitutes on street	no	15,100
	yes	11
People selling drugs	no	15,099
	yes	12

Exhibit 9.8 One Building in St. Jean's (2007) Study

residents and offenders, and also systematic social observation. St. Jean recorded neighborhood physical and social appearances with video cameras mounted in a van that was driven along neighbourhood streets. Pictures were then coded for the presence of neighbourhood disorder (see Exhibit 9.8 and the Student Study Site).

This study illustrates both the value of multiple methods and the technique of recording observations in a form from which quantitative data can be obtained. The systematic observations give us much greater confidence in the measurement of relative neighborhood disorder than we would have from unstructured descriptive reports or from responses of residents to survey questions. Interviews with residents and participant observation helped to identify the reasons that offenders chose particular locations when deciding where to commit crimes.

Intensive Interviewing

Intensive or depth interviewing is a qualitative method of finding out about people's experiences, thoughts, and feelings. Although intensive interviewing can be an important element in a participant observation study, it is often used by itself (Wolcott 1995:102–105). It shares with other qualitative research methods a commitment to learning about people in depth and on their own terms, and in the context of their situation.

Unlike the more structured interviewing that may be used in survey research (discussed in Chapter 8), intensive or depth interviewing relies on open-ended questions. Rather than asking standard questions in a fixed order, intensive interviewers may allow the specific content and order of questions to vary from one interviewee to another. Rather than presenting fixed responses that presume awareness of the range of answers that respondents might give, intensive interviewers expect respondents to answer questions in their own words.

What distinguishes intensive interviewing from less structured forms of questioning is consistency and thoroughness. The goal is to develop a comprehensive picture of the interviewee's background, attitudes, and actions, in his or her own terms; to "listen to people as they describe how they understand the worlds in which they live and work" (Rubin & Rubin 1995:3). For example, Spencer Moore and his colleagues (2004) sought through intensive interviewing of key community leaders "to elicit a more general discussion on county-wide events during the [Hurricane Floyd] flooding" (p. 209). The DRC researchers paid special attention to "first-hand personal accounts by individuals speaking about their own behavior" (Rodríguez et al. 2006:86).

Intensive interview studies do not reveal, as directly as does participant observation, the social context in which action is taken and opinions are formed. Nonetheless, intensive depth interviewers seek to take context into account. Jack D. Douglas (1985) made the point succinctly in *Creative Interviewing*:

> *Creative interviewing is purposefully situated interviewing.* Rather than denying or failing to see the situation of the interview as a determinant of what goes in the questioning and answering processes, creative interviewing embraces the immediate, concrete situation; tries to understand how it is affecting what is communicated; and, by understanding these effects, changes the interviewer's communication processes to increase the discovery of the truth about human beings. (p. 22)

So, similar to participant observation studies, intensive interviewing engages researchers more actively with subjects than standard survey research does. The researchers must listen to lengthy explanations, ask follow-up questions tailored to the preceding answers, and seek to learn about interrelated belief systems or personal approaches to things rather than measure a limited set of variables. As a result, intensive interviews are often much longer than standardized interviews, sometimes as long as 15 hours, conducted in several different sessions. The intensive interview becomes more like a conversation between partners than an interview between a researcher and a subject (Kaufman 1986:22–23). Some call it "a conversation with a purpose" (Rossman & Rallis 1998:126).

Intensive interviewers actively try to probe understandings and engage interviewees in a dialogue about what they mean by their comments. To prepare for this active interviewing, the interviewer should learn in advance about the setting to be studied. Preliminary discussion with key informants, inspection of written documents, and even a review of your own feelings about the setting can all help (Miller & Crabtree 1999c:94–96). Robert Bellah, Richard Madsen, William Sullivan, Ann Swidler, and Steven Tipton (1985) elaborate on this aspect of intensive interviewing in a methodological appendix to their national best-seller about American individualism, *Habits of the Heart:*

> We did not, as in some scientific version of "Candid Camera," seek to capture their beliefs and actions without our subjects being aware of us, rather, we sought to bring our preconceptions and questions into the conversation and to understand the answers we were receiving not only in terms of the language but also so far as we could discover, in the lives of those we were talking with. Though we did not seek to impose our ideas on those with whom we talked . . . , we did attempt to uncover assumptions, to make explicit what the person we were talking to might rather have left implicit. The interview as we employed it was active, Socratic. (p. 304)

The intensive interview follows a preplanned outline of topics. It may begin with a few simple questions that gather background information while building rapport. These are often followed by a few general **grand tour questions** that are meant to elicit lengthy narratives (Miller & Crabtree 1999c:96–99). Some projects may use relatively structured interviews, particularly when the focus is on developing knowledge about prior events or some narrowly defined topic. But more exploratory projects, particularly those aiming to learn about interviewees' interpretations of the world, may let each interview flow in a unique direction in response to the interviewee's experiences and interests (Kvale 1996:3–5; Rubin & Rubin 1995:6; Wolcott 1995:113–114). In either case, qualitative interviewers must adapt nimbly throughout the interview, paying attention to nonverbal cues, expressions with symbolic value, and the ebb and flow of the interviewee's feelings and interests. "You have to be free to follow your data where they lead" (Rubin & Rubin 1995:64).

> **Grand tour question** A broad question at the start of an interview that seeks to engage the respondent in the topic of interest.

Random selection is rarely used to select respondents for intensive interviews, but the selection method still must be considered carefully. If interviewees are selected in a haphazard manner, as by speaking just to those who happen to be available at the time when the researcher is on site, the interviews are likely to be of less value than when a more purposive selection strategy is used. Researchers should try to select interviewees who are knowledgeable about the subject of the interview, who are open to talking, and who represent the range of perspectives (Rubin & Rubin 1995:65–92). Selection of new interviewees should continue, if possible, at least until the **saturation point** is reached, the point when new interviews seem to yield little additional information (see Exhibit 9.9). As new issues are uncovered, additional interviewees may be selected to represent different opinions about these issues.

> **Saturation point** The point at which subject selection is ended in intensive interviewing, when new interviews seem to yield little additional information.

Establishing and Maintaining a Partnership

Because intensive interviewing does not engage researchers as participants in subjects' daily affairs, the problems of entering the field are much reduced. However, the social processes and logistics of arranging long periods for personal interviews can still be pretty complicated. Dag Nordanger (2007:177) made a social visit to his Tigrayan interviewees before the interviews to establish trust and provide information about the interview. It also is important to establish rapport with subjects by considering in advance how they will react

---

Exhibit 9.9 | **The Saturation Point in Intensive Interviewing**

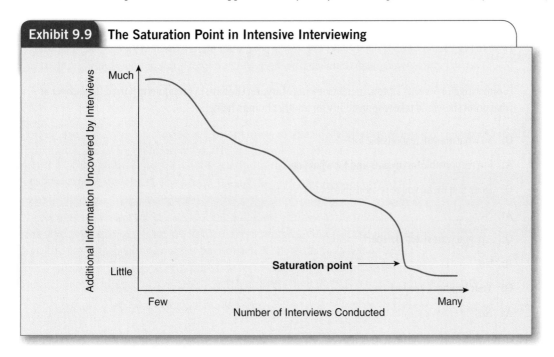

to the interview arrangements and by developing an approach that does not violate their standards for social behavior. Interviewees should be treated with respect, as knowledgeable partners whose time is valued (in other words, avoid coming late for appointments). A commitment to confidentiality should be stated and honored (Rubin & Rubin 1995).

But the intensive interviewer's relationship with the interviewee is not an equal partnership, because the researcher seeks to gain certain types of information and strategizes throughout to maintain an appropriate relationship (Kvale 1996:6). In the first few minutes of the interview, the goal is to show interest in the interviewee and to explain clearly what the purpose of the interview is (p. 128). During the interview, the interviewer should maintain an appropriate distance from the interviewee, one that doesn't violate cultural norms; the interviewer should maintain eye contact and not engage in distracting behavior. An appropriate pace is also important; pause to allow the interviewee to reflect, elaborate, and generally not feel rushed (Gordon 1992). When an interview covers emotional or otherwise stressful topics, the interviewer should give the interviewee an opportunity to unwind at the interview's end (Rubin & Rubin 1995:138).

More generally, intensive interviewers must be sensitive to the broader social context of their interaction with the interviewee and to the implications of their relationship in the way they ask questions and interpret answers. Tom Wengraf (2001) cautions new intensive interviewers to take account of their unconscious orientations to others based on prior experience:

> Your [prior] experience of being interviewed may lead you to behave and "come across" in your interviewing . . . like a policeman, or a parent, a teacher or academic, or any "authority" by whom you have been interviewed and from whom you learnt a way of handling stress and ambiguity. (p. 18)

Asking Questions and Recording Answers

Intensive interviewers must plan their main questions around an outline of the interview topic. The questions should generally be short and to the point. More details can then be elicited through nondirective probes (e.g., "Can you tell me more about that?" or "uh-huh," echoing the respondent's comment, or just maintaining a moment of silence). Follow-up questions can then be tailored to answers to the main questions.

Interviewers should strategize throughout an interview about how best to achieve their objectives while taking into account interviewees' answers. *Habits of the Heart* again provides a useful illustration (Bellah et al. 1985:304):

> [Coinvestigator Steven] Tipton, in interviewing Margaret Oldham [a pseudonym], tried to discover at what point she would take responsibility for another human being:
>
> **Q:** So what are you responsible for?
>
> **A:** I'm responsible for my acts and for what I do.
>
> **Q:** Does that mean you're responsible for others, too?
>
> **A:** No.
>
> **Q:** Are you your sister's keeper?
>
> **A:** No.
>
> **Q:** Your brother's keeper?
>
> **A:** No.

Q: Are you responsible for your husband?

A: I'm not. He makes his own decisions. He is his own person. He acts his own acts. I can agree with them, or I can disagree with them. If I ever find them nauseous enough, I have a responsibility to leave and not deal with it any more.

Q: What about children?

A: I . . . I would say I have a legal responsibility for them, but in a sense I think they in turn are responsible for their own acts.

Do you see how the interviewer actively encouraged the subject to explain what she meant by "responsibility"? This sort of active questioning undoubtedly did a better job of clarifying her concept of responsibility than a fixed set of questions would have.

The active questioning involved in intensive interviewing, without a fixed interview script, also means that the statements made by the interviewee can only be understood in the context of the interviewer's questions. Tom Wengraf's (2001:28–30) advice to a novice interviewer provides an example of how the interviewer's statements and questions can affect the interview process:

Interviewer:	Thank you for giving up this time for me.
Interviewee:	Well, I don't see it as giving up the time, more as contributing . . .
Interviewer:	Well, for giving me the time, contributing the time, thank you very much.
Wengraf:	*Stay silent, let him clarify whatever the point is he wishes to make.*
Later,	
Interviewer:	Ok, so you're anonymous [repeating an earlier statement], so you can say what you like.
Wengraf:	*Don't imply that he is slow on the uptake—it might be better to cut the whole sentence.*
Later,	
Interviewee:	I guess, it's difficult . . . being the breadwinner. Myself as a father, er . . . I'm not sure.
Wengraf:	*He's uncertain, hoping to be asked some more about "being a father himself today."*
Interviewer	
[Slightly desperately]:	Perhaps you could tell me a little about your own father.
Wengraf:	*Interviewer ignores implied request but moves eagerly on to her intended focus on him as a son.*

You can see in this excerpt how, at every step, the interviewer is actively constructing with the interviewee the text of the interview that will be analyzed. Becoming a good intensive interviewer means learning how to "get out of the way" as much as possible in this process. Becoming an informed critic of intensive interview studies means, in part, learning to consider how the social interaction between the interviewer and interviewee may have shaped in subtle ways what the interviewee said and to look for some report on how this interaction was managed.

Tape recorders commonly are used to record intensive and focus group interviews. Most researchers who have tape recorded interviews (including me) feel that they do not inhibit most interviewees and, in fact, are routinely ignored. The occasional respondent is very concerned with his or her public image and may therefore speak "for the tape recorder," but such individuals are unlikely to speak frankly in any research interview. In any case, constant note taking during an interview prevents adequate displays of interest and appreciation by the interviewer and hinders the degree of concentration that results in the best interviews.

Of course, there are exceptions to every rule. Fenno (1978) presents a compelling argument for avoiding the tape recorder when interviewing public figures who are concerned with their public image:

> My belief is that the only chance to get a nonroutine, nonreflexive interview [from many of the members of Congress] is to converse casually, pursuing targets of opportunity without the presence of a recording instrument other than myself. If [worse] comes to worst, they can always deny what they have said in person; on tape they leave themselves no room for escape. I believe they are not unaware of the difference. (p. 280)

Interviewing Online

Our social world now includes many connections initiated and maintained through e-mail and other forms of web-based communication, so it is only natural that interviewing has also moved online. Online interviewing can facilitate interviews with others who are separated by physical distance; it also is a means to conduct research with those who are only known through such online connections as a discussion group or an e-mail distribution list (James & Busher 2009:14).

Online interviews can be either synchronous—in which the interviewer and interviewee exchange messages as in online chatting—or asynchronous—in which the interviewee can respond to the interviewer's questions whenever it is convenient, usually through e-mail. Both styles of online interviewing have advantages and disadvantages (James & Busher 2009:13–16). Synchronous interviewing provides an experience more similar to an in-person interview, thus giving more of a sense of obtaining spontaneous reactions, but it requires careful attention to arrangements and is prone to interruptions. Asynchronous interviewing allows interviewees to provide more thoughtful and developed answers, but it may be difficult to maintain interest and engagement if the exchanges continue over many days. The online asynchronous interviewer should plan carefully how to build rapport as well as how to terminate the online relationship after the interview is concluded (King & Horrocks 2010:86–93).

Whether a synchronous or asynchronous approach is used, online interviewing can facilitate the research process by creating a written record of the entire interaction without the need for typed transcripts. The relative anonymity of online communications can also encourage interviewees to be more open and honest about their feelings than they would be if interviewed in person (James & Busher 2009:24–25). However, online interviewing lacks some of the most appealing elements of qualitative methods: The revealing subtleties of facial expression, intonation, and body language are lost, and the intimate rapport that a good intensive interviewer can develop in a face-to-face interview cannot be achieved. In addition, those who are being interviewed have much greater ability to present an identity that is completely removed from their in-person persona; for instance, basic characteristics such as age, gender, and physical location can be completely misrepresented.

▣ Focus Groups

Focus groups are groups of unrelated individuals that are formed by a researcher and then led into group discussion of a topic for 1 to 2 hours (Krueger & Casey 2009:2). The researcher asks specific questions and guides the discussion to ensure that group members address these questions, but the resulting information is qualitative and relatively unstructured. Focus groups do not involve representative samples; instead, a few individuals are recruited who have the time to participate, have some knowledge pertinent to the focus group topic, and share key characteristics with the target population. Focus group research projects usually involve several discussions involving similar participants.

Focus groups have their roots in the interviewing techniques developed in the 1930s by sociologists and psychologists who were dissatisfied with traditional surveys. Traditionally, in a questionnaire survey, subjects are directed to consider certain issues and particular response options in a predetermined order. The spontaneous exchange and development of ideas that characterize social life outside the survey situation is lost—and with it, some social scientists fear, the prospects for validity.

During World War II, the military used focus groups to investigate morale, and then the great American sociologist Robert K. Merton and two collaborators, Marjorie Fiske and Patricia Kendall, popularized them in *The Focused Interview* (1956). But marketing researchers were the first to adopt focus groups as a widespread methodology. Marketing researchers use focus groups to investigate likely popular reactions to possible advertising themes and techniques. Their success has prompted other social scientists to use focus groups to evaluate social programs and to assess social needs (Krueger & Casey 2009:3–4).

Focus groups are used to collect qualitative data, using open-ended questions posed by the researcher (or group leader). Thus, a focused discussion mimics the natural process of forming and expressing opinions. The researcher, or group moderator, uses an interview guide, but the dynamics of group discussion often require changes in the order and manner in which different topics are addressed (Brown 1999:120). No formal procedure exists for determining the generalizability of focus group answers, but the careful researcher should conduct at least several focus groups on the same topic and check for consistency in the findings. Some focus group experts advise conducting enough focus groups to reach the point of saturation, when an additional focus group adds little new information to that which already has been generated (Brown 1999:118). When differences in attitudes between different types of people are a concern, separate focus groups may be conducted that include these different types, and then the analyst can compare comments between them (Krueger & Casey 2009:21).

Most focus groups involve 5 to 10 people, a number that facilitates discussion by all in attendance (Krueger & Casey 2009:6). Participants usually do not know one another, although some studies in organized settings may include friends or coworkers. Opinions differ on the value of using homogeneous versus heterogeneous participants. Homogeneous groups may be more convivial and willing to share feelings, but heterogeneous groups may stimulate more ideas (Brown 1999:115–117). In any case, it is important to avoid having some participants who have supervisory or other forms of authority over other participants (Krueger & Casey 2009:22). It is also good to avoid focus groups when dealing with emotionally charged issues, when sensitive information is needed and when confidentiality cannot be ensured, or when the goal is to reach consensus (Krueger & Casey 2009:20).

Focus group moderators must begin the discussion by generating interest in the topic, creating the expectation that all will participate and making it clear that the researcher does not favor any particular perspective or participant (Smithson 2008:361). All questions should be clear, simple, and straightforward. The moderator should begin with easy to answer general factual questions and then, about one-quarter to halfway through the allotted time, shift to key questions on specific issues. In some cases, discussion may be stimulated by asking participants to make a list of concerns or experiences, to rate predefined items, or to choose among alternatives. If the question flow is successful, the participants should experience the focus group as an engaging and friendly discussion and the moderator should spend more time after the introductory period listening and guiding the discussion than asking questions. Disagreements should be handled carefully so that no participants feel uncomfortable and the discussion should be kept focused on the announced topic (Smithson 2008:361). The moderator may conclude the group discussion by asking participants for recommendations to policy makers or their further thoughts that they have not had a chance to express (Krueger & Casey 2009:36–48).

Keith Elder and his colleagues at the University of South Carolina and elsewhere (2007:S125) used focus groups to study the decisions by African Americans not to evacuate New Orleans before Hurricane Katrina. They conducted six focus groups with a total of 53 evacuees who were living in hotels in Columbia, South Carolina, between October 3 and October 14, 2005. African American women conducted the focus groups after American Red Cross relief coordinators announced them at a weekly "town hall" meeting.

One of the themes identified in the focus groups was the confusion resulting from inconsistent messages about the storm's likely severity. "Participants reported confusion about what to do because of inappropriate timing of mandatory evacuation orders and confusing recommendations from different authorities" (Elder et al. 2007:S126):

> The mayor did not say it was a mandatory evacuation at first. One or two days before the hurricane hit, he said it was mandatory. It was too late then.

> They didn't give us no warning. . . . When they said leave, it was already too late.

> After [the] levees broke the mayor said mandatory evacuation, before then he was not saying mandatory evacuation.

> Governor said on TV, you didn't want to go, you didn't have to go, cause it was no threat to us, she said.

Focus groups are now used extensively in political campaigns, as a quick means of generating insight into voter preferences and reactions to possible candidate positions. For example, Michigan Democratic legislators used focus groups to determine why voters were turning away from them in 1985. Elizabeth Kolbert (1992) found that white, middle-class Democrats were shifting to the Republican Party because of their feelings about race:

> These Democratic defectors saw affirmative action as a direct threat to their own livelihoods, and they saw the black-majority city of Detroit as a sinkhole into which their tax dollars were disappearing. . . . The participants listen[ed] to a quotation from Robert Kennedy exhorting whites to honor their "special obligation" to blacks. Virtually every participant in the four groups—37 in all—reacted angrily. (p. 21)

Focus group methods share with other field research techniques an emphasis on discovering unanticipated findings and exploring hidden meanings. They can be an indispensable aid for developing hypotheses and survey questions, for investigating the meaning of survey results, and for quickly assessing the range of opinion about an issue. The group discussion reveals the language participants used to discuss topics and think about their experiences (Smithson 2008:359). Because it is not possible to conduct focus groups with large, representative samples, it is always important to consider how recruitment procedures have shaped the generalizability of focus group findings. The issue of impact of interviewer style and questioning on intensive interview findings, which was discussed in the previous section, also must be considered when evaluating the results of focus groups.

🔲 **Participatory Action Research**

William Foote Whyte (1991) urged social researchers to engage with research participants throughout the research process. He formalized this recommendation into an approach he termed **participatory action research** (PAR). As the name implies, this approach encourages social researchers to get "out of the academic rut" and bring values into the research process (p. 285). Since Whyte's early call for this type of research, with a focus on research with organizational employees, PAR has become increasingly popular in disciplines ranging from public health to social work, as well as sociology (McIntyre 2008; Minkler 2000). Participatory action research is not itself a qualitative method, but PAR projects tend to use qualitative methods, which are more accessible to members of the lay public and which normally involve some of the same activities as in PAR: engaging with individuals in their natural settings and listening to them in their own words.

> **Participatory action research (PAR)** A type of research in which the researcher involves members of the population to be studied as active participants throughout the research process, from the selection of a research focus to the reporting of research results. Also termed *community-based participatory research*.

In PAR, also termed *community-based participatory research* (CBPR), the researcher involves as active participants some members of the setting studied. Both the members and the researcher are assumed to want to develop valid conclusions, to bring unique insights, and to desire change, but Whyte (1991) believed that these objectives were more likely to be obtained if the researcher collaborates actively with the persons being studied. For example, many academic studies have found that employee participation is associated with job satisfaction but not with employee productivity. After some discussion about this finding with employees and managers, Whyte realized that researchers had been using a general concept of employee participation that did not distinguish those aspects of participation that were most likely to influence productivity (pp. 278–279). For example, occasional employee participation in company meetings had not been distinguished from ongoing employee participation in and control of production decisions. When these and other concepts were defined more precisely, it became clear that employee participation in production decisions had substantially increased overall productivity, whereas simple meeting attendance had not. This discovery would not have occurred without the active involvement of company employees in planning the research.

Those who engage in PAR projects are making a commitment "to listen to, learn from, solicit and respect the contributions of, and share power, information, and credit for accomplishments with the groups that they are trying [to] learn about and help (Horowitz, Robinson, & Seifer 2009:2634). The emphasis on developing a partnership with the community is reflected in the characteristics of the method presented in a leading health journal (see Exhibit 9.10). Each characteristic (with the exception of "emphasis on multiple determinants of health") identifies a feature of the researcher's relationship with community members.

PAR can bring researchers into closer contact with participants in the research setting through groups that discuss and plan research steps and then take steps to implement research findings. Stephen Kemmis and Robin McTaggart (2005:563–568) summarize the key steps in the process of conducting a PAR project as creating "a spiral of self-reflecting cycles":

- Planning a change
- Acting and observing the process and consequences of the change
- Reflecting on these processes and consequences
- Replanning
- Acting and observing again

Exhibit 9.10 **Characteristics of CBPR**

Community members and researchers contribute equally and in all phases of research.

Trust, collaboration, shared decision making, and shared ownership of the research; findings and knowledge benefit all partners.

Researchers and community members recognize each other's expertise in bidirectional, colearning process.

Balance rigorous research and tangible community action.

Embrace skills, strength, resources, and assets of local individuals and organizations.

Community recognized as a unit of identity.

Emphasis on multiple determinants of health.

Partners commit to long-term research relationships.

Core elements include local capacity building, system development, empowerment, and sustainability.

In contrast with the formal reporting of results at the end of a research project, these cycles make research reporting an ongoing part of the research process. Community partners work with the academic researchers to make changes in the community reflecting the research findings. Publication of results is only part of the process.

Karen Hacker at Harvard and the Institute for Community Health in Cambridge, Massachusetts, collaborated with community partners in response to a public health emergency in the adjacent town of Somerville (Hacker et al. 2008). After a series of youth suicides and overdoses from 2000 to 2005, a PAR community coalition was formed with members from mental health service providers, school leaders, police, and community parents. After reviewing multiple statistics, the coalition concluded that the deaths represented a considerable increase over previous years. However, when mental health professionals attempted to interview family members of adolescents who had committed suicide in order to investigate the background to the suicides, they were rebuffed; in contrast, family members were willing to talk at length with PAR members from their community. The PAR team was then able to map out the relationships between the suicide victims. The process of using the results of this research to respond to the suicides included a candlelight vigil, a speak-out against substance abuse, the provision of crisis counseling, and programs to support families and educate the community. Subsequently, the suicide rate dropped back to its pre-2000 level.

🔲 Ethical Issues in Qualitative Research

No matter how hard the qualitative researcher strives to study the social world naturally, leaving no traces, the very act of research itself imposes something "unnatural" on the situation, so the qualitative researcher may have an impact that has ethical implications. Six ethical issues should be given particular attention:

1. *Voluntary participation:* Ensuring that subjects are participating in a study voluntarily is not often a problem with intensive interviewing and focus group research, but even in online interviews, all respondents must be given clear information about the research and provided the opportunity to decline to participate (King & Horrocks 2010:99). When qualitative interviews are conducted in multiple sessions, the opportunity

to withdraw consent should always be available to participants (King & Horrocks 2010:115). Maintaining the standard of voluntary participation can present more challenges in participant observation studies. Few researchers or institutional review boards are willing to condone covert participation because it offers no way to ensure that participation by the subjects is voluntary. Even when the researcher's role is more open, interpreting the standard of voluntary participation still can be difficult. Practically, much field research would be impossible if the participant observer were required to request permission of everyone having some contact, no matter how minimal, with a group or setting being observed. And should the requirement of voluntary participation apply equally to every member of an organization being observed? What if the manager consents, the workers are ambivalent, and the union says no? Requiring everyone's consent would limit participant observation research to settings without serious conflicts of interest.

2. *Subject well-being:* Every field researcher should consider carefully before beginning a project how to avoid harm to subjects. It is not possible to avoid every theoretical possibility of harm nor to be sure that any project will cause no adverse consequences whatsoever to any individual. Some of the Cornerville men read Whyte's book and felt discomfited by it (others found it enlightening). Some police accused Van Maanen of damaging their reputation with his studies. But such consequences could follow from any research, even from any public discourse. Direct harm to the reputations or feelings of particular individuals is what researchers must carefully avoid. They can do so, in part, by maintaining the confidentiality of research subjects. They also must avoid affecting adversely the course of events while engaged in a setting. Whyte (1955:335–337) found himself regretting having recommended that a particular politician be allowed to speak to a social club he was observing, because the speech led to serious dissension in the club and strains between Whyte and some of the club members. Some indigenous groups and disadvantaged communities now require that researchers seek approval for all their procedures and request approval before releasing findings, in order to prevent harm to their culture and interests (Lincoln 2009:160–161). Researchers should spend time in local settings before the research plan is finalized in order to identify a range of community stakeholders who can then be consulted about research plans (Bledsoe & Hopson 2009:400). These problems are less likely in intensive interviewing and focus groups, but even there, researchers should try to identify negative feelings and help distressed subjects cope with their feelings through debriefing or referrals for professional help. Online interviewing can create additional challenges for interviews in which respondents become inappropriately personal over time (King & Horrocks 2010:100–101). Plans for avoiding personal involvement and options for referring interviewees for assistance should be made in advance.

3. *Identity disclosure:* We already have considered the problems of identity disclosure, particularly in the case of covert participation. Current ethical standards require informed consent of research subjects, and most would argue that this standard cannot be met in any meaningful way if researchers do not disclose fully their identity. But how much disclosure about the study is necessary, and how hard should researchers try to make sure that their research purposes are understood? In field research on Codependents Anonymous, Leslie Irvine (1998) found that the emphasis on anonymity and expectations for group discussion made it difficult to disclose her identity. Less-educated subjects may not readily comprehend what a researcher is or be able to weigh the possible consequences of the research for themselves. Internet-based research can violate the principles of voluntary participation and identity disclosure when researchers participate in discussions and record and analyze text but do not identify themselves as researchers (Jesnadum 2000). The intimacy of the researcher-participant relationship in much qualitative research makes it difficult to inject reminders about the research into ongoing social interaction (Mabry 2008:221). Must researchers always inform everyone of their identity as researchers? Should researchers inform subjects if the study's interests and foci change while it is in progress? Can a balance be struck between the disclosure of critical facts and a coherent research strategy?

4. *Confidentiality:* Field researchers normally use fictitious names for the characters in their reports, but doing so does not always guarantee confidentiality to their research subjects. Individuals in the setting studied may be able to identify those whose actions are described and thus may become privy to some knowledge about their colleagues or neighbors that had formerly been kept from them. Therefore, researchers should make every effort to expunge possible identifying material from published information and to alter unimportant aspects of a description when necessary to prevent identity disclosure. In any case, no field research project should begin if some participants clearly will suffer serious harm by being identified in project publications. Focus groups create a particular challenge, since the researcher cannot guarantee that participants will not disclose information that others would like to be treated as confidential. This risk can be reduced at the start of the focus group by reviewing a list of "dos and don'ts" with participants. Focus group methodology simply should not be used for very personally sensitive topics (Smithson 2008:360–361).

5. *Appropriate boundaries:* This is an ethical issue that cuts across several of the others, including identity disclosure, subject well-being, and voluntary participation. You probably are familiar with this issue in the context of guidelines for professional practice: Therapists are cautioned to maintain appropriate boundaries with patients; teachers must maintain appropriate boundaries with students. This is a special issue in qualitative research because it often involves lessening the boundary between the "researcher" and the research "subject." Qualitative researchers may seek to build rapport with those they plan to interview by expressing an interest in their concerns and conveying empathy for their situation. Is this just "faking friendship" for the purpose of the research? Jean Duncombe and Julie Jessop (2002) posed the dilemma clearly in a book chapter entitled, "'Doing rapport' and the ethics of 'faking friendship.'" The long-term relationships that can develop in participant observation studies can make it seem natural for researchers to offer tangible help to research participants, such as helping take a child to school or lending funds. These involvements can in turn make it difficult to avoid becoming an advocate for the research participants, rather than a sympathetic observer.

> With deeper rapport, interviewees become more likely to explore their more intimate experiences and emotions. Yet they also become more likely to discover and disclose experiences and feelings which, upon reflection, they would have preferred to keep private from others . . . , or not to acknowledge even to themselves. (p. 112)

Participatory action research projects can heighten this dilemma, and make it difficult for researchers to disagree with the interpretations of community members about the implications of research findings. Focus groups in cross-cultural and cross-national research also create challenges for maintaining appropriate boundaries, since cultures differ in their openness to dissent and their sensitivity to public discussion of personal issues (Smithson 2008:365–366). Qualitative researchers need to be sensitive to the potential for these problems and respond flexibly and respectfully to the concerns of research participants.

6. *Researcher safety:* Research "in the field" about disasters or simply unfamiliar neighborhoods or nations should not begin until any potential risks to researcher safety have been evaluated. Qualitative methods may provide the only opportunity to learn about organized crime in Russian ports (Belousov et al. 2007) or street crime in the Dominican Republic (Gill 2004), but they should not be used if the risks to the researchers are unacceptably high. Safety needs to be considered at the time of designing the research, not as an afterthought on arriving in the research site. As Hannah Gill (2004) learned, such advance planning can require more investigation than just reading the local newspapers: "Due to the community's marginality, most crimes, including murders, were never reported in newspapers, making it impossible to have known the insecurity of the field site ahead of time" (p. 2).

But being realistic about evaluating risk does not mean simply accepting misleading assumptions about unfamiliar situations or communities. Reports of a widespread breakdown in law and order in New Orleans were broadcast repeatedly after Hurricane Katrina, but the DRC researchers found that most nontraditional behavior in that period was actually "prosocial," rather than antisocial (Rodríguez et al. 2004).

> One group named itself the "Robin Hood Looters." The core of this group consisted of eleven friends who, after getting their own families out of the area, decided to remain at some high ground and, after the floodwaters rose, commandeered boats and started to rescue their neighbors. . . . For about two weeks they kept searching in the area. . . . They foraged for food and water from abandoned homes, and hence their group name. Among the important norms that developed were that they were going to retrieve only survivors and not bodies and that group members would not carry weapons. The group also developed informal understandings with the police and the National Guard. (p. 91)

These ethical issues cannot be evaluated independently. The final decision to proceed must be made after weighing the relative benefits and risks to participants. Few qualitative research projects will be barred by consideration of these ethical issues, however, except for those involving covert participation. The more important concern for researchers is to identify the ethically troublesome aspects of their proposed research and resolve them before the project begins and to act on new ethical issues as they come up during the project.

回 Conclusions

Qualitative research allows the careful investigator to obtain a richer and more intimate view of the social world than is possible with more structured methods. It is not hard to understand why so many qualitative studies have become classics in the social science literature. And, the emphases in qualitative research on inductive reasoning and incremental understanding help stimulate and inform other research approaches. Exploratory research to chart the dimensions of previously unstudied social settings and intensive investigations of the subjective meanings that motivate individual action are particularly well served by the techniques of participant observation, intensive interviewing, and focus groups.

The very characteristics that make qualitative research techniques so appealing restrict their use to a limited set of research problems. It is not possible to draw representative samples for study using participant observation, and, for this reason, the generalizability of any particular field study's results cannot really be known. Only the accumulation of findings from numerous qualitative studies permits confident generalization, but here again, the time and effort required to collect and analyze the data make it unlikely that many field research studies will be replicated.

Even if qualitative researchers made more of an effort to replicate key studies, their notion of developing and grounding explanations inductively in the observations made in a particular setting would hamper comparison of findings. Measurement reliability is thereby hindered, as are systematic tests for the validity of key indicators and formal tests for causal connections.

In the final analysis, qualitative research involves a mode of thinking and investigating different from that used in experimental and survey research. Qualitative research is inductive and idiographic, whereas experiments and surveys tend to be conducted in a deductive, quantitative, and nomothetic framework. Both approaches can help social scientists learn about the social world; the proficient researcher must be ready to use either. Qualitative data are often supplemented with counts of characteristics or activities. And as you have already seen, quantitative data are often enriched with written comments and observations, while focus groups have become a common tool of survey researchers seeking to develop their questionnaires. Thus, the distinction between qualitative and quantitative research techniques is not always clear-cut, and combining methods is often a good idea.

Key Terms

Highlights

- Qualitative methods are most useful in exploring new issues, investigating hard-to-study groups, and determining the meaning people give to their lives and actions. In addition, most social research projects can be improved, in some respects, by taking advantage of qualitative techniques.

- Qualitative researchers tend to develop ideas inductively, try to understand the social context and sequential nature of attitudes and actions, and explore the subjective meanings that participants attach to events. They rely primarily on participant observation, intensive interviewing, and, in recent years, focus groups.

- Participant observers may adopt one of several roles for a particular research project. Each role represents a different balance between observing and participating. Many field researchers prefer a moderate role, participating as well as observing in a group but acknowledging publicly the researcher role. Such a role avoids the ethical issues that covert participation pose while still allowing the insights into the social world derived from participating directly in it. The role that the participant observer chooses should be based on an evaluation of the problems that are likely to arise from reactive effects and the ethical dilemmas of covert participation.

- Systematic observation techniques quantify the observational process to allow more systematic comparison between cases and greater generalizability.

- Field researchers must develop strategies for entering the field, developing and maintaining relations in the field, sampling, and recording and analyzing data. Selection of sites or other units to study may reflect an emphasis on typical cases, deviant cases, and/or critical cases that can provide more information than others. Sampling techniques commonly used within sites or in selecting interviewees in field research include theoretical sampling, purposive sampling, snowball sampling, quota sampling, and in special circumstances, random selection with the experience sampling method.

- Recording and analyzing notes is a crucial step in field research. Jottings are used as brief reminders about events in the field, while daily logs are useful to chronicle the researcher's activities. Detailed field notes should be recorded and analyzed daily. Analysis of the notes can guide refinement of methods used in the field and of the concepts, indicators, and models developed to explain what has been observed.

- Intensive interviews involve open-ended questions and follow-up probes, with specific question content and order varying from one interview to another. Intensive interviews can supplement participant observation data.

- Focus groups combine elements of participant observation and intensive interviewing. They can increase the validity of attitude measurement by revealing what people say when they present their opinions in a group context instead of in the artificial one-on-one interview setting.

- Four ethical issues that should be given particular attention in field research concern (1) voluntary participation, (2) subject well-being, (3) identity disclosure, and (4) confidentiality. Qualitative research conducted online, with discussion groups or e-mail traffic, raises special concerns about voluntary participation and identity disclosure.

- Adding qualitative elements to structured survey projects and experimental designs can enrich understanding of social processes.

STUDENT STUDY SITE

To assist in completing the web exercises, please access the study site at **www.sagepub.com/schuttisw7e**, where you will find the web exercises with accompanying links. You'll find other useful study materials such as self-quizzes and e-flashcards for each chapter, along with a group of carefully selected articles from research journals that illustrate the major concepts and techniques presented in the book.

Discussion Questions

1. You read in this chapter the statement by Maurice Punch (1994) that "the crux of the matter is that some deception, passive or active, enables you to get at data not obtainable by other means" (p. 91). What aspects of the social world would be difficult for participant observers to study without being covert? Are there any situations that would require the use of covert observation to gain access? What might you do as a participant observer to lessen access problems while still acknowledging your role as a researcher?

2. Review the experiments and surveys described in previous chapters. Pick one, and propose a field research design that would focus on the same research question but with participant observation techniques in a local setting. Propose the role that you would play in the setting, along the participant observation continuum, and explain why you would favor this role. Describe the stages of your field research study, including your plans for entering the field, developing and maintaining relationships, sampling, and recording

and analyzing data. Then, discuss what you would expect your study to add to the findings resulting from the study described in the book.

3. Intensive interviews are the core of many qualitative research designs. How do they differ from the structured survey procedures that you studied in the last chapter? What are their advantages and disadvantages over standardized interviewing? How does intensive interviewing differ from the qualitative method of participant observation? What are the advantages and disadvantages of these two methods?

4. Research on disasters poses a number of methodological challenges. In what ways are qualitative methods suited to disaster research? What particular qualitative methods would you have emphasized if you had been able to design research in New Orleans in the immediate aftermath of Hurricane Katrina? What unique challenges would you have confronted due to the nature of the disaster?

Practice Exercises

1. Conduct a brief observational study in a public location on campus where students congregate. A cafeteria, a building lobby, or a lounge would be ideal. You can sit and observe, taking occasional notes unobtrusively, without violating any expectations of privacy. Observe for 30 minutes. Write up field notes, being sure to include a description of the setting and a commentary on your own behavior and your reactions to what you observed.

2. Develop an interview guide that focuses on a research question addressed in one of the studies in this book. Using this guide, conduct an intensive interview with one person who is involved with the topic in some way. Take only brief notes during the interview, and then write up as complete a record of the interview as soon as you can immediately afterward. Turn in an evaluation of your

performance as an interviewer and note taker, together with your notes.

3. Devise a plan for using a focus group to explore and explain student perspectives on some current event. How would you recruit students for the group? What types of students would you try to include? How would you introduce the topic and the method to the group? What questions would you ask? What problems would you anticipate, such as discord between focus group members or digressions from the chosen topic? How would you respond to these problems?

4. Find the "Qualitative Research" lesson in the Interactive Exercises link on the study site. Answer the questions in this lesson to review the types of ethical issues that can arise in the course of participant observation research.

Ethics Questions

1. Should covert observation ever be allowed in social science research? Do you believe that social scientists should simply avoid conducting research on groups or individuals who refuse to admit researchers into their lives? Some have argued that members of privileged groups do not need to be protected from covert research by social scientists—that this restriction should only be applied to disadvantaged groups and individuals. Do you agree? Why or why not? Do you feel that Randy Alfred's (1976) covert participation observation in the Satanist group was unethical and should not be allowed? Why or why not?

2. Should any requirements be imposed on researchers who seek to study other cultures, to ensure that procedures are appropriate and interpretations are culturally sensitive? What practices would you suggest for cross-cultural researchers to ensure that ethical guidelines are followed? (Consider the wording of consent forms and the procedures for gaining voluntary cooperation.)

Web Exercises

1. Check your library's online holdings to see if it subscribes to the online version of the *Annual Review of Sociology*. If it does, go to their site and search for articles that use qualitative methods as the primary method of gathering data on any one of the following subjects: child development/socialization; gender/sex roles; aging/gerontology. Enter "Qualitative AND Methods" in the subject field to begin this search. Review at least five articles, and report on the specific method of field research used in each.

2. Go to the Intute site at www.intute.ac.uk/socialsciences/ and enter "Qualitative Methods" under "Search social sciences." Now choose three or four interesting sites to find out more about field research—either professional organizations of field researchers or journals that publish their work. Explore the sites to find out what information they provide regarding field research, what kinds of projects are being done that involve field research, and the purposes that specific field research methods are being used for.

3. You have been asked to do field research on the World Wide Web's impact on the socialization of children in today's world. The first part of the project involves your writing a compare and contrast report on the differences between how you and your generation were socialized as children and the way children today are being socialized. Collect your data by surfing the web "as if you were a kid." The web is your field, and you are the field researcher.

4. Using any of the major search engines (like Google), explore the web within the "Kids" or "Children" subject heading, keeping field notes on what you observe. Write a brief report based on the data you have collected. How has the web influenced child socialization in comparison with when you were a child?

SPSS Exercises

The cross-tabulations you examined in Chapter 8's SPSS exercises highlighted the strength of the association between attitudes related to race and support for capital punishment. In this chapter, you will explore related issues.

1. Examine the association between race and support for capital punishment. From the menu, click:

   Analyze/Descriptive Statistics/Crosstabs

In the Crosstabs window, set

Rows: CAPPUN

Columns: RACED

Cells: column percents

2. What is the association between race and support for capital punishment? How would you explain that association?

3. Now consider what might lead to variation in support for capital punishment among whites and blacks. Consider gun ownership (OWNGUNR), religious beliefs (FUND), attitudes about race (RACDIF1), education (EDUCR3), and political party identification (PARTYID3).

4. Generate crosstabs for the association of support for capital punishment with each of these variables, separately for minorities and whites. Follow the same procedures you used in Step 1, substituting the variables mentioned in Step 3 for RACED in Step 1. However, you must repeat the crosstab request for blacks and whites. To do this, before you choose Analyze, select black respondents only. From the menu above the Data Editor window, select Data, then

Select Cases. Then from the select Cases window, select If condition is satisfied and create this expression:

If . . . RACED=1

After you have generated the crosstabs, go back and repeat the data selection procedures, ending with RACED=2. When finished with the exercises, be sure to go back to Select Cases and select All Cases.

5. Are the bases of support for capital punishment similar among minorities and whites? Discuss your findings.

6. Propose a focus group to explore these issues further. Identify the setting and sample for the study, and describe how you would carry out your focus group.

Developing a Research Proposal

Add a qualitative component to your proposed study. You can choose to do this with a participant observation project or intensive interviewing. Pick the method that seems most likely to help answer the research question for the overall survey project (see Exhibit 2.14, #13 to #17).

1. For a participant observation component, propose an observational plan that would complement the overall survey project. Present in your proposal the following information about your plan: (a) Choose a site and justify its selection in terms of its likely value for the research; (b) choose a role along the participation-observation continuum and justify your choice; (c) describe access procedures and note any likely problems; (d) discuss how you will develop and maintain relations in the site; (e) review

any sampling issues; and (f) present an overview of the way in which you will analyze the data you collect.

2. For an intensive interview component, propose a focus for the intensive interviews that you believe will add the most to findings from the survey project. Present in your proposal the following information about your plan: (a) Present and justify a method for selecting individuals to interview; (b) write out three introductory biographical questions and five *grand tour* questions for your interview schedule; (c) list at least six different probes you may use; (d) present and justify at least two follow-up questions for one of your grand tour questions; and (e) explain what you expect this intensive interview component to add to your overall survey project.

Qualitative Data Analysis

*I was at lunch standing in line and he [another male student] came up to my face and started saying stuff
and then he pushed me. I said . . . I'm cool with you, I'm your friend and then he push me again and calling
me names. I told him to stop pushing me and then he push me hard and said something about my mom.
And then he hit me, and I hit him back. After he fell I started kicking him.*

—Morrill et al. (2000:521)

U nfortunately, this statement was not made by a soap opera actor but by a real student writing an in-class essay about conflicts in which he had participated. But then you already knew that such conflicts are common in many high schools, so perhaps it will be reassuring to know that this statement was elicited by a team of social scientists who were studying conflicts in high schools to better understand their origins and to inform prevention policies.

The first difference between qualitative and quantitative data analysis is that the data to be analyzed are text, rather than numbers, at least when the analysis first begins. Does it trouble you to learn that there are no variables and hypotheses in this qualitative analysis by Morrill et al. (2000)? This, too, is another difference between the typical qualitative and quantitative approaches to analysis, although there are some exceptions.

In this chapter, I present the features that most qualitative data analyses share, and I will illustrate these features with research on youth conflict and on being homeless. You will quickly learn that there is no one way to analyze textual data. To quote Michael Quinn Patton (2002), "Qualitative analysis transforms data into findings. No formula exists for that transformation. Guidance, yes. But no recipe. Direction can and will be offered, but the final destination remains unique for each inquirer, known only when—and if—arrived at" (p. 432).

I will discuss some of the different types of qualitative data analysis before focusing on computer programs for qualitative data analysis; you will see that these increasingly popular programs are blurring the distinctions between quantitative and qualitative approaches to textual analysis.

▣ Features of Qualitative Data Analysis

The distinctive features of qualitative data collection methods that you studied in Chapter 9 are also reflected in the methods used to analyze those data. The focus on text—on qualitative data rather than on numbers—is the most important feature of qualitative analysis. The "text" that qualitative researchers analyze is most often transcripts of interviews or notes from participant observation sessions, but text can also refer to pictures or other images that the researcher examines.

What can the qualitative data analyst learn from a text? Here qualitative analysts may have two different goals. Some view analysis of a text as a way to understand what participants "really" thought, felt, or did in some situation or at some point in time. The text becomes a way to get "behind the numbers" that are recorded in a quantitative analysis to see the richness of real social experience. Other qualitative researchers have adopted a hermeneutic perspective on texts—that is, a perspective that views a text as an interpretation that can never be judged true or false. The text is only one possible interpretation among many (Patton 2002:114).

The meaning of a text, then, is negotiated among a community of interpreters, and to the extent that some agreement is reached about meaning at a particular time and place, that meaning can only be based on consensual community validation.

From a hermeneutic perspective, a researcher is constructing a "reality" with his or her interpretations of a text provided by the subjects of research; other researchers, with different backgrounds, could come to markedly different conclusions.

You can see in this discussion about text that qualitative and quantitative data analyses also differ in the priority given to the prior views of the researcher and to those of the subjects of the research. Qualitative data analysts seek to describe their textual data in ways that capture the setting or people who produced this text on their own terms rather than in terms of predefined measures and hypotheses. What this means is that qualitative data analysis tends to be inductive—the analyst identifies important categories in the data, as well as patterns and relationships, through a process of discovery. There are often no predefined measures or hypotheses. Anthropologists term this an **emic focus**, which means representing the setting in terms of the participants and their viewpoint, rather than an **etic focus**, in which the setting and its participants are represented in terms that the researcher brings to the study.

> **Emic focus** Representing a setting with the participants' terms and from their viewpoint.
>
> **Etic focus** Representing a setting with the researchers' terms and from their viewpoint.

Good qualitative data analyses also are distinguished by their focus on the interrelated aspects of the setting, group, or person under investigation—the case—rather than breaking the whole into separate parts. The whole is always understood to be greater than the sum of its parts, and so the social context of events, thoughts, and actions becomes essential for interpretation. Within this framework, it doesn't really make sense to focus on two variables out of an interacting set of influences and test the relationship between just those two.

Qualitative data analysis is an iterative and reflexive process that begins as data are being collected rather than after data collection has ceased (Stake 1995). Next to her field notes or interview transcripts, the qualitative analyst jots down ideas about the meaning of the text and how it might relate to other issues. This process of reading through the data and interpreting them continues throughout the project. The analyst adjusts the data collection process itself when it begins to appear that additional concepts need to be investigated or new relationships explored. This process is termed **progressive focusing** (Parlett & Hamilton 1976).

> **Progressive focusing** The process by which a qualitative analyst interacts with the data and gradually refines her focus.

We emphasize placing an interpreter in the field to observe the workings of the case, one who records objectively what is happening but simultaneously examines its meaning and redirects observation to refine or substantiate those meanings. Initial research questions may be modified or even replaced in mid-study by the case researcher. The aim is to thoroughly understand [the case]. If early questions are not working, if new issues become apparent, the design is changed. (Stake 1995:9)

Elijah Anderson (2003) describes the progressive focusing process in his memoir about his study of Jelly's Bar.

Throughout the study, I also wrote conceptual memos to myself to help sort out my findings. Usually no more than a page long, they represented theoretical insights that emerged from my engagement with the data in my field notes. As I gained tenable hypotheses and propositions, I began to listen and observe selectively, focusing on those events that I thought might bring me alive to my research interests and concerns. This method of dealing with the information I was receiving amounted to a kind of a dialogue with the data, sifting out ideas, weighing new notions against the reality with which I was faced there on the streets and back at my desk (pp. 235–236).

Carrying out this process successfully is more likely if the analyst reviews a few basic guidelines when he or she starts the process of analyzing qualitative data (Miller & Crabtree 1999b:142–143):

- Know yourself, your biases, and preconceptions.

- Know your question.

- Seek creative abundance. Consult others and keep looking for alternative interpretations.

- S
  a
- C
  t
- S
- A
  f
- A

You'
tive data
data. You
and cou
better th
different

Exhibit I

1. C
2. C
3. C
4. C
   a
5. R

The
both, as
tion. Sim
make fr
"husban

An i
ships am
this proc
Her rese
with oth

The
prob
fami
anal

- Be flexible.

- Exhaust the data. Try to account for all the data in the texts, then publicly acknowledge the unexplained and remember the next principle.

- Celebrate anomalies. They are the windows to insight.

- Get critical feedback. The solo analyst is a great danger to self and others.

- Be explicit. Share the details with yourself, your team members, and your audiences.

Qualitative Data Analysis as an Art

If you find yourself longing for the certainty of predefined measures and deductively derived hypotheses, you are beginning to understand the difference between setting out to analyze data quantitatively and planning to do so with a qualitative approach in mind. Or, maybe you are now appreciating better the contrast between the positivist and interpretivist research philosophies that I summarized in Chapter 3. When it comes right down to it, the process of qualitative data analysis is even described by some as involving as much "art" as science—as a "dance," in the words of William Miller and Benjamin Crabtree (1999b) (Exhibit 10.1):

> Interpretation is a complex and dynamic craft, with as much creative artistry as technical exactitude, and it requires an abundance of patient plodding, fortitude, and discipline. There are many changing rhythms; multiple steps; moments of jubilation, revelation, and exasperation. . . . The dance of interpretation is a dance for two, but those two are often multiple and frequently changing, and there is always an audience, even if it is not always visible. Two dancers are the interpreters and the texts. (pp. 138–139)

Exhibit 10.1   Dance of Qualitative Analysis

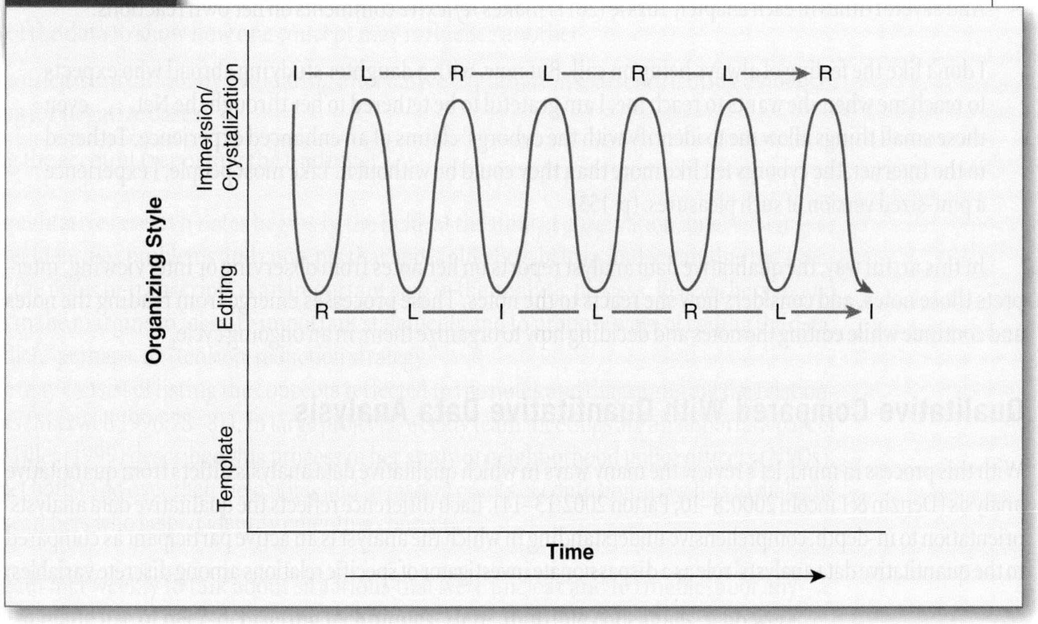

Conceptualization, Coding, and Categorizing

Identifying and refining important concepts is a key part of the iterative process of qualitative research. Sometimes, conceptualizing begins with a simple observation that is interpreted directly, "pulled apart," and then put back together more meaningfully. Robert Stake (1995) provides an example:

> When Adam ran a pushbroom into the feet of the children nearby, I jumped to conclusions about his interactions with other children: aggressive, teasing, arresting. Of course, just a few minutes earlier I had seen him block the children climbing the steps in a similar moment of smiling bombast. So I was aggregating, and testing my unrealized hypotheses about what kind of kid he was, not postponing my interpreting. . . . My disposition was to keep my eyes on him. (p. 74)

The focus in this conceptualization "on the fly" is to provide a detailed description of what was observed and a sense of why that was important.

More often, analytic insights are tested against new observations, the initial statement of problems and concepts is refined, the researcher then collects more data, interacts with the data again, and the process continues. Anderson (2003) recounts how his conceptualization of social stratification at Jelly's Bar developed over a long period of time:

> I could see the social pyramid, how certain guys would group themselves and say in effect, "I'm here and you're there." . . . I made sense of these crowds [initially] as the "respectables," the "nonrespectables," and the "near-respectables." . . . Inside, such non-respectables might sit on the crates, but if a respectable came along and wanted to sit there, the lower-status person would have to move. (pp. 225–226)

But this initial conceptualization changed with experience, as Anderson realized that the participants themselves used other terms to differentiate social status: *winehead, hoodlum,* and *regular* (Anderson 2003:230). What did they mean by these terms? The regulars basically valued "decency." They associated decency with conventionality but also with "working for a living," or having a "visible means of support" (Anderson 2003:231). In this way, Anderson progressively refined his concept as he gained experience in the setting.

Howard S. Becker (1958) provides another excellent illustration of this iterative process of conceptualization in his study of medical students:

> When we first heard medical students apply the term "crock" to patients, we made an effort to learn precisely what they meant by it. We found, through interviewing students about cases both they and the observer had seen, that the term referred in a derogatory way to patients with many subjective symptoms but no discernible physical pathology. Subsequent observations indicated that this usage was a regular feature of student behavior and thus that we should attempt to incorporate this fact into our model of student-patient behavior. The derogatory character of the term suggested in particular that we investigate the reasons students disliked these patients. We found that this dislike was related to what we discovered to be the students' perspective on medical school: the view that they were in school to get experience in recognizing and treating those common diseases most likely to be encountered in general practice. "Crocks," presumably having no disease, could furnish no such experience. We were thus led to specify connections between the student-patient relationship and the student's view of the purpose of this professional education. Questions concerning the genesis of this perspective led to discoveries about the organization of the student body and communication among students, phenomena which we had been assigning to another [segment of the larger theoretical model being developed]. Since "crocks" were also disliked because they gave the student no opportunity to assume medical responsibility, we were able to connect this aspect of the student-patient relationship with still another tentative model of the value system and hierarchical organization of the school, in which medical responsibility plays an important role. (p. 658)

This excerpt shows how the researcher first was alerted to a concept by observations in the field, then refined his understanding of this concept by investigating its meaning. By observing the concept's frequency of use, he came to realize its importance. Then he incorporated the concept into an explanatory model of student-patient relationships.

A well-designed chart, or **matrix**, can facilitate the coding and categorization process. Exhibit 10.4 shows an example of a coding form designed by Miles and Huberman (1994:93–95) to represent the extent to which

Exhibit 10.4 **Example of Checklist Matrix**

Presence of Supporting Conditions		
Condition	For Users	For Administrators
Commitment	*Strong*—"wanted to make it work."	*Weak* at building level. Prime movers in central office committed; others not.
Understanding	*"Basic"* ("felt I could do it, but I just wasn't sure how.") for teacher. *Absent* for aide ("didn't understand how we were going to get all this.")	*Absent* at building level and among staff. *Basic* for 2 prime movers ("got all the help we needed from developer.") *Absent* for other central office staff.
Materials	*Inadequate:* ordered late, puzzling ("different from anything I ever used"), discarded.	NA
Front-end training	*"Sketchy"* for teacher ("it all happened so quickly"); no demo class. *None* for aide ("totally unprepared. I had to learn along with the children.")	Prime movers in central office had training at developer site; none for others.
Skills	*Weak-adequate* for teacher. *"None"* for aide.	One prime mover (Robeson) skilled in substance; others unskilled.
Ongoing inservice	*None*, except for monthly committee meeting; no substitute funds.	*None*
Planning, coordination time	*None*: both users on other tasks during day; lab tightly scheduled, no free time.	*None*
Provisions for debugging	*None* systematized; spontaneous work done by users during summer.	*None*
School admin. support	*Adequate*	NA
Central admin. support	*Very strong* on part of prime movers.	Building admin. only acting on basis of central office commitment.
Relevant prior experience	*Strong* and useful in both cases: had done individualized instruction, worked with low achievers. But aide had no diagnostic experience.	*Present* and useful in central office, esp. Robeson (specialist).

Matrix A form on which can be recorded systematically particular features of multiple cases or instances that a qualitative data analyst needs to examine.

teachers and teachers' aides ("users") and administrators at a school gave evidence of various supporting conditions that indicate preparedness for a new reading program. The matrix condenses data into simple categories, reflects further analysis of the data to identify degree of support, and provides a multidimensional summary that will facilitate subsequent, more intensive analysis. Direct quotes still impart some of the flavor of the original text.

Examining Relationships and Displaying Data

Examining relationships is the centerpiece of the analytic process, because it allows the researcher to move from simple description of the people and settings to explanations of why things happened as they did with those people in that setting. The process of examining relationships can be captured in a matrix that shows how different concepts are connected, or perhaps what causes are linked with what effects.

Exhibit 10.5 displays a matrix used to capture the relationship between the extent to which stakeholders in a new program had something important at stake in the program and the researcher's estimate of their favorability toward the program. Each cell of the matrix was to be filled in with a summary of an illustrative case study. In other matrix analyses, quotes might be included in the cells to represent the opinions of these different stakeholders, or the number of cases of each type might appear in the cells. The possibilities are almost endless. Keeping this approach in mind will generate many fruitful ideas for structuring a qualitative data analysis.

Exhibit 10.5 **Coding Form for Relationships: Stakeholders' Stakes**

	Favorable	Neutral or Unknown	Antagonistic
High			
Moderate			
Low			

Note: Construct illustrative case studies for each cell based on fieldwork.

The simple relationships that are identified with a matrix like that shown in Exhibit 10.5 can be examined and then extended to create a more complex causal model. Such a model represents the multiple relationships among the constructs identified in a qualitative analysis as important for explaining some outcome. A great deal of analysis must precede the construction of such a model, with careful attention to identification of important variables and the evidence that suggests connections between them. Exhibit 10.6 provides an example of these connections from a study of the implementation of a school program.

Authenticating Conclusions

No set standards exist for evaluating the validity, or *authenticity,* of conclusions in a qualitative study, but the need to carefully consider the evidence and methods on which conclusions are based is just as great as with other types of research. Individual items of information can be assessed in terms of at least three criteria (Becker 1958):

1. *How credible was the informant?* Were statements made by someone with whom the researcher had a relationship of trust or by someone the researcher had just met? Did the informant have reason to lie? If the statements do not seem to be trustworthy as indicators of actual events, can they at least be used to help understand the informant's perspective?

2. *Were statements made in response to the researcher's questions, or were they spontaneous?* Spontaneous statements are more likely to indicate what would have been said had the researcher not been present.

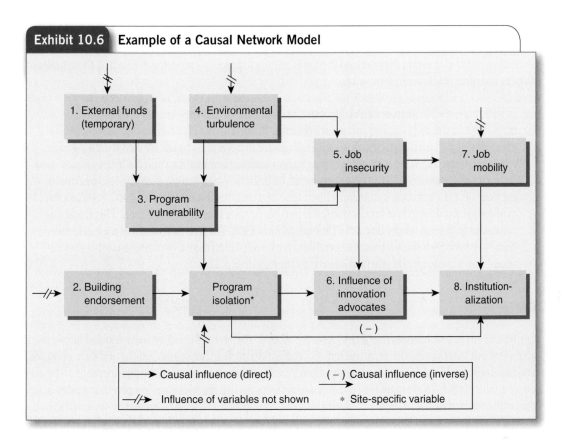

Exhibit 10.6   **Example of a Causal Network Model**

3. *How does the presence or absence of the researcher or the researcher's informant influence the actions and statements of other group members?* Reactivity to being observed can never be ruled out as a possible explanation for some directly observed social phenomenon. However, if the researcher carefully compares what the informant says goes on when the researcher is not present, what the researcher observes directly, and what other group members say about their normal practices, the extent of reactivity can be assessed to some extent (pp. 654–656).

A qualitative researcher's conclusions should also be assessed by his or her ability to provide a credible explanation for some aspect of social life. That explanation should capture group members' **tacit knowledge** of the social processes that were observed, not just their verbal statements about these processes. Tacit knowledge— "the largely unarticulated, contextual understanding that is often manifested in nods, silences, humor, and naughty nuances"—is reflected in participants' actions as well as their words and in what they fail to state but nonetheless feel deeply and even take for granted (Altheide & Johnson 1994:492–493). These features are evident in Whyte's (1955) analysis of Cornerville social patterns:

> **Tacit knowledge** In field research, a credible sense of understanding of social processes that reflects the researcher's awareness of participants' actions as well as their words, and of what they fail to state, feel deeply, and take for granted.

> The corner-gang structure arises out of the habitual association of the members over a long period of time. The nuclei of most gangs can be traced back to early boyhood. . . . Home plays a very small role in the group activities of the corner boy. . . . The life of the corner boy proceeds along regular and narrowly circumscribed channels. . . . Out of [social interaction within the group] arises a system of mutual obligations which is fundamental to group cohesion. . . . The code of the corner boy requires him to help his friends when he can and to refrain from doing anything to harm them. When life in the group runs smoothly, the obligations binding members to one another are not explicitly recognized. (pp. 255–257)

Comparing conclusions from a qualitative research project to those other researchers obtained while conducting similar projects can also increase confidence in their authenticity. Miller's (1999) study of NPOs found striking parallels in the ways they defined their masculinity to processes reported in research about males in nursing and other traditionally female jobs:

> In part, male NPOs construct an exaggerated masculinity so that they are not seen as feminine as they carry out the social-work functions of policing. Related to this is the almost defiant expression of heterosexuality, so that the men's sexual orientation can never truly be doubted even if their gender roles are contested. Male patrol officers' language—such as their use of terms like "pansy police" to connote neighborhood police officers—served to affirm their own heterosexuality. . . . In addition, the male officers, but not the women, deliberately wove their heterosexual status into conversations, explicitly mentioning their female domestic partner or spouse and their children. This finding is consistent with research conducted in the occupational field. The studies reveal that men in female-dominated occupations, such as teachers, librarians, and pediatricians, over-reference their heterosexual status to ensure that others will not think they are gay. (p. 222)

Reflexivity

Confidence in the conclusions from a field research study is also strengthened by an honest and informative account about how the researcher interacted with subjects in the field, what problems he or she encountered, and how these problems were or were not resolved. Such a "natural history" of the development of the evidence enables others to evaluate the findings and reflects the interpretivist philosophy that guides many qualitative researchers (see Chapter 3). Such an account is important first and foremost because of the evolving and variable nature of field research: To an important extent, the researcher "makes up" the method in the context of a particular investigation rather than applying standard procedures that are specified before the investigation begins.

Barrie Thorne (1993) provides a good example of this final element of the analysis:

> Many of my observations concern the workings of gender categories in social life. For example, I trace the evocation of gender in the organization of everyday interactions, and the shift from boys and girls as loose aggregations to "the boys" and "the girls" as self-aware, gender-based groups. In writing about these processes, I discovered that different angles of vision lurk within seemingly simple choices of language. How, for example, should one describe a group of children? A phrase like "six girls and three boys were chasing by the tires" already assumes the relevance of gender. An alternative description of the same event—"nine fourth-graders were chasing by the tires"—emphasizes age and downplays gender. Although I found no tidy solutions, I have tried to be thoughtful about such choices. . . . After several months of observing at Oceanside, I realized that my field notes were peppered with the words "child" and "children," but that the children themselves rarely used the term. "What do they call themselves?" I badgered in an entry in my field notes. The answer it turned out, is that children use the same practices as adults. They refer to one another by using given names ("Sally," "Jack") or language specific to a given context ("that guy on first base"). They rarely have occasion to use age-generic terms. But when pressed to locate themselves in an age-based way, my informants used "kids" rather than "children." (pp. 8–9)

Qualitative data analysts, more often than quantitative researchers, display real sensitivity to how a social situation or process is interpreted from a particular background and set of values and not simply based on the situation itself (Altheide & Johnson 1994). Researchers are only human, after all, and must rely on their own senses and process all information through their own minds. By reporting how and why they think they did what they did, they can help others determine whether, or how, the researchers' perspectives influenced their conclusions. "There should be clear 'tracks' indicating the attempt [to show the hand of the ethnographer] has been made" (Altheide & Johnson 1994:493).

Anderson's (2003) memoir about the Jelly's Bar research illustrates the type of "tracks" that an ethnographer makes as well as how the researcher can describe those tracks. Anderson acknowledges that his tracks began as a child:

> While growing up in the segregated black community of South Bend, from an early age, I was curious about the goings-on in the neighborhood, particularly the streets and more particularly the corner taverns where my uncles and my dad would go to hang out and drink. . . . Hence, my selection of a field setting was a matter of my background, intuition, reason, and a little bit of luck. (pp. 217–218)

After starting to observe at Jelly's, Anderson's (2003) tracks led to Herman:

> After spending a couple of weeks at Jelly's, I met Herman. I felt that our meeting marked an important step. We would come to know each other well . . . something of an informal leader at Jelly's. . . . We were becoming friends. . . . He seemed to genuinely like me, and he was one person I could feel comfortable with. (pp. 218–219)

So we learn that Anderson's observations were to be shaped, in part, by Herman's perspective, but we also find out that Anderson maintained some engagement with fellow students. This contact outside the bar helped shape his analysis: "By relating my experiences to my fellow students, I began to develop a coherent perspective, or a 'story' of the place that complemented the accounts I had detailed in my accumulating field notes" (Anderson 2003:220).

In this way, the outcome of Anderson's analysis of qualitative data resulted, in part, from the way in which he "played his role" as a researcher and participant, not just from the setting itself.

⊞ Alternatives in Qualitative Data Analysis

The qualitative data analyst can choose from many interesting alternative approaches. Of course, the research question under investigation should shape the selection of an analytic approach, but the researcher's preferences and experiences also will inevitably have an important influence on the method chosen. The alternative approaches I present here (ethnography, and its new online cousin, *netnography*); ethnomethodology; qualitative comparative analysis; narrative analysis; conversation analysis; case-oriented understanding; and grounded theory) give you a good sense of the different possibilities (Patton 2002).

Ethnography

Ethnography is the study of a culture or cultures that a group of people share (Van Maanen 1995:4). As a method, it is usually meant to refer to the process of participant observation by a single investigator who immerses himself or herself in the group for a long period of time (often one or more years), gradually establishing

> **Ethnography** The study of a culture or cultures that some group of people shares, using participant observation over an extended period of time.

trust and experiencing the social world as do the participants (Madden 2010:16). Ethnographic research can also be called *naturalistic,* because it seeks to describe and understand the natural social world as it really is, in all its richness and detail. This goal is best achieved when an ethnographer is fluent in the local language and spends enough time in the setting to know how people live, what they say about themselves and what they actually do, and what they value (Armstrong 2008:55).

As you learned in Chapter 9, anthropological field research has traditionally been ethnographic, and much sociological fieldwork shares these same characteristics. But there are no particular methodological techniques associated with ethnography, other than just "being there." The analytic process relies on the thoroughness and insight of the researcher to "tell us like it is" in the setting, as he or she experienced it.

Code of the Street, Anderson's (1999) award-winning study of Philadelphia's inner city, captures the flavor of this approach:

> My primary aim in this work is to render ethnographically the social and cultural dynamics of the interpersonal violence that is currently undermining the quality of life of too many urban neighborhoods. . . . How do the people of the setting perceive their situation? What assumptions do they bring to their decision making? (pp. 10–11)

The methods of investigation are described in the book's preface: participant observation, including direct observation and in-depth interviews, impressionistic materials drawn from various social settings around the city, and interviews with a wide variety of people. Like most traditional ethnographers, Anderson (1999) describes his concern with being "as objective as possible" and using his training as other ethnographers do, "to look for and to recognize underlying assumptions, their own and those of their subjects, and to try to override the former and uncover the latter" (p. 11).

From analysis of the data obtained in these ways, a rich description of life in the inner city emerges. Although we often do not "hear" the residents speak, we feel the community's pain in Anderson's (1999) description of "the aftermath of death":

> When a young life is cut down, almost everyone goes into mourning. The first thing that happens is that a crowd gathers about the site of the shooting or the incident. The police then arrive, drawing more of a crowd. Since such a death often occurs close to the victim's house, his mother or his close relatives and friends may be on the scene of the killing. When they arrive, the women and girls often wail and moan, crying out their grief for all to hear, while the young men simply look on, in studied silence. . . . Soon the ambulance arrives. (p. 138)

Anderson (1999) uses this description as a foundation on which he develops the key concepts in his analysis, such as "code of the street":

> The "code of the street" is not the goal or product of any individual's action but is the fabric of everyday life, a vivid and pressing milieu within which all local residents must shape their personal routines, income strategies, and orientations to schooling, as well as their mating, parenting, and neighbor relations. (p. 326)

Anderson's report on his Jelly's Bar study illustrates how his ethnographic analysis deepened as he became more socially integrated into the Jelly's Bar group. He thus became more successful at "blending the local knowledge one has learned with what we already know sociologically about such settings" (Anderson 2003:236):

I engaged the denizens of the corner and wrote detailed field notes about my experiences, and from time to time I looked for patterns and relationships in my notes. In this way, an understanding of the setting came to me in time, especially as I participated more fully in the life of the corner and wrote my field notes about my experiences; as my notes accumulated and as I reviewed them occasionally and supplemented them with conceptual memos to myself, their meanings became more clear, while even more questions emerged. (Anderson 2003:224)

A good ethnography like Anderson's is only possible when the ethnographer learns the subtleties of expression used in a group and the multiple meanings that can be given to statements or acts (Armstrong 2008:60–62). Good ethnographies also include some reflection by the researcher on the influence his or her own background has had on research plans, as well as on the impact of the research in the setting (Madden 2010:22–23).

Netnography

Communities can refer not only to people in a common physical location, but also to relationships that develop online. Online communities may be formed by persons with similar interests or backgrounds, perhaps to create new social relationships that location or schedules did not permit, or to supplement relationships that emerge in the course of work or school or other ongoing social activities. Like communities of people who interact face-to-face, online communities can develop a culture and become sources of identification and attachment (Kozinets 2010:14–15). And like physical communities, researchers can study online communities through immersion in the group for an extended period. **Netnography,** also termed *cyberethnography* and *virtual ethnography* (James & Busher 2009:34–35), is the use of ethnographic methods to study online communities.

In some respects, netnography is similar to traditional ethnography. The researcher prepares to enter the field by becoming familiar with online communities and their language and customs, formulating an exploratory research question about social processes or orientations in that setting, selecting an appropriate community to study. Unlike in-person ethnographies, netnographies can focus on communities whose members are physically distant and dispersed. The selected community should be relevant to the research question, involve frequent communication among actively engaged members, and have a number of participants who, as a result, generate a rich body of textual data (Kozinets 2010:89).

> **Netnography** The use of ethnographic methods to study online communities. Also termed *cyberethnography* and *virtual ethnography.*

The netnographer's self-introduction should be clear and friendly. Robert Kozinets (2010:93) provides the following example written about the online discussion space, alt.coffee:

I've been lurking here for a while, studying online coffee culture on alt.coffee, learning a lot, and enjoying it very much . . . I just wanted to pop out of lurker status to let you know I am here . . . I will be wanting to quote some of the great posts that have appeared here, and I will contact the individuals by personal e-mail who posted them to ask their permission to quote them. I also will be making the document on coffee culture available to any interested members of the newsgroup for their perusal and comments—to make sure I get things right.

A netnographer must keep both observational and reflective field notes, but unlike a traditional ethnographer can return to review the original data—the posted test—long after it was produced. The data can then be coded, annotated with the researcher's interpretations, checked against new data to evaluate the persistence of social patterns, and used to develop a theory that is grounded in the data.

Research in the News

READERS' ONLINE FEEDBACK CAN BE VICIOUS

After a woman published an article in an online magazine about postpartum post-traumatic stress disorder following a traumatic delivery experience with her baby boy, the nasty comments started to pour in to the area reserved for reader responses. She was told not to have any more babies and that she would be a bad mother. In a similar incident, an uninsured woman who had written of her inability to function after a car accident was told to "Get a minnie mouse bandage and go to sleep." Why do some people get so vicious on the Internet? One social scientist suggested that it is because of the lack of face-to-face interaction, which provides constant feedback about others' feelings through body language and gestures.

Source: Brodesser-Akner, Taffy. 2010. "E-Playgrounds Can Get Vicious (Online Feedback From Readers)." *The New York Times,* April 22:E8.

Ethnomethodology

Ethnomethodology A qualitative research method focused on the way that participants in a social setting create and sustain a sense of reality.

Ethnomethodology focuses on the way that participants construct the social world in which they live—how they "create reality"—rather than on describing the social world itself. In fact, ethnomethodologists do not necessarily believe that we can find an objective reality; it is the way that participants come to create and sustain a sense of reality that is of interest. In the words of Jaber F. Gubrium and James A. Holstein (1997), in ethnomethodology, as compared with the naturalistic orientation of ethnography,

the focus shifts from the scenic features of everyday life onto the ways through which the world comes to be experienced as real, concrete, factual, and "out there." An interest in members' methods of constituting their world supersedes the naturalistic project of describing members' worlds as they know them. (p. 41)

Unlike the ethnographic analyst, who seeks to describe the social world as the participants see it, the ethnomethodological analyst seeks to maintain some distance from that world. The ethnomethologist views a code of conduct like that described by Anderson (2003) not as a description of a real normative force that constrains social action, but as the way that people in the setting create a sense of order and social structure (Gubrium & Holstein 1997:44–45). The ethnomethodologist focuses on how reality is constructed, not on what it is.

Sociologist Harold Garfinkel (1967) developed ethnomethodology in the 1960s and first applied it to the study of gender. Focusing on a teenage male-to-female transsexual who he termed "Agnes," he described her "social achievement of gender" as

the tasks of securing and guaranteeing for herself the ascribed rights and obligations of an adult female by the acquisition and use of skills and capacities, the efficacious display of female appearances and performances, and the mobilizing of appropriate feelings and purposes. (p. 134)

The ethnomethodological focus on how the meaning of gender and other categories are socially constructed leads to a concern with verbal interaction. In recent years, this concern has led ethnomethodologists and others to develop a more formal approach, called *conversation analysis.*

Conversation Analysis

Conversation analysis is a specific qualitative method for analyzing the sequential organization and details of conversation. Like ethnomethodology, from which it developed, conversation analysis focuses on how reality is constructed, rather than on what it is. From this perspective, detailed analysis of conversational interaction is important because conversation is "sociological bedrock": . . ."a form of social organization through which the work of . . . institutions such as the economy, the polity, the family, socialization, etc." is accomplished (Schegloff 1996:4).

> . . . it is through conversation that we conduct the ordinary affairs of our lives. Our relationships with one another, and our sense of who we are to one another is generated, manifest, maintained, and managed in and through our conversations, whether face-to-face, on the telephone, or even by other electronic means. (Drew 2005:74)

Three premises guide conversation analysis (Gubrium & Holstein 2000:492):

1. Interaction is sequentially organized, and talk can be analyzed in terms of the process of social interaction rather than in terms of motives or social status.

2. Talk, as a process of social interaction, is contextually oriented—it is both shaped by interaction and creates the social context of that interaction.

3. These processes are involved in all social interaction, so no interactive details are irrelevant to understanding it.

Consider these premises as you read the following excerpt from Elizabeth Stokoe's (2006:479–480) analysis of the relevance of gender categories to "talk-in-interaction." The dialogue is between four first-year British psychology students who must write up a description of some photographs of people (Exhibit 10.7). Stokoe incorporates stills from the video recording of the interaction into her analysis of both the talk and embodied conduct in interaction. In typical conversation analysis style, the text is broken up into brief segments that capture shifts in meaning, changes in the speaker, pauses, nonspeech utterances and nonverbal actions, and emphases.

Can you see how the social interaction reinforces the link of "woman" and "secretary"? Here, in part, is how Elizabeth Stokoe (2006) analyzes this conversation:

> In order to meet the task demands, one member of the group must write down their ideas. Barney's question at the start of the sequence, "is somebody scribing" is taken up after a reformulation: "who's writin' it." Note that, through a variety of strategies, members of the group manage their responses such that they do not have to take on the role of scribe. At line 05, Neil's "Oh yhe:ah." treats Barney's turn as a proposal to be agreed with, rather than a request for action, and his subsequent nomination of Kay directs the role away from himself. . . . At line 08, Neil nominates Kay, his pointing gesture working in aggregate with the talk to accomplish the action ("She wants to do it."), whilst also attributing agency to Kay for taking up the role. A gloss [interpretation] might be "Secretaries in general are female, you're female, so you in particular are our secretary." (p. 481)

Exhibit 10.7 Conversation Analysis, Including Pictures

1. UT-23

01	N:		D' you reckon she's an instructor then.
02	N:		(0.2)
03	N:		Of some sort,
04	B:	→	Is somebody scribing. Who's writin' it. =
05	N:		=Oh yhe:ah.
06			(0.8)
07	M:		Well you can't [read my] =
08	N:		[((pointing to K)) She wants to do it.]

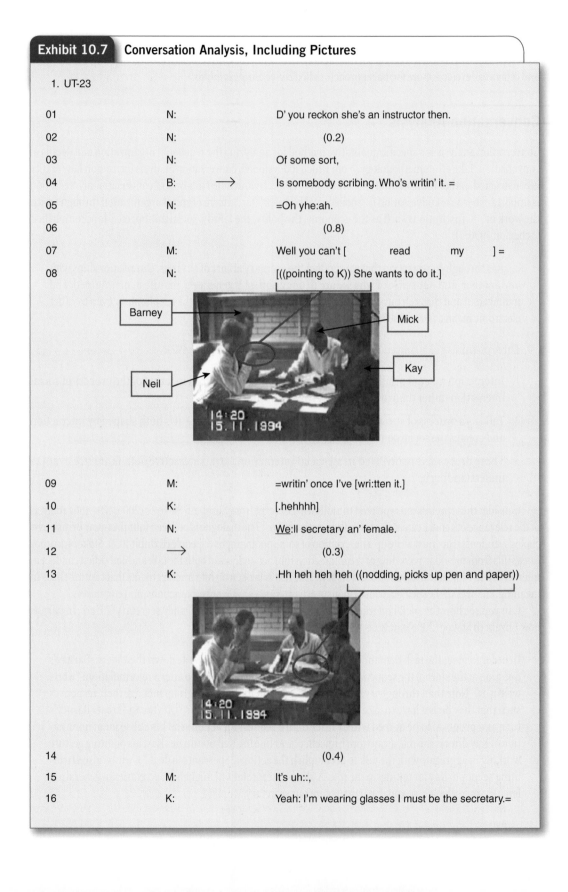

09	M:		=writin' once I've [wri:tten it.]
10	K:		[.hehhhh]
11	N:		We:ll secretary an' female.
12		→	(0.3)
13	K:		.Hh heh heh heh ((nodding, picks up pen and paper))
14			(0.4)
15	M:		It's uh::,
16	K:		Yeah: I'm wearing glasses I must be the secretary.=

Bethan Benwell and Elizabeth Stokoe (2006:61–62) used a conversation between three friends to illustrate key concepts in conversation analysis. The text is prepared for analysis by numbering the lines, identifying the speakers, and inserting ↑ symbols to indicate inflection and decimal numbers to indicate elapsed time.

104   **Marie:**   ↑ Has ↑ anyone- (0.2) has anyone got any really non:

105   sweaty stuff.

106   **Dawn:**   Dave has, but you'll smell like a ma:n,

107   (0.9)

108   **Kate:**   Eh [↑ huh heh]

109   **Marie:**   [Right has] anyone got any ↑ fe:minine non sweaty stuff.

The gap at line 107, despite being less than a second long, is nevertheless quite a long time in conversation, and indicates an interactional glitch or trouble. As Kate starts to laugh, Marie reformulates her request, from "↑ has ↑ anyone got any really non: sweaty stuff," to "right has anyone got any, ↑ fe:minine non sweaty stuff." The word *really* is replaced by *feminine,* and is produced with an audible increase in pitch and emphasis. This replacement, together with the addition of *right,* displays her understanding of the problem with her previous question. For these speakers, smelling like a man (when one is a woman) is treated as a trouble source, a laughable thing and something that needs attending to and fixing. In this way, conversation analysis can uncover meanings in interactions about which the participants are not fully aware (Antaki 2008:438).

Narrative Analysis

Narrative methods use interviews and sometimes documents or observations to "follow participants down their trails" (Riessman 2008:24). Unlike conversation analysis, which focuses attention on moment-by-moment interchange, narrative analysis seeks to put together the "big picture" about experiences or events as the participants understand them. **Narrative analysis** focuses on "the story itself" and seeks to preserve the integrity of personal biographies or a series of events that cannot adequately be understood in terms of their discrete elements (Riessman 2002:218). Narrative "displays the goals and intentions of human actors; it makes individuals, cultures, societies, and historical epochs comprehensible as wholes" (Richardson 1995:200). The coding for a narrative analysis is typically of the narratives as a whole, rather than of the different elements within them. The coding strategy revolves around reading the stories and classifying them into general patterns.

> **Narrative analysis** A form of qualitative analysis in which the analyst focuses on how respondents impose order on the flow of experience in their lives and thus make sense of events and actions in which they have participated.

For example, Calvin Morrill and his colleagues (2000:534) read through 254 conflict narratives written by the ninth graders they studied and found four different types of stories:

1. *Action tales,* in which the author represents himself or herself and others as acting within the parameters of taken-for-granted assumptions about what is expected for particular roles among peers.

2. *Expressive tales,* in which the author focuses on strong, negative emotional responses to someone who has wronged him or her.

3. *Moral tales,* in which the author recounts explicit norms that shaped his or her behavior in the story and influenced the behavior of others.

4. *Rational tales,* in which the author represents himself or herself as a rational decision maker navigating through the events of the story.

In addition to these dominant distinctions, Morrill et al. (2000:534–535) also distinguished the stories in terms of four stylistic dimensions: (1) plot structure (e.g., whether the story unfolds sequentially), (2) dramatic tension (how the central conflict is represented), (3) dramatic resolution (how the central conflict is resolved), and (4) predominant outcomes (how the story ends). Coding reliability was checked through a discussion between the two primary coders, who found that their classifications agreed for a large percentage of the stories.

The excerpt that begins this chapter exemplifies what Morrill et al. (2000) termed an *action tale.* Such tales

unfold in matter-of-fact tones kindled by dramatic tensions that begin with a disruption of the quotidian order of everyday routines. A shove, a bump, a look . . . triggers a response . . . Authors of action tales typically organize their plots as linear streams of events as they move briskly through the story's scenes . . . This story's dramatic tension finally resolves through physical fighting, but . . . only after an attempted conciliation. (p. 536)

You can contrast this action tale with the following narrative, which Morrill et al. (2000) classify as a *moral tale,* in which the students "explicitly tell about their moral reasoning, often referring to how normative commitments shape their decisionmaking" (p. 542):

I . . . got into a fight because I wasn't allowed into the basketball game. I was being harassed by the captains that wouldn't pick me and also many of the players. The same type of things had happened almost every day where they called me bad words so I decided to teach the ring leader a lesson. I've never been in a fight before but I realized that sometimes you have to make a stand against the people that constantly hurt you, especially emotionally. I hit him in the face a couple of times and I got [the] respect I finally deserved. (pp. 545–546)

Morrill et al. (2000:553) summarize their classification of the youth narratives in a simple table that highlights the frequency of each type of narrative and the characteristics associated with each of them (Exhibit 10.8). How does such an analysis contribute to our understanding of youth violence? Morrill et al. (2000) first emphasize that their narratives "suggest that consciousness of conflict among youths—like that among adults—is not a singular entity, but comprises a rich and diverse range of perspectives" (p. 551).

Theorizing inductively, Morrill et al. (2000:553–554) then attempt to explain why action tales were much more common than the more adult-oriented normative, rational, or emotionally expressive tales. One possibility is Gilligan's (1988) theory of moral development, which suggests that younger students are likely to limit themselves to the simpler action tales that "concentrate on taken-for-granted assumptions of their peer and wider cultures, rather than on more self-consciously reflective interpretation and evaluation" (Morrill et al. 2000:554). More generally, Morrill et al. (2000) argue, "We can begin to think of the building blocks of cultures as different narrative styles in which various aspects of reality are accentuated, constituted, or challenged, just as others are deemphasized or silenced" (p. 556).

In this way, Morrill et al.'s (2000) narrative analysis allowed an understanding of youth conflict to emerge from the youths' own stories while also informing our understanding of broader social theories and processes.

Narrative analysis can also use documents and observations and focus more attention on how stories are constructed, rather than on the resulting narrative (Hyvärinen 2008:452). Narrative analyst Catherine Kohler Riessman (2008:67–73) describes the effective combination of data from documents, interviews, and field observations to learn how members of Alcoholics Anonymous (AA) developed a group identity (Cain 1991). Propositions that Carole Cain (1991:228) identified repeatedly in the documents enter into stories as guidelines

Exhibit 10.8	Summary Comparison of Youth Narratives*			
Representation of	**Action Tales (N = 144)**	**Moral Tales (N = 51)**	**Expressive Tales (N = 35)**	**Rational Tales (N = 24)**
Bases of everyday conflict	Disruption of everyday routines & expectations	Normative violation	Emotional provocation	Goal obstruction
Decision making	Intuitive	Principled stand	Sensual	Calculative choice
Conflict handling	Confrontational	Ritualistic	Cathartic	Deliberative
Physical violence†	In 44% (N = 67)	In 27% (N = 16)	In 49% (N = 20)	In 29% (N = 7)
Adults in youth conflict control	Invisible or background	Sources of rules	Agents of repression	Institutions of social control

*Total N = 254.

†Percentages based on the number of stories in each category.

for describing the progression of drinking, the desire and inability to stop, the necessity of "hitting bottom" before the program can work, and the changes that take place in one's life after joining AA.

Cain then found that this same narrative was expressed repeatedly in AA meetings. She only interviewed three AA members but found that one who had been sober and in AA for many years told "his story" using this basic narrative, while one who had been sober for only 2 years deviated from the narrative in some ways. One interviewee did not follow this standard narrative at all as he told his story; he had attended AA only sporadically for 20 years and left soon after the interview. Cain (1991) explains,

> I argue that as the AA member learns the AA story model, and learns to place the events and experiences of his own life into the model, he learns to tell and to understand his own life as an AA life, and himself as an AA alcoholic. The personal story is a cultural vehicle for identity acquisition. (p. 215)

Grounded Theory

Theory development occurs continually in qualitative data analysis (Coffey & Atkinson 1996:23). Many qualitative researchers use a method of developing theory during their analysis that is termed **grounded theory**, which involves building up inductively a systematic theory that is *grounded* in, or based on, the observations. The grounded theorist first summarizes observations into conceptual categories, and tests the coherence of these categories directly in the research setting with more observations. Over time, as the researcher refines and links the conceptual categories, a theory evolves (Glaser & Strauss 1967; Huberman & Miles 1994:436). Exhibit 10.9 diagrams the grounded theory of a chronic illness "trajectory" developed by Strauss and Corbin (1990:221). Their notes suggested to them that conceptions of self, biography, and body are reintegrated after a process of grieving.

Grounded theory Systematic theory developed inductively, based on observations that are summarized into conceptual categories, reevaluated in the research setting, and gradually refined and linked to other conceptual categories.

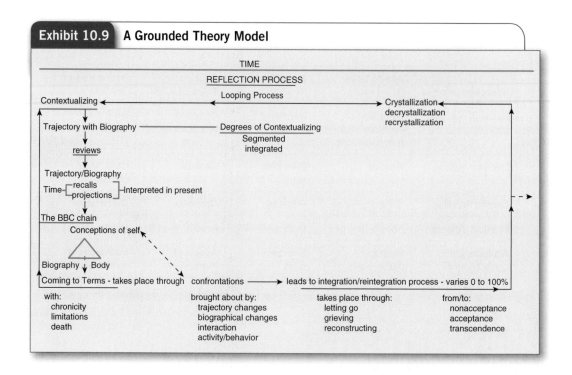

Exhibit 10.9   A Grounded Theory Model

As observation, interviewing, and reflection continue, grounded theory researchers refine their definitions of problems and concepts and select indicators. They can then check the frequency and distribution of phenomena: How many people made a particular type of comment? How often did social interaction lead to arguments? Social system models may then be developed, which specify the relationships among different phenomena. These models are modified as researchers gain experience in the setting. For the final analysis, the researchers check their models carefully against their notes and make a concerted attempt to discover negative evidence that might suggest that the model is incorrect.

Heidi Levitt, Rebecca Todd Swanger, and Jenny Butler (2008:435) used a systematic grounded method of analysis to understand the perspective of male perpetrators of violence on female victims. Research participants were recruited from programs the courts used in Memphis to assess and treat perpetrators who admitted to having physically abused a female intimate partner. All program participants were of low socioeconomic status, but in other respects Levitt and her colleagues (2008:436) sought to recruit a diverse sample.

The researchers (Levitt et al. 2008:437–438) began the analysis of their interview transcripts by dividing them up into "meaning units"—"segments of texts that each contain one main idea"—and labeling these units with terms like those used by participants. They then compared these labels and combined them into larger descriptive categories. This process continued until they had combined all the meaning units into seven different clusters. Exhibit 10.10 gives an example of two of their clusters and the four categories of meaning units combined within each (Levitt et al. 2008:439).

Here is how Levitt and her colleagues (2008) discuss the comments that were classified in Cluster 2, Category 3:

> Accordingly, when conflicts accumulated that could not be easily resolved, many of the men (5 of 12) thought that ending the relationship was the only way to stop violence from recurring. (p. 440)

Exhibit 10.10	Clusters and Categories in a Grounded Theory Analysis

Clusters (endorsement)	Categories (endorsement)
1. The arrest incident is a hurdle or a test from god that I alone have to deal with, although the responsibility for the abuse was not all my own. (10)	1. If alcohol or drugs had not been in the picture, we wouldn't have come to blows: Substance use is thought to increase the rate of IPV (2) 2. I don't want to get involved in conflict because I don't want to deal with its consequences (9) 3. Joint responsibility in conflict depends on who did more fighting (8) 4. How women cause IPV: Being treated as a child through nagging and being disrespected (5)
2. Passive avoidance and withdrawal from conflict is the best way to prevent aggression and to please god. (10)	1. DV thought to be "cured" by passively attending classes and learning anger management (6) 2. Religious interventions have been vague or guilt producing, we need explicit advice and aren't getting it (9) 3. Intimate partner violence can be stopped by cutting off relationships, but this can be a painful experience (5) 4. Should resolve conflict to create harmony and avoid depression—but conflict may increase as a result (10)

I don't deal with anybody so I don't have any conflicts. . . . It makes me feel bad because I be lonely sometime, but at the same time, it's the best thing going for me right now. I'm trying to rebuild me. I'm trying to put me on a foundation to where I can be a total leader. Like I teach my sons, "Be leaders instead of followers." (cited in Levitt et al. 2008:440)

Although this interviewee's choice to isolate himself was a strategy to avoid relational dependency and conflict, it left him without interpersonal support and it could be difficult for him to model healthy relationships for his children. (p. 440)

With procedures such as these, the grounded theory approach develops general concepts from careful review of text or other qualitative materials and can then suggest plausible relationships among these concepts.

Qualitative Comparative Analysis

Daniel Cress and David Snow (2000) asked a series of very specific questions about social movement outcomes in their study of homeless social movement organizations (SMOs). They collected qualitative data from about 15 SMOs in eight cities. A content analysis of newspaper articles indicated that these cities represented a range of outcomes, and the SMOs within them were also relatively accessible to Cress and Snow due to prior contacts. In each of these cities, Cress and Snow used a snowball sampling strategy to identify the homeless SMOs and the various supporters, antagonists, and significant organizational bystanders with whom they interacted. They then gathered information from representatives of these organizations, including churches, other activist organizations, police departments, mayors' offices, service providers, federal agencies, and, of course, the SMOs themselves.

To answer their research questions, Cress and Snow (2000) needed to operationalize each of the various conditions that they believed might affect movement outcomes, using coding procedures that were much more systematic than those often employed in qualitative research. For example, Cress and Snow defined "sympathetic allies" operationally as

the presence of one or more city council members who were supportive of local homeless mobilization. This was demonstrated by attending homeless SMO meetings and rallies and by taking initiatives to city agencies on behalf of the SMO. (Seven of the 14 SMOs had such allies.) (p. 1078)

Cress and Snow (2000) also chose a structured method of analysis, **qualitative comparative analysis (QCA)**, to assess how the various conditions influenced SMO outcomes. This procedure identifies the combination of factors that had to be present across multiple cases to produce a particular outcome (Ragin 1987). Cress and Snow (2000) explain why QCA was appropriate for their analysis:

QCA . . . is conjunctural in its logic, examining the various ways in which specified factors interact and combine with one another to yield particular outcomes. This increases the prospect of discerning diversity and identifying different pathways that lead to an outcome of interest and thus makes this mode of analysis especially applicable to situations with complex patterns of interaction among the specified conditions. (p. 1079)

Exhibit 10.11 summarizes the results of much of Cress and Snow's (2000) analysis. It shows that homeless SMOs that were coded as organizationally viable used disruptive tactics, had sympathetic political allies, and presented a coherent diagnosis and program in response to the problem they were protesting were very likely to achieve all four valued outcomes: (1) representation, (2) resources, (3) protection of basic rights, and (4) some form of tangible relief. Some other combinations of the conditions were associated with increased likelihood of achieving some valued outcomes, but most of these alternatives less frequently had positive effects.

The qualitative textual data on which the codes were based indicate how particular combinations of conditions exerted their influence. For example, one set of conditions that increased the likelihood of achieving increased protection of basic rights for homeless persons included avoiding disruptive tactics in cities that were more responsive to the SMOs. Cress and Snow (2000) use a quote from a local SMO leader to explain this process:

We were going to set up a picket, but then we got calls from two people who were the co-chairs of the Board of Directors. They have like 200 restaurants. And they said, "Hey, we're not bad guys, can we sit down and talk?" We had been set on picketing . . . Then we got to thinking, wouldn't it be better . . . if they co-drafted those things [rights guidelines] with us? So that's what we asked them to do. We had a work meeting, and we hammered out the guidelines. (p. 1089)

In Chapter 12, you will learn more about qualitative comparative analysis and see how this type of method can be used to understand political processes.

Case-Oriented Understanding

Like many qualitative approaches, a **case-oriented understanding** attempts to understand a phenomenon from the standpoint of the participants. The case-oriented understanding method reflects an interpretive research philosophy that is not geared to identifying causes but provides a different way to explain social phenomena. For example, Constance Fischer and Frederick Wertz (2002) constructed such an explanation of the effect of being criminally victimized. They first recounted crime victims' stories and then identified common themes in these stories.

Exhibit 10.11	Multiple Pathways to Outcomes and Level of Impact

Pathways	Outcomes	Impact
1. VIABLE * DISRUPT * ALLIES * DIAG * PROG	Representation, Resources, Rights, and Relief	Very strong
2. VIABLE * disrupt * CITY * DIAG * PROG	Representation and Rights	Strong
3. VIABLE * ALLIES * CITY * DIAG * PROG	Resources and Relief	Moderate
4. viable * DISRUPT * allies * diag * PROG.	Relief	Weak
5. viable * allies * city * diag * PROG	Relief	Weak
6. viable * disrupt * ALLIES * CITY * diag * prog	Resources	Weak

Note: Uppercase letters indicate presence of condition and lowercase letters indicate the absence of a condition. Conditions not in the equation are considered irrelevant. Multiplication signs (*) are read as "and."

Their explanation began with a description of what they termed the process of "living routinely" before the crime: **"he/she . . . feels that the defended against crime could never happen to him/her."** " . . . I said, 'nah, you've got to be kidding'" (pp. 288–289, emphasis in original).

In a second stage, "being disrupted," the victim copes with the discovered crime and fears worse outcomes: "You imagine the worst when it's happening . . . I just kept thinking my baby's upstairs." In a later stage, "reintegrating," the victim begins to assimilate the violation by taking some protective action: "But I clean out my purse now since then and I leave very little of that kind of stuff in there." (p. 289)

> **Case-oriented understanding** An understanding of social processes in a group, formal organization, community, or other collectivity that reflects accurately the standpoint of participants.

Finally, when the victim is "going on," he or she reflects on the changes the crime produced: "I don't think it made me stronger. It made me smarter." (p. 290)

You can see how Fischer and Wertz (2002:288–290) constructed an explanation of the effect of crime on its victims through this analysis of the process of responding to the experience. This effort to "understand" what happened in these cases gives us a much better sense of why things happened as they did.

Visual Sociology

For about 150 years, people have been creating a record of the social world with photography. This creates the possibility of "observing" the social world through photographs and films and of interpreting the resulting images as a "text." You have already seen in this chapter how Elizabeth Stokoe's conversation analysis of "gender talk" (2006) was enriched by her analysis of photographs. In the previous chapter, you learned how Robert Sampson and Stephen Raudenbush (1999) used systematic coding of videotaped observations to measure the extent of disorder in Chicago neighborhoods. Visual sociologists and other social researchers have been developing methods like these to learn how others "see" the social world and to create images for further study. Continuous video recordings help researchers unravel sequences of events and identify nonverbal expressions of feelings (Heath & Luff 2008:501). As in the analysis of written text, however, the visual sociologist must be sensitive to the way in which a photograph or film "constructs" the reality that it depicts.

An analysis by Eric Margolis (2004) of photographic representations of American Indian boarding schools gives you an idea of the value of analysis of photographs (Exhibit 10.12). On the left is a picture taken in 1886 of Chiricahua Apaches who had just arrived at the Carlisle Indian School in Carlisle, Pennsylvania. The school was run by a Captain Richard Pratt, who, like many Americans in that period, felt tribal societies were communistic, indolent, dirty, and ignorant, while Western civilization was industrious and individualistic. So Captain Pratt set out to acculturate American Indians to the dominant culture. The second picture shows the result: the same group of Apaches looking like Europeans, not Native Americans—dressed in standard uniforms, with standard haircuts, and with more standard posture.

Many other pictures display the same type of transformation. Are these pictures each "worth a thousand words"? They capture the ideology of the school management, but we can be less certain that they document accurately the "before and after" status of the students.

Captain Pratt "consciously used photography to represent the boarding school mission as successful" (Margolis 2004:79). While he clearly tried to ensure a high degree of conformity, there were accusations that the contrasting images were exaggerated to overemphasize the change (Margolis 2004:78). Reality was being constructed, not just depicted, in these photographs.

Darren Newbury (2005:1) cautioned the readers of his journal, *Visual Studies,* that "images cannot be simply taken of the world, but have to be made within it." **Photo voice** is a method of using photography to engage research participants in explaining how they have made sense of their social worlds. Rather than using images from other sources, the researcher directing a photo voice project distributes cameras to research participants and invites them to take pictures of their surroundings or everyday activities. The participants then meet with the researcher to present their pictures and discuss their meaning. In this way, researchers learn more about the participants' social worlds as they see it and react to it. The photo voice method also engages participants as part of the research team themselves, thus enriching the researcher's interpretations of the social world.

Photo voice A method in which research participants take pictures of their everyday surroundings with cameras the researcher distributes, and then meet in a group with the researcher to discuss the pictures' meaning.

Lisa Frohmann (2005) recruited 42 Latina and South Asian women from battered women's support groups in Chicago to participate in research about the meaning of violence in their lives. Frohman used photo voice methodology, so she gave each participant a camera. After they received some preliminary instruction, Frohmann invited participants to take about five to seven pictures weekly for 4 to 5 weeks. The photographs were to capture persons, places, and objects that represent

Exhibit 10.12 Pictures of Chiricahua Apache Children Before and After Starting Carlisle Indian School, Carlisle, Pennsylvania, 1886

the continuums of comfort-discomfort, happiness-sadness, safety-danger, security-vulnerability, serenity-anxiety, protection-exposure, strength-weakness, and love-hate (Exhibit 10.13). Twenty-nine women then returned to discuss the results.

With this very simple picture, one participant, Jenny, described how family violence affected her feelings:

This is the dining room table and I took this picture because the table is empty and I feel that although I am with my children, I feel that it is empty because there is no family harmony, which I think is the most important thing. (cited in Frohmann 2005:1407)

The image and narrative represent Jenny's concept of family: a husband and wife who love each other and their children. Food and eating together are important family activities. Part of being a mother and wife, of caring for her family, is preparing the food. Her concept of family is fractured (Frohmann 2005:1407).

Exhibit 10.13 **Picture in Photo Voice Project**

🖻 Mixed Methods

Different qualitative methods may be combined to take advantage of different opportunities for data collection and to enrich understanding of social processes. Researchers may also combine qualitative with quantitative methods in order to provide a more comprehensive analysis of different types of interrelated social processes. These combined approaches are all termed **mixed methods**, but they differ in the relative emphasis given to one or the other method and in the sequencing of their use in a research project (Cresswell 2010:57).

Combining Qualitative Methods

Qualitative researchers often combine one or more of these methods. Elif Kale-Lostuvali (2007) enriched his research by using a combination of qualitative methodologies—including participant observation and intensive interviewing—to study the citizen-state encounters after the İzmit earthquake.

One important concept that emerged from both the observations and the interviews was the distinction between a *mağdur* (sufferer) and a *depremzade* (son of the earthquake). This was a critical distinction, because a *mağdur* was seen as deserving of government assistance, while a *depremzade* was considered to be taking advantage of the situation for personal gain. Kale-Lostuvali (2007) drew on both interviews and participant observation to develop an understanding of this complex concept:

A prominent narrative that was told and retold in various versions all the time in the disaster area elaborated the contrast between *mağdur* (sufferer; that is, the truly needy) and *depremzades* (sons of the earthquake) on the other. The *mağdur* (sufferers) were the deserving recipients of the aid that was being distributed. However, they (1) were in great pain and could not pursue what they needed; or (2) were proud and could not speak of their need; or (3) were humble, always grateful for the little they got, and were certainly not after material gains; or (4) were characterized by a combination of the preceding. And because of these characteristics, they had not been receiving their rightful share of the aid and resources. In contrast, *depremzades* (sons of the earthquake) were people who took advantage of the situation. (p. 755)

The qualitative research by Spencer Moore and his colleagues (2004) on the social response to Hurricane Floyd demonstrates the interweaving of data from focus groups and from participant observation with relief workers.

Reports of heroic acts by rescuers, innumerable accounts of "neighbors helping neighbors," and the comments of HWATF [task force] participants suggest that residents, stranded motorists, relief workers, and rescuers worked and came together in remarkable ways during the relief and response phases of the disaster.

Like people get along better . . . they can talk to each other. People who hadn't talked before, they talk now, a lot closer. That goes, not only for the neighborhood, job-wise, organization-wise, and all that. . . . [Our] union sent some stuff for some of the families that were flooded out. (Focus Group #4) (pp. 210–211)

Combining Qualitative and Quantitative Methods

Conducting qualitative interviews can often enhance the value of a research design that uses primarily quantitative measurement techniques. Qualitative data can provide information about the quality of standardized case records and quantitative survey measures, as well as offer some insight into the meaning of particular fixed responses.

It makes sense to use official records to study the treatment of juveniles accused of illegal acts because these records document the critical decisions to arrest, to convict, or to release (Dannefer & Schutt 1982). But research based on official records can be only as good as the records themselves. In contrast to the controlled interview process in a research study, there is little guarantee that officials' acts and decisions were recorded in a careful and unbiased manner.

Case Study: Juvenile Court Records

Exhibit 10.14	Researchers' and Juvenile Court Workers' Discrepant Assumptions

Researcher Assumptions	Intake Worker Assumptions
• Being sent to court is a harsher sanction than diversion from court.	• Being sent to court often results in more lenient and less effective treatment.
• Screening involves judgments about individual juveniles.	• Screening centers on the juvenile's social situation.
• Official records accurately capture case facts.	• Records are manipulated to achieve the desired outcome.

Interviewing officials who create the records, or observing them while they record information, can strengthen research based on official records. A participant observation study of how probation officers screened cases in two New York juvenile court intake units shows how important such information can be (Needleman 1981). As indicated in Exhibit 10.14, Carolyn Needleman (1981) found that the concepts most researchers believe they are measuring with official records differ markedly from the meaning attached to these records by probation officers. Researchers assume that sending a juvenile case to court indicates a more severe disposition than retaining a case in the intake unit, but probation officers often diverted cases from court because they thought the court would be too lenient. Researchers assume that probation officers evaluate juveniles as individuals, but in these settings, probation officers often based their decisions on juveniles' current social situation (e.g., whether they were living in a stable home), without learning anything about the individual juvenile. Perhaps most troubling for research using case records, Needleman (1981) found that probation officers decided how to handle cases first and then created an official record that appeared to justify their decisions.

Case Study: Mental Health System

The same observation can be made about the value of supplementing fixed-choice survey questions with more probing, open-ended questions. For example, Renee Anspach (1991) wondered about the use of standard surveys to study the effectiveness of mental health systems. Instead of drawing a large sample and asking a set of closed-ended questions, Anspach used snowball sampling techniques to select some administrators, case managers, clients, and family members in four community mental health systems, and then asked these respondents a series of open-ended questions. When asked whether their programs were effective, the interviewees were likely to respond in the affirmative. Their comments in response to other questions, however, pointed to many program failings. Anspach concluded that the respondents simply wanted the interviewer (and others) to believe in the program's effectiveness, for several reasons: Administrators wanted to maintain funding and employee morale; case managers wanted to ensure cooperation by talking up the program with clients and their families; and case managers also preferred to deflect blame for problems to clients, families, or system constraints.

Case Study: Housing Loss in Group Homes

Ethnographic data complements quantitative data in my mixed-methods analysis of the value of group and independent living options for people who are homeless and have been diagnosed with severe mental illness. Exhibit 10.15 displays the quantitative association between lifetime substance abuse—a diagnosis recorded on a numerical scale that was made on the basis of an interview with a clinician—and housing

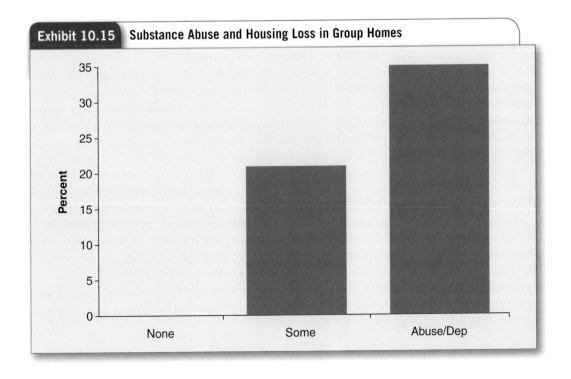

Exhibit 10.15 Substance Abuse and Housing Loss in Group Homes

loss—another quantitative indicator from service records (Schutt 2011:135). The ethnographic notes recorded in the group homes reveal orientations and processes that help to explain the substance abuse-housing loss association (Schutt 2011).

> . . . the time has come where he has to decide once and for all to drink or not. . . . Tom has been feeling "pinned to the bed" in the morning. He has enjoyed getting high with Sammy and Ben, although the next day is always bad. . . . Since he came back from the hospital Lisandro has been acting like he is taunting them to throw him out by not complying with rules and continuing to drink. . . . (pp. 131, 133)

In this way, my analysis of the quantitative data reveals *what* happened, while my analysis of the ethnographic data helps to understand *why*.

Computer-Assisted Qualitative Data Analysis

The analysis process can be enhanced in various ways by using a computer. Programs designed for qualitative data can speed up the analysis process, make it easier for researchers to experiment with different codes, test different hypotheses about relationships, and facilitate diagrams of emerging theories and preparation of research reports (Coffey & Atkinson 1996; Richards & Richards 1994). The steps involved in computer-assisted qualitative data analysis parallel those used traditionally to analyze text such as notes, documents, or interview transcripts: preparation, coding, analysis, and reporting. We use three of the most popular programs to illustrate these steps: HyperRESEARCH, QSR NVivo, and ATLAS.ti. (A free trial version of HyperRESEARCH and tutorials can be downloaded from the ResearchWare site, at http://www.researchware.com.)

> **Computer-assisted qualitative data analysis** Uses special computer software to assist qualitative analyses through creating, applying, and refining categories; tracing linkages between concepts; and making comparisons between cases and events.

Text preparation begins with typing or scanning text in a word processor or, with NVivo, directly into the program's rich text editor. NVivo will create or import a rich text file. HyperRESEARCH requires that your text be saved as a text file (as "ASCII" in most word processors) before you transfer it into the analysis program. HyperRESEARCH expects your text data to be stored in separate files corresponding to each unique case, such as an interview with one subject. These programs now allow multiple types of files, including pictures and videos as well as text. Exhibit 10.16 displays the different file types and how they are connected in the organization of a project (a "hermeneutic unit") with ATLAS.ti.

Coding the text involves categorizing particular text segments. This is the foundation of much qualitative analysis. Each program allows you to assign a code to any segment of text (in NVivo, you drag through the characters to select them; in HyperRESEARCH, you click on the first and last words to select text). You can make up codes as you go through a document and also assign codes that you have already developed to text segments. Exhibit 10.17 shows the screens that appear in HyperRESEARCH and NVivo at the coding stage, when a particular text ms "autocode" text by identifying a word or phrase that should always receive the same code, or, in NVivo, by coding each section identified by the style of the rich text document—for example, each question or speaker (of course, you should check carefully the results of autocoding). Both programs also let you examine the coded text "in context"—embedded in its place in the original document.

Exhibit 10.16 File Types and Unit Structure in ATLAS.ti

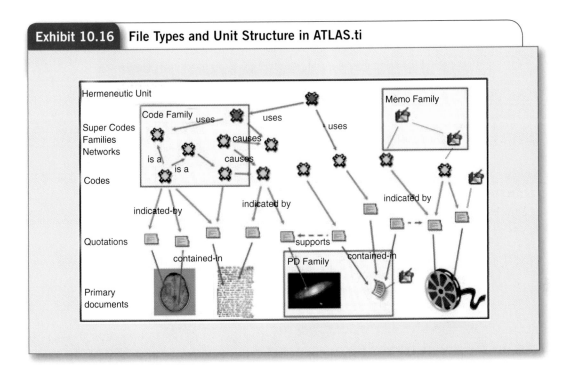

Exhibit 10.17a HyperRESEARCH Coding Stage

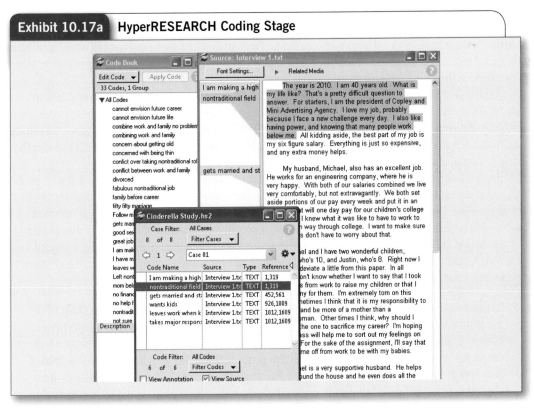

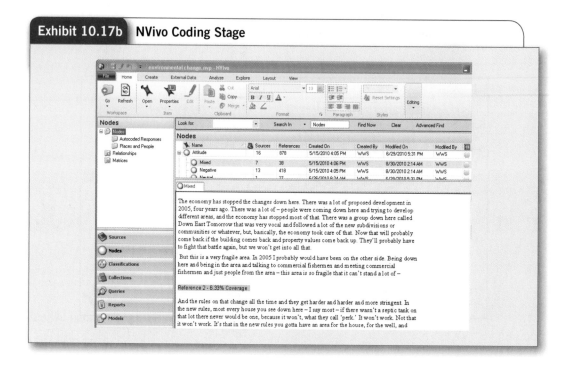

Exhibit 10.17b   NVivo Coding Stage

In qualitative data analysis, coding is not a one-time-only or one-code-only procedure. Each program allows you to be inductive and holistic in your coding: You can revise codes as you go along, assign multiple codes to text segments, and link your own comments ("memos") to text segments. You can work "live" with the coded text to alter coding or create new, more subtle categories. You can also place hyperlinks to other documents in the project or any multimedia files outside it.

Analysis focuses on reviewing cases or text segments with similar codes and examining relationships among different codes. You may decide to combine codes into larger concepts. You may specify additional codes to capture more fully the variation among cases. You can test hypotheses about relationships among codes and develop more free-form models (see Exhibit 10.18). You can specify combinations of codes that identify cases that you want to examine.

Reports from each program can include text to illustrate the cases, codes, and relationships that you specify. You can also generate counts of code frequencies and then import these counts into a statistical program for quantitative analysis. However, the many types of analyses and reports that can be developed with qualitative analysis software do not lessen the need for a careful evaluation of the quality of the data on which conclusions are based.

In reality, using a qualitative data analysis computer program is not always as straightforward as it appears. Scott Decker and Barrik Van Winkle (1996) describe the difficulty they faced in using a computer program to identify instances of the concept of *drug sales:*

> The software we used is essentially a text retrieval package . . . One of the dilemmas faced in the use of such software is whether to employ a coding scheme within the interviews or simply to leave them as unmarked text. We chose the first alternative, embedding conceptual tags at the appropriate points in the text. An example illustrates this process. One of the activities we were concerned with was drug sales. Our first chore (after a thorough reading of all the transcripts) was to use the software to "isolate" all of the transcript sections dealing with drug sales. One way to do this would be to search the transcripts for every instance in which the word "drugs" was used. However, such a strategy would have the disadvantages of providing information of too general a character while often missing important statements about drugs. Searching on the word "drugs" would have

Exhibit 10.18   A Free-Form Model in NVivo

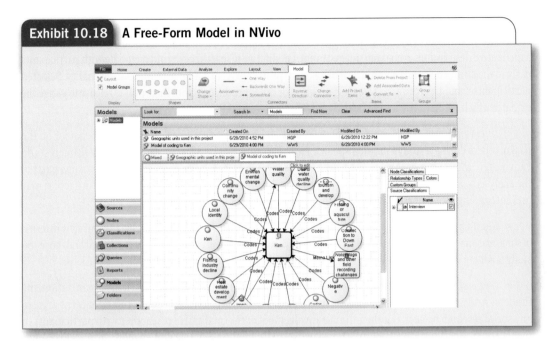

produced a file including every time the word was used, whether it was in reference to drug sales, drug use, or drug availability, clearly more information than we were interested in. However, such a search would have failed to find all of the slang used to refer to drugs ("boy" for heroin, "Casper" for crack cocaine) as well as the more common descriptions of drugs, especially rock or crack cocaine. (pp. 53–54)

Decker and Van Winkle (1996) solved this problem by parenthetically inserting conceptual tags in the text whenever talk of drug sales was found. This process allowed them to examine all the statements made by gang members about a single concept (drug sales). As you can imagine, however, this still left the researchers with many pages of transcript material to analyze.

▣ Ethics in Qualitative Data Analysis

The qualitative data analyst is never far from ethical issues and dilemmas. Data collection should not begin unless the researcher has a plan that others see as likely to produce useful knowledge. Relations developed with research participants and other stakeholders to facilitate data collection should also be used to keep these groups informed about research progress and findings. Research participants should be encouraged to speak out about emerging study findings (Lincoln 2009:154–155). Throughout the analytic process, the analyst must consider how the findings will be used and how participants in the setting will react. Miles and Huberman (1994:293–295) suggest several specific questions that are of particular importance during the process of data analysis:

- *Privacy, confidentiality, and anonymity:* "In what ways will the study intrude, come closer to people than they want? How will information be guarded? How identifiable are the individuals and organizations studied?" We have considered this issue already in the context of qualitative data collection, but it also must be a concern during the process of analysis. It can be difficult to present a rich description in a case study while at the same time not identifying the setting. It can be easy for participants in the study to identify each other in a qualitative description, even if outsiders cannot. Qualitative researchers should negotiate with participants early in the study the approach that will be taken to protect privacy and maintain confidentiality. Selected participants should also be asked to review reports or other products before their public release to gauge the extent to which they feel privacy has been appropriately preserved.

- *Intervention and advocacy:* "What do I do when I see harmful, illegal, or wrongful behavior on the part of others during a study? Should I speak for anyone's interests besides my own? If so, whose interests do I advocate?" Maintaining what is called *guilty knowledge* may force the researcher to suppress some parts of the analysis so as not to disclose the wrongful behavior, but presenting "what really happened" in a report may prevent ongoing access and violate understandings with participants.

- *Research integrity and quality:* "Is my study being conducted carefully, thoughtfully, and correctly in terms of some reasonable set of standards?" Real analyses have real consequences, so you owe it to yourself and those you study to adhere strictly to the analysis methods that you believe will produce authentic, valid conclusions.

- *Ownership of data and conclusions:* "Who owns my field notes and analyses: I, my organization, my funders? And once my reports are written, who controls their diffusion?" Of course, these concerns arise in any social research project, but the intimate involvement of the qualitative researcher with participants in the setting studied makes conflicts of interest between different stakeholders much more difficult to resolve. Working through the issues as they arise is essential.

- *Use and misuse of results:* "Do I have an obligation to help my findings be used appropriately? What if they are used harmfully or wrongly?" It is prudent to develop understandings early in the project with all major stakeholders that specify what actions will be taken to encourage appropriate use of project results and to respond to what is considered misuse of these results.

Netnographers must consider additional issues:

- *The challenge of online confidentiality:* While a netnographer should maintain the same privacy standards as other qualitative researchers, the retention of information on the web makes it difficult to protect confidentiality with traditional mechanisms. Anyone reading a distinctive quote based on online text can use a search engine to try to locate the original text. Distinctive online names intended to protect anonymity still identify individuals who may be the target of hurtful commentary (Kozinets 2010:143–145). Some users may think of their postings to some online community site as private (Markham 2007:274). For these reasons, netnographers who study sensitive issues should go to great lengths to disguise the identity of the community they have studied as well as of its participants who are quoted.

- *Gaining informed consent:* Since community members can disguise their identities, gaining informed consent creates unique challenges. Men may masquerade as women and children as adults, and yet the research should not proceed unless an effort is made to obtain informed and voluntary consent. Netnographers must make their own identity known, state clearly their expectations for participation, provide an explicit informed consent letter that is available as discussion participants come and go (Denzin & Lincoln 2008:274), and attempt to identify those whose identity is not credible through inconsistencies in their postings (Kozinets 2010:151–154).

Some indigenous peoples have established rules for outside researchers in order to preserve their own autonomy. These rules may require collaboration with an indigenous researcher, collective approval of admission to their culture, and review of all research products prior to publication. The primary ethical commitment is to the welfare of the community as a whole and the preservation of their culture, rather than to the rights of individuals (Lincoln 2009:162–163).

Conclusions

The variety of approaches to qualitative data analysis makes it difficult to provide a consistent set of criteria for interpreting their quality. Norman Denzin's (2002:362–363) "interpretive criteria" are a good place to start. Denzin suggests that at the conclusion of their analyses, qualitative data analysts ask the following questions about the materials they have produced. Reviewing several of them will serve as a fitting summary for your understanding of the qualitative analysis process.

- *Do they illuminate the phenomenon as lived experience?* In other words, do the materials bring the setting alive in terms of the people in that setting?

- *Are they based on thickly contextualized materials?* We should expect thick descriptions that encompass the social setting studied.

- *Are they historically and relationally grounded?* There must be a sense of the passage of time between events and the presence of relationships between social actors.

- *Are they processual and interactional?* The researcher must have described the research process and his or her interactions within the setting.

- *Do they engulf what is known about the phenomenon?* This includes situating the analysis in the context of prior research and also acknowledging the researcher's own orientation on first starting the investigation.

When an analysis of qualitative data is judged as successful in terms of these criteria, we can conclude that the goal of authenticity has been achieved.

As a research methodologist, you should be ready to use qualitative techniques, evaluate research findings in terms of these criteria, and mix and match specific analysis methods as required by the research problem to be investigated and the setting in which it is to be studied.

Key Terms

Case-oriented understanding 344
Computer-assisted qualitative data analysis 350
Emic focus 322
Ethnography 333

Ethnomethodology 336
Etic focus 322
Grounded theory 341
Matrix 329
Narrative analysis 339
Netnography 335

Photo voice 346
Progressive focusing 322
Qualitative comparative analysis (QCA) 344
Tacit knowledge 331

Highlights

- Qualitative data analysts are guided by an emic focus of representing persons in the setting on their own terms, rather than by an etic focus on the researcher's terms.

- Case studies use thick description and other qualitative techniques to provide a holistic picture of a setting or group.
- Ethnographers attempt to understand the culture of a group.

- Narrative analysis attempts to understand a life or a series of events as they unfolded, in a meaningful progression.

- Grounded theory connotes a general explanation that develops in interaction with the data and is continually tested and refined as data collection continues.

- Special computer software can be used for the analysis of qualitative, textual, and pictorial data. Users can record their notes, categorize observations, specify links between categories, and count occurrences.

STUDENT STUDY SITE

To assist in completing the web exercises, please access the study site at **www.sagepub.com/schuttisw7e**, where you will find the web exercise with accompanying links. You'll find other useful study materials such as self-quizzes and e-flashcards for each chapter, along with a group of carefully selected articles from research journals that illustrate the major concepts and techniques presented in the book.

Discussion Questions

1. List the primary components of qualitative data analysis strategies. Compare and contrast each of these components with those relevant to quantitative data analysis. What are the similarities and differences? What differences do these make?

2. Does qualitative data analysis result in trustworthy results—in findings that achieve the goal of authenticity? Why would anyone question its use? What would you reply to the doubters?

3. Narrative analysis provides the "large picture" of how a life or event has unfolded, while conversation analysis focuses on the details of verbal interchange? When is each method most appropriate? How could one method add to another?

4. Ethnography, grounded theory, and case-oriented understanding each refers to aspects of data analysis that are an inherent part of the qualitative approach. What do these approaches have in common? How do they differ? Can you identify elements of these three approaches in this chapter's examples of ethnomethodology, netnography, qualitative comparative analysis, and narrative analysis?

Practice Exercises

1. Attend a sports game as an ethnographer. Write up your analysis and circulate it for criticism.

2. Write a narrative in class about your first date, car, college course, or something else that you and your classmates agree on. Then collect all the narratives and analyze them in a "committee of the whole." Follow the general procedures discussed in the example of narrative analysis in this chapter.

3. Go forth and take pictures! Conduct a photo voice project with your classmates and write up your own review of the group's discussion of your pictures.

4. Review one of the articles on the book's study site, www.sagepub.com/schuttisw7e, that used qualitative methods. Describe the data that were collected, and identify the steps used in the analysis. What type of qualitative data analysis was this? If it is not one of the methods presented in this chapter, describe its similarities to and differences from one of these methods. How confident are you in the conclusions, given the methods of analysis used?

5. Review postings to an on-line discussion group. How could you study this group using netnography? What challenges would you encounter?

Ethics Questions

1. Pictures are worth a thousand words, so to speak, but is that 1,000 words too many? Should qualitative researchers (like yourself) feel free to take pictures of social interaction or other behaviors anytime, anywhere? What limits should an institutional review board place on researchers' ability to take pictures of others? What if the picture of the empty table in this chapter also included the abusive family member who is discussed? What if the picture was in a public park, rather than in a private residence?

2. Participants in social settings often "forget" that an ethnographer is in their midst, planning to record what they say and do, even when the ethnographer has announced his or her role. New participants may not have heard the announcement, and everyone may simply get used to the ethnographer as if he or she was just "one of us." What efforts should an ethnographer take to keep people informed about their work in the setting they are studying? Consider settings such as a sports team, a political group, and a book group.

Web Exercises

1. The *Qualitative Report* is an online journal about qualitative research. Inspect the table of contents for a recent issue at www.nova.edu/ssss/QR/index.html. Read one of the articles and write a brief article review.

2. Be a qualitative explorer! Go to the "Qualitative Page" website and see what you can find that enriches your understanding of qualitative research (www.qualitativeresearch.uga.edu/QualPage). Be careful to avoid textual data overload.

HyperRESEARCH Exercises

1. Eight essays written by college-age women in Hesse-Biber's (1989) "Cinderella Study" are saved on the book's study site. The essays touch on how young women feel about combining work and family roles. Download and install HyperRESEARCH from http://www.researchware.com/ and open the Cinderella Study (FinditatC:\ProgramFiles\HyperRESEARCH3.0\Documentation\Tutorials\Cinderella Study\Cinderella Study .hs2). Look over the preliminary code categories that have already been applied to each essay. Do you agree with the code categories/themes already selected? What new code categories would you add and why? Which would you delete? Why? What are some of the common themes/codes that cut across all eight cases concerning how young women think about what their life will be like in 20 years?

2. Work through the tutorial on HyperRESEARCH that was downloaded with the program. How does it seem that qualitative analysis software facilitates the analysis process? Does it seem to you that it might hinder the analysis process in some ways? Explain your answers.

Developing a Research Proposal

1. Which qualitative data analysis alternative is most appropriate for the qualitative data you proposed to collect for your project (Exhibit 2.14, #18)? Using the approach, develop a strategy for using the techniques of qualitative data analysis to analyze your textual data.

CHAPTER 11

Evaluation and Policy Research

Drug Abuse Resistance Education (DARE), as you probably know, is offered in elementary schools across America. For parents worried about drug abuse among youth and for many concerned citizens, the program has immediate appeal. It brings a special police officer into the schools once a week to talk to students about the hazards of drug abuse and to establish a direct link between local law enforcement and young people. You only have to check out bumper stickers or attend a few PTA meetings to learn that it's a popular program. It is one way many local governments have implemented antidrug policies.

And it is appealing. DARE seems to improve relations between the schools and law enforcement and to create a positive image of the police in the eyes of students.

> It's a very positive program for kids . . . a way for law enforcement to interact with children in a nonthreatening fashion . . . DARE sponsored a basketball game. The middle school jazz band played. . . . We had families there. . . . DARE officers lead activities at the [middle school]. . . . Kids do woodworking and produce a play. (Taylor 1999:1, 11)

For some, the positive police-community relationships created by the program are enough to justify its continuation (Birkeland, Murphy-Graham, & Weiss 2005:248), but most communities are concerned with its value in reducing drug abuse among children. Does DARE lessen the use of illicit drugs among DARE students? Does it do so while they are enrolled in the program or, more important, after they enter middle or high school? Unfortunately, evaluations of DARE using social science methods led to the conclusion that students who participated in DARE were no less likely to use illicit drugs than comparable students who did not participate in DARE (Ringwalt et al. 1994; West & O'Neal 2004).

If, like me, you have a child who enjoyed DARE, or were yourself a DARE student, this may seem like a depressing way to begin a chapter on evaluation research. Nonetheless, it drives home an important point: To know whether social programs work, or how they work, we have to evaluate them systematically and fairly, whether we personally like the programs or not. And there's actually an optimistic conclusion to this introductory story: Evaluation research can make a difference. After the accumulation of evidence that DARE programs were ineffective (West & O'Neal 2004), a "new" DARE program was designed that engaged students more actively (Toppo 2002).

> Gone is the old-style approach to prevention in which an officer stands behind a podium and lectures students in straight rows. New DARE officers are trained as "coaches" to support kids who are using research-based refusal strategies in high-stakes peer-pressure environments. (DARE 2008)

Of course, the "new DARE" is now being evaluated, too. Sorry to say, one early quasi-experimental evaluation in 17 urban schools, funded by DARE America, found no effect of the program on students' substance use (Vincus et al. 2010).

In this chapter, you will read about a variety of social program evaluations as I introduce the evaluation research process, illustrate the different types of evaluation research, highlight alternative approaches, and review ethical concerns. You will learn in this chapter that current major debates about social policies like health care, homelessness, domestic violence, and drug abuse are each being informed by important major evaluation projects. You should finish the chapter with a much better understanding of how the methods of applied social research can help improve society.

▣ History of Evaluation Research

Evaluation research is not a method of data collection, like survey research or experiments, nor is it a unique component of research designs, like sampling or measurement. Instead, evaluation research is social research that is conducted for a distinctive purpose: to investigate social programs (e.g., substance abuse treatment programs, welfare programs, criminal justice programs, or employment and training programs). For each project, an evaluation researcher must select a research design and a method of data collection that are useful for answering the particular research questions posed and appropriate for the particular program investigated.

The development of evaluation research as a major enterprise followed on the heels of the expansion of the federal government during the Great Depression and World War II. Large Depression-era government outlays for social programs stimulated interest in monitoring program output, and the military effort in World War II led to some of the necessary review and contracting procedures for sponsoring evaluation research. In the 1960s, criminal justice researchers began to use experiments to test the value of different policies

(Orr 1999:24). New government social programs of the 1960s often came with evaluation requirements attached, and more than 100 contract research and development firms began operation in the United States between 1965 and 1975 (Dentler 2002; Rossi & Freeman 1989:34).

The New Jersey Income Maintenance Experiment was the first large-scale randomized experiment to test social policy in action. Designed in 1967, the New Jersey Experiment randomly assigned 1,300 families to different income support levels to test the impact of cash transfers to the working poor on their work effort. It was soon followed by even larger experiments to test other income maintenance questions, most notably the Seattle-Denver Income Maintenance Experiment (Orr 1999:24–26). As Exhibit 11.1 illustrates, the number of social experiments like this continued to increase in subsequent years. The Campbell Collaboration research archive, a project begun by evaluation researchers, contained 10,449 reports on randomized evaluation studies at the end of the last millennium (Davies, Petrosino, & Chalmers 1999).

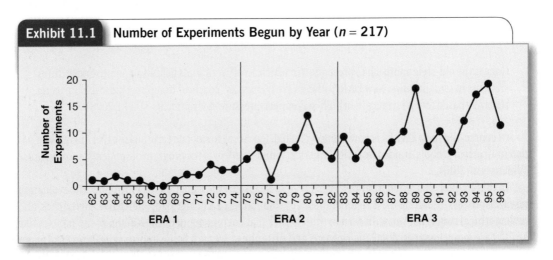

Exhibit 11.1 **Number of Experiments Begun by Year (*n* = 217)**

Government requirements and popular concern with the benefit derived from taxpayer-funded programs continue to stimulate evaluation research. The Community Mental Health Act Amendments of 1975 (Public Law 94–63) required quality assurance reviews, which often involved evaluation-like activities (Patton 2002:147–151), while the Government Performance and Results Act of 1993 required some type of evaluation of all government programs (U.S. Office of Management and Budget 2002). At century's end, the federal government was spending about $200 million annually on evaluating $400 billion in domestic programs, and the 30 major federal agencies had between them 200 distinct evaluation units (Boruch 1997). In 1999, the new Governmental Accounting Standards Board urged that more attention be given to "service efforts and accomplishments" in standard government fiscal reports (Campbell 2002). Universities and private organizations also often sponsor evaluation research projects.

Evaluation Basics

Exhibit 11.2 illustrates the process of evaluation research as a simple systems model. First, clients, customers, students, or some other persons or units—cases—enter the program as inputs. (You'll notice that this model treats programs like machines, with people functioning as raw materials to be processed.)

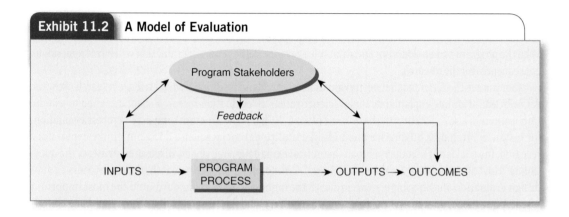

| Exhibit 11.2 | A Model of Evaluation |

Students may begin a new school program, welfare recipients may enroll in a new job training program, or crime victims may be sent to a victim advocate. The resources and staff a program requires are also program inputs.

Next, some service or treatment is provided to the cases. This may be attendance in a class, assistance with a health problem, residence in new housing, or receipt of special cash benefits. The **program process** may be simple or complicated, short or long, but it is designed to have some impact on the cases.

The direct product of the program's service delivery process is its **output**. Program outputs may include clients served, case managers trained, food parcels delivered, or arrests made. The program outputs may be desirable in themselves, but they primarily serve to indicate that the program is operating.

Program **outcomes** indicate the impact of the program on the cases that have been processed. Outcomes can range from improved test scores or higher rates of job retention to fewer criminal offenses and lower rates of poverty. Any social program is likely to have multiple outcomes, some intended and some unintended, some positive and others that are viewed as negative.

Variation in both outputs and outcomes, in turn, influences the inputs to the program through a **feedback** process. If not enough clients are being served, recruitment of new clients may increase. If too many negative side effects result from a trial medication, the trials may be limited or terminated. If a program does not appear to lead to improved outcomes, clients may go elsewhere.

Evaluation research is simply a systematic approach to feedback: It strengthens the feedback loop through credible analyses of program operations and outcomes. Evaluation research also broadens this loop to include connections to parties outside of the program itself. A funding agency or political authority may mandate the research, outside experts may be brought in to conduct the research, and the evaluation research findings may be released to the public, or at least the funders, in a formal report.

The evaluation process as a whole, and feedback in particular, can be understood only in relation to the interests and perspectives of program stakeholders. **Stakeholders** are those individuals and groups who have some basis of concern with the program. They might be clients, staff, managers, funders, or the public. The board of a program or agency, the parents or spouses of clients, the foundations that award program grants, the auditors who monitor program spending, the members of Congress—each is a potential stakeholder, and

Inputs The resources, raw materials, clients, and staff that go into a program.

Program process The complete treatment or service delivered by the program.

Outputs The services delivered or new products produced by the program process.

Outcomes The impact of the program process on the cases processed.

Feedback Information about service delivery system outputs, outcomes, or operations that is available to any program input.

Stakeholders Individuals and groups who have some basis of concern with the program.

each has an interest in the outcome of any program evaluation. Some may fund the evaluation; some may provide research data; and some may review, or even approve, the research report (Martin & Kettner 1996:3). Who the program stakeholders are and what role they play in the program evaluation will have tremendous consequences for the research.

Can you see the difference between evaluation research and traditional social science research (Posavac & Carey 1997)? Unlike explanatory social science research, evaluation research is not designed to test the implications of a social theory; the basic issue often is "What is the program's impact?" Process evaluation, for instance, often uses qualitative methods like traditional social science does, but unlike exploratory research, the goal is not to induce a broad theoretical explanation for what is discovered. Instead, the question is "How does the program do what it does?" Unlike social science research, the researchers cannot design evaluation studies simply in accord with the highest scientific standards and the most important research questions; instead, it is the program stakeholders who set the agenda. But there is no sharp boundary between the two. In their attempt to explain how and why the program has an impact, and whether the program is needed, evaluation researchers often bring social theories into their projects, but for immediately practical aims.

▣ Questions for Evaluation Research

Evaluation projects can focus on several questions related to the operation of social programs and the impact they have:

- Is the program needed?

- Can the program be evaluated?

- How does the program operate?

- What is the program's impact?

- How efficient is the program?

The specific methods used in an evaluation research project depend, in part, on which of these foci the project has.

Needs Assessment

Needs assessment A type of evaluation research that attempts to determine the needs of some population that might be met with a social program.

Is a new program needed or an old one still required? Is there a need at all? A needs assessment attempts to answer these questions with systematic, credible evidence. Need may be assessed by social indicators, such as the poverty rate or the level of home ownership; by interviews of local experts, such as school board members or team captains; by surveys of populations in need; or by focus groups composed of community residents (Rossi & Freeman 1989).

Research in the News

PREDICTING CRIMINAL PROPENSITY

In the News

Nowhere is needs assessment more needed than in predicting the risk of recidivism among applicants for parole. In the 1920s, sociologist Ernest Burgess studied previously released inmates' criminal histories and classified them on the basis of 22 variables as *hobos, ne'er-do-wells, farm boys, drug addicts, gangsters,* and *recent immigrants*. Illinois used a classification like this for 30 years, while other states relied on "clinical judgment." Now most states use a quantitative risk assessment tool in which risk predictions are based on identified correlates of future criminality, such as age at first arrest, job and relationship history, history of drug abuse and gang activity, and behavior while incarcerated.

Source: Neyfakh, Leon. 2011. "You Will Commit a Crime in the Future, True, False: Inside the Science of Predicting Violence." *Boston Sunday Globe,* February 20:K1, K4.

Needs assessment is not as easy as it sounds (Posavac & Carey 1997). Whose definitions or perceptions should be used to shape our description of the level of need? How will we deal with ignorance of need? How can we understand the level of need without understanding the social context from which that level of need emerges? (Short answer to that one: We can't!) What, after all, does *need* mean in the abstract? We won't really understand what the level of need is until we develop plans for implementing a program in response to the identified needs.

The results of the Boston McKinney Project reveal the importance of taking a multidimensional approach to the investigation of need. The Boston McKinney Project evaluated the merits of providing formerly homeless mentally ill persons with staffed group housing as compared with individual housing (Schutt 2011). In a sense, you can think of the whole experiment as involving an attempt to answer the question "What type of housing do these persons 'need'?" Our research team first examined this question at the start of the project, by asking each project participant which type of housing he or she wanted (Schutt & Goldfinger 1996) and by independently asking two clinicians to estimate which of the two housing alternatives would be best for each participant (Goldfinger & Schutt 1996).

Exhibit 11.3 displays the findings. The clinicians recommended staffed group housing for 69% of the participants, whereas most of the participants (78%) sought individual housing. In fact, there was no correspondence between the housing recommendations of the clinicians and the housing preferences of the participants (who did not know what the clinicians had recommended for them). So which perspective reveals the level of need for staffed group housing as opposed to individual housing?

Yet another perspective on housing needs is introduced by the project's outcomes. Individuals assigned to the group housing were somewhat more successful in retaining their housing than were those who were assigned to individual housing, and this differential success rate grew in the years after the project's end (Schutt 2011:247). Does this therefore reveal that these homeless mentally ill persons "needed" group housing more than they needed individual housing, in spite of their preference? What should we make of the fact that the participants who preferred individual housing were more likely to lose their housing during the project, whereas the participants whom the clinicians had rated as ready for independent living were less likely to lose their housing (Schutt 2011:248)? And what should we make of the fact that whether or not participants

Exhibit 11.3 Type of Residence: Preferred and Recommended

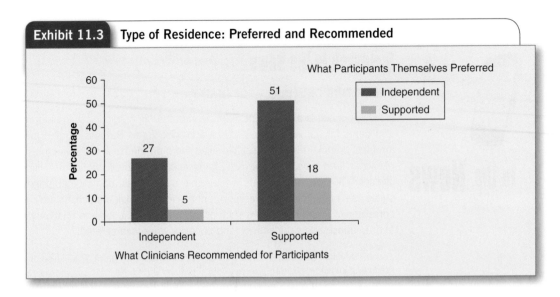

received the type of housing the clinicians recommended or that they themselves preferred made no difference to the likelihood of their losing their housing during the project (Schutt 2011:248)? Does this mean that neither initial preferences nor clinician recommendations tell us about the need for one or the other type of housing, only about the risk of losing whatever housing they were assigned to?

The methodological lesson here is that in needs assessment, as in other forms of evaluation research, it is a good idea to use multiple indicators. You can also see that there is no absolute definition of *need* in this situation, nor is there likely to be in any but the most simplistic evaluation projects. A good evaluation researcher will do his or her best to capture different perspectives on need and then help others make sense of the results.

A wonderful little tale, popular with evaluation researchers, reveals the importance of thinking creatively about what people need:

> The manager of a 20-story office building had received many complaints about the slowness of the elevators. He hired an engineering consultant to propose a solution. The consultant measured traffic flow and elevator features and proposed replacing the old with new ones, which could shave 20 seconds off the average waiting time. The only problem: It cost $100,000. A second consultant proposed adding 2 additional elevators, for a total wait time reduction of 35 seconds and a cost of $150,000. Neither alternative was affordable. A third consultant was brought in. He looked around for a few days and announced that the problem was not really the waiting times, but boredom. For a cost of less than $1000, the manager had large mirrors installed next to the elevators so people could primp and observe themselves while waiting for an elevator. The result: no more complaints. Problem solved. (Witkin & Altschuld 1995:38)

Evaluability Assessment

Evaluability assessment A type of evaluation research conducted to determine whether it is feasible to evaluate a program's effects within the available time and resources.

Evaluation research will be pointless if the program itself cannot be evaluated. Yes, some type of study is always possible, but a study specifically to identify the effects of a particular program may not be possible within the available time and resources. So researchers may conduct an **evaluability assessment** to learn this in advance, rather than expend time and effort on a fruitless project.

Why might a social program not be evaluable? Several factors may affect this:

- Management only wants to have its superior performance confirmed and does not really care whether the program is having its intended effects. This is a very common problem.

- Staff are so alienated from the agency that they don't trust any attempt sponsored by management to check on their performance.

- Program personnel are just "helping people" or "putting in time" without any clear sense of what the program is trying to achieve.

- The program is not clearly distinct from other services delivered from the agency and so can't be evaluated by itself (Patton 2002:164).

Because evaluability assessments are preliminary studies to "check things out," they often rely on qualitative methods. Program managers and key staff may be interviewed in depth, or program sponsors may be asked about the importance they attach to different goals. The evaluators may then suggest changes in program operations to improve evaluability. They may also use the evaluability assessment to "sell" the evaluation to participants and sensitize them to the importance of clarifying their goals and objectives. If the program is judged to be evaluable, knowledge gleaned through the evaluability assessment is used to shape evaluation plans.

The President's Family Justice Center (FJC) Initiative was initiated in the administration of President George W. Bush to plan and implement comprehensive domestic violence services that would provide "one stop shopping" for victims in need of services. In 2004, the National Institute of Justice contracted with Abt Associates in Cambridge, Massachusetts to assess the evaluability of 15 pilot service programs that been awarded a total of $20 million and to develop an evaluation plan. In June 2005, Abt researchers Meg Townsend, Dana Hunt, Caity Baxter, and Peter Finn reported on their evaluability assessment.

Abt's assessment began with conversations to collect background information and perceptions of program goals and objectives from those who had designed the program. These conversations were followed by a review of the grant applications submitted by each of the 15 sites and phone conversations with site representatives. Site-specific data collection focused on the project's history at the site, its stage of implementation, staffing plans and target population, program activities and stability, goals identified by the site's director, apparent contradictions between goals and activities, and the state of data systems that could be used in the evaluation. Exhibit 11.4 shows the resulting logic model that illustrates the intended activities, outcomes and impacts for the Alameda County, California program. Although they had been able to begin the evaluability assessment process, Meg Townsend, Dana Hunt, and their colleagues concluded that in the summer of 2005, none of the 15 sites were far enough along with their programs to complete the assessment.

Process Evaluation

What actually happens in a social program? The New Jersey Income Maintenance Experiment was designed to test the effect of some welfare recipients receiving higher payments than others (Kershaw & Fair 1976). Did that occur? In the Minneapolis experiment on the police response to domestic violence (Sherman & Berk 1984), police officers were to either arrest or warn individuals accused of assaulting their spouses on the basis of a random selection protocol, unless they concluded that they must override the experimental assignment to minimize the risk of repeat harm. Did the police officers follow this protocol? How often did they override it due to concerns about risk? Questions like these about program implementation must be answered before it

Exhibit 11.4 Alameda Family Justice Center Logic Model

Inputs	Activities	Outcomes	Impacts	Goals

• On-site partners
• Intake systems
• Client management process
• Space design
• Site location
	FJC
	• Case management
	• Assistance with restraining orders
	• Assistance with police reports
	• Legal assistance
	• Advocacy
	• Medical care
	• Forensic exams
	• Assessments and referral for treatment
	• Counseling
	• Safety planning
	• Emergency food/cash/ transportation
	• Referral for shelter and other on-going care
	• Assistance with public assistance
	• 24-hour helpline
	• Parenting classes
	• Child care
	• Rape crises services
	• Faith-based services
	• Job training
	• Translation services
		Victims
		• Increase likelihood to access services
		• Increase demand for services
		• Increase usage of services
		• Increase frequency of cross-referrals or use of multiple services
			Victims
			• Reduce tendency to blame oneself for abuse
			• Reduce conditions prevent women from leaving
			• Increase likelihood of reporting incident
			• Increase likelihood of request for temporary/ permanent restraining orders
			• Increase likelihood of participating in prosecution
				• Decrease incidents of DV
				  ○ Decreased repeat victimizations
				  ○ Decreased seriousness
				• Hold offenders accountable
				  ○ Decrease repeat offenders
				• Break cycle of violence

	Community • Early intervention and prevention programming • FJC informational materials	**Community** • Increase knowledge of DV/SA/Elder Abuse • Increase awereness of services available	**Community** • Increase awereness of FJC • Decrease social tolerance for VAW	

Inputs	Activities	Outcomes	Impacts	Goals
	Systems • Collaboration between government and non-gov't providers • Improve access to batterer information	**Systems** • Improve DV policies and procedures • Increase understanding of each other's services • Increase coordination of services	**Systems** • Improve institutional response to DV • Decrease secondary trauma • Increase assurance of victim safety • Increase the number of successful criminal legal actions • Increase the number of successful civil legal actions	

is possible to determine whether the program's key elements had the desired effect. Answers to such program implementation questions are obtained through **process evaluation**—research to investigate the process of service delivery.

Process evaluation is even more important when more complex programs are evaluated. Many social programs comprise multiple elements and are delivered over an extended period of time, often by different providers in different areas. Due to this complexity, it is quite possible that the program as delivered is not the same for all program recipients or consistent with the formal program design.

> **Process evaluation** Evaluation research that investigates the process of service delivery.

The evaluation of DARE by Research Triangle Institute researchers Christopher Ringwalt and colleagues (1994:7) included a process evaluation with three objectives:

1. To assess the organizational structure and operation of representative DARE programs nationwide

2. To review and assess the factors that contribute to the effective implementation of DARE programs nationwide

3. To assess how DARE and other school-based drug prevention programs are tailored to meet the needs of specific populations

The process evaluation (they called it an *implementation assessment*) was an ambitious research project in itself, with site visits, informal interviews, discussions, and surveys of DARE program coordinators and advisors. These data indicated that DARE was operating as designed and was running relatively smoothly. As shown in Exhibit 11.5, drug prevention coordinators in DARE school districts rated the program components as much more satisfactory than did coordinators in school districts with other types of alcohol and drug prevention programs.

Process evaluation also can be used to identify the specific aspects of the service delivery process that have an impact. This, in turn, will help explain why the program has an effect and which conditions are required for these effects. (In Chapter 6, I described this as identifying the causal mechanism.) Implementation problems identified in site visits included insufficient numbers of officers to carry out the program as planned and a lack of Spanish-language DARE books in a largely Hispanic school. Classroom observations indicated engaging presentations and active student participation (Ringwalt et al. 1994:58).

Process analysis of this sort can also help show how apparently unambiguous findings may be incorrect. The apparently disappointing results of the Transitional Aid Research Project (TARP) provide an instructive lesson of this sort. TARP was a social experiment designed to determine whether financial aid during the transition from prison to the community would help released prisoners find employment and avoid returning to crime. Two thousand participants in Georgia and Texas were randomized to receive either a particular level of benefits over a particular period of time or no benefits at all (the control group). Initially, it seemed that the payments had no effect: The rate of subsequent arrests for both property and nonproperty crimes was not affected by the TARP treatment condition.

But this wasn't all there was to it. Peter Rossi tested a more elaborate causal model of TARP effects, summarized in Exhibit 11.6 (Chen 1990). Participants who received TARP payments had more income to begin with and so had more to lose if they were arrested; therefore, they were less likely to commit crimes. However, TARP payments also created a disincentive to work and therefore increased the time available in which to

Exhibit 11.5	Components of DARE and Other Alcohol and Drug Prevention Programs Rated as Very Satisfactory (%)	
Components	**DARE Program (N = 222)**	**Other AOD Programs (N = 406)**
Curriculum	67.5	34.2
Teaching	69.7	29.8
Administrative Requirements	55.7	23.1
Receptivity of Students	76.5	34.6
Effects on Students	63.2	22.8

Exhibit 11.6 Model of TARP Effects

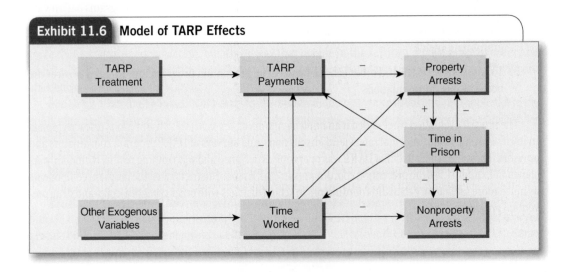

commit crimes. Thus, the positive direct effect of TARP (more to lose) was cancelled out by its negative indirect effect (more free time).

The term **formative evaluation** may be used instead of process evaluation when the evaluation findings are used to help shape and refine the program (Rossi & Freeman 1989). Formative evaluation procedures that are incorporated into the initial development of the service program can specify the treatment process and lead to changes in recruitment procedures, program delivery, or measurement tools (Patton 2002:220).

> **Formative evaluation** Process evaluation that is used to shape and refine program operations.

You can see the formative element in the following excerpt from the report on Ohio's Assisted Living Medicaid Waiver Program by a research team from the Scripps Gerontology Center and Miami University (Applebaum et al. 2007). The program is designed to allow disabled elderly persons to move into assisted living facilities that maximize autonomy rather than more expensive nursing homes.

> In order for the state to develop a viable assisted living program, a plan to increase provider participation is critical. . . . The perspectives of residents, and their families, providers, case managers, and representatives of the Department of Aging and Health will be needed to refine the program as it develops. . . . Every program needs to evolve and therefore a solid structure to identify and implement necessary changes will be crucial for long-term program success. (p. 31)

Process evaluation can employ a wide range of indicators. Program coverage can be monitored through program records, participant surveys, community surveys, or utilizers versus dropouts and ineligibles. Service delivery can be monitored through service records completed by program staff, a management information system maintained by program administrators, or reports by program recipients (Rossi & Freeman 1989).

Qualitative methods are often a key component of process evaluation studies because they can be used to elucidate and understand internal program dynamics—even those that were not anticipated (Patton 2002:159; Posavac & Carey 1997). Qualitative researchers may develop detailed descriptions of how program participants engage with each other, how the program experience varies for different people, and how the program changes and evolves over time.

Impact Analysis

The core questions of evaluation research are "Did the program work?" and "Did it have the intended result?" This part of the research is variously called *impact analysis, impact evaluation,* or *summative evaluation.* Formally speaking, impact analysis compares what happened after a program with what would have happened had there been no program.

> **Impact evaluation (or analysis)** Analysis of the extent to which a treatment or other service has an effect. Also known as **summative evaluation**.

Think of the program—a new strategy for combating domestic violence, an income supplement, whatever—as an independent variable and the result it seeks as a dependent variable. The DARE program (independent variable), for instance, tries to reduce drug use (dependent variable). When the program is present, we expect less drug use. In a more elaborate study, we might have multiple values of the independent variable; for instance, we might look at *no program, DARE program,* and *other drug/alcohol education* conditions and compare the results of each.

Elizabeth J. D'Amico and Kim Fromme's (2002) study of a new Risk Skills Training Program (RSTP) is a good example of a more elaborate study. They compared the impact of RSTP on children 14 to 19 years of age with that of an abbreviated version of DARE and with results for a control group. The impacts they examined included positive and negative alcohol expectancies (the anticipated effects of drinking) as well as perception of peer risk taking and actual alcohol consumption. They found that negative alcohol expectancies increased for the RSTP group in the posttest but not for the DARE group or the control group, while weekly drinking and positive expectancies for drinking outcomes actually *increased* for the DARE group and/or the control group by the 6-month follow-up but not for the RSTP group (pp. 568–570; see Exhibit 11.7).

Exhibit 11.7 Impact of RSTP, DARE-A

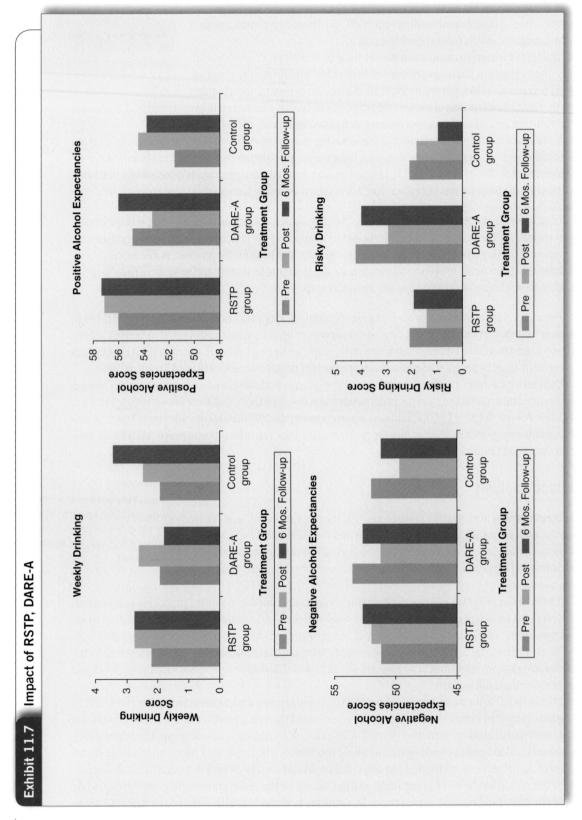

As in other areas of research, an experimental design is the preferred method for maximizing internal validity—that is, for making sure your causal claims about program impact are justified. Cases are assigned randomly to one or more experimental treatment groups and to a control group so that there is no systematic difference between the groups at the outset (see Chapter 7). The goal is to achieve a fair, unbiased test of the program itself, so that differences between the types of people who are in the different groups do not influence judgment about the program's impact. It can be a difficult goal to achieve, because the usual practice in social programs is to let people decide for themselves whether they want to enter a program or not, and also to establish eligibility criteria that ensure that people who enter the program are different from those who do not (Boruch 1997). In either case, a selection bias is introduced.

Impact analyses that do not use an experimental design can still provide useful information and may be all that is affordable, conceptually feasible, or ethically permissible in many circumstances. Evaluation of the State Children's Health Insurance Program (SCHIP) provides an example. The U.S. Congress enacted SCHIP in 1997 to expand health insurance coverage for low-income uninsured children. The federal Centers for Medicare & Medicaid Services (CMS) then contracted with Mathematica Policy Research to evaluate SCHIP (Rosenbach et al. 2007:1).

Given the nature of SCHIP, children could not be assigned randomly to participate. Instead, the Mathematica researchers tracked the growth of enrollment in SCHIP and changes in the percentage of children without health insurance. You can see in Exhibit 11.8 that the percentage of uninsured children declined from 25.2 to 20.1 from 1997 to 2003, during which time enrollment in SCHIP grew from .7 million to 6.2 million children (Rosenbach et al. 2007:ES.2). SCHIP, Rosenbach et al. (2007) concluded,

> Provided a safety net for children whose families lost employer-sponsored coverage during the economic downturn [of 2000–2003] . . . [even as] nonelderly adults experienced a significant 2 percentage point increase in their overall uninsured rate. . . . in the absence of SCHIP . . . the number of uninsured children would have grown by 2.7 million, rather than declining by 0.4 million. (p. ES.7)

Of course, program impact may also be evaluated with quasi-experimental designs (see Chapter 7) or survey or field research methods. But if current participants who are already in a program are compared with nonparticipants, it is unlikely that the treatment group will be comparable with the control group. Participants will probably be a selected group, different at the outset from nonparticipants. As a result, causal conclusions about program impact will be on much shakier ground. For instance, when a study at New York's maximum-security prison for women concluded, "Income Education [i.e., classes] Is Found to Lower Risk of New Arrest," the findings were immediately suspect: The research design did not ensure that the women who enrolled in the prison classes were similar to those who had not enrolled in the classes, "leaving open the possibility that the results were due, at least in part, to self-selection, with the women most motivated to avoid reincarceration being the ones who took the college classes" (Lewin 2001a:A18).

Rigorous evaluations often lead to the conclusion that a program does not have the desired effect (Patton 2002:154). A program depends on political support for achieving its goals, and such evaluations may result in efforts to redesign the program (as with DARE) or reduction or termination of program funding. The latter outcome occurred with the largest U.S. federal training program for the disadvantaged, the Job Training Partnership Act (JTPA). In 1995, in part due to negative results of evaluation research, Congress both restructured JTPA and cut its funding by more than 80% (Heckman, Hohmann, & Smith 2000:651).

But the JTPA experience also raises an important caution about impact analysis. The National JTPA Study used a rigorous experimental design to evaluate the JTPA program and concluded that it did not have the desired impact. However, a subsequent, more intensive analysis by the economist James Heckman and his colleagues (2000) indicated that the initial evaluation had overlooked two critical issues: (1) between 27% and 40% of control group members found some other training program to participate in while the research was being conducted and (2) between 49% and 59% of treatment group members dropped out of the program

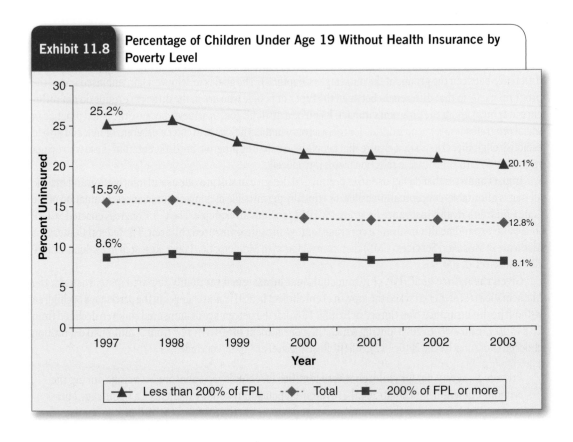

Exhibit 11.8 Percentage of Children Under Age 19 Without Health Insurance by Poverty Level

and so did not actually receive the classroom training (p. 660). After Heckman et al. took account of these problems of control group "substitution bias" and treatment group dropout, they concluded that classroom training in JTPA had "a large positive effect on monthly earnings" for those completing training (p. 688). So, it is important to design research carefully and consider all possible influences on program impact before concluding that a program should be terminated due to poor results.

Efficiency Analysis

Whatever the program's benefits, are they sufficient to offset the program's costs? Are the taxpayers getting their money's worth? What resources are required by the program? These efficiency questions can be the primary reason why funders require evaluation of the programs they fund. As a result, **efficiency analysis**, which compares program effects with costs, is often a necessary component of an evaluation research project.

A **cost-benefit analysis** must identify the specific program costs and the procedures for estimating the economic value of specific program benefits. This type of analysis also requires that the analyst identify whose perspective will be used to determine what can be considered a benefit rather than a cost. Program clients will have a different perspective on these issues than taxpayers or program staff. Exhibit 11.9 lists the factors that can be considered as costs or benefits in an employment and training program, from the standpoint of program participants, the rest of society, and society as a whole (the combination of program participants

Efficiency analysis A type of evaluation research that compares program costs with program effects. It can be either a cost-benefit analysis or a cost-effectiveness analysis.

Cost-benefit analysis A type of evaluation research that compares program costs with the economic value of program benefits.

| Exhibit 11.9 | Conceptual Framework for Cost-Benefit Analysis of an Employment and Training Program |

Costs/Benefits	Perspective of Program Participants	Perspective of Rest of Society	Perspective of Entire Society*
Costs			
Operational costs of the program	0	−	−
Forgone leisure and home production	−	0	−
Benefits			
Earnings gains	+	0	+
Reduced costs of nonexperimental services	0	+	+
Transfers			
Reduced welfare benefits	−	+	0
Wage subsidies	+	−	0
Net benefits	±	±	±

Note: − = program costs; + = program benefits; ± = program costs and benefits; 0 = no program costs or benefits.

*Entire society = program participants + rest of society.

and the rest of society) (Orr 1999:224). Note that some anticipated impacts of the program, on welfare benefits and wage subsidies, are considered a cost to one group and a benefit to another group, whereas some are not relevant to one of the groups.

> **Cost-effectiveness analysis** A type of evaluation research that compares program costs with actual program outcomes.

A **cost-effectiveness analysis** focuses attention directly on the program's outcomes rather than on the economic value of those outcomes. In a cost-effectiveness analysis, the specific costs of the program are compared with the program's outcomes, such as the number of jobs obtained, the extent of improvement in reading scores, or the degree of decline in crimes committed. For example, one result might be an estimate of how much it cost the program for each job obtained by a program participant.

Social science training often doesn't devote much attention to cost-benefit analysis, so it can be helpful to review possible costs and benefits with an economist or business school professor or student. Once the potential costs and benefits have been identified, they must be measured. This is a need highlighted in new government programs (Campbell 2002):

> The Governmental Accounting Standards Board's (GASB) mission is to establish and improve standards of accounting and financial reporting for state and local governments in the United States. In June 1999, the GASB issued a major revision to current reporting requirements ("Statement 34"). The new reporting will provide information that citizens and other users can utilize to gain an understanding of the financial position and cost of programs for a government and a descriptive management's discussion and analysis to assist in understanding a government's financial results. (p. 1)

In addition to measuring services and their associated costs, a cost-benefit analysis must be able to make some type of estimation of how clients benefited from the program. Normally, this will involve a comparison of some indicators of client status before and after clients received program services or between clients who received program services and a comparable group that did not.

A recent study of therapeutic communities (TCs) provides a clear illustration. A therapeutic community is a method for treating substance abuse in which abusers participate in an intensive, structured living experience with other addicts who are attempting to stay sober. Because the treatment involves residential support as well as other types of services, it can be quite costly. Are those costs worth it?

Sacks et al. (2002) conducted a cost-benefit analysis of a modified TC. In the study, 342 homeless, mentally ill chemical abusers were randomly assigned to either a TC or a "treatment-as-usual" comparison group. Employment status, criminal activity, and utilization of health care services were each measured for the 3 months prior to entering treatment and the 3 months after treatment. Earnings from employment in each period were adjusted for costs incurred by criminal activity and utilization of health care services.

Was it worth it? The average cost of TC treatment for a client was $20,361. In comparison, the economic benefit (based on earnings) to the average TC client was $305,273, which declined to $273,698 after comparing postprogram with preprogram earnings, but it was still $253,337 even after adjustment for costs. The resulting benefit-cost ratio was 13:1, although this ratio declined to only 5.2:1 after further adjustments (for cases with extreme values). Nonetheless, the TC program studied seems to have had a substantial benefit relative to its costs.

Design Decisions

Once we have decided on, or identified, the goal or focus of a program evaluation, there are still important decisions to be made about how to design the specific evaluation project. The most important decisions are the following:

- *Black box or program theory:* Do we care how the program gets results?
- *Researcher or stakeholder orientation:* Whose goals matter the most?
- *Quantitative or qualitative methods:* Which methods provide the best answers?
- *Simple or complex outcomes:* How complicated should the findings be?

Black Box Evaluation or Program Theory

The "meat and potatoes" of most evaluation research involves determining whether a program has the intended effect. If the effect occurred, the program has "worked"; if the effect didn't occur, then, some would say, the program should be abandoned or redesigned. In **black box evaluation**, the process by which a program has an effect on outcomes is often treated as a black box—that is, the focus of the evaluation researcher is on whether cases seem to have changed as a result of their exposure to the program, between the time they entered the program as inputs and when they exited the program as outputs (Chen 1990). The assumption is that program evaluation requires only the test of a simple input/output model, like that shown in Exhibit 11.2. There may be no attempt to open the black box of the program process.

Black box evaluation This type of evaluation occurs when an evaluation of program outcomes ignores, and does not identify, the process by which the program produced the effect.

But there is good reason to open the black box and investigate how the process works (or why it doesn't work). Consider recent research on welfare-to-work programs. The Manpower Demonstration Research

Corporation reviewed findings from research on these programs in Florida, Minnesota, and Canada. In each location, adolescents with parents in a welfare-to-work program were compared with a control group of teenagers whose parents were on welfare but were not enrolled in welfare-to-work. In all three locations, teenagers in the welfare-to-work families actually did worse in school than those in the control group—troubling findings.

But why? Why did requiring welfare mothers to work hurt their children's schoolwork? Unfortunately, because the researchers had not investigated the program process—had not "opened the black box"—we can't know for sure. Martha Zaslow, an author of the resulting research report, speculated,

> Parents in the programs might have less time and energy to monitor their adolescents' behavior once they were employed. . . . Under the stress of working, they might adopt harsher parenting styles . . . the adolescents' assuming more responsibilities at home when parents got jobs was creating too great a burden. (cited in Lewin 2001b:A16)

But as Zaslow admitted, "We don't know exactly what's causing these effects, so it's really hard to say, at this point, what will be the long-term effects on these kids" (cited in Lewin 2001b:A16).

If an investigation of program process is conducted, a **program theory** may be developed. A program theory describes what has been learned about how the program has its effect. When a researcher has sufficient knowledge of this before the investigation begins, outlining a program theory can help guide the investigation of program process in the most productive directions. This is termed a **theory-driven evaluation**.

A program theory specifies how the program is expected to operate and identifies which program elements are operational (Chen 1990:32). In addition, a program theory specifies how a program is to produce its effects and thus improves understanding of the relationship between the independent variable (the program) and the dependent variable (the outcome or outcomes). For example, Exhibit 11.10 illustrates the theory for an alcoholism treatment program. It shows that persons entering the program are expected to respond to the combination of motivational interviewing and peer support. A program theory can also decrease the risk of failure when the program is transported to other settings, because it will help identify the conditions required for the program to have its intended effect.

> **Program theory** A descriptive or prescriptive model of how a program operates and produces effects.
>
> **Theory-driven evaluation** A program evaluation that is guided by a theory that specifies the process by which the program has an effect.

Program theory can be either descriptive or prescriptive (Chen 1990). Descriptive theory specifies what impacts are generated and how they occur. It suggests a causal mechanism, including intervening factors, and the necessary context for the effects. Descriptive theories are generally empirically based. On the other hand, prescriptive theory specifies what the program ought to do, and is not actually tested. Prescriptive theory specifies how to design or implement the treatment, what outcomes should be expected, and how performance should be judged. Comparison of the descriptive and prescriptive theories of the program can help identify implementation difficulties and incorrect understandings that can be corrected (Patton 2002:162–164).

Researcher or Stakeholder Orientation

Whose prescriptions specify how the program should operate, what outcomes it should try to achieve, or who it should serve? Most social science research assumes that the researcher specifies the research questions, the applicable theory or theories, and the outcomes to be investigated. Social science research results are most often reported in a professional journal or at professional conferences, where scientific standards determine how the research is received. In program evaluation, however, the research question is often set by the program sponsors or the government agency that is responsible for reviewing the program. In consulting projects for businesses, the client—a manager, perhaps, or a division president—decides what question researchers will

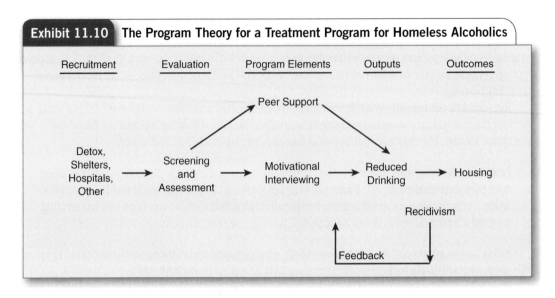

Exhibit 11.10 The Program Theory for a Treatment Program for Homeless Alcoholics

study. It is to these authorities that research findings are reported. Most often, these authorities also specify the outcomes to be investigated. The first evaluator of the evaluation research is the funding agency, then, rather than the professional social science community. Evaluation research is research for a client, and its results may directly affect the services, treatments, or even punishments (e.g., in the case of prison studies) that program users receive. In this case, the person who pays the piper gets to call the tune.

Should the evaluation researcher insist on designing the evaluation project and specifying its goals, or should he or she accept the suggestions and adopt the goals of the funding agency? What role should the preferences of program staff or clients play? What responsibility does the evaluation researcher have to politicians and taxpayers when evaluating government-funded programs? The different answers that various evaluation researchers have given to these questions are reflected in different approaches to evaluation (Chen 1990:66–68).

Stakeholder approaches encourage researchers to be responsive to program stakeholders (so this approach is also termed **responsive evaluation**). Issues for study are to be based on the views of people involved with the program, and reports are to be made to program participants (Shadish, Cook, & Leviton 1991:275–276). The program theory is developed by the researcher to clarify and develop the key stakeholders' theory of the program (Shadish et al. 1991:254–255). In one stakeholder approach, termed *utilization-focused evaluation,* the evaluator forms a task force of program stakeholders, who help to shape the evaluation project so that they are most likely to use its results (Patton 2002:171–175). In evaluation research termed *action research* or *participatory research* (discussed in Chapter 9), program participants are engaged with the researchers as coresearchers and help to design, conduct, and report the research. One research approach that has been termed *appreciative inquiry* eliminates the professional researcher altogether in favor of a structured dialogue about needed changes among program participants themselves (Patton 2002:177–185).

In their book, *Fourth Generation Evaluation,* Egon Guba and Yvonna Lincoln (1989) argue for evaluations oriented toward stakeholders:

> The stakeholders and others who may be drawn into the evaluation are welcomed as equal partners in every aspect of design, implementation, interpretation, and resulting action of an evaluation—that is, they are accorded a full measure of political parity and control . . . determining what questions are to be asked and what information is to be collected on the basis of stakeholder inputs. (p. 11)

Stakeholder approach An orientation to evaluation research that expects researchers to be responsive primarily to the people involved with the program. Also termed **responsive evaluation**.

Because different stakeholders may differ in their reports about or assessment of the program, there is not likely to be one conclusion about program impact. The evaluators are primarily concerned with helping participants understand the views of other stakeholders and with generating productive dialogue. Tineke A. Abma (2005) took this approach in a study of an injury prevention program at a dance school in the Netherlands:

> The evaluators acted as facilitators, paying deliberate attention to the development of trust and a respectful, open and comfortable climate. . . . Furthermore, the evaluation stimulated a public discourse about issues that were taboo, created a space for reflection fostered dynamics and motivated participants to think about ways to improve the quality of their teaching practice. (pp. 284–285)

Social science approaches emphasize the importance of researcher expertise and maintenance of some autonomy in order to develop the most trustworthy, unbiased program evaluation. It is assumed that "evaluators cannot passively accept the values and views of the other stakeholders" (Chen 1990:78). Evaluators who adopt this approach derive a program theory from information they obtain on how the program operates and extant social science theory and knowledge, not from the views of stakeholders. In one somewhat extreme form of this approach, *goal-free evaluation,* researchers do not even permit themselves to learn what goals the program stakeholders have for the program. Instead, the researcher assesses and then compares the needs of participants with a wide array of program outcomes (Scriven 1972b). The goal-free evaluator wants to see the unanticipated outcomes and to remove any biases caused by knowing the program goals in advance.

> **Social science approach** An orientation to evaluation research that expects researchers to emphasize the importance of researcher expertise and maintenance of autonomy from program stakeholders.

Of course, there are disadvantages in both stakeholder and social science approaches to program evaluation. If stakeholders are ignored, researchers may find that participants are uncooperative, that their reports are unused, and that the next project remains unfunded. On the other hand, if social science procedures are neglected, standards of evidence will be compromised, conclusions about program effects will likely be invalid, and results are unlikely to be generalizable to other settings. These equally undesirable possibilities have led to several attempts to develop more integrated approaches to evaluation research.

Integrative approaches attempt to cover issues of concern to both stakeholders and evaluators and to include stakeholders in the group from which guidance is routinely sought (Chen & Rossi 1987:101–102). The emphasis given to either stakeholder or social science concerns is expected to vary with the specific project circumstances. Integrated approaches seek to balance the goal of carrying out a project that is responsive to stakeholder concerns with the goal of objective, scientifically trustworthy and generalizable results. When the research is planned, evaluators are expected to communicate and negotiate regularly with key stakeholders and to take stakeholder concerns into account. Findings from preliminary inquiries are reported back to program decision makers so that they can make improvements in the program before it is formally evaluated. When the actual evaluation is conducted, the evaluation research team is expected to operate more autonomously, minimizing intrusions from program stakeholders.

> **Integrative approach** An orientation to evaluation research that expects researchers to respond to the concerns of people involved with the program—stakeholders—as well as to the standards and goals of the social scientific community.

Many evaluation researchers now recognize that they must take account of multiple values in their research and be sensitive to the perspectives of different stakeholders, in addition to maintaining a commitment to the goals of measurement validity, internal validity, and generalizability (Chen 1990). Ultimately, evaluation research takes place in a political context, in which program stakeholders may be competing or collaborating to increase program funding or to emphasize particular program goals. It is a political process that creates social programs, and it is a political process that determines whether these programs are evaluated and what is done with the evaluation findings (Weiss 1993:94). Developing supportive relations with stakeholder groups will increase the odds that political processes will not undermine evaluation practice. You don't want to find out after you are finished that "people operating ineffective programs who depend on them for their jobs" are able to prevent an evaluation report from having any impact ("'Get Tough' Youth Programs Are Ineffective, Panel Says" 2004:25).

Quantitative or Qualitative Methods

Evaluation research that attempts to identify the effects of a social program typically is quantitative: Did the response times of emergency personnel tend to decrease? Did the students' test scores increase? Did housing retention improve? Did substance abuse decline? It's fair to say that when there's an interest in comparing outcomes between an experimental and a control group or tracking change over time in a systematic manner, quantitative methods are favored.

But qualitative methods can add much to quantitative evaluation research studies, including more depth, detail, nuance, and exemplary case studies (Patton 2002). Perhaps the greatest contribution qualitative methods can make in many evaluation studies is investigating program process—finding out what is "inside the black box." Although it is possible to track service delivery with quantitative measures such as frequency of staff contact and number of complaints, finding out what is happening to clients and how clients experience the program can often best be accomplished by observing program activities and interviewing staff and clients intensively.

For example, Patton (2002:160) describes a study in which process analysis in an evaluation of a prenatal clinic's outreach program led to program changes. The process analysis revealed that the outreach workers were spending a lot of time responding to immediate problems, such as needs for rat control, protection from violence, and access to English classes. As a result, the outreach workers were recruiting fewer community residents for the prenatal clinic. New training and recruitment strategies were adopted to lessen this deviation from program goals.

Another good reason for using qualitative methods in evaluation research is the importance of learning how different individuals react to the treatment. For example, a quantitative evaluation of student reactions to an adult basic skills program for new immigrants relied heavily on the students' initial statements of their goals. However, qualitative interviews revealed that most new immigrants lacked sufficient experience in America to set meaningful goals; their initial goal statements simply reflected their eagerness to agree with their counselors' suggestions (Patton 2002:177–181).

Qualitative methods can also help reveal how social programs actually operate. Complex social programs have many different features, and it is not always clear whether it is the combination of those features or some particular features that are responsible for the program's effect—or for the absence of an effect. Lisbeth B. Schorr, Director of the Harvard Project on Effective Interventions, and Daniel Yankelovich, President of Public Agenda, put it this way: "Social programs are sprawling efforts with multiple components requiring constant midcourse corrections, the involvement of committed human beings, and flexible adaptation to local circumstances" (Schorr & Yankelovich 2000:A19).

The more complex the social program is, the more value qualitative methods can add to the evaluation process. Schorr and Yankelovich (2000) point to the Ten Point Coalition, an alliance of black ministers that helped reduce gang warfare in Boston through multiple initiatives, "ranging from neighborhood probation patrols to safe havens for recreation" (p. A19). Qualitative methods would help describe a complex, multifaceted program such as this. A skilled qualitative researcher will be flexible and creative in choosing methods for program evaluation and will often develop mixed methods, so that the evaluation benefits from the advantages of both qualitative and quantitative techniques.

Simple or Complex Outcomes

Does the program have only one outcome? Unlikely. How many outcomes are anticipated? How many might be unintended? Which are direct consequences of program action, and which are indirect effects that occur as a result of the direct effects (Mohr 1992)? Do the longer-term outcomes follow directly from the immediate program outputs? Does the output (e.g., the increase in test scores at the end of the preparation course) result surely in the desired outcomes (i.e., increased rates of college admission)? Due to these and other possibilities, the selection of outcome measures is a critical step in evaluation research.

The decision to focus on one outcome rather than another, on a single outcome, or on several can have enormous implications. When Sherman and Berk (1984) evaluated the impact of an immediate arrest policy in cases of domestic violence in Minneapolis, they focused on recidivism as the key outcome. Similarly, the reduction of recidivism was the single desired outcome of prison "boot camps" opened in the 1990s. Boot camps are military-style programs for prison inmates that provide tough, highly regimented activities and harsh punishment for disciplinary infractions, with the goal of scaring inmates "straight." They were quite a rage in the 1990s, and the researchers who evaluated their impact understandably focused on criminal recidivism.

But these single-purpose programs turned out to be not quite so simple to evaluate. The Minneapolis researchers found that there was no adequate single source for records of recidivism in domestic violence cases, so they had to hunt for evidence from court and police records, follow-up interviews with victims, and family member reports (Sherman & Berk 1984). More easily measured variables, such as partners' ratings of the accused's subsequent behavior, eventually received more attention. Boot camp researchers soon concluded that the experience did not reduce recidivism: "Many communities are wasting a great deal of money on those types of programs" (Robert L. Johnson, cited in "'Get Tough' Youth Programs Are Ineffective, Panel Says" 2004:25). However, some participants felt that the study had missed something (Latour 2002):

> [A staff member] saw things unfold that he had never witnessed among inmates and their caretakers. Those experiences profoundly affected the drill instructors and their charges, who still call to talk to the guards they once saw as torturers. Graduation ceremonies routinely reduced inmates, relatives, and sometimes even supervisors to tears. (p. B7)

A former boot camp superintendent, Michael Corsini, compared the Massachusetts boot camp with other correctional facilities and concluded, "Here, it was a totally different experience" (Latour 2002:B7).

Some now argue that the failure of boot camps to reduce recidivism was due to the lack of postprison support rather than failure of the camps to promote positive change in inmates. Looking only at recidivism rates would be to ignore some important positive results.

So in spite of the additional difficulties introduced by measuring multiple outcomes, most evaluation researchers attempt to do so (Mohr 1992). The result usually is a much more realistic, and richer, understanding of program impact.

Some of the multiple outcomes measured in the evaluation of Project New Hope appear in Exhibit 11.11. Project New Hope was an ambitious experimental evaluation of the impact of guaranteeing jobs to poor persons (DeParle 1999). It was designed to answer the following question: If low-income adults are given a job at a sufficient wage, above the poverty level, with child care and health care assured, how many would ultimately prosper?

The project involved 677 low-income adults in Milwaukee who were offered a job involving work for 30 hours a week, along with child care and health care benefits. The outcome? Only 27% stuck with the job long enough to lift themselves out of poverty, and their earnings as a whole were only slightly higher than those of a control group that did not receive guaranteed jobs. Levels of depression were not decreased, or self-esteem increased, by the job guarantee. But there were some positive effects: The number of people who never worked at all declined, and rates of health insurance and use of formal child care increased. Perhaps most important, the classroom performance and educational hopes of participants' male children increased, with the boys' test scores rising by the equivalent of 100 points on the SAT and their teachers ranking them as better behaved.

| Exhibit 11.11 | Outcomes in Project New Hope |

Income and Employment (2nd program year)	New Hope	Control Group
Earnings	$6,602	$6,129
Wage subsidies	1,477	862
Welfare income	1,716	1,690
Food stamp income	1,418	1,242
Total income	11,213	9,915
% above poverty level	27%	19%
% continuously unemployed for 2 years	6%	13%
Hardships and Stress	**New Hope**	**Control Group**
% reporting:		
Unmet medical needs	17%	23%
Unmet dental needs	27%	34%
Periods without health insurance	49%	61%
Living in overcrowded conditions	14%	15%
Stressed much or all of the time	45%	50%
Satisfied or very satisfied with standard of living	65%	67%

So, did the New Hope program "work"? Clearly it didn't live up to initial expectations, but it certainly showed that social interventions can have some benefits. Would the boys' gains continue through adolescence? Longer-term outcomes would be needed. Why didn't girls (who were already performing better than the boys) benefit from their parents' enrollment in New Hope just as the boys did? A process analysis would have added a great deal to the evaluation design. The long and short of it is that collection of multiple outcomes gave a better picture of program impact.

Of course, there is a potential downside to the collection of multiple outcomes. Policy makers may choose to publicize only those outcomes that support their own policy preferences and ignore the rest. Often, evaluation researchers themselves have little ability to publicize a more complete story.

Groups or Individuals

Robert G. St. Pierre and the late Peter H. Rossi (2006) urge evaluation researchers to consider randomizing groups rather than individuals to alternative programs in an impact analysis. For example, a study of the effectiveness of a new educational program could randomly assign classes to either the new program or an alternative program rather than randomly assign individual students.

Randomization of groups to different treatments may be preferable when the goal is to compare alternative programs, so that different groups each receive some type of program. It can be easier to implement this approach if there are already different programs available and parents or other constituents are concerned that they (or their children) receive *some* type of program. Using group randomization also makes it easier to determine whether some characteristics of different sites (that offer different programs) influence program impact. However, this approach requires a larger number of participants and, often, cooperation across many governmental or organizational units (St. Pierre & Rossi 2006:667–675).

In a sense, all these choices (black box or program theory, researcher or stakeholder interests, and so on) hinge on (1) what your real goals are in doing the project and (2) how able you will be in a "research for hire" setting to achieve those goals. Not every agency really wants to know if its programs work, especially if the answer is no. Dealing with such issues, and the choices they require, is part of what makes evaluation research both scientifically and politically fascinating.

🔳 Ethics in Evaluation

Evaluation research can make a difference in people's lives while the research is being conducted as well as after the results are reported. Job opportunities, welfare requirements, housing options, treatment for substance abuse, and training programs are each potentially important benefits, and an evaluation research project can change both the type and the availability of such benefits. This direct impact on research participants and, potentially, their families, heightens the attention that evaluation researchers have to give to human subject concerns (Wolf, Turner, & Toms 2009:171). Although the particular criteria that are at issue and the decisions that are judged most ethical vary with the type of evaluation research conducted and the specifics of a particular project, there are always serious ethical as well as political concerns for the evaluation researcher (Boruch 1997:13; Dentler 2002:166).

It is when program impact is the focus that human subject considerations multiply. What about assigning persons randomly to receive some social program or benefit? One justification for this given by evaluation researchers has to do with the scarcity of these resources. If not everyone in the population who is eligible for a program can receive it, due to resource limitations, what could be a fairer way to distribute the program benefits than through a lottery? Random assignment also seems like a reasonable way to allocate potential program benefits when a new program is being tested with only some members of the target recipient population. However, when an ongoing entitlement program is being evaluated and experimental subjects would normally be eligible for program participation, it may not be ethical simply to bar some potential participants from the programs. Instead, evaluation researchers may test alternative treatments or provide some alternative benefit while the treatment is being denied.

There are many other ethical challenges in evaluation research:

- How can confidentiality be preserved when the data are owned by a government agency or are subject to discovery in a legal proceeding?

- Who decides what level of burden an evaluation project may tolerably impose on participants?

- Is it legitimate for research decisions to be shaped by political considerations?

- Must evaluation findings be shared with stakeholders rather than only with policy makers?

- Is the effectiveness of the proposed program improvements really uncertain?

- Will a randomized experiment yield more defensible evidence than the alternatives?

- Will the results actually be used?

The Health Research Extension Act of 1985 (Public Law 99–158) mandated that the Department of Health and Human Services require all research organizations receiving federal funds to have an institutional review board to assess all research for adherence to ethical practice guidelines. We have already reviewed the federally mandated criteria (Boruch 1997:29–33):

- Are risks minimized?

- Are risks reasonable in relation to benefits?

- Is the selection of individuals equitable? (randomization implies this)

- Is informed consent given?

- Are the data monitored?

- Are privacy and confidentiality assured?

Evaluation researchers must consider whether it will be possible to meet each of these criteria long before they even design a study.

The problem of maintaining subject confidentiality is particularly thorny, because researchers, in general, are not legally protected from the requirements that they provide evidence requested in legal proceedings, particularly through the process known as "discovery." However, it is important to be aware that several federal statutes have been passed specifically to protect research data about vulnerable populations from legal disclosure requirements. For example, the Crime Control and Safe Streets Act (28 CFR Part 11) includes the following stipulation:

> Copies of [research] information [about persons receiving services under the act or the subjects of inquiries into criminal behavior] shall be immune from legal process and shall not, without the consent of the persons furnishing such information, be admitted as evidence or used for any purpose in any action, suit, or other judicial or administrative proceedings. (Boruch 1997:60)

Ethical concerns must also be given special attention when evaluation research projects involve members of vulnerable populations as subjects. To conduct research on children, parental consent usually is required before the child can be approached directly about the research. Adding this requirement to an evaluation research project can dramatically reduce participation, because many parents simply do not bother to respond to mailed consent forms.

Tricia Leakey and her colleagues (2004:511) demonstrated that this problem can be overcome in their evaluation of Project SPLASH (Smoking Prevention Launch Among Students in Hawaii). When the project began in the seventh grade, the researchers gave students project information and a consent card to take home to their parents. A pizza party was then held in every class where at least 90% of the students returned a signed consent card. In subsequent follow-ups in the eighth grade, a reminder letter was sent to parents whose children had previously participated. Classes with high participation rates also received a candy thank you. As you can see in Exhibit 11.12, the result was a very high rate of participation.

When it appears that it will be difficult to meet the ethical standards in an evaluation project, at least from the perspective of some of the relevant stakeholders, modifications should be considered in the study design. Several steps can be taken to lessen any possibly detrimental program impact (Boruch 1997: 67–68):

- Alter the group allocation ratios to minimize the number in the untreated control group.

- Use the minimum sample size required to be able to adequately test the results.

- Test just parts of new programs rather than the entire program.

- Compare treatments that vary in intensity (rather than presence or absence).

- Vary treatments between settings rather than among individuals within a setting.

| **Exhibit 11.12** | **Parental Consent Response Rates and Outcomes** |

Survey	Population Size[a]	Consent to Parents[b]	Consent Returned[c]	Refused Consent[d]	Consent to Participate	Student Assent "Yes"[e]
7th grade baseline[f]	4,741	4,728	89.5% (n = 4,231)	7.3% (n = 310)	92.7% (n = 3,921)	99.4% (n = 3,716)
8th grade baseline[g]	4,222	421	58.0% (n = 244)	11.9% (n = 29)	88.1% (n = 215)	99.0% (n = 3,235)
8th grade follow-up[g]	3,703	177	41.8% (n = 74)	5.4% (n = 4)	94.6% (n = 70)	98.7% (n = 2,999)

Note: Parents who had refused participation at the previous survey point were again contacted for permission at the next survey point.

[a]Number of students who were enrolled in the program, varies over time depending on enrollment and teacher participation rates.

[b]Number of consent forms that were handed out at each time period to new students.

[c]Out of the total number of consent forms distributed.

[d]Out of the total number of consent forms returned.

[e]Out of all students who had parental consent and were present on the day of the survey.

[f]Project staff explained and distributed consent on site.

[g]Teachers explained and distributed consent.

Essentially, each of these approaches limits the program's impact during the experiment and so lessens any potential adverse effects on human subjects. It is also important to realize that it is costly to society and potentially harmful to participants to maintain ineffective programs. In the long run, at least, it may be more ethical to conduct an evaluation study than to let the status quo remain in place.

▣ Conclusions

Hopes for evaluation research are high: Society could benefit from the development of programs that work well, that accomplish their policy goals, and that serve the people who genuinely need them. At least that is the hope. Unfortunately, there are many obstacles to realizing this hope (Posavac & Carey 1997):

- Because social programs and the people who use them are complex, evaluation research designs can easily miss important outcomes or aspects of the program process.

- Because the many program stakeholders all have an interest in particular results from the evaluation, researchers can be subjected to an unusual level of cross-pressures and demands.

- Because the need to include program stakeholders in research decisions may undermine adherence to scientific standards, research designs can be weakened.

- Because some program administrators want to believe that their programs really work well, researchers may be pressured to avoid null findings or, if they are not responsive, may find their research report ignored. Plenty of well-done evaluation research studies wind up in a recycling bin or hidden away in a file cabinet.

- Because the primary audience for evaluation research reports are program administrators, politicians, or members of the public, evaluation findings may need to be overly simplified, distorting the findings.

The rewards of evaluation research are often worth the risks, however. Evaluation research can provide social scientists with rare opportunities to study a complex social process, with real consequences, and to contribute to the public good. Although they may face unusual constraints on their research designs, most evaluation projects can result in high-quality analysis and publications in reputable social science journals. In many respects, evaluation research is an idea whose time has come. We may never achieve Donald Campbell's (Campbell & Russo 1999) vision of an "experimenting society," in which research is consistently used to evaluate new programs and to suggest constructive changes, but we are close enough to continue trying.

Key Terms

Black box evaluation   374
Cost-benefit analysis   372
Cost-effectiveness
    analysis   373
Efficiency analysis   372
Evaluability assessment   364
Feedback   361
Formative evaluation   369

Impact evaluation
    (or analysis)   369
Inputs   361
Integrative approach   377
Needs assessment   362
Outcomes   361
Outputs   361
Process evaluation   367

Program process   361
Program theory   375
Responsive evaluation   376
Social science approach   377
Stakeholder approach   376
Stakeholders   361
Summative evaluation   369
Theory-driven evaluation   375

Highlights

- Evaluation research is social research that is conducted for a distinctive purpose: to investigate social programs.
- The development of evaluation research as a major enterprise followed on the heels of the expansion of the federal government during the Great Depression and World War II.
- The evaluation process can be modeled as a feedback system, with inputs entering the program, which generates outputs and then outcomes, which feed back to program stakeholders and affect program inputs.
- The evaluation process as a whole, and the feedback process in particular, can be understood only in relation to the interests and perspectives of program stakeholders.
- The process by which a program has an effect on outcomes is often treated as a "black box," but there is good reason to open the black box and investigate the process by which the program operates and produces, or fails to produce, an effect.
- A program theory may be developed before or after an investigation of the program process is completed. It may be either descriptive or prescriptive.

- Evaluation research is done for a client, and its results may directly affect the services, treatments, or punishments that program users receive. Evaluation researchers differ in the extent to which they attempt to orient their evaluations to program stakeholders.
- Qualitative methods are useful in describing the process of program delivery.
- Multiple outcomes are often necessary to understand program effects.
- There are five primary types of program evaluation: (1) needs assessment, (2) evaluability assessment, (3) process evaluation (including formative evaluation), (4) impact evaluation (also termed as summative evaluation), and (5) efficiency (cost-benefit) analysis.
- Evaluation research raises complex ethical issues because it may involve withholding desired social benefits.

STUDENT STUDY SITE

To assist in completing the web exercises, please access the study site at **www.sagepub.com/schuttisw7e**, where you will find the web exercises with accompanying links. You'll find other useful study materials, such as self-quizzes and e-flashcards for each chapter, along with a group of carefully selected articles from research journals that illustrate the major concepts and techniques presented in the book.

Discussion Questions

1. Would you prefer that evaluation researchers use a stakeholder or a social science approach? Compare and contrast these perspectives, and list at least four arguments for the one you favor.

2. Propose a randomized experimental evaluation of a social, medical, or educational program with which you are familiar. Possibilities could range from a job training program to a community health center, or a even a college. Include in your proposal a description of the program and its intended outcomes. Discuss the strengths and weaknesses of your proposed design.

3. How would you describe the contents of the "black box" of program operations? What "program theory" would specify how the program (in Question 2) operates?

4. What would be the advantages and disadvantages of using qualitative methods to evaluate this program? What would be the advantages and disadvantages of using quantitative methods? Which approach would you prefer and why?

Practice Exercises

1. Read and summarize an evaluation research report published in the journal *Evaluation and Program Planning*. Be sure to identify the type of evaluation research that is described. Discuss the strengths and weaknesses of the design.

2. Identify the key stakeholders in a local social or educational program. Interview several stakeholders to determine what their goals for the program are and what tools they use to assess goal achievement. Compare and contrast the views of each stakeholder and try to account for any differences you find.

3. Review the "Evaluation Research" lesson in Interactive Exercises link on the book's study site to learn more about the language and logic of evaluation research.

4. Identify an article that reports an evaluation research study on the book's website, at www.sagepub.com/schuttisw7e. What type of evaluation research does this study represent? What alternatives did the author(s) select when designing the research? After reading the entire article, do you agree with the author's choices? Why or why not?

Ethics Questions

1. Imagine that you are evaluating a group home for persons with serious mental illness and learn that a house resident has been talking about cutting himself. Would you immediately inform house staff about this? What if the resident asked you not to tell anyone? In what circumstances would you feel it is ethical to take action to prevent the likelihood of a subject harming himself or herself or others?

2. Is it ethical to assign people to receive some social benefit on a random basis? Form two teams, and debate the ethics of the TARP randomized evaluation of welfare payments described in this chapter.

Web Exercises

1. Inspect the website maintained by the Governmental Accounting Standards Board (GASB), www.gasb.org and particularly the section at www.gasb.org/project_pages/index.html. Read and report on the GASB effort to improve performance measurement in government as described at www.gasb.org/plain-language_documents/SEA_Plain-Language_Article.pdf.

2. Describe the resources available for evaluation researchers at one of the following three websites: www.wmich.edu/evalctr, http://www.resources4evaluators.info/CommunitiesOf Evaluators.html, and www.worldbank.org/oed.

3. You can check out the latest information regarding the DARE program at www.dare.com. What is the current approach? Can you find information on the web about current research on DARE?

4. Evaluation research is a big industry! Two examples are provided by Mathematica Policy Research, www.mathematica-mpr.com/publications/PDFs/evalabstinence.pdf, and the Policy Evaluation and Research Center at Educational Testing Services, http://www.ets.org/research/perc/pic/reports/. Summarize their work.

SPSS Exercises

1. Neighborhood and school integration has often been a focus of government social policy. Does the racial composition of a neighborhood have any association with attitudes related to racial issues? Although we cannot examine the effects of social policies or programs directly in the General Social Survey (GSS) data, we can consider the association between neighborhood racial composition and attitudes related to race. The variable RACLIVE indicates whether the respondent lives in a racially integrated neighborhood. Request its frequency distribution as well as those for several attitudes related to race: RACOPEN, AFFRMACT, WRKWAYUP, HELPBLK, and CLOSEBLK3.

2. Do attitudes vary with the experience of living in a racially integrated neighborhood? Request the cross-tabulation of the variables used in Step 1, RACOPEN to CLOSEBLK3 by RACLIVE (request percentages on the column totals). Read the tables and explain what they tell us about attitudes and neighborhoods. Does the apparent effect of racial integration vary with the different attitudes? How would you explain this variation in these "multiple outcomes"?

3. What other attitudes differ between whites who live in integrated and segregated neighborhoods? Review the GSS2010x variable list to identify some possibilities and request cross-tabulations for these variables. Do you think that these differences are more likely to be a consequence of a racially integrated neighborhood experience or a cause of the type of neighborhood that people choose to live in? Explain.

Developing a Research Proposal

If you plan an evaluation research project, you will have to revisit the decisions about research designs (Exhibit 2.14, #13 to #17).

1. Develop a brief model for a program that might influence the type of attitude or behavior in which you are interested. List the key components of this model.

2. Design a program evaluation to test the efficacy of your program model, using an impact analysis approach.

3. Add to your plan a discussion of a program theory for your model. In your methodological plan, indicate whether you will use qualitative or quantitative techniques and simple or complex outcomes.

4. Who are the potential stakeholders for your program? How will you relate to them before, during, and after your evaluation?

Historical and Comparative Research

A lthough the United States and several European nations have maintained democratic systems of governance for more than 100 years, democratic rule has more often been brief and unstable, when it has occurred at all. What explains the presence of democratic practices in one country and their absence in another? Are democratic politics a realistic option for every nation? What about Libya? Egypt? Iraq? Are there some prerequisites in historical experience, cultural values, or economic resources? (Markoff 2005:384–386). A diverse set of methodological tools allow us to investigate social processes at other times and in other places, when the actual participants in these processes are not available.

Historical and comparative research methods can generate new insights into social processes due to their ability to focus on aspects of the social world beyond recent events in one country. They involve several different approaches and a diverse set of techniques, and they may have qualitative and/or quantitative components. These methods provide ways to investigate topics that usually cannot be studied with experiments, participant observations, or surveys. However, because this broader focus involves collecting data from records on the past or from other nations, the methods used in historical and comparative investigations present unique challenges to social researchers.

In this chapter, I review the major unobtrusive methods social scientists use to understand historical processes and to compare different societies or regions. I also introduce oral histories, a qualitative tool for historical investigations, as well as demographic methods, which can strengthen both historical and comparative studies. Throughout the chapter, I will draw many examples from research on democracy and the process of democratization.

Overview of Comparative and Historical Research Methods

Historical events research
Research in which social events are studied at one past time period.

Historical process research
Research in which historical processes are studied over a long period of time.

Cross-sectional comparative research Research comparing data from one time period between two or more nations.

Comparative historical research
Research comparing data from more than one time period in more than one nation.

The central insight behind historical and comparative research methods is that we can improve our understanding of social process when we make comparisons to other times and places. Max Weber's comparative study of world religions (Bendix 1962), Emile Durkheim's (1984) historical analysis of the division of labor, and Seymour Martin Lipset's (1990) contrast of U.S. and Canadian politics affirm the value of this insight. Beyond this similarity, however, historical and comparative methods are a diverse collection of approaches. Research may be historical, comparative, or both historical and comparative. Historical and comparative methods can be quantitative or qualitative, or a mixture of both. Both nomothetic and idiographic approaches to establishing causal effects can be used.

There are no hard-and-fast rules for determining how far in the past the focus of research must be in order to consider it historical or what types of comparisons are needed to warrant calling research comparative. In practice, research tends to be considered historical when it focuses on a period prior to the experience of most of those conducting research (Abbott 1994:80). Research involving different nations is usually considered comparative, but so are studies of different regions within one nation if they emphasize interregional comparison. In recent years, the globalization of U.S. economic ties and the internationalization of scholarship have increased the use of unobtrusive methods for comparative research across many different countries (Kotkin 2002).

Distinguishing research in terms of a historical and/or comparative focus results in four basic types of research: **historical events research**, **historical process research**, **cross-sectional comparative research**, and **comparative historical research**. Research that focuses on events in one short historical period is historical events research, whereas longitudinal research that traces a sequence of events over a number of years is historical process research (cf. Skocpol 1984:359). There are also two types of comparative research, the first involving cross-sectional comparisons and the second comparing longitudinal data about historical processes between multiple cases. The resulting four types of research are displayed in Exhibit 12.1.

Exhibit 12.1 Types of Historical and Comparative Research

	Cross-Sectional	Longitudinal
Single Case	Historical Events Research	Historical Process Research
Multiple Cases	Cross-Sectional Comparative Research	Comparative Historical Research

Historical Social Science Methods

Both historical events research and historical process research investigate questions concerning past times. These methods are used increasingly by social scientists in sociology, anthropology, political science, and economics, as well as by many historians (Monkkonen 1994). The late 20th and early 21st centuries have seen so much change in so many countries that many scholars have felt a need to investigate the background of these changes and to refine their methods of investigation (Hallinan 1997; Robertson, 1993). The accumulation of large bodies of data about the past has not only stimulated more historically oriented research but has also led to the development of several different methodologies.

Much historical (and comparative) research is qualitative. This style of historical social science research tends to have several features that are similar to those used in other qualitative methodologies. Qualitative historical research has the following characteristics:

> **Case-oriented research** Research that focuses attention on the nation or other unit as a whole.
>
> **Narrative explanation** An idiographic causal explanation that involves developing a narrative of events and processes that indicate a chain of causes and effects.

- *Case oriented:* **Case-oriented research** focuses on the nation or other unit as a whole, rather than only on different parts of the whole in isolation from each other (Ragin 2000:68). This could be considered the most distinctive feature of qualitative research on historical processes.

- *Holistic:* Qualitative historical research is concerned with the context in which events occurred and the interrelations among different events and processes: "how different conditions or parts fit together" (Ragin 1987:25–26).

- *Conjunctural:* Qualitative historical research is conjunctural because, it is argued, "no cause ever acts except in complex conjunctions with others" (Abbott 1994:101).

- *Temporal:* Qualitative historical research becomes temporal by taking into account the related series of events that unfold over time.

- *Historically specific:* Qualitative historical research is likely to be limited to the specific time(s) and place(s) studied, like traditional historical research.

- *Narrative:* Qualitative historical research researches a story involving specific actors and other events occurring at the same time (Abbott 1994:102) or one that takes account of the position of actors and events in time and in a unique historical context (Griffin 1992). (You can think of this as a combination of the previous two features.) **Narrative explanations** involve idiographic causal reasoning (see Chapter 6).

- *Inductive:* Qualitative historical research develops an explanation for what happened from the details discovered about the past.

The focus on the past presents special methodological challenges:

- Documents and other evidence may have been lost or damaged.

- Available evidence may represent a sample biased toward more newsworthy figures.

- Written records will be biased toward those who were more prone to writing.

- Feelings of individuals involved in past events may be hard, if not impossible, to reconstruct.

Before you judge historical social science research as credible, you should look for convincing evidence that each of these challenges has been addressed.

Historical Events Research

Research on past events that does not follow processes for some long period of time is historical events research rather than historical process research. Historical events research basically uses a cross-sectional, rather than longitudinal, design. Investigations of past events may be motivated by the belief that they had a critical impact on subsequent developments or because they provide opportunities for testing the implications of a general theory (Kohn 1987).

> **Event-structure analysis** A systematic method of developing a causal diagram showing the structure of action underlying some chronology of events; the end result is an idiographic causal explanation.

Event-Structure Analysis

One technique useful in historical events research, as well as in other types of historical and comparative research, is **event-structure analysis**. Event-structure is a qualitative approach that relies on a systematic coding of key events or national characteristics to identify the underlying structure of action in a chronology of events. The codes are then used to construct event sequences, make comparisons between cases, and develop an idiographic causal explanation for a key event.

An event-structure analysis consists of the following steps:

1. Classifying historical information into discrete events

2. Ordering events into a temporal sequence

3. Identifying prior steps that are prerequisites for subsequent events

4. Representing connections between events in a diagram

5. Eliminating from the diagram connections that are not necessary to explain the focal event

Larry Griffin (1993) used event-structure analysis to explain a unique historical event, a lynching in the 1930s in Mississippi. According to published accounts and legal records, the lynching occurred after David Harris, an African American who sold moonshine from his home, was accused of killing a white tenant farmer. After the killing was reported, the local deputy was called and a citizen search party was formed. The deputy did not intervene as the search party trailed Harris and then captured and killed him. Meanwhile, Harris's friends killed another African American who had revealed Harris's hiding place. This series of events is outlined in Exhibit 12.2.

Which among the numerous events occurring between the time that the tenant farmer confronted Harris and the time that the mob killed Harris had a causal influence on that outcome? To identify these idiographic causal links (see Chapter 6), Griffin identified plausible counterfactual possibilities—events that might have occurred but did not—and considered whether the outcome might have been changed if a counterfactual had occurred instead of a particular event.

> If, contrary to what actually happened, the deputy had attempted to stop the mob, might the lynching have been averted? . . . Given what happened in comparable cases and the Bolivar County deputy's clear knowledge of the existence of the mob and of its early activities, his forceful intervention to prevent the lynching thus appears an objective possibility. (Griffin 1993:1112)

So, Griffin concluded that nonintervention by the deputy had a causal influence on the lynching.

Exhibit 12.2 **Event-Structure Analysis: Lynching Incident in the 1930s**

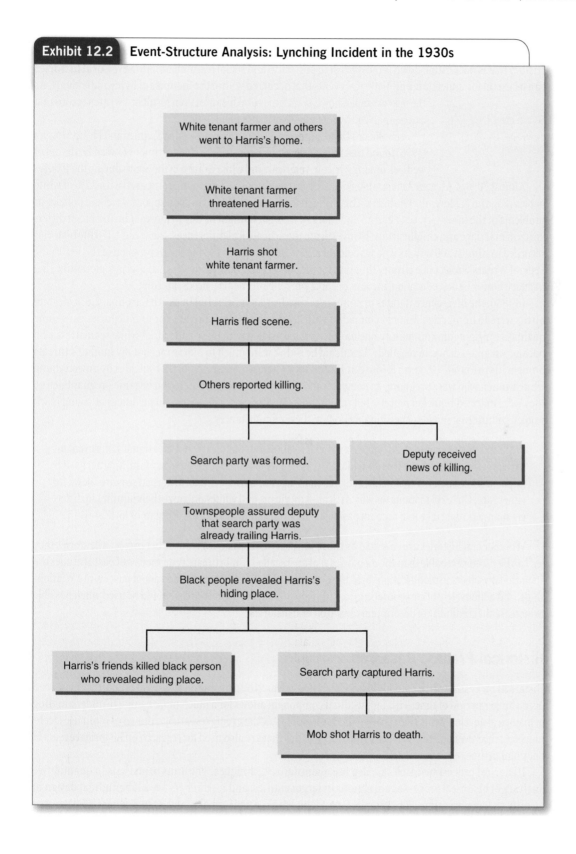

Oral History

History that is not written down is mostly lost to posterity (and social researchers). However, **oral histories** can be useful for understanding historical events that occurred within the lifetimes of living individuals. As the next example shows, sometimes oral histories even result in a written record that can be analyzed by researchers at a later point in time.

> **Oral history** Data collected through intensive interviews with participants in past events.

Thanks to a Depression-era writers' project, Deanna Pagnini and Philip Morgan (1996) found that they could use oral histories to study attitudes toward births out of wedlock among African American and white women in the South during the 1930s.

Almost 70% of African American babies are born to unmarried mothers, compared with 22% of white babies (Pagnini & Morgan 1996:1696). This difference often is attributed to contemporary welfare policies or problems in the inner city, but Pagnini and Morgan thought it might be due to more enduring racial differences in marriage and childbearing. To investigate these historical differences, they read 1,170 life histories recorded by almost 200 writers who worked for a New Deal program during the Depression of the 1930s, the Federal Writers' Project Life History Program for the Southeast. The interviewers had used a topic outline that included family issues, education, income, occupation, religion, medical needs, and diet.

In 1936, the divergence in rates of nonmarital births was substantial in North Carolina: 2.6% of white births were to unmarried women, compared with 28.3% of nonwhite births. The oral histories gave some qualitative insight into community norms that were associated with these patterns. A white seamstress who became pregnant at age 16 recalled, "I'm afraid he didn't want much to marry me, but my mother's threats brought him around" (Pagnini & Morgan 1996:1705). There were some reports of suicides by unwed young white women who were pregnant. In comparison, African American women who became pregnant before they were married reported regrets, but rarely shame or disgrace. There were no instances of young black women committing suicide or getting abortions in these circumstances.

> We found that bearing a child outside a marital relationship was clearly not the stigmatizing event for African-Americans that it was for whites. . . . When we examine contemporary family patterns, it is important to remember that neither current marriage nor current childbearing patterns are "new" for either race. Our explanations for why African-Americans and whites organize their families in different manners must take into account past behaviors and values. (Pagnini & Morgan 1996:1714–1715)

Whether oral histories are collected by the researcher or obtained from an earlier project, the stories they tell can be no more reliable than the memories that are recalled. Unfortunately, memories of past attitudes are "notoriously subject to modifications over time" (Banks 1972:67), as are memories about past events, relationships, and actions. Use of corroborating data from documents or other sources should be used when possible to increase the credibility of descriptions based on oral histories.

Historical Process Research

Historical process research extends historical events research by focusing on a series of events that happened over a longer period of time. This longitudinal component allows for a much more complete understanding of historical developments than is often the case with historical events research, although it often uses techniques such as event history analysis and oral histories that are also used for research on historical events at one point in time.

Historical process research can also use quantitative techniques. The units of analysis in quantitative analyses of historical processes are nations or larger entities, and researchers use a longitudinal design to identify changes over time. For example, David John Frank, Ann Hironaka, and Evan Schofer (2000) treated the entire world as their "case" for their deductive test of alternative explanations for the growth of national activities to protect the natural environment during the 20th century. Were environmental protection activities a response to environmental degradation and economic affluence within nations, as many had theorized? Or, instead, were they the result of a "top-down" process in which a new view of national responsibilities was spread by international organizations? Their measures of environmental protectionism included the number of national parks among all countries in the world and memberships in international environmental organizations; one of their indicators of global changes was the cumulative number of international agreements (see Exhibit 12.3 for a list of some of their data sources).

Exhibit 12.4a charts the growth of environmental activities identified around the world. Compare the pattern in this exhibit with the pattern of growth in the number of international environmental agreements and national environmental laws shown in Exhibit 12.4b, and you can see that environmental protectionism at the national level was rising at the same time that it was becoming more the norm in international relations. In more detailed analyses, Frank and colleagues (2000) attempt to show that the growth in environmental protectionism was not explained by increasing environmental problems or economic affluence within nations. As in most research that relies on historical and/or comparative data, however, some variables that would indicate alternative influences (such as the strength of national environmental protest movements) could not be measured (Buttel 2000). Therefore, further research is needed.

One common measurement problem in historical research projects is the lack of data from some historical periods (Rueschemeyer, Stephens, & Stephens 1992:4; Walters, James, & McCammon 1997). For example, the widely used U.S. Uniform Crime Reporting System did not begin until 1930 (Rosen 1995). Sometimes, alternative sources of documents or estimates for missing quantitative data can fill in gaps (Zaret 1996), but even when measures can be created for key concepts, multiple measures of the same concepts are likely to be out of the question; as a result, tests of reliability and validity may not be feasible. Whatever the situation, researchers must assess the problem honestly and openly (Bollen, Entwisle, & Alderson 1993; Paxton 2002).

Those measures that are available are not always adequate. What is included in the historical archives may be an unrepresentative selection of materials that still remain from the past. At various times, some

---

Exhibit 12.3 Variables for Historical Analysis of Environmental Protectionism

Dependent Variables	Definition	Data Source(s)	Period of Analysis
National parks and protected areas	Annual cumulative numbers of parks per nation-state	IUCN (1990)	1900–1990
Country chapters of international environmental nongovernmental associations	Annual numbers of chapters per nation-state	Fried (1905–1911); League of Nations (1921, 1938); UIA (1948–1990)	1900–1988
Nation-state memberships in intergovernmental environmental organizations	Annual numbers of memberships per nation-state	Fried (1905–1911); League of Nations (1921, 1938); UIA (1948–1990)	1900–1984
Environmental impact assessment laws	Year of founding	Wood (1995)	1966–1992
National environmental ministries	Year of founding	Europa Year Book (1970–1995)	1970–1995

Exhibit 12.4   International Environmental Activity

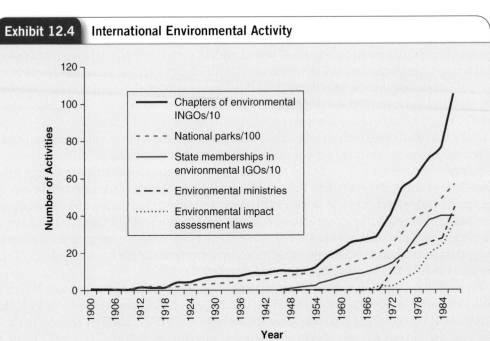

(a) Cumulative Numbers of Five National Environment Activities, 1900 to 1988

Note: INGOs are international nongovernment organizations; IGOs are intergovernmental organizations.

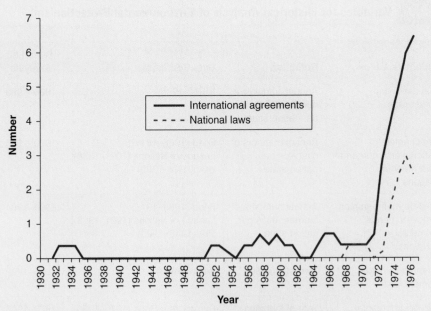

(b). Environmental Impact Assessment: International Agreements and National Laws, 1930 to 1977

documents could have been discarded, lost, or transferred elsewhere for a variety of reasons. *Original* documents may be transcriptions of spoken words or handwritten pages and could have been modified slightly in the process; they could also be outright distortions (Erikson 1966:172, 209–210; Zaret 1996). When relevant data are obtained from previous publications, it is easy to overlook problems of data quality, but this simply makes it all the more important to evaluate the primary sources. It is very important to develop a systematic plan for identifying and evaluating relevant documents.

▣ Comparative Social Science Methods

The limitations of single-case historical research have encouraged many social scientists to turn to comparisons between nations. These studies allow for a broader vision about social relations than is possible with cross-sectional research limited to one country or other unit. From 1985 to 1990, more than 80 research articles in top sociology journals and 200 nonedited books were published in which the primary purpose was the comparison of two or more nations (Bollen et al. 1993). About half of this research used cross-sectional data rather than longitudinal data collected over a period of time.

Research in the News

SHOULD THE STATUS OF TEACHERS IN THE UNITED STATES BE RAISED?

Countries in which students receive the highest average scores on international achievement tests recruit high-performing college graduates to be teachers, provide considerable mentoring and classroom support, and actively seek to raise teachers' status. The United States, which scored 15th in reading and 19th in science on the Program for International Student Assessment, lacks all of this. Comparative research indicates that raising the status of teachers is the single most important step to take. Although the U.S. spends more than almost any other country on education, much of that money goes to bus transportation and sports facilities rather than to actions that could improve student academic performance.

Source: Dillon, Sam. 2011. "Study: U.S. Must Raise Status of Its Teachers." *The New York Times,* March 16:A22.

Cross-Sectional Comparative Research

Comparisons between countries during one time period can help social scientists identify the limitations of explanations based on single-nation research. Such comparisons can suggest the relative importance of universal factors in explaining social phenomena as compared to unique factors rooted in specific times and places (de Vaus 2008:251). These comparative studies may focus on a period in either the past or the present.

Historical and comparative research that is quantitative may obtain data from national statistics or other sources of published data; if it is contemporary, such research may rely on cross-national surveys. Like other

Variable-oriented research
Research that focuses attention on variables representing particular aspects of the cases studied and then examines the relations among these variables across sets of cases.

types of quantitative research, quantitative historical and comparative research can be termed **variable-oriented research**, with a focus on variables representing particular aspects of the units studied (Demos 1998).

Causal reasoning in quantitative historical and comparative research is nomothetic, and the approach is usually deductive, testing explicit hypotheses about relations among these variables (Kiser & Hechter 1991). For example, Clem Brooks and Jeff Manza (2006:476–479) deduce from three theories about welfare states—national values, power resources, and path dependency theory—the hypothesis that voters' social policy preferences will influence welfare state expenditures. Using country-level survey data collected by the International Social Survey Program (ISSP) in 15 democracies in five different years and expenditure data from the Organization for Economic Cooperation and Development (OECD), Brooks and Manza were able to identify a consistent relationship between popular preferences for social welfare spending and the actual national expenditures (see Exhibit 12.5).

Exhibit 12.5 **Interrelationship of Policy Preferences and Welfare State Output**

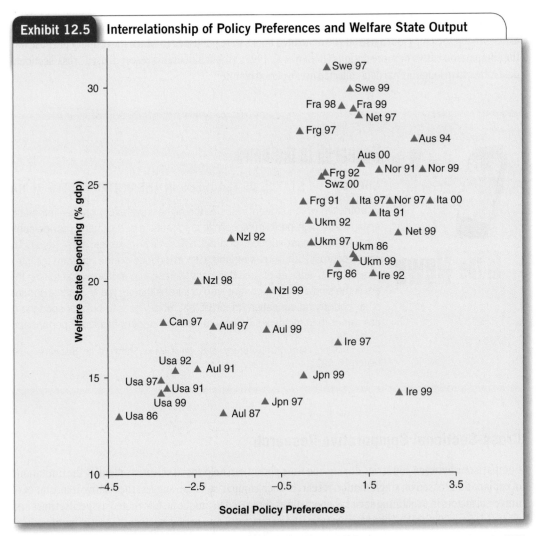

Note: Scattergram shows data for policy preferences and welfare state spending in 15 OECD democracies. Data are from the ISSP/OECD (International Social Survey Program/Organization for Economic Cooperation and Development).

Popular preferences are also important factors in political debates over immigration policy. Christopher A. Bail (2008) asked whether majority groups in different European countries differ in the way that they construct "symbolic boundaries" that define "us" versus an immigrant "them." For his cross-sectional comparative investigation, he drew on 333,258 respondents in the 21-country European Social Survey. The key question about immigrants in the ESS was: "Please tell me how important you think each of these things should be in deciding whether someone born, brought up and living outside [country] should be able to come and live here." The "things" whose importance they were asked to rate were six individual characteristics: being (1) white, (2) well educated, (3) from a Christian background, (4) speaking the official national language, (5) being committed to the country's way of life, and (6) having work skills needed in the country. Bail then calculated the average importance rating in each country for each of these characteristics and used a statistical procedure to cluster the countries in terms of the extent to which their ratings and other characteristics were similar.

Bail's (2008:54–56) analysis identified the countries as falling into three clusters (see Exhibit 12.6). Cluster A countries are on the periphery of Europe and have only recently experienced considerable immigration; their populations tend to draw boundaries in terms of race and religion. Cluster B countries are in the core of Western Europe (except Slovenia), have a sizable and long-standing immigrant population, and their populations tend to base their orientations toward immigrants on linguistic and cultural differences. Countries in Cluster C are in Scandinavia, have a varied but relatively large immigrant population, and attach

Exhibit 12.6 | **Symbolic Boundaries Against Immigrants in 21 European Countries**

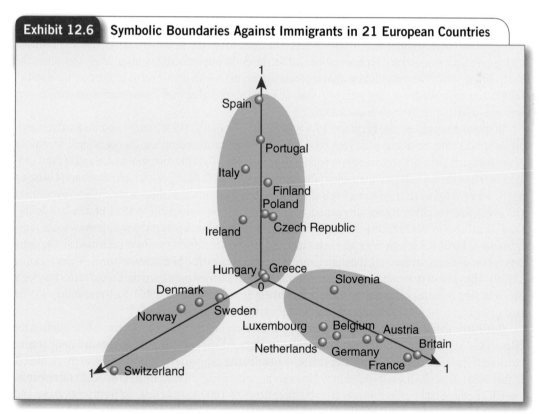

Notes: A country's membership scores describe how closely its configuration of symbolic boundaries resembles the sets described in Table 7. Countries with high membership in a given set are prototypical; those with low membership are simply atypical. Ellipses depict the "crisp" clustering solution: the major diameter, or length, describes the range of membership scores within each crisp cluster whereas the minor diameter, or width, describes overlap between them.

Fuzzy Membership Scores in Three Sets

much less importance to any of the six symbolic boundaries than those in the other countries. Bail (2008:56) encourages longitudinal research to determine the extent to which these different symbolic boundaries are the product or the source of social inequality in these countries.

Cross-sectional comparative research has also helped to explain variation in voter turnout. This research focuses on a critical issue in political science: Although free and competitive elections are a defining feature of democratic politics, elections themselves cannot orient governments to popular sentiment if citizens do not vote (LeDuc, Niemi, & Norris 1996). As a result, the low levels of voter participation in U.S. elections have long been a source of practical concern and research interest.

International data give our first clue for explaining voter turnout: The historic rate of voter participation in the United States (48.3%, on average) is much lower than it is in many other countries that have free, competitive elections; for example, Italy has a voter turnout of 92.5%, on average, since 1945 (Exhibit 12.7).

Is this variation due to differences among voters in knowledge and wealth? Do media and political party get-out-the-vote efforts matter? Mark Franklin's (1996:219–222) analysis of international voting data indicates that neither explanation accounts for much of the international variation in voter turnout. Instead, it is the structure of competition and the importance of issues that are influential. Voter turnout is maximized where structural features maximize competition: compulsory voting (including, in Exhibit 12.7, Austria, Belgium, Australia, and Greece), mail and Sunday voting (including the Netherlands and Germany), and multiday voting. Voter turnout also tends to be higher where the issues being voted on are important and where results are decided by proportional representation (as in Italy and Israel, in Exhibit 12.7) rather than on a winner-take-all basis (as in U.S. presidential elections)—so individual votes are more important.

Franklin concludes that it is these characteristics that explain the low level of voter turnout in the United States, not the characteristics of individual voters. The United States lacks the structural features that make voting easier, the proportional representation that increases the impact of individuals' votes, and, often, the sharp differences between candidates that are found in countries with higher turnout. Because these structural factors generally do not vary within nations, we would never realize their importance if our analysis was limited to data from individuals in one nation.

In spite of the unique value of comparative analyses like Franklin's (1996), such cross-national research also confronts unique challenges (de Vaus 2008:255). The meaning of concepts and the operational definitions of variables may differ between nations or regions (Erikson 1966:xi), so the comparative researcher must consider how best to establish measurement equivalence (Markoff 2005:402). For example, the concept of being a *good son or daughter* refers to a much broader range of behaviors in China than in most Western countries (Ho 1996). Rates of physical disability cannot be compared among nations due to a lack of standard definitions (Martin & Kinsella 1995:364–365). Individuals in different cultures may respond differently to the same questions (Martin & Kinsella 1995:385). Alternatively, different measures may have been used for the same concepts in different nations, and the equivalence of these measures may be unknown (van de Vijver & Leung 1997:9). The value of statistics for particular geographic units such as counties in the United States may vary over time simply due to changes in the boundaries of these units (Walters et al. 1997). Such possibilities should be considered, and any available opportunity should be taken to test for their effects.

Qualitative data can also be used as a primary tool for comparative research. The Human Relations Area Files (HRAF) Collection of Ethnography provides an extraordinary resource for qualitative comparative cross-sectional research (and, to a lesser extent, for qualitative comparative historical research) (Ember & Ember 2005). The HRAF was founded in 1949 as a corporation designed to facilitate cross-cultural research. The HRAF ethnography collection now contains more than 1,000,000 pages of material from publications and other reports from about 400 different cultural, ethnic, religious, and national groups all over the world. The information is indexed by topic, in 710 categories, and now made available electronically (if your school pays to maintain access to the HRAF). Exhibit 12.8 is an example of a page from an HRAF document that has been indexed for easy retrieval.

Most of the significant literature published on the chosen groups is included in the HRAF and used to prepare a standard summary about the group. Researchers can use these summaries and systematic

Exhibit 12.7	Average Percentage of Voters Who Participated in Presidential or Parliamentary Elections, 1945–1998[*]		

Country	Vote %	Country	Vote %
Italy	92.5	St. Kitts and Nevis	58.1
Cambodia	90.5	Morocco	57.6
Seychelles	96.1	Cameroon	56.3
Iceland	89.5	Paraguay	56.0
Indonesia	88.3	Bangladesh	56.0
New Zealand	86.2	Estonia	56.0
Uzbekistan	86.2	Gambia	55.8
Albania	85.3	Honduras	55.3
Austria	85.1	Russia	55.0
Belgium	84.9	Panama	53.4
Czech	84.8	Poland	52.3
Netherlands	84.8	Uganda	50.6
Australia	84.4	Antigua and Barbuda	50.2
Denmark	83.6	Burma/Myanmar	50.0
Sweden	83.5	Switzerland	49.3
Mauritius	82.8	USA	48.3
Portugal	82.4	Mexico	48.1
Mongolia	82.3	Peru	48.0
Tuvalu	81.9	Brazil	47.9
Western Samoa	81.9	Nigeria	47.6
Andorra	81.3	Thailand	47.4
Germany	80.9	Sierra Leone	46.8
Slovenia	80.6	Botswana	46.5
Aruba	80.4	Chile	45.9
Namibia	80.4	Senegal	45.6
Greece	80.3	Ecuador	44.7
Guyana	80.3	El Salvador	44.3
Israel	80.0	Haiti	42.9
Kuwait	79.6	Ghana	42.4
Norway	79.5	Pakistan	41.8
San Marino	79.1	Zambia	40.5
Finland	79.0	Burkina Faso	38.3
Suriname	77.7	Nauru	37.3
Malta	77.6	Yemen	36.8
Bulgaria	77.5	Colombia	36.2
Romania	77.2	Niger	35.6

*Based on entire voting-age population in countries that held at least two elections during these years. Only countries with highest and lowest averages are shown.

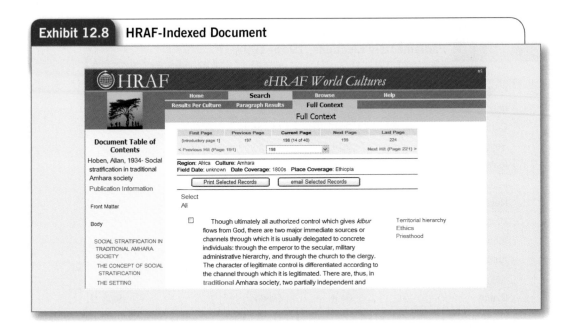

Exhibit 12.8 **HRAF-Indexed Document**

searches for specific index terms to answer many questions about other social groups with the HRAF files, such as "What percentage of the world's societies practice polygyny?" and "Does punitive child training affect the frequency of warfare?" (Ember & Ember 2005:n.p.).

Comparative Historical Research

The combination of historical analysis with comparisons between nations or other units often leads to the most interesting results. Historical social scientists may use comparisons between cases "to highlight the particular features of each case" (Skocpol 1984:370) or to identify general historical patterns across nations. A study of processes within one nation may therefore be misleading if important processes within the nation have been influenced by social processes that operate across national boundaries (Markoff 2005:403). For example, comparisons between nations may reveal that differences in political systems are much more important than voluntary decisions by individual political leaders (Rueschemeyer et al. 1992:31–36).

Comparative historical research can also help identify the causal processes at work within the nations or other entities (Lipset 1968:34; Skocpol 1984:374–386). Comparative historical research can result in historically conditional theory, in which the applicability of general theoretical propositions is limited to particular historical circumstances—for example, what explains the development of capitalism in Turkey may apply to some societies but not others (Paige 1999).

The comparative historical approach focuses on sequences of events rather than on some single past (or current) occurrence that might have influenced an outcome in the present. Comparisons of these sequences may be either quantitative or qualitative. Some studies collect quantitative longitudinal data about a number of nations and then use these data to test hypotheses about influences on national characteristics. (Theda Skocpol [1984:375] terms this *analytic historical sociology*.) Others compare the histories or particular historical experiences of nations in a narrative form, noting similarities and differences and inferring explanations for key national events (*interpretive historical sociology* in Skocpol's terminology [1984:368]).

There are several stages for a systematic qualitative comparative historical study (Ragin 1987:44–52; Rueschemeyer et al. 1992:36–39):

1. Specify a theoretical framework and identify key concepts or events that should be examined to explain a phenomenon.

2. Select cases (such as nations) that vary in terms of the key concepts or events.

3. Identify similarities and differences between the cases in terms of these key concepts or events and the outcome to be explained.

4. Propose a causal explanation for the historical outcome and check it against the features of each case. The criterion of success in this method is to explain the outcome for each case, without allowing deviations from the proposed causal pattern.

Rueschemeyer et al. (1992) used a method such as this to explain why some nations in Latin America developed democratic politics, whereas others became authoritarian or bureaucratic-authoritarian states. First, Rueschemeyer et al. developed a theoretical framework that gave key attention to the power of social classes, state (government) power, and the interaction between social classes and the government. They then classified the political regimes in each nation over time (Exhibit 12.9). Next, they noted how each nation varied over time in terms of the variables they had identified as potentially important for successful democratization.

Their analysis identified several conditions for initial democratization: consolidation of state power (ending overt challenges to state authority); expansion of the export economy (reducing conflicts over resources); industrialization (increasing the size and interaction of middle and working classes); and some agent of political articulation of the subordinate classes (which could be the state, political parties, or mass movements). Historical variation in these conditions was then examined in detail.

The great classical sociologists also used comparative methods, although their approach was less systematic. For example, Max Weber's comparative sociology of religions contrasted Protestantism in the West, Confucianism and Taoism in China, Hinduism and Buddhism in India, and Ancient Judaism. As Reinhard Bendix (1962) explained,

> His [Weber's] aim was to delineate religious orientations that contrasted sharply with those of the West, because only then could he specify the features that were peculiar to Occidental [Western] religiosity and hence called for an explanation. . . . to bring out the distinctive features of each historical phenomenon. (p. 268)

So, for example, Weber concluded that the rise of Protestantism, with its individualistic approach to faith and salvation, was an important factor in the development of capitalism.

When geographic units such as nations are sampled for comparative purposes, it is assumed that the nations are independent of each other in terms of the variables examined. Each nation can then be treated as a separate case for identifying possible chains of causes and effects. However, in a very interdependent world, this assumption may be misplaced—nations may develop as they do because of how other nations are developing (and the same can be said of cities and other units). As a result, comparing the particular histories of different nations may overlook the influence of global culture, international organizations, or economic dependency—just the type of influence identified in Frank et al.'s study of environmental protectionism (Skocpol 1984:384; cf. Chase-Dunn & Hall 1993). These common international influences may cause the same pattern of changes to emerge in different nations; looking within the history of these nations for the explanatory influences would lead to spurious conclusions (de Vaus 2008:258). The possibility of such complex interrelations should always

Exhibit 12.9   Classification of Regimes Over Time

	Constitutional Oligarchic	Authoritarian; Traditional, Populist, Military, or Corporatist	Restricted Democrat	Fully Democratic	Bureaucratic-Authoritarian
Argentina	before 1912	1930–46 1951–55 1955–58 1962–63	1958–62 1963–66	1912–30 1946–51 1973–76 1983–90	1966–73 1976–83
Brazil	before 1930	1930–45	1945–64 1985–90		1964–85
Bolivia	before 1930	1930–52 1964–82	1982–90	1952–64	
Chile	before 1920	1924–32	1920–24 1932–70 1990	1970–73	1973–89
Colombia	before 1936	1949–58	1936–49 1958–90		
Ecuador	1916–25	before 1916 1925–48 1961–78	1948–61 1978–90		
Mexico Paraguay		up to 1990 up to 1990			
Peru		before 1930 1930–39 1948–56 1962–63 1968–80	1939–48 1956–62 1963–68	1980–90	
Uruguay		before 1903 1933–42	1903–19	1919–33 1942–73 1984–90	1973–84
Venezuela		before 1935 1935–45	1958–68	1945–48 1968–90	

Method of agreement A method proposed by John Stuart Mill for establishing a causal relation, in which the values of cases that agree on an outcome variable also agree on the value of the variable hypothesized to have a causal effect, while they differ in terms of other variables.

be considered when evaluating the plausibility of a causal argument based on a comparison between two apparently independent cases (Jervis 1996).

Comparative Case Study Designs

Some comparative researchers use a systematic method for identifying causes that owes its origins to the English philosopher John Stuart Mill (1872). One approach that Mill developed was called the **method of agreement**. The core of this approach is the comparison of nations (*cases*) in terms of similarities and differences on potential causal variables and the phenomenon to be explained. As comparative historian Theda Skocpol (1979:36) explains, researchers who use this method should "try to establish that several cases having in common the phenomenon one is trying to explain also have in common a set of causal factors, although they vary in other ways that might have seemed causally relevant."

For example, suppose three countries that have all developed democratic political systems are compared in terms of four socioeconomic variables hypothesized by different theories to influence democratization (Exhibit 12.10). If the countries differ in terms of three of the variables but are similar in terms of the fourth, this is evidence that the fourth variable influences democratization. In Exhibit 12.10, the method of agreement would lead the analyst to conclude that an expanding middle class was a cause of the democratization

Exhibit 12.10 **John Stuart Mill's Method of Agreement (hypothetical cases and variables)**

Variable	Case 1	Case 2	Case 3
Importance of peasant agriculture	Different	Different	Different
Expanding industrial base	Different	Same	Same
Rising educational levels	Different	Different	Different
Expanding middle class	Same	Same	Same
Democratization (outcome)	**Same**	**Same**	**Same**

experienced in all three countries. Since the focus of the method of agreement is actually on identifying a similarity among cases that differ in many respects, this approach is also called the *most different case studies* method.

The second approach John Stuart Mill developed was the **method of difference**. Again, in the words of comparative historian Theda Skocpol (1979),

> One can contrast the cases in which the phenomenon to be explained and the hypothesized causes are present to other cases in which the phenomenon and the causes are both absent, but which are otherwise as similar as possible to the positive cases. (p. 36)

The method of difference approach is diagrammed in Exhibit 12.11. In this example, "moderate income disparities" are taken to be the cause of democratization, since the country that didn't democratize differs in this respect from the country that did democratize. These two countries are similar with respect to other potential influences on democratization. The argument could be improved by adding more positive and negative cases. Since the focus of the method of difference is actually on identifying a difference among cases that are similar in other respects, this approach is also called the *most similar case studies* method.

Method of difference A method proposed by John Stuart Mill for establishing a causal relation, in which the values of cases that differ on an outcome variable also differ on the value of the variable hypothesized to have a causal effect, while they agree in terms of other variables.

Exhibit 12.11 **John Stuart Mill's Method of Difference (hypothetical cases and variables)**

Country A (Positive Case)	Country B (Negative Case)
Economic development	Economic development
Two-party system begun	Two-party system begun
Proportional representation	Proportional representation
Moderate income disparities	Extreme income disparities
Democratization	**No democratization**

The method of agreement and method of difference approaches can also be combined, "by using at once several positive cases along with suitable negative cases as contrasts" (Skocpol 1979:37). This is the approach that Theda Skocpol (1979) used in her classic book about the French, Russian, and Chinese revolutions, *States and Social Revolutions.* Exhibit 12.12 summarizes part of her argument about the conditions for peasant insurrections, based on a careful historical review. In this exhibit, Skocpol (1979:156) shows how the three countries that experienced revolutions (France, Russia, and China) tended to have more independent peasants and more autonomy in local politics than did three contrasting countries (Prussia/Germany, Japan, and England) that did not experience social revolutions.

Cautions for Comparative Analysis

Of course, ambitious methods that compare different countries face many complications. The features of the cases selected for comparison have a large impact on the researcher's ability to identify influences. Cases should be chosen for their difference in terms of key factors hypothesized to influence the outcome of interest and their similarity on other, possibly confounding, factors (Skocpol 1984:383). For example, to understand how industrialization influences democracy, you would need to select cases for comparison that differ in industrialization, so that you could then see if they differ in democratization (King et al. 1994:148–152). Nonetheless, relying on just a small number of cases for comparisons introduces uncertainty into the conclusions (de Vaus 2008:256).

And what determines whether cases are similar and different in certain respects? In many comparative analyses, the values of continuous variables are dichotomized. For example, nations may be coded as

Exhibit 12.12	**Methods of Agreement and Difference Combined: Conditions for Peasant Insurrections**

Country	Agrarian Class Structures	Local Politics
France	Peasant smallholders own 30% to 40% of land; work 80% in small plots. Individual property established, but peasant community opposes seigneurs, who collect dues.	Villages relatively autonomous under supervision of royal officials.
Russia	Peasants own 60%+ and rent more; control process of production on small plots; pay rents and redemption payments. Strong community based on collective ownership.	Village sovereign under control of tzarist bureaucracy.
China	Peasants own 50% and work virtually all land in small plots. Pay rents to gentry. No peasant community.	Gentry landlords, usurers, and literati dominate local organizational life; cooperate with Imperial officials.
Contrasts		
Prussia/ Germany	West of Elbe: resembles France. East of Elbe: large estates worked by laborers and peasants with tiny holdings and no strong communities.	Junker landlords are local agents of bureaucratic state; dominate local administration and policing.
Japan	Communities dominated by rich peasants.	Strong bureaucratic controls over local communities.
England	Landed class owns 70%. Peasantry polarizing between yeomen farmers and agricultural laborers. No strong peasant community.	Landlords are local agents of monarchy; dominate administration and policing.

democratic or *not democratic* or as having *experienced revolution* or *not experienced revolution.* The methods of agreement and difference that I have just introduced presume these types of binary (dichotomous) distinctions. However, variation in the social world often involves degrees of difference, rather than all or none distinctions (de Vaus 2008:255). Some countries may be partially democratic and some countries may have experienced a limited revolution. At the individual level, you know that distinctions such as *rich* and *poor* or *religious* and *not religious* reflect differences on underlying continua of wealth and religiosity. So, the use of dichotomous distinctions in comparative analyses introduces an imprecise and somewhat arbitrary element into the analysis (Lieberson 1991). On the other hand, for some comparisons, qualitative distinctions such as *simple majority rule* or *unanimity required* may capture the important differences between cases better than quantitative distinctions. We don't want to simply ignore important categorical considerations such as this in favor of *degree of majority rule* or some other underlying variable (King et al. 1994:158–163). Careful discussion of the bases for making distinctions is an important first step in any comparative historical research (also see Ragin 2000).

The focus on comparisons between nations may itself be a mistake for some analyses. National boundaries often do not correspond to key cultural differences, so comparing subregions within countries or larger cultural units that span multiple countries may make more sense for some analyses (de Vaus 2008:258). Comparing countries that have fractured along cultural or religious divides simply in terms of average characteristics would obscure many important social phenomena.

With cautions such as these in mind, the combination of historical and comparative methods allows for rich descriptions of social and political processes in different nations or regions as well as for causal inferences that reflect a systematic, defensible weighing of the evidence. Data of increasingly good quality are available on a rapidly expanding number of nations, creating many opportunities for comparative research. We cannot expect one study comparing the histories of a few nations to control adequately for every plausible alternative causal influence, but repeated investigations can refine our understanding and lead to increasingly accurate causal conclusions (King et al. 1994:33).

▣ Demographic Analysis

The social processes that are the focus of historical and comparative research are often reflected in and influenced by changes in the makeup of the population being studied. For example, the plummeting birthrates in European countries will influence the politics of immigration in those countries, their living standards, the character of neighborhoods, and national productivity (Bruni 2002). **Demography** is the field that studies these dynamics. Demography is the statistical and mathematical study of the size, composition, and spatial distribution of human populations and how these features change over time. Demographers explain population change in terms of five processes: (1) fertility, (2) mortality, (3) marriage, (4) migration, and (5) social mobility (Bogue 1969:1).

> **Demography** The statistical and mathematical study of the size, composition, and spatial distribution of human populations and how these features change over time.

Demographers obtain data from a census of the population (see Chapter 5) and from registries—records of events such as births, deaths, migrations, marriages, divorces, diseases, and employment (Anderton, Barrett, & Bogue 1997: 54–79; Baum, 1993). They compute various statistics from these data to facilitate description and analysis (Wunsch & Termote 1978). To use these data, you need to understand how they are calculated and the questions they answer. Four concepts are key to understanding and using demographic methods: population change, standardization of population numbers, the demographic bookkeeping equation, and population composition.

Population change is a central concept in demography. The absolute population change is calculated simply as the difference between the population size in one census minus the population size in an earlier census. This measure of absolute change is of little value, however, because it does not take into account the total size of the population that was changing (Bogue 1969:32–43). A better measure is the *intercensal percent change*, which is the absolute change in population between two censuses divided by the population size in the earlier census (and multiplied by 100 to obtain a percentage). With the percent change statistic, we can meaningfully compare the growth in two or more nations that differ markedly in size (as long as the intercensal interval does not vary between the nations) (White 1993:1–2).

Standardization of population numbers, as with the calculation of intercensal percent change, is a key concern of demographic methods (Gill, Glazer, & Thernstrom 1992:478–482; Rele 1993). To make meaningful comparisons between nations and over time, numbers that describe most demographic events must be adjusted for the size of the population at risk for the event. For example, the fertility rate is calculated as the ratio of the number of births to women of childbearing age to the total number of women in this age range (multiplied by 1,000). Unless we make such adjustments, we will not know if a nation with a much higher number of births or deaths in relation to its total population size simply has more women in the appropriate age range or has more births per "eligible" woman.

The *demographic bookkeeping* (or *balancing*) *equation* is used to identify the four components of population growth during a time interval ($P_2 P_1$): births (B), deaths (D), and in-migration (M_i) and out-migration (M_o). The equation is written as follows: $P_2 = P_1 + (B—D) + (M_{i\_} M_o)$. That is, population at a given point in time is equal to the population at an earlier time plus the excess of births over deaths during the interval and the excess of in-migration over out-migration (White 1993:1–4). Whenever you see population size or change

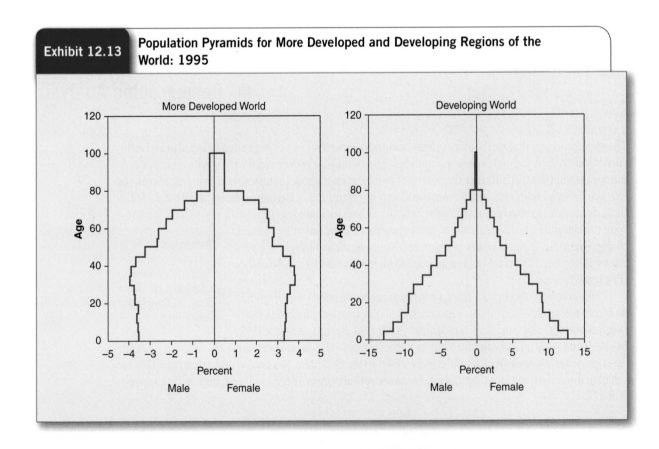

Exhibit 12.13 **Population Pyramids for More Developed and Developing Regions of the World: 1995**

statistics used in a comparative analysis, you will want to ask yourself whether it is also important to know which component in the equation was responsible for the change over time or for the difference between countries (White 1993:1–4).

Population composition refers to a description of a population in terms of basic characteristics such as age, race, sex, or marital status (White 1993:1–7). Descriptions of population composition at different times or in different nations can be essential for understanding social dynamics identified in historical and comparative research. For example, Exhibit 12.13 compares the composition of the population in more developed and developing regions of the world by age and sex in 1995, using United Nations data. By comparing these *population pyramids,* we see that children comprise a much greater proportion of the population in less developed regions. The more developed regions' population pyramid also shows the greater proportion of women at older ages and the post–World War II baby boom bulge in the population.

Demographic analysis can be an important component of historical research (Bean, Mineau, & Anderton 1990), but problems of data quality must be carefully evaluated (Vaessen 1993). The hard work that can be required to develop demographic data from evidence that is hundreds of years old does not always result in worthwhile information. The numbers of people for which data are available in particular areas may be too small for statistical analysis; data that are easily available (e.g., a list of villages in an area) may not provide the information that is important (e.g., population size); and lack of information on the original data-collection procedures may prevent assessment of data quality (Hollingsworth 1972:77).

▣ Ethical Issues in Historical and Comparative Research

Analysis of historical documents or quantitative data collected by others does not create the potential for harm to human subjects that can be a concern when collecting primary data. It is still important to be honest and responsible in working out arrangements for data access when data must be obtained from designated officials or data archivists, but, of course, many data are available easily in libraries or on the web. Researchers who conclude that they are being denied access to public records of the federal government may be able to obtain the data by filing a Freedom of Information Act (FOIA) request. The FOIA stipulates that all persons have a right to access all federal agency records unless the records are specifically exempted (Riedel 2000:130–131). Researchers who review historical or government documents must also try to avoid embarrassing or otherwise harming named individuals or their descendants by disclosing sensitive information.

Ethical concerns are multiplied when surveys are conducted or other data are collected in other countries. If the outside researcher lacks much knowledge of local norms, values, and routine activities, the potential for inadvertently harming subjects is substantial. For this reason, cross-cultural researchers should spend time learning about each of the countries in which they plan to collect primary data and strike up collaborations with researchers in those countries (Hantrais & Mangen 1996). Local advisory groups may also be formed in each country so that a broader range of opinion is solicited when key decisions must be made. Such collaboration can also be invaluable when designing instruments, collecting data, and interpreting results.

Cross-cultural researchers who use data from other societies have a particular obligation to try to understand the culture and norms of those societies before they begin secondary data analyses. It is a mistake to assume that questions asked in other languages or cultural contexts will have the same meaning as when asked in the researcher's own language and culture, so a careful, culturally sensitive process of review by knowledgeable experts must precede measurement decisions in these projects. Ethical standards themselves may vary between nations and cultures, so cross-cultural researchers should consider collaborating with others in the places to be compared and take the time to learn about cultural practices and ethical standards (Stake & Rizvi 2009:527).

▣ Conclusions

Historical and comparative social science investigations use a variety of techniques that range from narrative histories having much in common with qualitative methods to analyses of secondary data that are in many respects like traditional survey research. Each of these techniques can help the researchers gain new insights into processes such as democratization. They encourage intimate familiarity with the cause of development of the nations studied and thereby stimulate inductive reasoning about the interrelations among different historical events. Systematic historical and comparative techniques can be used to test deductive hypotheses concerning international differences as well as historical events.

Most historical and comparative methods encourage causal reasoning. They require the researcher to consider systematically the causal mechanism, or historical sequences of events, by which earlier events influence later outcomes. They also encourage attention to causal context, with a particular focus on the ways in which different cultures and social structures may result in different effects of other variables. There is much to be gained by learning and continuing to use and develop these methods.

Key Terms

Case-oriented research 389
Comparative historical
  research 388
Cross-sectional comparative
  research 388

Demography 405
Event-structure analysis 390
Historical events research 388
Historical process
  research 388

Method of agreement 402
Method of difference 403
Narrative explanation 389
Oral history 392
Variable-oriented research 396

Highlights

- The central insight behind historical and comparative methods is that we can improve our understanding of social processes when we make comparisons to other times and places.

- There are four basic types of historical and comparative research methods: (1) historical events research, (2) historical process research, (3) cross-sectional comparative research, and (4) comparative historical research. Historical events research and historical process research are likely to be qualitative, whereas comparative studies are often quantitative; however, research of each type may be either quantitative or qualitative.

- Qualitative historical process research uses a narrative approach to causal explanation, in which historical events are treated as part of a developing story. Narrative explanations are temporal, holistic, and conjunctural.

- Methodological challenges for comparative and historical research include missing data, variation in the meaning of words and phrases and in the boundaries of geographic units across historical periods and between cultures, bias or inaccuracy of historical documents, lack of measurement equivalence, the need for detailed knowledge of the cases chosen, a limited number of cases, case selection on an availability basis, reliance on dichotomous categorization of cases, and interdependence of cases selected.

- Central concepts for demographic research are population change, standardization of population numbers, the demographic bookkeeping equation, and population composition.

- Oral history provides a means of reconstructing past events. Data from other sources should be used whenever possible to evaluate the accuracy of memories.

STUDENT STUDY SITE

To assist in completing the web exercises, please access the study site at **www.sagepub.com/schuttisw7e**, where you will find the web exercises with accompanying links. You'll find other useful study materials such as self-quizzes and e-flashcards for each chapter, along with a group of carefully selected articles from research journals that illustrate the major concepts and techniques presented in the book.

Discussion Questions

1. Review the differences between case-oriented, historically specific, inductive explanations and those that are more variable oriented, theoretically general, and deductive. List several arguments for and against each approach. Which is more appealing to you and why?

2. What historical events have had a major influence on social patterns in the nation? The possible answers are too numerous to list, ranging from any of the wars to major internal political conflicts, economic booms and busts, scientific discoveries, and legal changes. Pick one such event in your own nation for this exercise. Find one historical book on this event and list the sources of evidence used. What additional evidence would you suggest for a social science investigation of the event?

3. Consider the comparative historical research by Rueschemeyer et al. (1992) on democratic politics in Latin America. What does comparison among nations add to the researcher's ability to develop causal explanations?

4. Olzak, Shanahan, and McEneaney (1996) developed a nomothetic causal explanation of variation in racial rioting in the United States over time, whereas Griffin's (1993) explanation of a lynching can be termed *idiographic*. Discuss the similarities and differences between these types of causal explanation. Use these two studies to illustrate the strengths and weaknesses of each.

5. Select a major historical event or process, such as the Great Depression, World War II, the civil rights movement, or the war in Iraq. Why do you think this event happened? Now, select one of the four major types of historical and comparative methods that you think could be used to test your explanation. Why did you choose this method? What type of evidence would support your proposed explanation? What problems might you face in using this method to test your explanation?

Practice Exercises

1. The journals *Social Science History* and *Journal of Social History* report many studies of historical processes. Select one article from a recent journal issue about a historical process used to explain some event or other outcome. Summarize the author's explanation. Identify any features of the explanation that are temporal, holistic, and conjunctural. Prepare a chronology of the important historical events in that process. Do you agree with the author's causal conclusions? What additional evidence would strengthen the author's argument?

2. Exhibit 12.14 identifies voting procedures and the level of turnout in one election for 10 countries. Do voting procedures appear to influence turnout in these countries? To answer this question using Mill's methods, you will first have

to decide how to dichotomize the values of variables that have more than two values (postal voting, proxy voting, and turnout). You must also decide what to do about missing values. Apply Mill's method of agreement to the pattern in the table. Do any variables emerge as likely causes? What additional information would you like to have for your causal analysis?

3. Using your library's government documents collection or the U.S. Census site on the web, select one report by the U.S. Census Bureau about the population of the United States or some segment of it. Outline the report and list all the tables included in it. Summarize the report in two paragraphs. Suggest a historical or comparative study for which this report would be useful.

Exhibit 12.14 Voting Procedures in 10 Countries

	Voting Age	Number of Days Polling Booth Open	Voting Day on Work Day or Rest Day	Postal Voting	Proxy Voting	Constituency Transfer	Advance Voting	Voter Turnout (in %)	Year (P=presidential, L = legislative election)
Switzerland	20	2	Rest day	Automatic for armed forces, otherwise by application 4 days before voting	Varies by canton	No	No	46	1991L
Taiwan	20	1	Rest day					72	1992L
Thailand	20	1	Rest day	No				62	1995L
Turkey	20	1	Rest day	No	No	Special polling stations at border posts for citizens residing abroad	No	80	1991L
Ukraine	18	1	Rest day					71.6	1994P
United Kingdom	18	1	Work day	On application	On application	No	No	77.8	1992L
United States	18	1	Work day	By application; rules vary across states	In some states for blind and disabled	No		51.5	1992P
Uruguay	18	1	Rest day	No	No	No	No	89.4	1994P
Venezuela	18	1	Rest day	No	Assisted voting for blind and disabled	No	No	60	1993P
Zambia		1	Work day		No			50	1991P

Note: P = Presidential election; L = Legislative election.

4. Find a magazine or newspaper report on a demographic issue, such as population change or migration. Explain how one of the key demographic concepts could be used or was used to improve understanding of this issue.

5. Review the Interactive Exercises on the study site for a lesson that will help you master the terms used in historical and comparative research.

6. Select an article from the book's study site, at www.sagepub.com/schuttisw7e, that used a historical/comparative design. Which specific type of design was used? What were the advantages of this design for answering the research question posed?

Ethics Questions

1. Oral historians can uncover disturbing facts about the past. What if a researcher were conducting an oral history project such as the Depression Writer's Project and learned from an interviewee about his previously undisclosed involvement in a predatory sex crime many years ago? Should the researcher report what he learned to a government attorney who might decide to bring criminal charges? What about informing the victim and/or her surviving relatives? Would it matter if the statute of limitations had expired, so that the offender could not be prosecuted any longer? Would it matter if the researcher were subpoenaed to testify before a grand jury?

2. In this chapter's ethics section, I recommended that researchers who conduct research in other cultures form an advisory group of local residents to provide insight into local customs and beliefs. What are some other possible benefits of such a group for cross-cultural researchers? What disadvantages might arise from use of such a group?

Web Exercises

1. The World Bank offers numerous resources that are useful for comparative research. Visit the World Bank website at www.worldbank.org. Click on the "Countries" link at the top of the site and then select one region, such as "Africa." Now, choose a specific country and topic that interests you and write a brief summary of the reported data. Then, compare these data with those for another country in the same region, and summarize the differences and similarities you have identified between the countries.

2. The U.S. Bureau of Labor Statistics (BLS) website provides extensive economic indicator data for regions, states, and cities. Go to the BLS web page that offers statistics by location: http://stats.bls.gov/eag. Now, click on a region and explore the types of data that are available. Write out a description of the steps you would have to take to conduct a comparative analysis using the data available from the BLS website.

3. The U.S. Census Bureau's home page can be found at www.census.gov. This site contains extensive reporting of census data, including population data, economic indicators, and other information acquired through the U.S. Census. This website allows you to collect information on numerous subjects and topics. This information can then be used to make comparisons between different states or cities. Comparative analysis is facilitated by the "State and County Quick Facts" option, which can be accessed directly at http://quickfacts.census.gov/qfd. Now, choose your own state and the county in which you live and copy down several statistics of interest. Repeat this process for other counties in your state. Use the data you have collected to compare your county with other counties in the state. Write a one-page report summarizing your findings.

SPSS Exercises

1. In this exercise, you will use Mill's method of agreement to examine international differences in egalitarianism (belief that the government should reduce income differences). For this cross-sectional comparative analysis, you will use the

ISSP data set on Family and Gender Roles III 2002. It is on the study site and contains results of an international survey involving respondents in more than 25 countries.

a. First, examine the labels and response choices for each of the following variables:

GOVDIFF, UNION, DEGREE

Which, if any, do you believe varies between nations? State a hypothesis specifying which countries you expect to be relatively egalitarian (high level of agreement with GOVDIFF) and which countries you expect to be relatively inegalitarian (low level of agreement with GOVDIFF). Explain your reasoning.

b. Now, request the distributions for the above variables, by country. You can do this by requesting the cross-tabulation of each variable by COUNTRY. Summarize what you have found in a table that indicates, for each country, whether it is "high" or "low" in terms of each variable, in comparison with the other countries. From the menu, click

Analyze/Descriptive Statistics/Crosstabs

In the Crosstabs window, set:

Row(s): GOVDIFF (and the other variables)

Column(s): COUNTRY

Cells: COLUMN %

c. Review your table of similarities and differences between countries. See if on this basis you can develop a tentative explanation of international variation in support for egalitarian government policies using Mill's method of agreement.

d. Discuss the possible weaknesses in the type of explanation you have constructed, following John Stuart Mill. Propose a different approach for a comparative historical analysis.

2. How do the attitudes of immigrants to the United States compare with those of people born in the United States? Use the GSS2010x file and request the cross-tabulations (in percentage form) of POLVIEWS3, BIBLE, SPKATH by BORN (with BORN as the column variable). Inspect the output. Describe the similarities and differences you have found.

3. Because the GSS file is cross-sectional, we cannot use it to conduct historical research. However, we can develop some interesting historical questions by examining differences in the attitudes of Americans in different birth cohorts.

a. Inspect the distributions of the same set of variables. Would you expect any of these attitudes and behaviors to have changed over the 20th century? State your expectations in the form of hypotheses.

b. Request a cross-tabulation of these variables by birth COHORTS. What appear to be the differences among the cohorts? Which differences do you think are due to historical change, and which do you think are due to the aging process? Which attitudes and behaviors would you expect to still differentiate the baby-boom generation and the post-Vietnam generation in 20 years?

Developing a Research Proposal

Add a historical or comparative dimension to your proposed study (Exhibit 2.14, #13 to #17).

1. Consider which of the four types of comparative/historical methods would be most suited to an investigation of your research question. Think of possibilities for qualitative and quantitative research on your topic with the method you prefer. Will you conduct a variable-oriented or case-oriented study? Write a brief statement justifying the approach you choose.

2. Review the possible sources of data for your comparative/historical project. Search the web and relevant government, historical, and international organization sites or publications. Search the social science literature for similar studies and read about the data sources that they used.

3. Specify the hypotheses you will test or the causal sequences you will investigate. Describe what your cases will be (nations, regions, years, etc.). Explain how you will select cases. List the sources of your measures, and describe the specific type of data you expect to obtain for each measure. Discuss how you will evaluate causal influences, and indicate whether you will take a nomothetic or idiographic approach to causality.

4. Review the list of potential problems in comparative/historical research, and discuss those that you believe will be most troublesome in your proposed investigation. Explain your reasoning.

Secondary Data Analysis and Content Analysis

I rish researchers Richard Layte (Economic and Social Research Institute) and Christopher T. Whelan (University College Dublin) sought to improve understanding of poverty in Europe. Rather than design their own data collection effort, they turned to five waves of data from the European Community Household Panel Survey, which were available to them from Eurostat, the Statistical Office of the European Communities (Eurostat 2003). The data they obtained represented the years from 1994 to 1998, thus allowing Layte and Whelan (2003) to investigate whether poverty tends to persist more in some countries than in others and what factors influence this persistence in different countries. Their investigation of "poverty dynamics" found a tendency for individuals and households to be "trapped" in poverty, but this phenomenon varied with the extent to which countries provided social welfare supports.

Secondary data analysis The method of using pre-existing data in a different way or to answer a different research question than intended by those who collected the data.

Secondary data Previously collected data that are used in a new analysis.

Secondary data analysis is the method of using preexisting data in a different way or to answer a different research question than intended by those who collected the data. The most common sources of **secondary data**—previously collected data that are used in a new analysis—are social science surveys and data collected by government agencies, often with survey research methods. It is also possible to reanalyze data that have been collected in experimental studies or with qualitative methods. Even a researcher's reanalysis of data that he or she collected previously qualifies as secondary analysis if it is employed for a new purpose or in response to a methodological critique.

Thanks to the data collected by social researchers, governments, and organizations over many years, secondary data analysis has become the research method used by many contemporary social scientists to investigate important research questions. Why consider secondary data? (1) Data collected in previous investigations is available for use by other social researchers on a wide range of topics. (2) Available data sets often include many more measures and cases and reflect more rigorous research procedures than another researcher will have the time or resources to obtain in a new investigation. (3) Much of the groundwork involved in creating and testing measures with the data set has already been done. (4) Most important, most funded social science research projects collect data that can be used to investigate new research questions that the primary researchers who collected the data did not consider. Analyzing secondary data, then, is nothing like buying "used goods"!

Content analysis A research method for systematically analyzing and making inferences from recorded human communication, including books, articles, poems, constitutions, speeches, and songs.

Content analysis is similar to secondary data analysis in its use of information that has already been collected. Therefore, like secondary data analysis, content analysis can be called an *unobtrusive method* that does not need to involve interacting with live people. In addition, most content analyses, like most secondary data analyses, use quantitative analysis procedures and you will find some data sets resulting from content analyses in collections of secondary data sets. Content analyses can even be used to code data collected in surveys, so you can find content analysis data included in some survey data sets. However, content analysis methods usually begin with text, speech broadcasts, or visual images, and not data already collected by social scientists. The content analyst develops procedures for coding various aspects of the textual, aural (spoken), or visual material and then analyzes this coded content.

I will first review the procedures involved in secondary data analysis, identify many of the sources for secondary data sets and explain how to obtain data from these sources. I will give special attention to some easy-to-overlook problems with the use of secondary data. I then present basic procedures for content analysis and several examples of this method. The chapter concludes with some ethical cautions related to use of both methods.

▣ Secondary Data Sources

Secondary data analysis has been an important social science methodology since the earliest days of social research, whether when Karl Marx (1967) reviewed government statistics in the Reading Room of the British Library or Emile Durkheim (1966) analyzed official government cause-of-death data for his study of suicide rates throughout Europe. With the advent of modern computers and, even more important, the Internet, secondary data analysis has become an increasingly accessible social research method. Literally, thousands of large-scale data sets are now available for the secondary data analyst, often with no more effort than the few commands required to download the data set; a number of important data sets can even be analyzed directly on the web by users who lack their own statistical software.

There are many sources of data for secondary analysis within the United States and internationally. These sources range from data compiled by governmental units and private organizations for administrative purposes, which are subsequently made available for research purposes, to data collected by social researchers for one purpose that are then made available for reanalysis. Many important data sets are collected for the specific purpose of facilitating secondary data analysis. Government units from the the Census Bureau to the U.S. Department of Housing and Urban Development; international organizations such as the United Nations, the Organization for Economic Co-Operation and Development (OECD), and the World Bank; and internationally involved organizations such as the CIA sponsor a substantial amount of social research that is intended for use by a broader community of social scientists. The National Opinion Research Corporation (NORC), with its General Social Survey (GSS), and the University of Michigan, with its Detroit Area Studies, are examples of academically based research efforts that are intended to gather data for social scientists to use in analyzing a range of social science research questions.

Many social scientists who have received funding to study one research question have subsequently made the data they collect available to the broader social science community for investigations of other research questions. Many of these data sets are available from a website maintained by the original research organization, often with some access restrictions. Examples include the Add Health study conducted at the University of North Carolina Population Center, the University of Michigan's Health and Retirement Study as well as its Detroit Area Studies, and the United Nations University's World Inequality.

What makes secondary data analysis such an exciting and growing option today are the considerable resources being devoted to expanding the amount of secondary data and to making it available to social scientists. For example, the National Data Program for the Social Sciences, funded in part by the National Science Foundation, sponsors the ongoing GSS to make current data on a wide range of research questions available to social scientists. Since 1985, the GSS has participated in an International Social Survey Program that generates comparable data from 47 countries around the world (www.issp.org). Another key initiative is the Data Preservation Alliance for the Social Sciences (Data-PASS), funded by the Library of Congress in 2004 as a part of the National Digital Preservation Program (www.icpsr.umich.edu/icpsrweb/DATAPASS/This project is designed to ensure the preservation of digitized social science data. Led by the Inter-University Consortium for Political and Social Research (ICPSR) at the University of Michigan, it combines the efforts of other major social research organizations, including the Roper Center for Public Opinion Research at the University of Connecticut; the Howard W. Odum Institute for Research in Social Sciences at the University of North Carolina, Chapel Hill; the Henry A. Murray Research Archive and the Harvard-MIT Data Center at Harvard University; and the Electronic and Special Media Records Service Division of the U.S. National Archives and Records Administration.

Fortunately, you do not have to google your way around the web to find all these sources on your own. There are many websites that provide extensive collections of secondary data. Chief among these is the ICPSR at the University of Michigan. The University of California at Berkeley's Survey Documentation and Analysis (SDA) archive provides several data sets from national omnibus surveys, as well as from U.S. Census microdata, from surveys on racial attitudes and prejudice, and from several labor and health surveys. The National Archive of Criminal Justice Data is an excellent source of data in the area of criminal justice, although, like many other data collections, including key data from the U.S. Census, it is also available through the ICPSR. Much of the statistical data collected by U.S. federal government agencies can be accessed through the consolidated FedStats website, www.fedstats.gov.

In this section, I describe several sources of online data in more detail. The decennial population census by the U.S. Census Bureau is the single most important governmental data source, but many other data sets are collected by the U.S. Census and by other government agencies, including the U.S. Census Bureau's Current Population Survey and its Survey of Manufactures or the Bureau of Labor Statistics' Consumer Expenditure Survey. These government data sets typically are quantitative; in fact, the term *statistics*—state-istics—is derived from this type of data.

U.S. Census Bureau

The U.S. government has conducted a census of the population every 10 years since 1790; since 1940, this census has also included a census of housing (see also Chapter 5). This decennial Census of Population and Housing is a rich source of social science data (Lavin 1994). The Census Bureau's monthly *Current Population Survey (CPS)* provides basic data on labor force activity that is then used in U.S. Bureau of Labor Statistics reports. The Census Bureau also collects data on agriculture, manufacturers, construction and other business, foreign countries, and foreign trade.

The U.S. Census of Population and Housing aims to survey one adult in every household in the United States. The basic *complete-count* census contains questions about household composition as well as ethnicity and income. More questions are asked in a longer form of the census that is administered to a sample of the households. A separate census of housing characteristics is conducted at the same time (Rives & Serow 1988:15). Participation in the census is required by law, and confidentiality of the information obtained is mandated by law for 72 years after collection. Census data are reported for geographic units, including states, metropolitan areas, counties, census tracts (small, relatively permanent areas within counties), and even blocks (see Exhibit 13.1). These different units allow units of analysis to be tailored to research questions. Census data are used to apportion seats in the U.S. House of Representatives and to determine federal and state legislative district boundaries, as well as to inform other decisions by government agencies.

The U.S. Census website (www.census.gov) provides much information about the nearly 100 surveys and censuses that the Census Bureau directs each year, including direct access to many statistics for particular geographic units. An interactive data retrieval system, American FactFinder, is the primary means for distributing results from the 2010 Census: You can review its organization and download data at http://factfinder2.census.gov/main.html. The catalog of the ICPSR (www.icpsr.umich.edu/icpsrweb/ICPSR/) also lists many census reports. Many census files containing microdata—records from persons, households, or housing units—are available online, while others can be purchased on CD-ROM or DVD from the Customer Services

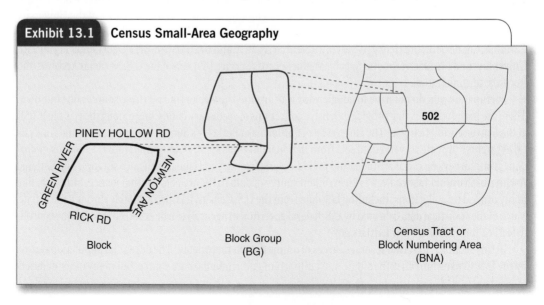

Exhibit 13.1 Census Small-Area Geography

PINEY HOLLOW RD

GREEN RIVER

NEWTON AVE

RICK RD

502

Block

Block Group
(BG)

Census Tract or
Block Numbering Area
(BNA)

Center at (301) 763-INFO (4636); census data can also be inspected online or downloaded for various geographic levels, including counties, cities, census tracts, and even blocks using the DataFerrett application (Federated Electronic Research, Review, Extract, and Tabulation Tool). You can download, install, and use this tool at http://dataferrett.census.gov. This tool also provides access to data sets collected by other federal agencies. An even more accessible way to use U.S. Census data is through the website maintained by the Social Science Data Analysis Network, at www.ssdan.net/. Check out the DataCounts! options.

States also maintain census bureaus and may have additional resources. Some contain the original census data collected in the state 100 or more years ago. The ambitious historical researcher can use these returns to conduct detailed comparative studies at the county or state level (Lathrop 1968:79).

Integrated Public Use Microdata Series

Individual-level samples from U.S. Census data for the years 1850 to 2000, as well as historical census files from several other countries, are available through the Integrated Public Use Microdata Series (IPUMS) at the University of Minnesota's Minnesota Population Center (MPC). These data are prepared in an easy-to-use format that provides consistent codes and names for all the different samples.

This exceptional resource offers 39 samples of the American population selected from 15 federal censuses, as well as results of the Census Bureau's annual American Community Survey from 2000 to 2006. Each sample is independently selected, so that individuals are not linked between samples. In addition to basic demographic measures, variables in the U.S. samples include educational, occupational, and work indicators; respondent income; disability status; immigration status; veteran status; and various household characteristics, including family composition and dwelling characteristics. The international samples include detailed characteristics from hundreds of thousands of individuals in countries ranging from France and Mexico to Kenya and Vietnam. You can view these resources at www.ipums.umn.edu. You must register to download data, but the registration is free.

Bureau of Labor Statistics (BLS)

Another good source of data is the BLS of the U.S. Department of Labor, which collects and analyzes data on employment, earnings, prices, living conditions, industrial relations, productivity and technology, and occupational safety and health (U.S. Bureau of Labor Statistics 1991, 1997b). Some of these data are collected by the U.S. Census Bureau in the monthly *CPS;* other data are collected through surveys of establishments (U.S. Bureau of Labor Statistics 1997a).

The *CPS* provides a monthly employment and unemployment record for the United States, classified by age, sex, race, and other characteristics. The *CPS* uses a stratified random sample of about 60,000 households (with separate forms for about 120,000 individuals). Detailed questions are included to determine the precise labor force status (whether they are currently working or not) of each household member over the age of 16. Statistical reports are published each month in the BLS's *Monthly Labor Review* and can also be inspected at its website (http://stats.bls.gov). Data sets are available on computer tapes and disks from the BLS and services like the ICPSR.

Other U.S. Government Sources

Many more data sets useful for historical and comparative research have been collected by federal agencies and other organizations. The National Technical Information Service (NTIS) of the U.S. Department of

Commerce maintains a Federal Computer Products Center that collects and catalogs many of these data sets and related reports.

By 2008, more than 2,000,000 data sets and reports were described in the NTIS Database. The NTIS Database is the essential source of information about the data sets and can be searched at www.ntis.gov. Data set summaries can be searched in the database by either subject or agency. Government research reports cataloged by NTIS and other agencies can be searched online at the NTIS website, www.fedworld.gov.

Independent Investigator Data Sources

Many researchers who have received funding to investigate a wide range of research topics make their data available on websites where they can be downloaded by other researchers for secondary data analyses. One of the largest, introduced earlier, is the Add Health study, funded at the University of North Carolina by the National Institute of Child Health and Human Development (NICHD) and 23 other agencies and foundations to investigate influences on adolescents' health and risk behaviors (www.cpc.unc.edu/projects/addhealth). The study began in 1994–95 with a representative sample of more than 90,000 adolescents who completed questionnaires in school and more than 20,000 who were interviewed at home. This first wave of data collection has been followed by three more, resulting in longitudinal data for more than 10 years. Another significant data source, the Health and Retirement Study (HRS), began in 1992 with funding from the National Institute on Aging (NIA) (http://hrsonline.isr.umich.edu/). The University of Michigan oversees HRS interviews every 2 years with more than 22,000 Americans over the age of 50. To investigate family experience change, researchers at the University of Wisconsin designed the National Survey of Families and Households (www.soc.wisc.edu/nsfh). With funding from both NICHD and NIA, members of more than 10,000 households were interviewed in three waves, from 1987 to 2002. Another noteworthy example, among many, is the Detroit Area Studies, with annual surveys between 1951 and 2004 on a wide range of personal, political, and social issues (www.icpsr.umich.edu/icpsrwewb/detroitareastudies/).

Inter-University Consortium for Political and Social Research

The University of Michigan's ICPSR is the premier source of secondary data useful to social science researchers. ICPSR was founded in 1962 and now includes more than 640 colleges and universities and other institutions throughout the world. ICPSR archives the most extensive collection of social science data sets in the United States outside the federal government: More than 7,990 studies are represented in more than 500,000 files from 130 countries and from sources that range from U.S. government agencies such as the Census Bureau to international organizations such as the United Nations, social research organizations such as the National Opinion Research Center, and individual social scientists who have completed funded research projects.

The data sets archived by ICPSR are available for downloading directly from the ICPSR website, www.icpsr.umich.edu. ICPSR makes data sets obtained from government sources available directly to the general public, but many other data sets are available only to individuals at the colleges and universities around the world that have paid the fees required to join ICPSR. The availability of some data sets is restricted due to confidentiality issues (see section in this chapter on research ethics); in order to use them, researchers must sign a contract and agree to certain conditions (see www.icpsr.umich.edu/icpsrweb/ICPSR/help/datausers/index.jsp).

Survey data sets obtained in the United States and in many other countries that are stored at the ICPSR provide data on topics ranging from elite attitudes to consumer expectations. For example, data collected in the British Social Attitudes Survey in 1998, designated by the University of Chicago's National Opinion Research Center, are available through the ICPSR (go to the ICPSR website, www.icpsr.umich.edu, and search for study no. 3101). Data collected in a monthly survey of Spaniards' attitudes, by the Center for Research on Social Reality (Spain) Survey, are also available (see study no. 6964). Survey data from Russia, Germany, and other countries can also be found in the ICPSR collection.

Do you have an interest in events and interactions between nations, such as threats of military force? A data set collected by Charles McClelland includes characteristics of 91,240 such events (study no. 5211). The history of military interventions in nations around the world between 1946 and 1988 is coded in a data set developed by Frederic Pearson and Robert Baumann (study no. 6035). This data set identifies the intervener and target countries, the starting and ending dates of military intervention, and a range of potential motives (such as foreign policies, related domestic disputes, and pursuit of rebels across borders).

Census data from other nations are also available through the ICPSR, as well as directly through the Internet. In the ICPSR archives, you can find a data set from the Statistical Office of the United Nations on the 1966 to 1974 population of 220 nations throughout the world (study no. 7623). More current international population data are available through data sets available from a variety of sources, such as the study of indicators of globalization from 1975 to 1995 (study no. 4172). (See also the preceding description of the Eurobarometer Survey Series.) More than 3,000 data sets from countries outside of the United States are available through ICPSR's International Data Resource Center.

Obtaining Data From ICPSR

You begin a search for data in the ICPSR archives at www.icpsr.umich.edu/icpsrweb/ICPSR/access/index.jsp. Exhibit 13.2 shows the search screen as I began a search for data from studies involving a subject of domestic

Exhibit 13.2   **Search Screen: Domestic Violence**

violence. You can also see in this screen that you can search the data archives for specific studies, identified by study number or title, as well as for studies by specific investigators (this would be a quick way to find the data set contributed by Richard A. Berk and Lawrence W. Sherman from their research, discussed in Chapter 2, on the police response to domestic violence).

Exhibit 13.3 displays the results of my search: a list of 63 data sets that involved research on domestic violence and that are available through ICPSR. For most data sets, you can obtain a description, the files that are available for downloading, and a list of "related literature"—that is, reports and articles that use the listed data set. Some data sets are made available in collections on a CD-ROM; the CD-ROM's contents are described in detail on the ICPSR site, but you have to place an order to receive the CD-ROM itself.

When you click on the "Download" option, you are first asked to enter your e-mail address and password. What you enter will determine which data sets you can access; if you are not at an ICPSR member institution, you will be able to download only a limited portion of the data sets—mostly those from government sources. If you are a student at a member institution, you will be able to download most of the data sets directly, although you may have to be using a computer that is physically on your campus to do so. Exhibit 13.4 displays the ICPSR download screen after I selected files I wanted to download from the study by Lisa Newmark, Adele Harrell, and Bill Adams on victim ratings of police response in New York and Texas. Because I wanted to analyze the data with the SPSS statistical package, I downloaded the data set in the form of an "SPSS Portable File." The files downloaded in a *zip* file, so I had to use the WinZip program to unzip them. After unzipping the SPSS portable file, I was able to start my data analysis with the SPSS program. If you'd like to learn how to analyze data with the SPSS statistical program, jump ahead to Chapter 14 and Appendix D (on the student study site).

Exhibit 13.3   Search Screen: Domestic Violence Search Results

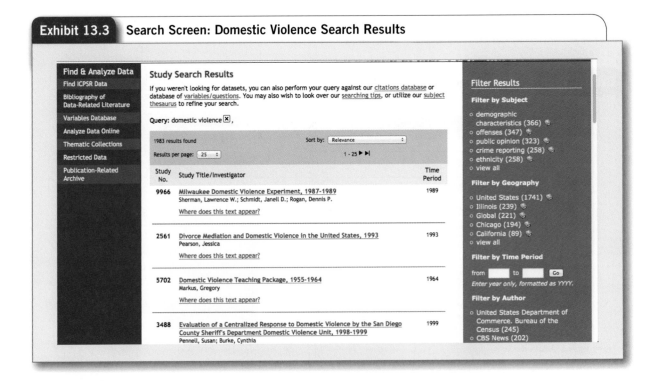

If you prepare your own paper based on an analysis of ICPSR data, be sure to include a proper citation. Here's an example from the ICPSR itself (www.icpsr.umich.edu/icpsrweb/ICPSR/citations/):

Reif, Karlheinz, and Anna Melich. *Euro-Barometer 39.0: European Community Policies and Family Life, March-April 1993* [Computer file]. Conducted by INRA (Europe), Brussels. ICPSR06195-v4. Ann Arbor, MI: Inter-University Consortium for Political and Social Research [producer], 1995. Koeln, Germany: Zentralarchiv fuer Empirische Sozialforschung/Ann Arbor, MI: Inter-University Consortium for Political and Social Research [distributors], 1997.

You can also search the entire ICPSR database for specific variables and identify the various studies in which they have appeared. Exhibit 13.4 displays one segment of the results of searching for variables related to "victimization." A total of 5,554 names of variables in a multitude of studies were obtained. Reviewing some of these results can suggest additional search strategies and alternative databases to consider.

Some of the data sets are also offered with the option of "online analysis." If you have this option, you can immediately inspect the distributions of responses to each question in a survey and examine the relation between variables, without having any special statistical programs of your own. At the bottom of Exhibit 13.5, you'll find the wording reported in the study "codebook" for a question used in the study of a collaborative health care and criminal justice intervention in Texas, as well as, in the top portion, the available statistical options. After choosing one or more variables from the codebook, you can request the analysis.

Exhibit 13.4   ICPSR Variables Related to Victimization

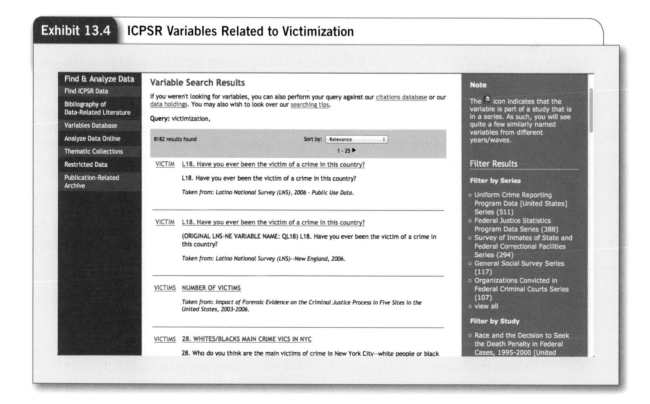

| Exhibit 13.5 | ICPSR Online Analysis: Codebook Information and Statistical Options |

My analysis began with a chart of the distribution of victims' responses to a question about their current relationship with the abuser. As you can see in Exhibit 13.6, about half had left the relationship, but half were still married or living as married with the abuser. This approach to analysis with secondary data can get you jumpstarted in your work. An online analysis option is also starting to appear at other websites that offer secondary data.

ICPSR also catalogs reports and publications containing analyses that have used ICPSR data sets since 1962—more than 58,000 citations were in this archive on May 2, 2011. This superb resource provides an excellent starting point for the literature search that should precede a secondary data analysis. In most cases, you can learn from detailed study reports a great deal about the study methodology, including the rate of response in a sample survey and the reliability of any indexes constructed. Published articles provide not only examples of how others have described the study methodology but also research questions that have already been studied with the data set and issues that remain to be resolved. You can search this literature at the ICPSR site simply by entering the same search terms that you used to find data sets or by entering the specific study number of the data set on which you have focused (see Exhibit 13.7). Don't start a secondary analysis without reviewing such reports and publications.

Even if you are using ICPSR, you shouldn't stop your review of the literature with the sources listed on the ICPSR site. Conduct a search in Sociological Abstracts or another bibliographic database to learn about related studies that used different databases (see Chapter 2).

Institute for Quantitative Social Science

Harvard University's Henry A. Murray Research Archive (www.murray.harvard.edu) has developed a remarkable collection of social science research data sets which are now made available through a larger collaborative secondary data project as part of its Institute for Quantitative Social Science (IQSS) (http://dvn

---

Exhibit 13.6   **ICPSR Online Analysis Frequency Distribution**

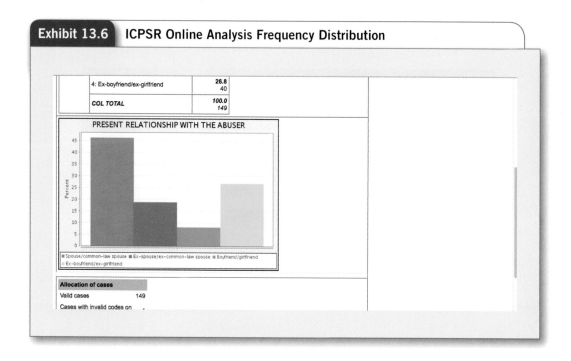

.iq.harvard.edu/dvn/). As of May 2011, IQSS provided information on about 39,285 studies, cross-referencing those in the ICPSR archives. You can search data sets in the IQSS collection by title, abstract, keywords, and other fields; if you identify a data set that you would like to analyze, you must then submit an application in order to be given access.

International Data Sources

Comparative researchers and those conducting research in other countries can find data sets on the population characteristics, economic and political features, and political events of many nations. Some of these are available from U.S. government agencies. For example, the Social Security Administration reports on the characteristics of social security throughout the world (Wheeler 1995). This comprehensive source classifies nations in terms of their type of social security program and provides detailed summaries of the characteristics of each nation's programs. Current information is available online at www.ssa.gov/policy/docs/progdesc/ssptw/index.html. More recent data are organized by region. A broader range of data is available in the *World Handbook of Political and Social Indicators,* with political events and political, economic, and social data coded from 1948 to 1982 (www.icpsr.umich.edu, study no. 7761) (Taylor & Jodice 1986).

The European Commission administers the Eurobarometer Survey Series at least twice yearly across all the member states of the European Union. The survey monitors social and political attitudes and reports are published regularly online at www.gesis.org/en/services/data/survey-data/eurobarometer-data-service/. Case-level Eurobarometer survey data are stored at the ICPSR. The United Nations University makes available a World Income Inequality Database from ongoing research on income inequality in developed, developing,

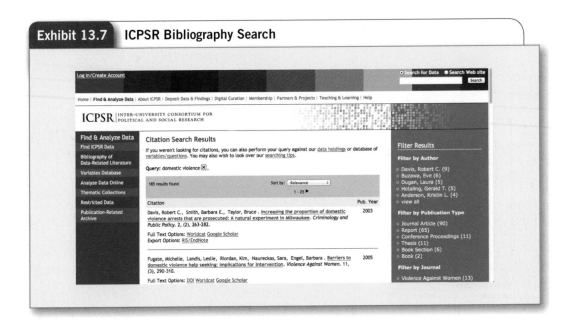

Exhibit 13.7 ICPSR Bibliography Search

and transition countries (www.wider.unu.edu/resarch/Database/en_GB/database/). ICPSR also maintains an International Data Resource Center that provides access to many other data sets from around the world (www.icpsr.umich.edu/icpsrweb/IDRC/index.jsp).

Both the Council of European Social Science Data Archives (CESSDA) (www.cessda.org/) and the International Federation of Data Organisations (IFDO) (www.ifdo.org/) maintain lists of data archives upheld by a wide range of nations (Dale, Wathan, & Higgins 2008:521). CESSDA makes available more than 25,000 data sets from European countries to European researchers (as of May 2011), while IFDO provides an overview of social science data sets collected throughout the world; access procedures vary, but some data sets can be downloaded directly from the IFDO site.

Qualitative Data Sources

Far fewer qualitative data sets are available for secondary analysis, but the number is growing. European countries, particularly England, have been in the forefront of efforts to promote archiving of qualitative data. The United Kingdom's Economic and Social Research Council established the Qualitative Data Archiving Resource Center at the University of Essex in 1994 (Heaton 2008:507). Now part of the Economic and Social Data Service, ESDS Qualidata provides access to data from more than 335 qualitative research projects (Universities of Essex and Manchester 2011). After registering at the ESDS Qualidata site, interview transcripts and other materials from many qualitative studies can be browsed or searched directly online, but access to many studies is restricted to users in the United Kingdom or according to other criteria.

In the United States, the ICPSR collection includes an expanding number of studies containing at least some qualitative data or measures coded from qualitative data (494 such studies as of May 2011). Studies range from transcriptions of original handwritten and published materials relating to infant and child care from the turn of the century to World War II (LaRossa 1995) to transcripts of open-ended interviews with high school students involved in violent incidents (Lockwood 1996). Harvard University's Institute for

Quantitative Social Science has archived at least 100 studies which contain at least some qualitative data (as of May 2011).

The most unique source of qualitative data available for researchers in the United States is the Human Relations Area Files (HRAF) at Yale University, described in Chapter 12. The HRAF has made anthropological reports available for international cross-cultural research since 1949 and currently contains more than 1,000,000 pages of information on more than 400 different cultural, ethnic, religious, and national groups (Ember & Ember 2011). If you are interested in cross-cultural research, it is well worth checking out the HRAF and exploring access options (reports can be accessed and searched online by those at affiliated institutions).

The University of Southern Maine's Center for the Study of Lives (usm.maine.edu/olli/national/lifestory center/) collects interview transcripts that record the life stories of people of diverse ages and backgrounds. As of March 2011, their collection included transcripts from over 400 life stories, representing more than 35 different ethnic groups, experiences of historical events ranging from the Great Depression to the Vietnam War, and including reports on dealing with health problems such as HIV/AIDS. These qualitative data are available directly online without any registration or fee.

There are many other readily available sources, including administrative data from hospitals, employers, and other organizations; institutional research data from university offices that collect such data; records of transactions from businesses; and data provided directly by university-based researchers (Hakim 1982:6).

▣ Challenges for Secondary Data Analyses

The use of the method of secondary data analysis has the following clear advantages for social researchers (Rew et al. 2000:226):

- It allows analyses of social processes in other inaccessible settings.

- It saves time and money.

- It allows the researcher to avoid data collection problems.

- It facilitates comparison with other samples.

- It may allow inclusion of many more variables and a more diverse sample than otherwise would be feasible.

- It may allow data from multiple studies to be combined.

The secondary data analyst also faces some unique challenges. The easy availability of data for secondary analysis should not obscure the fundamental differences between a secondary and a primary analysis of social science data. In fact, a researcher who can easily acquire secondary data may be tempted to minimize the limitations of the methods used to collect the data as well as insufficient correspondence between the measures in the data set and the research questions that the secondary analyst wants to answer.

So, the greatest challenge faced in secondary data analysis results from the researcher's inability to design data collection methods that are best suited to answer his or her research question. The secondary data analyst also cannot test and refine the methods to be used on the basis of preliminary feedback from the population or processes to be studied. Nor is it possible for the secondary data analyst to engage in the iterative process of making observations, developing concepts, or making more observations and refining the concepts. This last problem is a special challenge for those seeking to conduct secondary analyses of qualitative data, since an inductive process of developing research questions and refining observation and interview strategies is a hallmark of much qualitative methodology (Heaton 2008:511).

These limitations mean that it may not be possible for a secondary data analyst to focus on the specific research question of original interest or to use the most appropriate sampling or measurement approach for studying that research question. Secondary data analysis inevitably involves a trade-off between the ease with which the research process can be initiated and the specific hypotheses that can be tested and methods that can be used. If the primary study was not designed to measure adequately a concept that is critical to the secondary analyst's hypothesis, the study may have to be abandoned until a more adequate source of data can be found. Alternatively, hypotheses, or even the research question itself, may be modified to match the analytic possibilities presented by the available data (Riedel 2000:53).

Data quality is always a concern with secondary data, even when the data are collected by an official government agency. Government actions result, at least in part, from political processes that may not have as their first priority the design or maintenance of high-quality data for social scientific analysis. For example, political opposition over the British Census's approach to recording ethnic origin led to changes in the 1991 Census that rendered its results inconsistent with prior years and that demonstrated the "tenuous relationship between enumeration [Census] categories and possible social realities" (Fenton 1996:155).

It makes sense to use official records to study the treatment of juveniles accused of illegal acts because these records document the critical decisions to arrest, to convict, or to release (Dannefer & Schutt 1982). But research based on official records can be only as good as the records themselves. In contrast to the controlled interview process in a research study, there is little guarantee that the officials' acts and decisions were recorded in a careful and unbiased manner. The same is true for data collected by employees of private and nonprofit organizations. For example, research on the quality of hospital records has created, at best, mixed support for the validity of the key information they contain (Iezzoni 1997:391).

This one example certainly does not call into question all legal records or all other types of official records. It does, however, highlight the value of using multiple methods, particularly when the primary method of data collection is analysis of records generated by **street-level bureaucrats**—officials who serve clients and have a high degree of discretion (Lipsky 1980). When officials make decisions and record the bases for their decisions without much supervision, records may diverge considerably from the decisions they are supposed to reflect.

> **Street-level bureaucrats** Officials who serve clients and have a high degree of discretion.

More generally, it is always important to learn how people make sense of the social world when we want to describe their circumstances and explain their behavior (Needleman 1981).

The basis for concern is much greater in research across national boundaries, because different data collection systems and definitions of key variables may have been used (Glover 1996). Census counts can be distorted by incorrect answers to census questions as well as by inadequate coverage of the entire population (Rives & Serow 1988:32–35). National differences in the division of labor between genders within households can confuse the picture when comparing household earnings between nations without taking these differences into account (Jarvis 1997:521).

Reanalyzing qualitative data someone else collected also requires setting aside the expectation that qualitative research procedures and interpretations will be informed by intimate familiarity with the context in which the data were collected and with those from whom the data were obtained (Heaton 2008:511). Instead, the secondary analyst of qualitative data must seek opportunities for carrying on a dialogue with the original researchers.

Many of these problems can be lessened by seeking conscientiously to review data features and quality before deciding to develop an analysis of secondary data (Riedel 2000:55–69; Stewart & Kamins 1993:17–31) and then developing analysis plans that maximize the value of the available data. Replicating key analyses with alternative indicators of key concepts, testing for the stability of relationships across theoretically meaningful subsets of the data, and examining findings of comparable studies conducted with other data sets can each strengthen confidence in the findings of a secondary analysis.

Any secondary analysis will improve if the analyst—yourself or the author of the work that you are reviewing—answers several questions before deciding to develop an analysis of secondary data in the first place and then continues to develop these answers as the analysis proceeds (adapted from Riedel 2000:55–69; Stewart & Kamins 1993:17–31):

1. What were the agency's or researcher's goals in collecting the data?

The goals of the researcher and/or research and/or research sponsor influence every step in the process of designing a research project, analyzing the resulting data, and reporting the results. Some of these goals will be stated quite explicitly while others may only be implicit—reflected in the decisions made but not acknowledged in the research report or other publications. When you consider whether to use a data set for a secondary analysis, you should consider whether your own research goals are similar to those of the original investigator and sponsor. The data collected are more likely to include what is necessary for achieving your own research goals if the original investigator and/or sponsor had similar goals. When your research question or other goals diverge from those of the original investigator, you should consider how this divergence may have affected the course of the primary research project and whether this affects your ability to use the resulting data for a different purpose.

For example, Pamela Paxton (2002) studied the role of secondary organizations in democratic politics in a sample of 101 countries, but found that she could only measure the prevalence of international nongovernmental associations (INGOs) because comparable figures on purely national associations were not available. She cautioned that "INGOs represent only a specialized subset of all the associations present in a country" (Paxton 2002:261). We need to take this limitation into account when interpreting the results of her secondary analysis.

2. What data were collected, and what were they intended to measure?

You should develop a clear description of how data enter the data collection system, for what purpose, and how cases leave the system and why. Try to obtain the guidelines that agency personnel are supposed to follow in processing cases. Have there been any changes in these procedures during the period of investigation (Riedel 2000:57–64)?

3. When was the information collected?

Both historical and comparative analyses can be affected. For example, the percentage of the U.S. population not counted in the U.S. Census appears to have declined since 1880 from about 7% to 1%, but undercounting continues to be more common among poorer urban dwellers and recent immigrants (King & Magnuson 1995; see also Chapter 5). The relatively successful 2000 U.S. Census reduced undercounting (Forero 2000b) but still suffered from accusations of shoddy data collection procedures in some areas (Forero 2000a).

4. What methods were used for data collection? Who was responsible for data collection, and what were their qualifications? Are they available to answer questions about the data? Each step in the data collection process should be charted and the involved personnel identified. The need for concern is much greater in research across national boundaries, because different data collection systems and definitions of key variables may have been used (Glover 1996). Incorrect answers to census questions as well as inadequate coverage of the entire population can distort census counts (see Chapter 5; Rives & Serow 1988:32–35). Copies of the forms used for data collection should be obtained, specific measures should be inspected, and the ways in which these data are processed by the agency/agencies should be reviewed.

5. How is the information organized (by date, event, etc.)? Are there identifiers that are used to identify the different types of data available (computer tapes, disks, paper files) (Riedel 2000:58–61)?

Answers to these questions can have a major bearing on the work that will be needed to carry out the study.

6. What is known about the success of the data collection effort? How are missing data indicated? What kind of documentation is available? How consistent are the data with data available from other sources?

The U.S. Census Bureau provides extensive documentation about data quality, including missing data, and it also documents the efforts it makes to improve data quality. The Census 2000 Testing, Experimentation, and Evaluation Program was designed to improve the next decennial census in 2010, as well as other Census Bureau censuses and surveys. This is an ongoing effort, since 1950, with tests of questionnaire design and other issues. You can read more about it at www.census.gov/pred/www/Intro.htm.

Answering these questions helps ensure that the researcher is familiar with the data he or she will analyze and can help identify any problems with it. It is unlikely that you or any secondary data analyst will be able to develop complete answers to all these questions prior to starting an analysis, but it still is critical to make the attempt to assess what you know and don't know about data quality before deciding whether to conduct the analysis. If you uncover bases for real concern after checking documents, the other publications with the data, information on websites, and perhaps by making some phone calls, you may have to decide to reject the analytic plan and instead search for another data set. If your initial answers to these six questions give sufficient evidence that the data can reasonably be used to answer your research question, you should still keep seeking to add in missing gaps in your initial answers to the six questions; through this ongoing process, you will develop the fullest possible understanding of the quality of your data. This understanding can lead you to steer your analysis in the most productive directions and can help you write a convincing description of the data set's advantages and limitations.

This seems like a lot to ask, doesn't it? After all, you can be married for life after answering only one question; here, I'm encouraging you to attempt to answer six questions before committing yourself to a brief relationship with a data set. Fortunately, the task is not normally so daunting. If you acquire a data set for analysis from a trusted source, many of these questions will already have been answered for you. You may need to do no more than read through a description of data available on a website to answer the secondary data questions and consider yourself prepared to use the data for your own purposes. If you are going to be conducting major analyses of a data set, you should take more time to read the complete study documents, review other publications with the data, and learn about the researchers who collected the data.

Exhibit 13.8 contains the description of a data set available from the ICPSR. Read through it and see how many of the secondary data questions it answers.

You will quickly learn that this data set represents one survey conducted as part of the ongoing Detroit Area Studies, so you'll understand the data set better if you also read a general description of that survey project (Exhibit 13.9).

Exhibit 13.8 **ICPSR Data Set Description**

Description—Study No. 4120

Bibliographic Description

ICPSR Study No.:	4120
Title:	Detroit Area Study, 1997: Social Change in Religion and Child Rearing
Principal Investigator(s):	Duane Alwin, University of Michigan
Series:	*Detroit Area Studies Series*
Bibliographic Citation:	Alwin, Duane. DETROIT AREA STUDY, 1997: SOCIAL CHANGE IN RELIGION AND CHILD REARING [Computer file]. ICPSR04120-v1. Ann Arbor, MI: Detroit Area Studies [producer], 1997. Ann Arbor, MI: Inter-university Consortium for Political and Social Research [distributor], 2005-06-02.
Scope of Study *Summary:*	For this survey, respondents from three counties in the Detroit, Michigan, area were queried about their work, health, marriage and family, finances, political views, religion, and child rearing. With respect to finances, respondent views were elicited on credit card purchases, recording expenditures, and savings and investments. Regarding political views, respondents were. . . .
Subject Term(s):	*abortion, Atheism, Bible, birth control, Catholicism, Catholics, child rearing, children, Christianity, church attendance, communism, Creationism, credit card use, divorce, drinking behavior, economic behavior, educational background, employment, ethnicity, families, . . .*
Geographic Coverage:	Detroit, Michigan, United States
Time Period:	1997
Date(s) of Collection:	1997
Universe:	Residents 21 years and older in the tri-county area (Wayne, Oakland, and Macomb) of Michigan.
Data Type:	survey data
Methodology *Sample:*	A random-digit dialing sample of residential telephone numbers in the Michigan counties of Wayne, Oakland, and Macomb. The sample was restricted to adults 21 years of age and older.
Mode of Data Collection:	telephone interview
Extent of Processing:	CDBK.ICPSR/ DDEF.ICPSR/ REFORM.DATA
Access & Availability *Extent of Collection:*	1 data file + machine-readable documentation (PDF) + SAS setup file + SPSS setup file + Stata setup file
Data Format:	Logical Record Length with SAS, SPSS, and Stata setup files, SPSS portable file, and Stata system file
Original ICPSR Release:	2005-06-02

Note: Detailed file-level information (such as LRECL, case count, and variable count) may be found in the *file manifest.*

In an environment in which so many important social science data sets are instantly available for reanalysis, the method of secondary data analysis should permit increasingly rapid refinement of social science knowledge, as new hypotheses can be tested and methodological disputes clarified if not resolved quickly. Both the necessary technology and the supportive ideologies required for this rapid refinement

Exhibit 13.9 ICPSR Description of Detroit Area Studies

Detroit Area Studies Series

- View studies in the series
- Related Literature

The Detroit Area Studies series was initiated in 1951 at the University of Michigan and has been carried out nearly every year till the present. The Department of Sociology and the Survey Research Center of the Institute for Social Research are associated with the development of the series. It was initially supported by funds from the Ford Foundation, but since 1988 the University of Michigan has provided primary financial support for the series, with supplemental funding obtained frequently from outside sources. The purpose of these surveys is to provide practical social research training for graduate students and reliable data on the Greater Detroit community. Each survey probes a different aspect of personal and public life, economic and political behavior, political attitudes, professional and family life, and living experiences in the Detroit metropolitan area. The different specific problems investigated each year are selected by the executive committee of the project.

have spread throughout the world. Social science researchers now have the opportunity to take advantage of this methodology as well as the responsibility to carefully and publicly delineate and acknowledge the limitations of the method.

Content Analysis

How are medical doctors regarded in American culture? Do newspapers use the term *schizophrenia* in a way that reflects what this serious mental illness actually involves? Does the portrayal of men and women in video games reinforce gender stereotypes? Are the body images of male and female college students related to their experiences with romantic love? If you are concerned with understanding culture, attitudes toward mental illness, or gender roles, you'll probably find these to be important research questions. You now know that you could probably find data about each of these issues for a secondary data analysis, but in this section, I would like to introduce procedures for analyzing a different type of data that awaits the enterprising social researcher. Content analysis is "the systematic, objective, quantitative analysis of message characteristics" and is a method particularly well suited to the study of popular culture and many other issues concerning human communication (Neuendorf 2002:1).

The goal of content analysis is to develop inferences from human communication in any of its forms, including books, articles, magazines, songs, films, and speeches (Weber 1990:9). You can think of content analysis as a "survey" of some documents or other records of communication—a survey with fixed-choice responses that produce quantitative data. This method was first applied to the study of newspaper and film content and then developed systematically for the analysis of Nazi propaganda broadcasts in World War II. Since then, content analysis has been used to study historical documents, records of speeches, and other "voices from the past" as well as media of all sorts (Neuendorf 2002:31–37). The same techniques can now be used to analyze blog sites, wikis, and other text posted on the Internet (Gaiser & Schreiner 2009:81–90). Content analysis techniques are also used to analyze responses to open-ended survey questions.

Research in the News

GOOGLE'S PROJECT OXYGEN TRIES TO IMPROVE MANAGEMENT PERFORMANCE

Google content analyzed more than 10,000 observations about managers gleaned from performance reviews, feedback surveys, and award nominations. Comments were coded in terms of more than 100 variables and then statistically analyzed to identify patterns. Good management behaviors identified included being a good coach and empowering the team and not micromanaging them, while pitfalls included spending too little time managing and communicating. Providing training based on these findings resulted in significant improvements in manager quality for 75% of Google's worst-performing managers.

Source: Bryant, Adam. 2011. "The Quest to Build a Better Boss." *The New York Times,* March 13: BU1, BU8.

In the News

Content analysis bears some similarities to qualitative data analysis, because it involves coding and categorizing text and identifying relationships among constructs identified in the text. However, since it usually is conceived as a quantitative procedure, content analysis overlaps with qualitative data analysis only at the margins—the points where qualitative analysis takes on quantitative features or where content analysis focuses on qualitative features of the text. This distinction becomes fuzzy, however, because content analysis techniques can be used with all forms of messages, including visual images, sounds, and interaction patterns, as well as written text (Neuendorf 2002:24–25).

Kimberly Neuendorf's (2002:3) analysis of medical prime-time network television programming introduces the potential of content analysis. As Exhibit 13.10 shows, medical programming has been dominated by noncomedy shows, but there have been two significant periods of comedy medical shows—during the 1970s and early 1980s and then again in the early 1990s. It took a qualitative analysis of medical show content to reveal that the 1960s shows represented a very distinct "physician-as-God" era, which shifted to a more human view of the medical profession in the 1970s and 1980s. This era has been followed, in turn, by a mixed period that has had no dominant theme.

Content analysis is useful for investigating many questions about the social world. To illustrate its diverse range of applications, I will use in the next sections Neuendorf's (2002) analysis of TV programming, Kenneth Duckworth et al.'s (2003) (and my) analysis of newspaper articles, Karen Dill and Kathryn Thill's (2007) analysis of video game characters, and Suman Ambwani and Jaine Strauss's (2007) analysis of student responses to open-ended survey questions. These examples will demonstrate that the units that are "surveyed" in a content analysis can range from newspapers, books, or TV shows to persons referred to in other communications, themes expressed in documents, or propositions made in different statements.

Content analysis proceeds through several stages.

Identify a Population of Documents or Other Textual Sources

This population should be selected so that it is appropriate to the research question of interest. Perhaps the population will be all newspapers published in the United States, college student newspapers, nomination speeches at political party conventions, or "state of the nation" speeches by national leaders. Books or films are

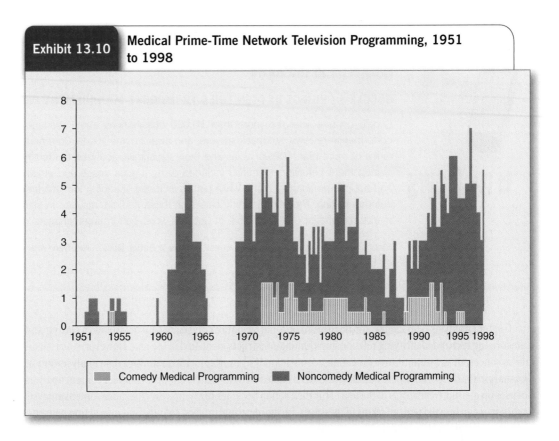

Exhibit 13.10 Medical Prime-Time Network Television Programming, 1951 to 1998

Comedy Medical Programming   Noncomedy Medical Programming

also common sources for content analysis projects. Often, a comprehensive archive can provide the primary data for the analysis (Neuendorf 2002:76–77). For a fee, the LexisNexis service makes a large archive of newspapers available for analysis. For her analysis of prime-time programming since 1951, Neuendorf (2002:3–4) used a published catalog of all TV shows. For my analysis with Duckworth and others (2003:1402) of newspapers' use of the terms *schizophrenia* and *cancer,* I requested a sample of articles from the LexisNexis national newspaper archive. Dill and Thill (2007:855–856) turned to video game magazines for their analysis of the depiction of gender roles in video games. For their analysis of gender differences in body image and romantic love, Ambwani and Strauss (2007:15) surveyed students at a small midwestern liberal arts college.

Determine the Units of Analysis

These could be items such as newspaper articles, whole newspapers, speeches, or political conventions, or they could be more microscopic units such as words, interactions, time periods, or other bits of a communication (Neuendorf 2002:71). The content analyst has to decide what units are most appropriate to the research question and how the communication content can be broken up into those units. If the units are individual issues of a newspaper, in a study of changes in news emphases, this step may be relatively easy. However, if the units are most appropriately the instances of interaction between characters in a novel or a movie, in a study of conflict patterns between different types of characters, it will require a careful process of testing to determine how to define operationally the specific units of interaction (Weber 1990:39–40).

Units of analysis varied across the four content analysis projects I have introduced. The units of analysis for Neuendorf (2002:2) were "the individual medically oriented TV program"; for Duckworth et al. (2003:1403), they were newspaper articles; for Dill and Thill (2007:856), they were images appearing in magazine articles; while for Ambwani and Strauss (2007:15), they were individual students.

Select a Sample of Units From the Population

The simplest strategy might be a simple random sample of documents. However, a stratified sample might be needed to ensure adequate representation of community newspapers in large and in small cities, or of weekday and Sunday papers, or of political speeches during election years and in off years (see Chapter 4) (Weber 1990:40–43). Nonrandom sampling methods have also been used in content analyses when the entire population of interest could not be determined (Neuendorf 2002:87–88).

The selected samples in our four content analysis projects were diverse. In fact, Neuendorf (2002:2) included the entire population of medically oriented TV programs between 1951 and 1998. For my content analysis with Ken Duckworth (Duckworth et al. 2003), I had my student, Chris Gillespie, draw a stratified random sample of 1,802 articles published

> in the five U.S. newspapers with the highest daily circulation in 1996 to 1997 in each of the four regions identified in the LexisNexis database, as well as the two high-circulation national papers in the database, *The New York Times* and *USA Today.* (pp. 1402–1403)

Because individual articles cannot be sampled directly in the LexisNexis database, a random sample of days was drawn first. All articles using the terms *schizophrenia* or *cancer* (or several variants of these terms) were then selected from the chosen newspapers on these days. Dill and Thill (2007:855–856) used all images in the current issues (as of January 2006) of the six most popular video game magazines sold on Amazon.com. Ambwani and Strauss (2007:15) used an availability sampling strategy, with 220 students from Introductory Psychology and a variety of other sources.

Design Coding Procedures for the Variables to Be Measured

This requires deciding what variables to measure, using the unit of text to be coded such as words, sentences, themes, or paragraphs. Then, the categories into which the text units are to be coded must be defined. These categories may be broad such as *supports democracy* or narrow such as *supports universal suffrage.* Reading or otherwise reviewing some of the documents or other units to be coded is an essential step in thinking about variables that should be coded and in developing coding procedures. Development of clear instructions and careful training of coders is essential.

As an example, Exhibit 13.11 is a segment of the coding form that I developed for a content analysis of union literature that I collected during a mixed-methods study of union political processes (Schutt 1986). My sample was of 362 documents: all union newspapers and a stratified sample of union leaflets given to members during the years of my investigation. My coding scheme included measures of the source and target for the communication, as well as measures of concepts that my theoretical framework indicated were important in organizational development: types of goals, tactics for achieving goals, organizational structure, and forms of participation. The analysis documented a decline in concern with client issues and an increase in focus on organizational structure, which were both trends that also emerged in interviews with union members.

Developing reliable and valid coding procedures deserves special attention in a content analysis, for it is not an easy task. The meaning of words and phrases is often ambiguous. Homographs create special problems (words such as *mine* that have different meanings in different contexts), as do many phrases that have special meanings (such as *point of no return*) (Weber 1990:29–30). As a result, coding procedures cannot simply categorize and count words; text segments in which the words are embedded must also be inspected before codes are finalized. Because different coders may perceive different meanings in the same text segments, explicit coding rules are required to ensure coding consistency. Special dictionaries can be developed to keep track of how the categories of interest are defined in the study (Weber 1990:23–29).

Exhibit 13.11 Union Literature Coding Form*

I. Preliminary Codes

1. Document # _____

2. Date _____

         mo   yr

3. Length of text _____ pp. (round up to next 1/4 page; count legal size as 1.25)

4. Literature Type

    1. General leaflet for members/employees
    2. Newspaper/Newsletter article
    3. Rep Council motions
    4. Other material for Reps, Stewards, Delegates (e.g., budget, agenda)
    5. Activity reports of officers, President's Report
    6. Technical information-filing grievances, processing forms
    7. Buying plans/Travel packages
    8. Survey Forms, Limited Circulation material (correspondence)
    9. Non-Union
    10. Other _____ (specify)

4A. If newspaper article

Position

    1. Headline story
    2. Other front page
    3. Editorial
    4. Other

4B. If Rep Council motion

Sponsor

    1. Union leadership
    2. Office
    3. Leadership faction
    4. Opposition faction
    5. Other

5. Literature content-Special issues

    1. First strike (1966)
    2. Second strike (1967)
    3. Collective bargaining (1977)
    4. Collective bargaining (1979)
    5. Election/campaign literature
    6. Affiliation with AFSCME/SEIU/other national union
    7. Other

II. Source and Target

6. Primary source (code in terms of those who prepared this literature for distribution).

    1. Union-newspaper (Common Sense; IUPAE News)
    2. Union-newsletter (Info and IUPAE Bulletin)
    3. Union-unsigned
    4. Union officers
    5. Union committee
    6. Union faction (the Caucus; Rank-and-Filers; Contract Action, other election slate; PLP News; Black Facts)
    7. Union members in a specific work location/office
    8. Union members-other
    9. Dept. of Public Aid/Personnel

10. DVR/DORS
11. Credit Union
12. Am. Buyers' Assoc.
13. Other nonunion

7. Secondary source (use for lit. at least in part reprinted from another source, for distribution to members)

1. Newspaper-general circulation
2. Literature of other unions, organizations
3. Correspondence of union leaders
4. Correspondence from DPA/DVR-DORS/Personnel
5. Correspondence from national union
6. Press release
7. Credit Union, Am. Buyers'
8. Other _____ (specify)
9. None

8. Primary target (the audience for which the literature is distributed)

1. Employees-general (if mass-produced and unless otherwise stated)
2. Employees-DVR/DORS
3. Union members (if refers only to members or if about union elections)
4. Union stewards, reps, delegates committee
5. Non-unionized employees (recruitment lit, etc.)
6. Other _____ (specify)
7. Unclear

III. *Issues*

A. Goal

B. Employee conditions/benefits (Circle up to 5)

1. Criteria for hiring
2. Promotion
3. Work out of Classification, Upgrading
4. Step increases
5. Cost-of-living, pay raise, overtime pay, "money"
6. Layoffs (nondisciplinary); position cuts
7. Workloads, Redeterminations, "30 for 40," GA Review
8. Office physical conditions, safety
9. Performance evaluations
10. Length of workday
11. Sick benefits/leave—holidays, insurance, illness, vacation, voting time
12. Educational leave
13. Grievances—change in procedures
14. Discrimination (race, sex, age, religion, national origin)
15. Discipline—political (union-related)
16. Discipline—performance, other
17. Procedures with clients, at work
18. Quality of work, "worthwhile jobs"—other than relations with clients

*Coding instruction available from author.

After coding procedures are developed, their reliability should be assessed by comparing different coders' codes for the same variables. Computer programs for content analysis can enhance reliability by facilitating the consistent application of text-coding rules (Weber 1990:24–28). Validity can be assessed with a construct validation approach by determining the extent to which theoretically predicted relationships occur (see Chapter 4).

Neuendorf's (2002:2) analysis of medical programming measured two variables that did not need explicit coding rules: length of show in minutes and the year(s) the program was aired. She also coded shows as comedies or noncomedies, as well as medical or not, but she does not report the coding rules for these distinctions. We provided a detailed description of coding procedures in our analysis of newspaper articles that used the terms *schizophrenia* or *cancer* (Duckworth et al. 2003). This description also mentions our use of a computerized text-analysis program and procedures for establishing measurement reliability.

> Content coding was based on each sentence in which the key term was used. Review of the full text of an article resolved any potential ambiguities in proper assignment of codes. Key terms were coded into one of eight categories: metaphor, obituary, first person or human interest, medical news, prevention or education, incidental, medically inappropriate, and charitable foundation. Fifty-seven of the 913 articles that mentioned schizophrenia, but none of those that mentioned cancer, were too ambiguous to be coded into one of these defined categories and so were excluded from final comparisons. Coding was performed by a trained graduate assistant using QSR's NUD*IST program. In questionable cases, final decisions were made by consensus of the two psychiatrist coauthors. A random subsample of 100 articles was also coded by two psychiatry residents, who, although blinded to our findings, assigned the same primary codes to 95 percent of the articles. (p. 1403)

Dill and Thill (2007) used two coders and a careful training procedure for their analysis of the magazine images about video games.

> One male and one female rater, both undergraduate psychology majors, practiced on images from magazines similar to those used in the current investigation. Raters discussed these practice ratings with each other and with the first author until they showed evidence of properly applying the coding scheme for all variables. Progress was also checked part way through the coding process, as suggested by Cowan (2002). Specifically, the coding scheme was re-taught by the first author, and the two raters privately discussed discrepancies and then independently assessed their judgments about their ratings of the discrepant items. They did not resolve discrepancies, but simply reconsidered their own ratings in light of the coding scheme refresher session. Cowan (2002) reports that this practice of reevaluating ratings criteria is of particular value when coding large amounts of violent and sexual material because, as with viewers, coders suffer from desensitization effects. (p. 856)

Ambwani and Strauss (2007) also designed a careful training process to achieve acceptable levels of reliability before their raters coded the written answers to their open-ended survey questions.

> We developed a coding scheme derived from common themes in the qualitative responses; descriptions of these themes appear in the Appendix. We independently developed lists of possible coding themes by reviewing the responses and then came to a consensus regarding the most frequently emerging themes. Next, a team of four independent raters was trained in the coding scheme through oral, written, and group instruction by the first author. The raters first coded a set of hypothetical responses; once the raters reached an acceptable level of agreement on the coding for the hypothetical responses, they began to code the actual data. These strategies, recommended by Orwin (1994), were

employed to reduce error in the data coding. All four independent coders were blind to the hypotheses. Each qualitative response could be coded for multiple themes. Thus, for example, a participant who described the influence of body image in mate selection and quality of sexual relations received positive codes for both *choice* and *quality of sex*. (p. 16)

Develop Appropriate Statistical Analyses

The content analyst creates variables for analysis by counting occurrences of particular words, themes, or phrases and then tests relations between the resulting variables. These analyses could use some of the statistics that will be introduced in Chapter 14, including frequency distributions, measures of central tendency and variation, cross-tabulations, and correlation analysis (Weber 1990:58–63). Computer-aided qualitative analysis programs, like those you learned about in Chapter 10 and like the one I selected for the preceding newspaper article analysis, can help, in many cases, develop coding procedures and then to carry out the content coding.

The simple chart that Neuendorf (2002:3) used to analyze the frequency of medical programming appears in Exhibit 13.10. Duckworth et al.'s (2003) primary analysis was simply a comparison of percentages showing that 28% of the articles mentioning schizophrenia used it as a metaphor, compared with only 1% of the articles mentioning cancer. We also presented examples of the text that had been coded into different categories. For example, *the nation's schizophrenic perspective on drugs* was the type of phrase coded as a metaphorical use of the term *schizophrenia* (p. 1403). Dill and Thill (2007:858) presented percentages and other statistics that showed that, among other differences, female characters were much more likely to be portrayed in sexualized ways in video game images than were male characters. Ambwani and Strauss (2007:16) used other statistics that showed that body esteem and romantic love experiences are related, particularly for women. They also examined the original written comments and found further evidence for this relationship. For example, one woman wrote, "[My current boyfriend] taught me to love my body. Now I see myself through his eyes, and I feel beautiful" (p. 17).

The criteria for judging quantitative content analyses of text are the same standards of validity applied to data collected with other quantitative methods. We must review the sampling approach, the reliability and validity of the measures, and the controls used to strengthen any causal conclusions.

The various steps in a content analysis are represented in the flowchart in Exhibit 13.12. Note that the steps are comparable to the procedures in quantitative survey research. Use this flowchart as a checklist when you design or critique a content analysis project.

🔲 Ethical Issues in Secondary Data Analysis and Content Analysis

Analysis of data collected by others, as well as content analysis of text, does not create the same potential for harm as does the collection of primary data, but neither ethical nor related political considerations can be ignored. First and foremost, because in most cases the secondary researchers did not collect the data, a key ethical obligation is to cite the original, principal investigators, as well as the data source, such as the ICPSR. Researchers who seek access to data sets available through the CESSDA (Center for European Social Science Data Archives) must often submit a request to the national data protection authority in the country (or countries) of interest (Johnson & Bullock 2009:214).

Exhibit 13.12 **Flowchart for the Typical Process of Content Analysis Research**

1. *Theory and rationale:* What content will be examined, and *why*? Are there certain *theories* or perspectives that indicate that this particular message content is important to study? Library work is needed here to conduct a good literature review. Will you be using an integrative model, linking content analysis with other data to show relationships with source or receiver characteristics? Do you have *research questions? Hypotheses?*

2. *Conceptualizations:* What *variables* will be used in the study, and how do you define them *conceptually* (i.e., with dictionary-type definitions)? Remember, you are the boss! There are many ways to define a given construct, and there is no one right way. You may want to screen some examples of the content you're going to analyze, to make sure you've covered everything you want.

3. *Operationalizations (measures):* Your measures should match your conceptualizations . . . What *unit of data collection* will you use? You may have more than one unit (e.g., a by-utterance coding scheme and a by-speaker coding scheme). Are the variables measured well (i.e., at a high *level of measurement,* with categories that are *exhaustive and mutually exclusive)?* An *a priori* coding scheme describing all measures must be created. Both face validity and content validity may also be assessed at this point.

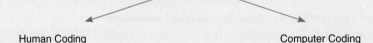

Human Coding   Computer Coding

4a. *Coding schemes:* You need to create the following materials:

  a. *Codebook* (with all variable measures *fully* explained)

  b. *Coding form*

4b. *Coding schemes:* With computer text content analysis, you still need a codebook of sorts—a full explanation of your *dictionaries* and method of applying them. You may use standard dictionaries (e.g., those in Hart's program, *Diction*) or originally created dictionaries. When creating custom dictionaries, be sure to first generate a frequencies list from your text sample and examine for key words and phrases.

Human Coding   Computer Coding

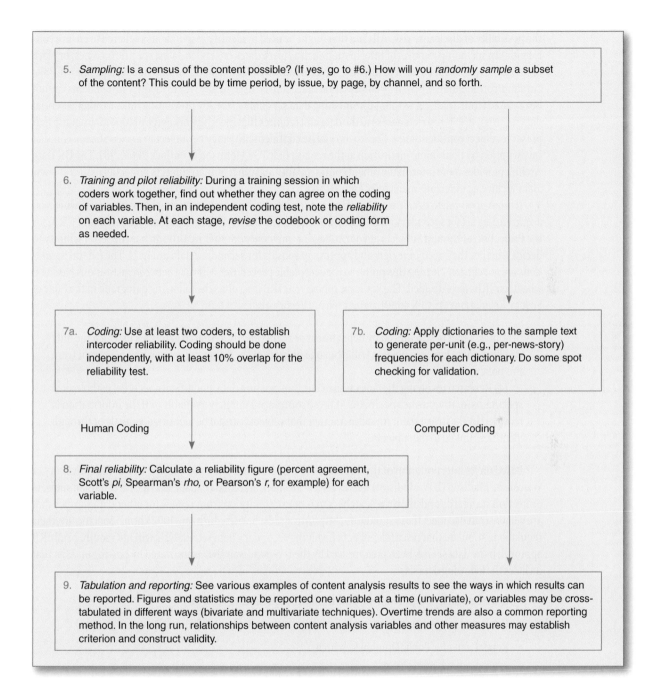

5. *Sampling:* Is a census of the content possible? (If yes, go to #6.) How will you *randomly sample* a subset of the content? This could be by time period, by issue, by page, by channel, and so forth.

6. *Training and pilot reliability:* During a training session in which coders work together, find out whether they can agree on the coding of variables. Then, in an independent coding test, note the *reliability* on each variable. At each stage, *revise* the codebook or coding form as needed.

7a. *Coding:* Use at least two coders, to establish intercoder reliability. Coding should be done independently, with at least 10% overlap for the reliability test.

7b. *Coding:* Apply dictionaries to the sample text to generate per-unit (e.g., per-news-story) frequencies for each dictionary. Do some spot checking for validation.

Human Coding

Computer Coding

8. *Final reliability:* Calculate a reliability figure (percent agreement, Scott's *pi*, Spearman's *rho,* or Pearson's *r,* for example) for each variable.

9. *Tabulation and reporting:* See various examples of content analysis results to see the ways in which results can be reported. Figures and statistics may be reported one variable at a time (univariate), or variables may be cross-tabulated in different ways (bivariate and multivariate techniques). Overtime trends are also a common reporting method. In the long run, relationships between content analysis variables and other measures may establish criterion and construct validity.

Subject confidentiality is a key concern when original records are analyzed. Whenever possible, all information that could identify individuals should be removed from the records to be analyzed so that no link is possible to the identities of living subjects or the living descendants of subjects (Huston & Naylor 1996:1698). When you use data that have already been archived, you need to find out what procedures were used to preserve subject confidentiality. The work required to ensure subject confidentiality probably will have been done for you by the data archivist. For example, the ICPSR examines carefully all data deposited in the archive for the possibility of disclosure risk. All data that might be used to identify respondents is altered to ensure confidentiality, including removal of information such as birth dates or service dates, specific incomes, or place of residence that could be used to identify subjects indirectly (see www.icpsr.umich.edu/icpsrweb/ICPSR/curation/confidentiality.jsp). If all information that could be used in any way to identify respondents cannot be removed from a data set without diminishing data set quality (e.g., by preventing links to other essential data records), ICPSR restricts access to the data and requires that investigators agree to conditions of use that preserve subject confidentiality. Those who violate confidentiality may be subject to a scientific misconduct investigation by their home institution at the request of ICPSR (Johnson & Bullock 2009:218). The UK Data Archive provides more information about confidentiality and other human subjects protection issues at www.data-archive.ac.uk/create-manage/consent-ethics.

It is not up to you to decide whether there are any issues of concern regarding human subjects when you acquire a data set for secondary analysis from a responsible source. The Institutional Review Board (IRB) for the Protection of Human Subjects at your college or university or other institution has the responsibility to decide whether they need to review and approve proposals for secondary data analysis. The federal regulations are not entirely clear on this point, so the acceptable procedures will vary between institutions based on what their IRBs have decided. The key issue is how your IRB interprets the following paragraph in the Code of Federal Regulations (45 CFR 46.101, paragraph b, category 4):

> (b) Unless otherwise required by department or agency heads, research activities in which the only involvement of human subjects will be in one or more of the following categories are exempt from this policy: . . .
>
> (4) Research involving the collection or study of existing data, documents, records, pathological specimens, or diagnostic specimens, if these sources are publicly available or if the information is recorded by the investigator in such a manner that subjects cannot be identified, directly or through identifiers linked to the subjects.

Based on its interpretation of this section in the federal code, for example, the IRB at the University of Wisconsin-Madison (UW-Madison) waives review for specific secondary data sets that meet human subjects protection standards and allows researchers to request exemption from review for other data sources that are demonstrated to meet these standards (http://my.gradsch.wisc.edu/hrpp/10023.htm). Specifically, their regulations stipulate that research projects involving secondary data set analysis will *not* require prior IRB approval, if the data set has been preapproved by the UW-Madison IRB as indicated by posting on a list that includes the following data sets:

- Inter-University Consortium for Political and Social Research (ICPSR)
- University of Wisconsin Data and Information Services Center (DISC)
- Roper Center for Public Opinion Research
- U.S. Census Bureau
- National Center for Health Statistics
- National Center for Education Statistics
- National Election Studies

Data sets that may qualify for inclusion on UW-Madison's list of approved data sources include the following:

- Public use data sets posted on the Internet that include a responsible use statement or other confidentiality agreement for authors to protect human subjects (e.g., see the ICPSR's responsible use statement).

- Survey data distributed by UW principal investigators who can certify that (1) the data collection procedures were approved by a qualified IRB meeting the Common Rule criteria for IRB and (2) the data set and documentation as distributed do not contain information that could be used to identify individual research participants.

Note: Research projects that merge more than one data set in such a way that individuals may be identified are not covered by this policy, and require prior IRB approval.

Data quality is always a concern with secondary data, even when the data are collected by an official government agency. Researchers who rely on secondary data inevitably make trade-offs between their ability to use a particular data set and the specific hypotheses they can test. If a concept that is critical to a hypothesis was not measured adequately in a secondary data source, the study might have to be abandoned until a more adequate source of data can be found. Alternatively, hypotheses or even the research question itself may be modified to match the analytic possibilities presented by the available data (Riedel 2000:53).

Political concerns intersect with ethical practice in secondary data analyses. How are *race* and *ethnicity* coded in the U.S. Census? You learned in Chapter 4 that changing conceptualizations of race have affected what questions are asked in the census to measure race. This data collection process reflects, in part, the influence of political interest groups, and it means that analysts using the census data must understand why the proportion of individuals choosing "other" as their race and the proportion in a "multiracial" category has changed. The same types of issues influence census and other government statistics collected in other countries. Britain's Census first asked about ethnic group in 1991. British researcher Steve Fenton (1996) reports that the design of the specific questions and categories used to measure ethnic groups "was clearly based on a conception of ethnic minorities as constituted by Black Caribbean and Asian populations" (p. 156). Respondents were asked to classify themselves only as White or Black (in several subcategories), Indian, Pakistani, Bangladeshi, Chinese, or "any other ethnic group."

Other concerns can be much greater in research across national boundaries, because different data collection systems and definitions of key variables may have been used (Glover 1996). Government funding decisions can affect the availability of government statistics on particular social issues (Levitas & Guy 1996:3). Census counts can be distorted by inadequate coverage of the entire population (see Chapter 5; Rives & Serow 1988:32–35). Social and political pressures may influence the success of a census in different ways in different countries. Some Mexicans were concerned that the results of Mexico's 2000 Census would be "used against them" by the government, and nearly 200,000 communities were inaccessible for follow-up except by a day's mule travel (Burke 2000). In rural China, many families who had flouted the government's official one-child policy sought to hide their "extra" children from census workers (Rosenthal 2000).

🔲 Conclusions

The easy availability for secondary analyses of data sets collected in thousands of social science investigations is one of the most exciting features of social science research in the 21st century. You can often find a previously collected data set that is suitable for testing new hypotheses or exploring new issues of interest. Moreover, the research infrastructure that has developed at ICPSR and other research consortia, both in the United States and internationally, ensures that a great many of these data sets have been carefully checked for quality and archived in a form that allows easy access.

Many social scientists now review available secondary data before they consider collecting new data with which to investigate a particular research question. Even if you do not leave this course with a plan to become a social scientist yourself, you should now have the knowledge and skills required to find and use secondary data or existing textual data to answer your own questions about the social world.

Key Terms

Content analysis   414

Secondary data   414

Secondary data analysis   414

Street-level bureaucrats   426

Highlights

- Secondary data analysts should have a good understanding of the research methods used to collect the data they analyze. Data quality is always a concern, particularly with historical data.

- Secondary data for historical and comparative research are available from many sources. The ICPSR provides the most comprehensive data archive.

- Content analysis is a tool for systematic quantitative analysis of documents and other textual data. It requires careful testing and control of coding procedures to achieve reliable measures.

STUDENT STUDY SITE

To assist in completing the web exercises, please access the study site at **www.sagepub.com/schuttisw7e,** where you will find the web exercises with accompanying links. You'll find other useful study materials such as self-quizzes and e-flashcards for each chapter, along with a group of carefully selected articles from research journals that illustrate the major concepts and techniques presented in the book.

Discussion Questions

1. What are the strengths and weaknesses of secondary data analysis? Do you think it's best to encourage researchers to try to address their research questions with secondary data if at all possible?

2. What are the similarities and differences between content analysis, secondary data analysis, and qualitative data analysis? Do you feel one of these three approaches is more likely to yield valid conclusions? Explain your answer.

3. In a world of limited resources and time constraints, should social researchers be required to include in their proposals to collect new data an explanation of why they cannot investigate their proposed research question with secondary data? Such a requirement might include a systematic review of data that already is available at ICPSR and other sources. Discuss the merits and demerits of such a requirement. If such a requirement were to be adopted, what specific rules would you recommend?

Practice Exercises

1. Using your library's government documents collection or the U.S. Census site on the web, select one report by the U.S. Census Bureau about the population of the United States or some segment of it. Outline the report and list all the tables included in it. Summarize the report in two paragraphs. Suggest a historical or comparative study for which this report would be useful.

2. Review the survey data sets available through the ICPSR, using their Internet site (www.icpsr.umich.edu/icpsrweb/ICPSR/). Select two data sets that might be used to study a research question in which you are interested. Use the information ICPSR reports about them to answer Questions 1 to 6 from the "Challenges for Secondary Data Analyses" section of this chapter. Is the information adequate to answer these questions? What are the advantages and disadvantages of using one of these data sets to answer your research question compared with designing a new study?

3. Select a current social or political topic that has been the focus of news articles. Propose a content analysis strategy for this topic, using newspaper articles or editorials as your units of analysis. Your strategy should include a definition of the population, selection of the units of analysis, a sampling plan, and coding procedures for key variables. Now find an article on this topic and use it to develop your coding procedures. Test and refine your coding procedures with another article on the same topic.

Ethics Questions

1. Reread the University of Wisconsin's IRB statement about secondary data analysis. Different policies about secondary analyses have been adopted by different IRBs. Do you agree with UW-Madison's policy? Would you recommend exempting all secondary analyses from IRB review, just some of them, or none of them? Explain your reasoning.

2. A controversy between the United States and Mexico over race erupted in 2005 when the Mexican government issued a stamp commemorating cartoon character Memín Pinguín, whose features were viewed in the United States as a racist caricature. The "Memingate Affair" drew attention to differences in the history

and understanding of the concept of race in Mexico and the United States. It also demonstrated the difficulty of applying coding rules and other content analysis procedures in different cultures. What procedures would you suggest that content analysts follow when developing research on other countries? What mistakes might they make by simply using the same coding procedures cross-culturally? You might reread the "Consider Translation" section of the survey chapter (Chapter 8) to get some ideas about how to deal with differences in meaning. You can read more about Mimingate at http://bostonreview.net/BR30.6/lomnitz.php.

Web Exercises

1. Explore the ICPSR website. Start by browsing their list of subject headings and then writing a short summary of the data sets available about one subject. You can start at www.icpsr.umich.edu/icpsrweb/ICPSR/

2. Try an online analysis. Go to one of the websites that offer online analysis of secondary data, like the ICPSR site.

Review the list of variables available for online analysis of one data set. Now form a hypothesis involving two of these variables, request frequency distributions for both, and generate the crosstab in percentage form to display their relationship.

SPSS Exercises

1. This is the time to let your social scientific imagination run wild, since any analysis of the GSS or ISSP data sets will qualify as a secondary data analysis. Review the list of variables in either the GSS or the ISSP, formulate one hypothesis involving two of these variables *that have no more than five categories,* and test the hypotheses using the cross-tabulation procedure.

(Specify the independent variable as the column variable in the selection window, specify the dependent variable as the row variable, and then choose the Option for your table of column percents. See if the percentage distribution of the dependent variable varies across the categories of the independent variable.)

Developing a Research Proposal

If you plan a secondary data analysis research project, you will have to revisit at least one of the decisions about research designs (Exhibit 2.14, #15).

1. Convert your proposal for research using survey data, in Chapter 8, into a proposal for a secondary analysis of survey data. Begin by identifying an appropriate data set available through ICPSR. Be sure to include a statement about the limitations of your approach, in which you note any differences between what you proposed to study initially and what you are actually able to study with the available data.

2. Identify a potential source of textual data that would be appropriate for a content analysis related to your research question. Develop a complete content analysis plan by answering each of the questions posed in Exhibit 13.12.

CHAPTER 14

Quantitative Data Analysis

T his chapter introduces several common statistics used in social research and highlights the factors that must be considered when using and interpreting statistics. Think of it as a review of fundamental social statistics, if you have already studied them, or as an introductory overview, if you have not. Two preliminary sections lay the foundation for studying statistics. In the first, I discuss the role of statistics in the research process, returning to themes and techniques with which you are already familiar. In the second preliminary section, I outline the process of preparing data for statistical analysis. In the rest of the chapter, I explain how to describe the distribution of single variables and the relationship between variables. Along the way, I address ethical issues related to data analysis. This chapter will have been successful if it encourages you to use statistics responsibly, to evaluate statistics critically, and to seek opportunities for extending your statistical knowledge.

Although many colleges and universities offer social statistics in a separate course, and for good reason (there's a *lot* to learn), I don't want you to think of this chapter as something that deals with a different topic than the rest of this book. Data analysis is an integral component of research methods, and it's important that any proposal for quantitative research include a plan for the data analysis that will follow data collection. (As you learned in Chapter 10, data analysis in qualitative projects often occurs in tandem with data collection.) You have to anticipate your data analysis needs if you expect your research design to secure the requisite data.

▣ Introducing Statistics

Statistics play a key role in achieving valid research results, in terms of measurement, causal validity, and generalizability. Some statistics are useful primarily to describe the results of measuring single variables and to construct and evaluate multi-item scales. These statistics include frequency distributions, graphs, measures of central tendency and variation, and reliability tests. Other statistics are useful primarily in achieving causal validity, by helping us describe the association among variables and to control for, or otherwise take account of, other variables. Cross-tabulation is the technique for measuring association and controlling other variables that is introduced in this chapter. All these statistics are termed **descriptive statistics** because they are used to describe the distribution of, and relationship among, variables.

> **Descriptive statistics** Statistics used to describe the distribution of, and relationship among, variables.

You have already learned in Chapter 5 that it is possible to estimate the degree of confidence that can be placed in generalization from a sample to the population from which the sample was selected. The statistics used in making these estimates are termed *inferential statistics*. In this chapter, I introduce the use of inferential statistics for testing hypotheses involving sample data.

Social theory and the results of prior research should guide our statistical choices, as they guide the choice of other research methods. There are so many particular statistics and so many ways for them to be used in data analysis that even the best statistician can be lost in a sea of numbers if he or she does not use prior research and theorizing to develop a coherent analysis plan. It is also important to choose statistics that are appropriate to the level of measurement of the variables to be analyzed. As you learned in Chapter 4, numbers used to represent the values of variables may not actually signify different quantities, meaning that many statistical techniques will be inapplicable for some variables.

Case Study: The Likelihood of Voting

In this chapter, I use for examples data from the 2010 General Social Survey (GSS) on voting and the variables associated with it, and I will focus on a research question about political participation: What influences the likelihood of voting? Prior research on voting in both national and local settings provides a great deal of support for one hypothesis: The likelihood of voting increases with social status (Manza, Brooks, & Sauder 2005:208; Milbrath & Goel 1977:9295; Salisbury 1975:326; Verba & Nie 1972:892). Research suggests that social status influences the likelihood of voting through the intervening variable of perceived political efficacy, or the feeling that one's vote matters (see Exhibit 14.1). But some research findings on political participation are inconsistent with the social status–voting hypothesis. For example, African Americans participate in politics at higher rates than do white Americans of similar social status—at least when there is an African American candidate for whom to vote (Manza et al. 2005:209; Verba & Nie 1972; Verba, Nie, & Kim 1978). This discrepant finding suggests that the impact of social status on voting and other forms of political participation varies with the social characteristics of potential participants.

The remarkable 2008 presidential primary season sharply reversed years of decline in turnout for presidential primaries (Exhibit 14.2) (Gans 2008), and the marked upsurge of participation among African Americans in the Democratic primaries supported prior research findings that African Americans vote at a higher rate when there is an African American candidate (Associated Press [AP] 2008). Turnout also rose among young people and women and in some geographic areas (AP 2008).

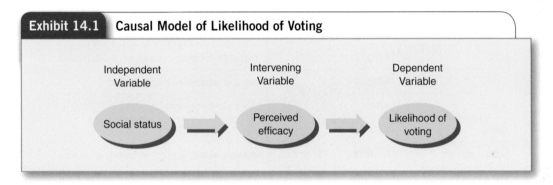

Exhibit 14.1 Causal Model of Likelihood of Voting

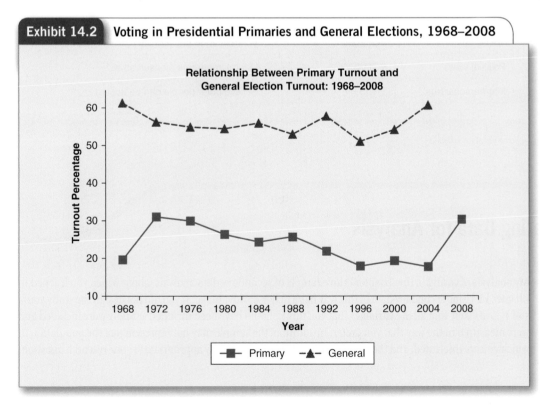

Exhibit 14.2 Voting in Presidential Primaries and General Elections, 1968–2008

If we are guided by prior research, a test of the hypothesis that likelihood of voting increases with social status should also take into account political efficacy and some social characteristics, such as race. We can find indicators for each of these variables, except political efficacy, in the 2010 GSS (see Exhibit 14.3). We will substitute the variable *interpersonal trust* for political efficacy. I will use the variables to illustrate particular statistics throughout this chapter, drawing on complete 2010 GSS data. You can replicate my analysis with the subset of the 2010 GSS data that is posted on the study site for this book (and on the CD-ROM included with the book, if you purchased the SPSS Studentware version).

Exhibit 14.3   List of GSS 2010 Variables for Analysis of Voting

Variable[a]	SPSS Variable Name	Description
Social Status		
Family income	INCOME4R	Family income (in categories)
Education	EDUCR6	Years of education completed (6 categories)
	EDUC4	Years of education completed (4 categories)
	EDUC3	Years of education, trichotomized
Age	AGE4	Years old (categories)
Gender	SEX	Sex
Marital status	MARITAL	Married, never married, widowed, divorced
Race	RACED	White, minority
Politics	PARTYID3	Political party affiliation
Voting	VOTE08D	Voted in 2004 presidential election (yes/no)
Political views	POLVIEWS3	Liberal, moderate, conservative
Interpersonal trust	TRUSTD	Believe other people can be trusted

a. Some variables recoded.

▣ Preparing Data for Analysis

My analysis of voting in this chapter is an example of secondary data analysis, about which you learned in Chapter 13. Using secondary data in this way has a major disadvantage: If you did not design the study yourself, it is unlikely that all the variables that you think should have been included actually were included and were measured in the way that you prefer. In addition, the sample may not represent just the population in which you are interested, and the study design may be only partially appropriate to your research question.

For example, because it is a survey of individuals, the GSS lacks measures of political context (such as the dominant party in an area). Because the survey sample is selected only from the United States and because the questions concern just one presidential election, we will not be able to address directly the larger issues of political context that are represented in cross-national and longitudinal research (for more on cross-national and longitudinal research, see Verba et al. 1978).

It is the availability of secondary data that makes their use preferable for many purposes. As you learned in Chapter 13, a great many high-quality data sets are available for reanalysis from the Inter-University Consortium for Political and Social Research (ICPSR) at the University of Michigan; many others can be obtained from the government, individual researchers, and other research organizations. Many of these data sets are stored online and can be downloaded directly for analysis; some can even be analyzed directly online. Others are available on CD-ROM. For a great many research problems, therefore, a researcher should first review the available data sets pertaining to that topic (and determine whether his or her university or other employer is a member of ICPSR). As a result of this, a lot of time and resources can be saved.

If you have conducted your own survey or experiment, your quantitative data must be prepared in a format suitable for computer entry. Several options are available. Questionnaires or other data entry forms can be designed for scanning or direct computer entry (see Exhibit 14.4). Once the computer database software is programmed to recognize the response codes, the forms can be fed through a scanner and the data will then be entered directly into the database. If responses or other forms of data have been entered on nonscannable paper forms, a computer data entry program should be used that will allow the data to be entered into the databases by clicking on boxes corresponding to the response codes. Alternatively, if a data entry program is not used, responses can be typed directly into a computer database. If data entry is to be done this way, the questionnaires or other forms should be precoded. Precoding means that a number represents every response choice, and respondents are instructed to indicate their response to a question by checking a number. It will then be easier to type in the strings of numbers than to type in the responses themselves.

Whatever data entry method is used, the data must be checked carefully for errors—a process called **data cleaning**. The first step in data cleaning is to check responses before they are entered into the database to make sure that one and only one valid answer code has been clearly circled or checked for each question (unless multiple responses are allowed or a skip pattern was specified). Written answers can be assigned their own numerical codes. The next step in data cleaning is to make sure that no invalid codes have been entered. Invalid codes are codes that fall outside the range of allowable values for a given variable and those that represent impossible combinations of responses to two or more questions. (For example, if a respondent says that he or she did not vote in an election, a response to a subsequent question indicating whom that person voted for would be invalid.) Most survey research organizations now use a database management program to control data entry. The program prompts the data entry clerk for each response code, checks the code to ensure that it represents a valid response for that variable, and saves the response code in the data file. This process reduces sharply the possibility of data entry errors.

> **Data cleaning** The process of checking data for errors after the data have been entered in a computer file.

If data are typed into a text file or entered directly through the data sheet of a statistics program, a computer program must be written to "define the data." A data definition program identifies the variables that are coded in each column or range of columns, attaches meaningful labels to the codes, and distinguishes values representing missing data. The procedures for doing so vary with the specific statistical package used. I used the Statistical Package for the Social Sciences (SPSS) for the analysis in this chapter; you will find examples of SPSS commands for defining and analyzing data in Appendix D (on the study site at www.sagepub.com/schuttisw7e). More information on using SPSS is contained in SPSS manuals and in the SAGE Publications volume, *Using IBM SPSS Statistics for Social Statistics and Research Methods,* 3rd edition, by William E. Wagner, III (2011).

Exhibit 14.4 Form for Direct Data Entry

OMB Control No: 6691-0001
Expiration Date: 04/30/07

Bureau of Economic Analysis
Customer Satisfaction Survey

1. Which data products do you use?	Frequently (every week)	Often (every month)	Infrequently	Rarely	Never	Don't know or not applicable
GENERAL DATA PRODUCTS	(On a scale of 1-5, please circle the appropriate answer.)					
Survey of Current Business	5	4	3	2	1	N/A
CD-ROMs	5	4	3	2	1	N/A
BEA Web site (www.bea.gov)	5	4	3	2	1	N/A
STAT-USA Web site (www.stat-usa.gov)	5	4	3	2	1	N/A
Telephone access to staff	5	4	3	2	1	N/A
E-Mail access to staff	5	4	3	2	1	N/A
INDUSTRY DATA PRODUCTS						
Gross Product by Industry	5	4	3	2	1	N/A
Input-Output Tables	5	4	3	2	1	N/A
Satellite Accounts	5	4	3	2	1	N/A
INTERNATIONAL DATA PRODUCTS						
U.S. International Transactions . . . . (Balance of Payments)	5	4	3	2	1	N/A
U.S. Exports and Imports of Private Services . .	5	4	3	2	1	N/A
U.S. Direct Investment Abroad	5	4	3	2	1	N/A
Foreign Direct Investment in the United States . .	5	4	3	2	1	N/A
U.S. International Investment Position	5	4	3	2	1	N/A
NATIONAL DATA PRODUCTS						
National Income and Product Accounts (GDP)	5	4	3	2	1	N/A
NIPA Underlying Detail Data	5	4	3	2	1	N/A
Capital Stock (Wealth) and Investment . . . . by Industry	5	4	3	2	1	N/A
REGIONAL DATA PRODUCTS						
State Personal Income	5	4	3	2	1	N/A
Local Area Personal Income	5	4	3	2	1	N/A
Gross State Product by Industry	5	4	3	2	1	N/A
RIMS II Regional Multipliers	5	4	3	2	1	N/A

回 Displaying Univariate Distributions

The first step in data analysis is usually to display the variation in each variable of interest. For many descriptive purposes, the analysis may go no further. Graphs and frequency distributions are the two most popular approaches; both allow the analyst to display the distribution of cases across the categories of a variable. Graphs have the advantage of providing a picture that is easier to comprehend, although frequency distributions are preferable when exact numbers of cases having particular values must be reported and when many distributions must be displayed in a compact form.

Whichever type of display is used, the primary concern of the data analyst is to display accurately the distribution's shape, that is, to show how cases are distributed across the values of the variable. Three features of shape are important: **central tendency, variability**, and **skewness** (lack of symmetry). All three features can be represented in a graph or in a frequency distribution.

These features of a distribution's shape can be interpreted in several different ways, and they are not all appropriate for describing every variable. In fact, all three features of a distribution can be distorted if graphs, frequency distributions, or summary statistics are used inappropriately.

A variable's level of measurement is the most important determinant of the appropriateness of particular statistics. For example, we cannot talk about the skewness (lack of symmetry) of a variable measured at the nominal level. If the values of a variable cannot be ordered from lowest or highest—if the ordering of the values is arbitrary—we cannot say that the distribution is not symmetric because we could just reorder the values to make the distribution more (or less) symmetric. Some measures of central tendency and variability are also inappropriate for variables measured at the nominal level.

The distinction between variables measured at the ordinal level and those measured at the interval or ratio level should also be considered when selecting statistics for us, but social researchers differ in just how much importance they attach to this distinction. Many social researchers think of ordinal variables as imperfectly measured interval-level variables and believe that, in most circumstances, statistics developed for interval level variables also provide useful summaries for ordinal variables. Other social researchers believe that variation in ordinal variables will often be distorted by statistics that assume an interval level of measurement. We will touch on some of the details in the following sections on particular statistical techniques.

We will now examine graphs and frequency distributions that illustrate these three features of shape. Summary statistics used to measure specific aspects of central tendency and variability are presented in a separate section. There is a summary statistic for the measurement of skewness, but it is used only rarely in published research reports and will not be presented here.

> **Central tendency** The most common value (for variables measured at the nominal level) or the value around which cases tend to center (for a quantitative variable).
>
> **Variability** The extent to which cases are spread out through the distribution or clustered in just one location.
>
> **Skewness** The extent to which cases are clustered more at one or the other end of the distribution of a quantitative variable rather than in a symmetric pattern around its center. Skew can be positive (a right skew), with the number of cases tapering off in the positive direction, or negative (a left skew), with the number of cases tapering off in the negative direction.

Graphs

A picture often is worth some unmeasurable quantity of words. Even for the uninitiated, graphs can be easy to read, and they highlight a distribution's shape. They are useful particularly for exploring data because they show the full range of variation and identify data anomalies that might be in need of further study. And good, professional-looking graphs can now be produced relatively easily with software available for personal computers. There are many types of graphs, but the most common and most useful are bar charts, histograms, and frequency polygons. Each has two axes, the vertical axis (the *y*-axis) and the horizontal axis (the *x*-axis), and labels to identify the variables and the values, with tick marks showing where each indicated value falls along the axis.

A **bar chart** contains solid bars separated by spaces. It is a good tool for displaying the distribution of variables measured at the nominal level because there is, in effect, a gap between each of the categories. The bar chart of marital status in Exhibit 14.5 indicates that almost half of adult Americans were married at the time of the survey. Smaller percentages were divorced, separated, widowed, while more than one-quarter had never married. The most common value in the distribution is married, so this would be the distribution's central tendency. There is a moderate amount of variability in the distribution, because the half who are not married are spread across the categories of widowed, divorced, separated, and never married. Because marital status is not a quantitative variable, the order in which the categories are presented is arbitrary, and so skewness is not relevant.

Bar chart A graphic for qualitative variables in which the variable's distribution is displayed with solid bars separated by spaces.

Histogram A graphic for quantitative variables in which the variable's distribution is displayed with adjacent bars.

Histograms, in which the bars are adjacent, are used to display the distribution of quantitative variables that vary along a continuum that has no necessary gaps. Exhibit 14.6 shows a histogram of years of education from the 2010 GSS data. The distribution has a clump of cases at 12 years. The distribution is skewed, because the number of cases declines to below the mean much more quickly than above the mean.

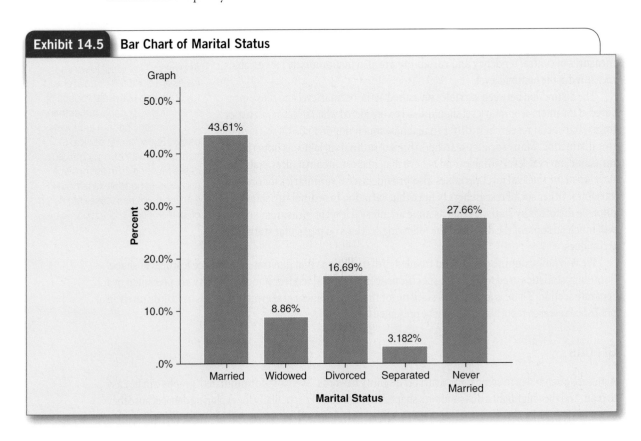

Exhibit 14.5   Bar Chart of Marital Status

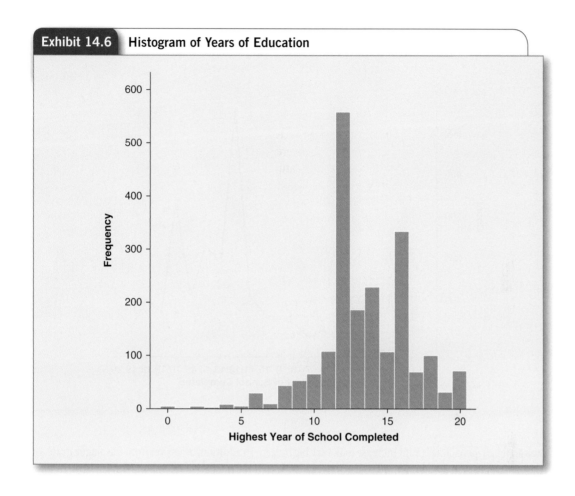

Exhibit 14.6 Histogram of Years of Education

In a **frequency polygon**, a continuous line connects the points representing the number or percentage of cases with each value. The frequency polygon is an alternative to the histogram when the distribution of a quantitative variable must be displayed; this alternative is particularly useful when the variable has a wide range of values. It is easy to see in the frequency polygon of years of education in Exhibit 14.7 that the most common value is 12 years, high school completion, and that this value also seems to be at the center of the distribution. There is moderate variability in the distribution, with many cases having more than 12 years of education and almost one-third having completed at least 4 years of college (16 years). The distribution is highly skewed in the negative direction, with few respondents reporting less than 10 years of education.

> **Frequency polygon** A graphic for quantitative variables in which a continuous line connects data points representing the variable's distribution.

If graphs are misused, they can distort, rather than display, the shape of a distribution. Compare, for example, the two graphs in Exhibit 14.8. The first graph shows that high school seniors reported relatively stable rates of lifetime use of cocaine between 1980 and 1985. The second graph, using exactly the same numbers, appeared in a 1986 *Newsweek* article on the coke plague (Orcutt & Turner 1993). Looking at this graph, you would think that the rate of cocaine usage among high school seniors had increased dramatically during this period. But, in fact, the difference between the two graphs is due simply to changes in how the graphs are drawn. In the plague graph, the percentage scale on the vertical axis begins at 15 rather than at 0, making what

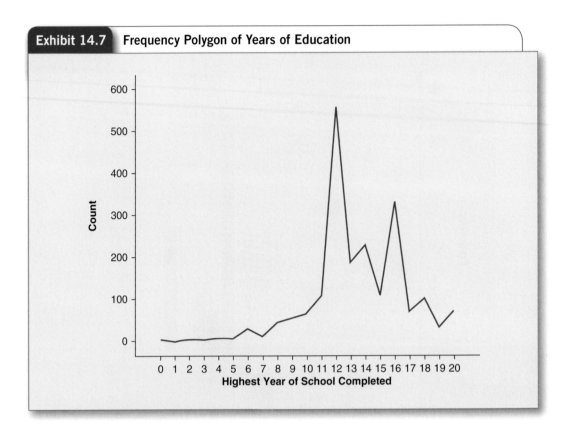

Exhibit 14.7 Frequency Polygon of Years of Education

was about a 1-percentage point increase look very big indeed. In addition, omission from the plague graph of the more rapid increase in reported usage between 1975 and 1980 makes it look as if the tiny increase in 1985 were a new, and thus more newsworthy, crisis.

Adherence to several guidelines (Tufte 1983; Wallgren et al. 1996) will help you spot these problems and avoid them in your own work:

- The difference between bars can be exaggerated by cutting off the bottom of the vertical axis and displaying less than the full height of the bars. Instead, begin the graph of a quantitative variable at 0 on both axes. It may be reasonable, at times, to violate this guideline, as when an age distribution is presented for a sample of adults, but in this case be sure to mark the break clearly on the axis.

- Bars of unequal width, including pictures instead of bars, can make particular values look as if they carry more weight than their frequency warrants. Always use bars of equal width.

- Either shortening or lengthening the vertical axis will obscure or accentuate the differences in the number of cases between values. The two axes usually should be of approximately equal length.

- Avoid chart junk that can confuse the reader and obscure the distribution's shape (a lot of verbiage or umpteen marks, lines, lots of cross-hatching, etc.).

Frequency distribution Numerical display showing the number of cases, and usually the percentage of cases (the relative frequencies), corresponding to each value or group of values of a variable.

Base number (*N*) The total number of cases in a distribution.

Frequency Distributions

A **frequency distribution** displays the number of, percentage (the relative frequencies) of, or both cases corresponding to each of a variable's values or group of values. The components of the frequency distribution should be clearly labeled, with a title,

| Exhibit 14.8 | Two Graphs of Cocaine Usage |

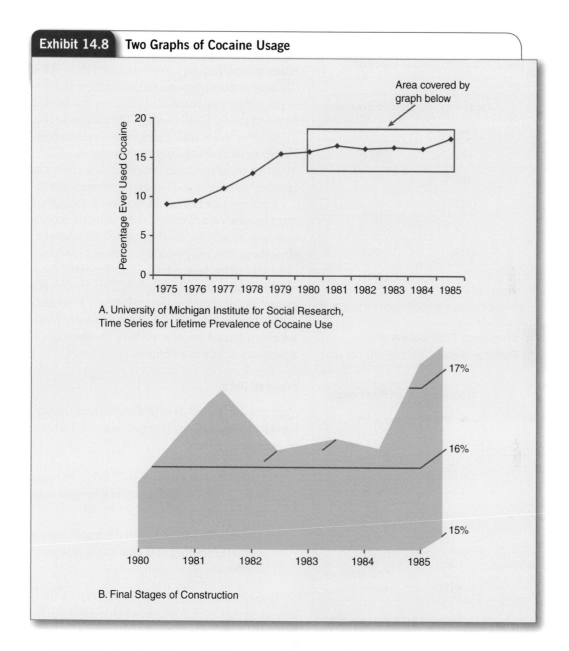

A. University of Michigan Institute for Social Research,
Time Series for Lifetime Prevalence of Cocaine Use

B. Final Stages of Construction

a stub (labels for the values of the variable), a caption (identifying whether the distribution includes frequencies, percentages, or both), and perhaps the number of missing cases. If percentages, rather than frequencies, are presented (sometimes both are included), the total number of cases in the distribution (the base number N) should be indicated (see Exhibit 14.9).

Ungrouped Data

Constructing and reading frequency distributions for variables with few values is not difficult. The frequency distribution of voting in Exhibit 14.9, for example, shows that 72.4% of the respondents eligible to vote said they voted, and 27.6% reported they did not vote. The total number of respondents to this question was 1,920, although 2,044 actually were interviewed. The rest were ineligible to vote, said they did not know whether they had voted or not, or gave no answer.

Exhibit 14.9	Frequency Distribution of Voting in the 2008 Presidential Election	
Value	**Frequency**	**Valid Percentage**
Voted	1390	72.4
Did not vote	530	27.6
Ineligible	103	
Don t know	12	
No answer	9	
Total	2044	100.0
N		(1920)

Exhibit 14.10	Frequency Distribution of Political Views	
Value	**Frequency**	**Valid Percentage**
Extremely liberal	76	3.9%
Liberal	259	13.1
Slightly liberal	232	11.8
Moderate	746	37.8
Slightly conservative	265	13.4
Conservative	315	16.0
Extremely conservative	80	4.1
Total	1973	100.0%

Political ideology was measured with a question having seven response choices, resulting in a longer but still relatively simple frequency distribution (see Exhibit 14.10). The most common response was moderate, with 37.8% of the sample that responded choosing this label to represent their political ideology. The distribution has a symmetric shape, with somewhat more respondents identifying themselves as conservative rather than liberal. About 4% of the respondents identified themselves as extremely conservative and about 4% as extremely liberal.

If you compare Exhibits 14.10 and 14.6, you can see that a frequency distribution (Exhibit 14.10) can provide more precise information than a graph (Exhibit 14.6) about the number and percentage of cases in a variable's categories. Often, however, it is easier to see the shape of a distribution when it is graphed. When the goal of a presentation is to convey a general sense of a variable's distribution, particularly when the presentation is to an audience that is not trained in statistics, the advantages of a graph outweigh those of a frequency distribution.

Grouped Data

Many frequency distributions (and graphs) require grouping of some values after the data are collected. There are two reasons for grouping:

1. There are more than 15 to 20 values to begin with, a number too large to be displayed in an easily readable table.

2. The distribution of the variable will be clearer or more meaningful if some of the values are combined.

Inspection of Exhibit 14.11 should clarify these reasons. In the first distribution, which is only a portion of the entire ungrouped GSS age distribution, it is very difficult to discern any shape, much less the central tendency. In the second distribution, age is grouped in the familiar 10-year intervals (except for the first, abbreviated category), and the distribution's shape is immediately clear.

Once we decide to group values, or categories, we have to be sure that in doing so, we do not distort the distribution. Adhering to the following guidelines for combining values in a frequency distribution will prevent many problems:

- Categories should be logically defensible and preserve the distribution's shape.

- Categories should be mutually exclusive and exhaustive, so that every case should be classifiable in one and only one category.

Exhibit 14.11 Grouped Versus Ungrouped Frequency Distributions

Ungrouped		Grouped	
Age	Percentage	Age	Percentage
18	0.5%	18–19	1.7%
19	1.2	20–29	16.7
20	1.2	30–39	17.8
21	1.7	40–49	18.0
22	.9	50–59	18.5
23	1.9	60–69	14.8
24	1.5	70–79	7.1
25	2.4	80–89	5.4
26	1.4		100.0%
27	2.1		(2041)
28	1.5		
29	2.2		
30	2.0		
31	2.1		
32	1.6		
33	1.8		
34	1.6		
35	2.2		
36	1.5		
37	1.9		
38	1.5		
39	1.7		
40	1.8		
41	1.9		
42	1.7		
43	2.1		
44	1.9		
45	1.8		
46	1.9		
. . .	. . .		

Exhibit 14.12	Years of Education Completed	
Years of Education	**Percentage**	
Less than 8	5.7%	
8–11	11.2	
12	27.3	
13–15	25.6	
16	16.3	
17 or more	13.8	
	100.0	
	(2044)	

Violating these two guidelines is easier than you might think. If you were to group all the ages above 59 together, as 60 or higher, it would create the appearance of a bulge at the high end of the age distribution, with 27.3% of the cases. Combining other categories so that they include a wide range of values could create the same type of misleading impression. In some cases, however, the most logically defensible categories will vary in size. A good example would be grouping years of education as less than 8 (did not finish grade school), 8 to 11 (finished grade school), 12 (graduated high school), 13 to 15 (some college), 16 (graduated college), and 17 or more (some postgraduate education). Such a grouping captures the most meaningful distinctions in the educational distribution and preserves the information that would be important for many analyses (see Exhibit 14.12).

It is also easy to imagine how the requirement that categories be mutually exclusive can be violated. You sometimes see frequency distributions or categories in questionnaires that use such overlapping age categories as 20 to 30, 30 to 40, and so on instead of mutually exclusive categories such as those in Exhibit 14.11. The problem is that we then can't tell which category to place someone in who is age 30, 40, and so on.

Combined and Compressed Distributions

In a **combined frequency display**, the distributions for a set of conceptually similar variables having the same response categories are presented together. Exhibit 14.13 is a combined display reporting the frequency distributions in percentage form for 13 variables that indicate GSS respondents' level of confidence in American institutions. The different variables are identified in the leftmost column, and their values are labeled along the top. By looking at the table, you can see quickly that confidence is greatest in the military, the scientific community, and medicine; educational institutions and the Supreme Court are regarded with only slightly less confidence. Smaller portions of the American public have much confidence in major companies, organized labor, and the executive branch. Banks and financial institutions, the U.S. Congress, the press, and television elicit the least confidence. Note that the specific variables are ordered in decreasing order of confidence to make it easier to see the pattern, and the source for the data is cited in the table's footnote. The number of cases on which the distributions are based is included for each variable.

Combined frequency display
A table that presents together the distributions for a set of conceptually similar variables having the same response categories; common headings are used for the responses.

Compressed frequency display
A table that presents cross classification data efficiently by eliminating unnecessary percentages, such as the percentage corresponding to the second value of a dichotomous variable.

Compressed frequency displays can also be used to present cross-tabular data and summary statistics more efficiently, by eliminating unnecessary percentages (such as those corresponding to the second value of a dichotomous variable) and by reducing the need for repetitive labels. Exhibit 14.14 presents a compressed display of agreement that abortion should be allowed given particular conditions. Note that this display presents (in parentheses) the number of cases on which the percentages are based.

Combined and compressed statistical displays facilitate the presentation of a large amount of data in a relatively small space. They should be used with caution, however, because they may baffle people who are not used to them.

Exhibit 14.13 Confidence in Institutions

Confidence in . . .	A Great Deal	(%) Only Some	(%) Hardly Any	(%) Total	(%) n
Congress	9.3	47.1	43.6	100.0	1,347
Press	10.3	45.8	43.8	100.0	1,355
Banks and financial institutions	10.6	48.1	41.3	100.0	1,364
Organized labor	11.2	60.1	28.7	100.0	1,300
Television	12.1	49.7	38.2	100.0	1,363
Major companies	13.0	63.3	23.7	100.0	1,339
Executive branch of federal government	16.9	46.1	37.1	100.0	1,346
Organized religion	20.5	54.6	24.9	100.0	1,319
Education	26.9	58.2	14.9	100.0	1,363
U.S. Supreme Court	30.0	52.9	17.2	100.0	1,335
Medicine	40.9	47.5	11.6	100.0	1,363
Scientific community	41.7	52.3	6.0	100.0	1,306
Military	53.1	37.4	9.5	100.0	1,356

Exhibit 14.14 Conditions When Abortion Should Be Allowed

Statement	% Agree	N
Woman's health seriously endangered	87.1	(1,224)
Pregnant as result of rape	79.9	(1,235)
Strong chance of serious defect	75.2	(1,225)
Married and wants no more children	48.0	(1,223)
Low income—can't afford more children	45.5	(1,231)
Abortion if woman wants for any reason	43.7	(1,230)
Not married	42.7	(1,228)

▣ Summarizing Univariate Distributions

Summary statistics focus attention on particular aspects of a distribution and facilitate comparison among distributions. For example, if your purpose were to report variation in income by state in a form that is easy for most audiences to understand, you would usually be better off presenting average incomes; many people would find it difficult to make sense of a display containing 50 frequency distributions, although they could readily comprehend a long list of average incomes. A display of average incomes would also be preferable to multiple frequency distributions if your only purpose were to provide a general idea of income differences among states.

Research in the News

GENERAL SOCIAL SURVEY SHOWS INFIDELITY ON THE RISE

Since 1972, about 12% of married men and 7% of married women have said each year that they have had sex outside their marriage. However, the lifetime rate of infidelity for men over age 60 increased from 20% in 1991 to 28% in 2006, while for women in this age group it increased from 5% to 15%. Infidelity has also increased among those under age 35: from 15% to 20% among young married men and from 12% to 15% among young married women. On the other hand, couples appear to be spending slightly more time with each other.

Source: Parker-Pope, Tara. 2008. "Love, Sex, and the Changing Landscape of Infidelity." *The New York Times,* October 28:D1.

Of course, representing a distribution in one number loses information about other aspects of the distribution's shape and so creates the possibility of obscuring important information. If you need to inform a discussion about differences in income inequality among states, for example, measures of central tendency and variability would miss the point entirely. You would either have to present the 50 frequency distributions or use some special statistics that represent the unevenness of a distribution. For this reason, analysts who report summary measures of central tendency usually also report a summary measure of variability and sometimes several measures of central tendency, variability, or both.

Measures of Central Tendency

Central tendency is usually summarized with one of three statistics: the mode, the median, or the mean. For any particular application, one of these statistics may be preferable, but each has a role to play in data analysis. To choose an appropriate measure of central tendency, the analyst must consider a variable's level of measurement, the skewness of a quantitative variable's distribution, and the purpose for which the statistic is used. In addition, the analyst's personal experiences and preferences inevitably will play a role.

Mode

The **mode** is the most frequent value in a distribution. It is also termed the **probability average** because, being the most frequent value, it is the most probable. For example, if you were to pick a case at random from the distribution of political views (refer back to Exhibit 14.10), the probability of the case being a moderate would be .378 out of 1, or about 38%—the most probable value in the distribution.

> **Mode** The most frequent value in a distribution; also termed the **probability average**.

The mode is used much less often than the other two measures of central tendency because it can so easily give a misleading impression of a distribution's central tendency. One problem with the mode occurs when a distribution is **bimodal**, in contrast to being **unimodal**. A bimodal (or trimodal, etc.) distribution has two or more categories with an equal number of cases and with more cases than any of the other categories. There is no single mode. Imagine that a particular distribution has two categories, each having just about the same number of cases (and these are the two most frequent categories). Strictly speaking, the mode would be the one with more cases, even though the other frequent category had only slightly fewer cases. Another potential problem with the mode is that it might happen to fall far from the main clustering of cases in a distribution. It would be misleading in most circumstances to say simply that the variable's central tendency was whatever the modal value was.

> **Bimodal** A distribution that has two nonadjacent categories with about the same number of cases, and these categories have more cases than any others.
>
> **Unimodal** A distribution of a variable in which there is only one value that is the most frequent.

Nevertheless, there are occasions when the mode is very appropriate. Most important, the mode is the only measure of central tendency that can be used to characterize the central tendency of variables measured at the nominal level. We can't say much more about the central tendency of the distribution of marital status in Exhibit 14.5 than that the most common value is married. The mode also is often referred to in descriptions of the shape of a distribution. The terms *unimodal* and *bimodal* appear frequently, as do descriptive statements such as "The typical [most probable] respondent was in her 30s." Of course, when the issue is what the most probable value is, the mode is the appropriate statistic. Which ethnic group is most common in a given school? The mode provides the answer.

Median

The **median** is the position average, or the point that divides the distribution in half (the 50th percentile). The median is inappropriate for variables measured at the nominal level because their values cannot be put in order, and so there is no meaningful middle position. To determine the median, we simply array a distribution's

> **Median** The position average or the point that divides a distribution in half (the 50th percentile).

values in numerical order and find the value of the case that has an equal number of cases above and below it. If the median point falls between two cases (which happens if the distribution has an even number of cases), the median is defined as the average of the two middle values and is computed by adding the values of the two middle cases and dividing by 2.

The median in a frequency distribution is determined by identifying the value corresponding to a cumulative percentage of 50. Starting at the top of the years of education distribution in Exhibit 14.12, for example, and adding up the percentages, we find that we have reached 44.2% in the first 12 years category and then 69.8% in the 13 to 15 years category. The median is therefore 13 to 15.

With most variables, it is preferable to compute the median from ungrouped data because that method results in an exact value for the median, rather than an interval. In the grouped age distribution in Exhibit 14.11, for example, the median is in the 40s interval. But if we determine the median from the ungrouped data, we can state that the exact value of the median is 46.

Mean The arithmetic, or weighted, average, computed by adding up the value of all the cases and dividing by the total number of cases.

Mean

The **mean**, or arithmetic average, takes into account the values of each case in a distribution—it is a weighted average. The mean is computed by adding up the value of all the cases and dividing by the total number of cases, thereby taking into account the value of each case in the distribution:

Mean = Sum of value of cases/Number of cases

In algebraic notation, the equation is $\bar{Y} = \sum Y_i / N$ For example, to calculate the mean of eight cases, we add the values of all the cases ($\sum Y_i$) and divide by the number of cases (N):

$$(28 + 117 + 42 + 10 + 77 + 51 + 64 + 55)/8 = 444/8 = 55.5$$

Because computing the mean requires adding up the values of the cases, it makes sense to compute a mean only if the values of the cases can be treated as actual quantities—that is, if they reflect an interval or ratio level of measurement, or if they are ordinal and we assume that ordinal measures can be treated as interval. It would make no sense to calculate the mean religion. For example, imagine a group of four people in which there were two Protestants, one Catholic, and one Jew. To calculate the mean, you would need to solve the equation (Protestant + Protestant + Catholic + Jew) + 4 = ? Even if you decide that Protestant = 1, Catholic = 2, and Jew = 3 for data entry purposes, it still doesn't make sense to add these numbers because they don't represent quantities of religion.

Median or Mean?

Both the median and the mean are used to summarize the central tendency of quantitative variables, but their suitability for a particular application must be carefully assessed. The key issues to be considered in this assessment are the variable's level of measurement, the shape of its distribution, and the purpose of the statistical summary. Consideration of these issues will sometimes result in a decision to use both the median and the mean and will sometimes result in neither measure being seen as preferable. But in many other situations, the choice between the mean and median will be clear-cut as soon as the researcher takes the time to consider these three issues.

Level of measurement is a key concern because to calculate the mean, we must add up the values of all the cases—a procedure that assumes the variable is measured at the interval or ratio level. So even though we know that coding *Agree* as 2 and *Disagree* as 3 does not really mean that *Disagree* is 1 unit more of disagreement than *Agree,* the mean assumes this evaluation to be true. Because calculation of the median requires only that we order the values of cases, we do not have to make this assumption. Technically speaking, then, the mean is simply an inappropriate statistic for variables measured at the ordinal level (and you already know that it is completely meaningless for variables measured at the nominal level). In practice, however, many social researchers use the mean to describe the central tendency of variables measured at the ordinal level, for the reasons outlined earlier.

The shape of a variable's distribution should also be taken into account when deciding whether to use the median or mean. When a distribution is perfectly symmetric, so that the distribution of values below the median is a mirror image of the distribution of values above the median, the mean and median will be the same. But the values of the mean and median are affected differently by skewness, or the presence of cases with extreme values on one side of the distribution but not the other side. Because the median takes into account only the number of cases above and below the median point, not the value of these cases, it is not affected in any way by extreme values. Because the mean is based on adding the value of all the cases, it will be pulled in the direction of exceptionally high (or low) values. When the value of the mean is larger than the value of the median, we know that the distribution is skewed in a positive direction, with proportionately more cases with higher than lower values. When the mean is smaller than the median, the distribution is skewed in a negative direction.

This differential impact of skewness on the median and mean is illustrated in Exhibit 14.15. On the first balance beam, the cases (bags) are spread out equally, and the median and mean are in the same location. On the second and third balance beams, the median corresponds to the value of the middle case, but the mean is pulled toward the value of the one case with an extremely high value. For this reason, the mean age (47.97) for the 2,041 cases represented partially in the detailed age distribution in Exhibit 14.11 is higher than the median age (47). Although in the distribution represented in Exhibit 14.11 the difference is small, in some distributions the two measures will have markedly different values, and in such instances the median may be preferred.

The single most important influence on the choice of the median or the mean for summarizing the central tendency of quantitative variables should be the purpose of the statistical summary. If the purpose is to report the middle position in one or more distributions, then the median is the appropriate statistic, whether or not the distribution is skewed. For example, with respect to the age distribution from the GSS, you could report that half the American population is younger than 47 years old, and half the population is older than that. But if the purpose is to show how likely different groups are to have age-related health problems, the measure of

---

Exhibit 14.15   **The Mean as a Balance Point**

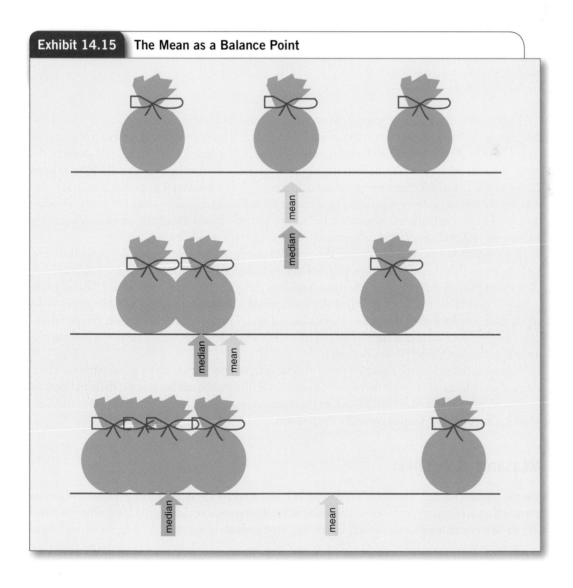

central tendency for these groups should take into account people's ages, not just the number of people who are older and younger than a particular age. For this purpose, the median would be inappropriate, because it would not distinguish between two distributions that have the same median but with different numbers at older people. In one distribution, everyone is between the ages of 35 and 45, with a median of 41. In the other distribution, the median is still 41, but half of the cases have ages above 60. The higher mean in the second distribution reflects the fact that it has a greater number of older people.

Exhibit 14.16	Insensitivity of Median to Variation at End of Distribution		
Level of Measurement	**Most Appropriate MCT**	**Potentially Useful MCT**	**Definitely Inappropriate MCT**
Nominal	Mode	None	Median, mean
Ordinal	Median	Mean	None
Interval, ratio	Mean	Median, mode	None

Note: MCT, measure of central tendency.

Keep in mind that it is not appropriate to use either the median or the mean as a measure of central tendency for variables measured at the nominal level, because at this level, the different attributes of a variable cannot be ordered as higher or lower (as reflected in the above correspondence list). Technically speaking, the mode should be used to measure the central tendency of variables measured at the nominal level (and it can also be used with variables measured at the ordinal, interval, and ratio levels). The median is most suited to measure the central tendency of variables measured at the ordinal level (and it can also be used to measure the central tendency of variables measured at the interval and ratio levels). Finally, the mean is only suited to measure the central tendency for variables measured at the interval and ratio levels.

It is not entirely legitimate to represent the central tendency of a variable measured at the ordinal level with the mean: Calculation of the mean requires summing the values of all cases, and at the ordinal level, these values indicate only order, not actual numbers. Nonetheless, many social scientists use the mean with ordinal-level variables and find that this is potentially useful for comparisons among variables and as a first step in more complex statistical analyses. The median and mode can also be useful as measures of central tendency for variables measured at the interval and ratio levels, when the goal is to indicate middle position (the median) or the most frequent value (the mode).

In general, the mean is the most commonly used measure of central tendency for quantitative variables, both because it takes into account the value of all cases in the distribution and because it is the foundation for many other more advanced statistics. However, the mean's very popularity results in its use in situations for which it is inappropriate. Keep an eye out for this problem.

Measures of Variation

You already have learned that central tendency is only one aspect of the shape of a distribution—the most important aspect for many purposes but still just a piece of the total picture. A summary of distributions based only on their central tendency can be very incomplete, even misleading. For example, three towns might have the same median income but still be very different in their social character due to the shape of their income distributions. As illustrated in Exhibit 14.17, Town A is a homogeneous middle-class community; Town B is very heterogeneous; and Town C has a polarized, bimodal income distribution, with mostly very poor and very rich people and few in between. However, all three towns have the same median income.

The way to capture these differences is with statistical measures of variation. Four popular measures of variation are the range, the interquartile range, the variance, and the standard deviation (which is the most popular measure of variability). To calculate each of these measures, the variable must be at the interval or ratio level (but many would argue that, like the mean, they can be used with ordinal-level measures, too). Statistical measures of variation are used infrequently with variables measured at the nominal level, so these measures will not be presented here.

It's important to realize that measures of variability are summary statistics that capture only part of what we need to be concerned with about the distribution of a variable. In particular, they do not tell us about the extent to which a distribution is skewed, which we've seen is very important for interpreting measures of central tendency. Researchers usually evaluate the skewness of distributions just by eyeballing them.

Exhibit 14.17 | **Distributions Differing in Variability but Not Central Tendency**

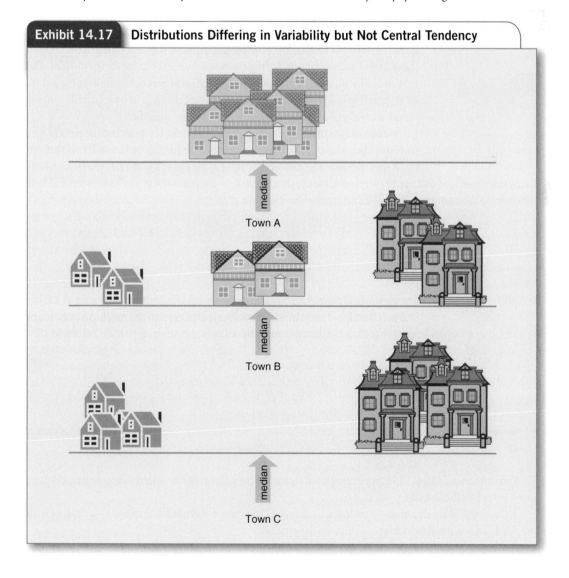

Range

Range The true upper limit in a distribution minus the true lower limit (or the highest rounded value minus the lowest rounded value, plus one).

Outlier An exceptionally high or low value in a distribution.

The **range** is a simple measure of variation, calculated as the highest value in a distribution minus the lowest value:

Range = Highest value − Lowest value

It often is important to report the range of a distribution to identify the whole range of possible values that might be encountered. However, because the range can be drastically altered by just one exceptionally high or low value (termed an **outlier**), it does not do an adequate job of summarizing the extent of variability in a distribution.

Interquartile Range

Interquartile range The range in a distribution between the end of the first quartile and the beginning of the third quartile.

Quartiles The points in a distribution corresponding to the first 25% of the cases, the first 50% of the cases, and the first 75% of the cases.

A version of the range statistic, the **interquartile range**, avoids the problem created by outliers. **Quartiles** are the points in a distribution corresponding to the first 25% of the cases, the first 50% of the cases, and the first 75% of the cases. You already know how to determine the second quartile, corresponding to the point in the distribution covering half of the cases—it is another name for the median. The first and third quartiles are determined in the same way but by finding the points corresponding to 25% and 75% of the cases, respectively. The interquartile range is the difference between the first quartile and the third quartile.

We can use the distribution of age for an example. If you add up the percentages corresponding to each value of age (ungrouped) in Exhibit 14.11, you'll find that you reach the first quartile (25% of the cases) at the age value of 34. If you were to continue, you would find that age 59 corresponds to the third quartile—the point where you have covered 75% of the cases. So the interquartile range for age, in the GSS 2010 data, is 30:

Third quartile − First quartile = Interquartile range
61 − 33 = 30.

Variance

Variance A statistic that measures the variability of a distribution as the average squared deviation of each case from the mean.

The **variance** is the average squared deviation of each case from the mean, so it takes into account the amount by which each case differs from the mean. An example of how to calculate the variance, using the following formula, appears in Exhibit 14.18:

$$\sigma^2 = \frac{\sum (Y_i - \overline{Y}_i)^2}{N}$$

Symbol key: $\overline{Y}$ = mean; N = number of cases; $\sum$ = sum over all cases; Y_i = value of variable Y for case i; σ^2 = variance.

You can see in Exhibit 14.18 two examples of summing over all cases, the operation represented by the Greek letter $\sum$ in the formula.

The variance is used in many other statistics, although it is more conventional to measure variability with the closely related standard deviation than with the variance.

Exhibit 14.18 Calculation of the Variance

Case #	Score (Y$_i$)	Y$_i$ – Ȳ	(Y$_i$ – Ȳ)²
1	21	–3.27	10.69
2	30	5.73	32.83
3	15	–9.27	85.93
4	18	–6.27	39.31
5	25	0.73	0.53
6	32	7.73	59.75
7	19	–5.27	27.77
8	21	–3.27	10.69
9	23	–1.27	1.61
10	37	12.73	162.05
11	26	1.73	2.99
			434.15

Standard Deviation

The **standard deviation** is simply the square root of the variance. It is the square root of the average squared deviation of each case from the mean:

$$\sigma = \sqrt{\frac{\sum (Y_i - \overline{Y}_i)^2}{N}}$$

> **Standard deviation** The square root of the average squared deviation of each case from the mean.

Symbol key: $\overline{Y}$ = mean; N = number of cases; $\sum$ = sum over all cases; Y_i = value of variable Y for case i; $\sqrt{}$ = square root; σ = standard deviation.

When the standard deviation is calculated from sample data, the denominator is supposed to be $N-1$, rather than N, an adjustment that has no discernible effect when the number of cases is reasonably large. You also should note that the use of *squared* deviations in the formula accentuates the impact of relatively large deviations, because squaring a large number makes that number count much more.

The standard deviation has mathematical properties that increase its value for statisticians. You already learned about the **normal distribution** in Chapter 5. A normal distribution is a distribution that results from chance variation around

> **Normal distribution** A symmetric, bell-shaped distribution that results from chance variation around a central value.

the mean price of homes in Community B to that for Community A (one with a homogeneous mid-priced set of homes) and therefore makes Community B look much better. In truth, the higher mean in Community B reflects a very skewed, lopsided distribution of property values—most residents own small, cheap homes. A median would provide a better basis for comparison.

You have already seen that it is possible to distort the shape of a distribution by ignoring some of the guidelines for constructing graphs and frequency distributions. Whenever you need to group data in a frequency distribution or graph, you can reduce the potential for problems by inspecting the ungrouped distributions and then using a grouping procedure that does not distort the distribution's basic shape. When you create graphs, be sure to consider how the axes you choose may change the distribution's apparent shape.

Cross-Tabulating Variables

Most data analyses focus on relationships among variables in order to test hypotheses or just to describe or explore relationships. For each of these purposes, we must examine the association among two or more variables. Cross-tabulation (crosstab) is one of the simplest methods for doing so. A **cross-tabulation**, or **contingency table**, displays the distribution of one variable for each category of another variable; it can also be termed a *bivariate distribution*. You can also display the association between two variables in a graph; we will see an example in this section. In addition, crosstabs provide a simple tool for statistically controlling one or more variables while examining the associations among others. In the next section, you will learn how crosstabs used in this way can help test for spurious relationships and evaluate causal models. We will examine several trivariate tables.

Cross-tabulation (crosstab) or **contingency table** In the simplest case, a bivariate (two-variable) distribution, showing the distribution of one variable for each category of another variable; can be elaborated using three or more variables.

Constructing Contingency Tables

The exhibits throughout this section are based on the 2010 GSS data. You can learn in the SPSS exercises at the end of this chapter and online how to use the SPSS program to generate cross-tabulations from a data set that has been prepared for analysis with SPSS. But let's now briefly see how you would generate a cross-tabulation without a computer. Exhibit 14.21 shows the basic procedure, based on results of responses in a survey about income and voting:

Exhibit 14.21	Converting Data From Case Records (raw data) Into a Crosstab

1. Record each respondent's answers to both these questions.
2. Create the categories for the table.
3. Tally the number of respondents whose answers fall in each table category.
4. Convert the tallies to frequencies and add up the row and column totals.

Simple, isn't it? Now, you're ready to shift to inspecting crosstabs produced with the SPSS computer program for the 2010 GSS data. We'll consider in just a bit the procedures for converting frequencies in a table into percentages and for interpreting, or "reading," crosstabs.

Cross-tabulation is only a useful method for examining the relationship between variables when they have only a few categories. For most analyses, 10 categories is a reasonable upper limit, but even 10 is too many unless you have a pretty large number of cases (more than 100). If you wish to include in a crosstab a variable with many categories, or that varies along a continuum with many values, you should first recode the values of that variable to a smaller number. For example, you might recode the values of a continuous index to just *high, medium,* and *low.* You might recode the numerical values of age to 10-year intervals. Exhibit 14.22 provides several examples.

Exhibit 14.23 displays the cross-tabulation of voting by family income, using the 2010 GSS data, so that we can test the hypothesis that likelihood of voting increases with this one social status indicator. The table is presented first with frequencies and then again with percentages. In both tables, the *body* of the table is the part between the row and column labels and the row and column totals. The cells of the table are defined by combinations of row and column values. Each cell represents cases with a unique combination of values of the two variables, corresponding to that particular row and column. The **marginal distributions** of the table are on the right (the *row marginals*) and underneath (the *column marginals*). These are just the frequency distributions for the two variables (in number of cases, percentages, or both), considered separately. (The column marginals in Exhibit 14.23 are for family income; the row marginals are for the distribution of voting.) The independent variable is usually the column variable; the dependent variable then is the row variable.

The table in the upper panel of Exhibit 14.23 shows the number of cases with each combination of values of voting and family income. In it, you can see that 228 of those earning less than $20,000 per year voted, whereas 187 did not. On the other hand, 404 of those whose family earned $75,000 or more voted, whereas 56 of these high-income respondents did not vote. It often is hard to look at a table in this form, with just the numbers of cases in each cell, and determine whether there is a relationship between the two variables. We need to convert the cell frequencies into **percentages**, as in the table in the lower panel of Exhibit 14.23. This table presents the data as percentages within the categories of the independent variable (the column variable, in this case). In other words, the cell frequencies have been converted into percentages of the column totals (the *n* in each column). For example, in Exhibit 14.23 the number of people in families earning less than $20,000 who voted is 228 out of 415, or 54.9%. Because the cell frequencies have been converted to percentages of the column totals, the numbers add up to 100 in each column, but not across the rows.

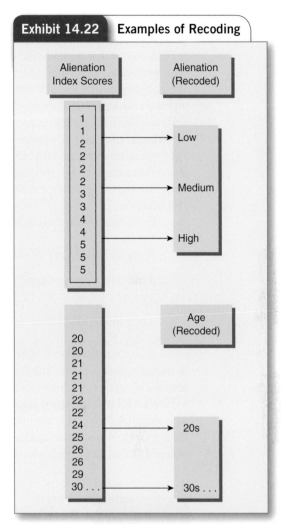

Exhibit 14.22 Examples of Recoding

Alienation Index Scores	Alienation (Recoded)
1, 1, 2, 2, 2, 2	Low
3, 3	Medium
4, 4, 5, 5, 5, 5	High

	Age (Recoded)
20, 20, 21, 21, 21, 22, 22, 24, 25, 26, 26, 29	20s
30 . . .	30s . . .

Marginal distribution The summary distributions in the margins of a cross-tabulation that correspond to the frequency distribution of the row variable and of the column variable.

Percentages Relative frequencies, computed by dividing the frequency of cases in a particular category by the total number of cases and then multiplying by 100.

To read the percentage table (the bottom panel of Exhibit 14.23), compare the percentage distribution of voting across the columns, starting with the lowest income category (in the left column) and moving from left to right. You see that as income increases, the percentage who voted also rises, from 55% (rounding off) of those with annual family incomes less than $20,000 (in the first cell in the first column), to 68% of those with family incomes between $20,000 and $39,999, then 75% of those with family incomes from $40,000 to $74,999, and then up to 88% of those with family incomes of $75,000 or more (the last cell in the body of the table in the first row). This result is consistent with the hypothesis.

When a table is converted to percentages, usually just the percentage in each cell should be presented, not the number of cases in each cell. Include 100% at the bottom of each column (if the independent variable is the column variable) to indicate that the percentages add up to 100, as well as the *base number* (*n*) for each column (in parentheses). If the percentages add up to 99 or 101 due to rounding error, just indicate this in a footnote.

Follow these rules when you create and then read a percentage table:

1. Make the independent variable the column variable and the dependent variable the row variable.

2. Percentage the table column by column, on the column totals. The percentages should add to 100 (or perhaps 99 or 101, if there has been rounding error) in each column.

3. Compare the distributions of the dependent variable (the row variable) across each column.

Skip ahead to the bottom panel of Exhibit 14.34 and try your hand at the table reading process (as described by Rule 3, above) with this larger table. The table in the bottom panel describes the relationship between education and family income. Examine the distribution of family income for those with only a grade-school education (first column). More than half (52.7%) reported a family income under $20,000, whereas just 4.2% reported a family income of $75,000 or more. Then, examine the distribution of family income for the respondents who had finished high school but had gone no further. Here, the distribution of family income has shifted upward, with just 26.6% reporting a family income under $20,000 and 15.5%

| Exhibit 14.23 | Cross-Tabulation of Voting in 2008 by Family Income: Cell Counts and Percentages |

Voting	<$20,000	$20,000–$39,999	$40,000–$74,999	$75,000+
Cell Counts	Family Income			
Voted	228	266	334	404
Did not vote	187	124	109	56
Total (n)	(415)	(390)	(443)	(460)
Percentages				
Voted	55	68	75	88
Did not vote	45	32	25	12
Total	100	100	100	100

reporting family incomes of $75,000 or more—that's more than three times the percentage in that category that we saw for those with a grade-school education. You can see there are also more respondents in the $40,000 to $74,999 category than there were for the grade schoolers. Now, examine the column representing those who had completed some college. The percentage with family incomes under $20,000 has dropped again, to 23.6%, whereas the percentage in the highest income category has risen to 24%. The college graduates have a much higher family income distribution than those with less education, with just 10.1% reporting family incomes of less than $20,000 and half (50.3%) reporting family incomes of $75,000 or more. If you step back and compare the income distributions across the four categories of education, you see that incomes increase markedly and consistently. The relationship is positive (fortunately, for students like you who are working so hard to finish college!).

But the independent variable does not *have* to be the column variable; what is critical is to be consistent within a report or paper. You will find in published articles and research reports some percentage tables in which independent variable and dependent variable positions are reversed. If the independent variable is the row variable, we percentage the table on the row totals (the *n* in each row), and so the percentages add up to 100 across the rows. Let's examine Exhibit 14.24, which is percentaged on the row variable: age. When you read the table in Exhibit 14.24, you find that 55.9% of those in their 20s voted, compared with 63.6% of those in their 30s, 76.7% of those in their 40s, and 75.3% to 85.0% of those between ages 50 and 80 or older.

Graphing Association

Graphs provide an efficient tool for summarizing relationships among variables. Exhibit 14.25 displays the relationship between race and region in graphic form. It shows that the percentage of the population that is black is highest in the South, whereas persons of "other" races are most common on the coasts.

Another good example of the use of graphs to show relationships is provided by a Bureau of Justice Statistics report on criminal victimization (Rand, Lynch, & Cantor 1997:1). Exhibit 14.26, taken from that report, shows how the rates of different violent crimes have varied over time: Most rates fell in the late 1980s, rose in the early 1990s, and then fell again by 1995. Because the four crime rates displayed in this graph as well as year are measured at the interval-ratio level, the graph can represent the variation over time with continuous lines.

Exhibit 14.24 Cross-Tabulation of Voting in 2008 by Age

Age	Voting			
	Yes (%)	No (%)	Total (%)	(n)
20—29	55.9	44.1	100.0	(315)
30—39	63.6	36.4	100.0	(335)
40—49	76.7	23.3	100.0	(344)
50—59	75.3	24.7	100.0	(364)
60—69	82.4	17.6	100.0	(295)
70—79	85.3	14.7	100.0	(143)
80 or older	85.0	15.0	100.0	(107)

Exhibit 14.25 Race by Region of the United States

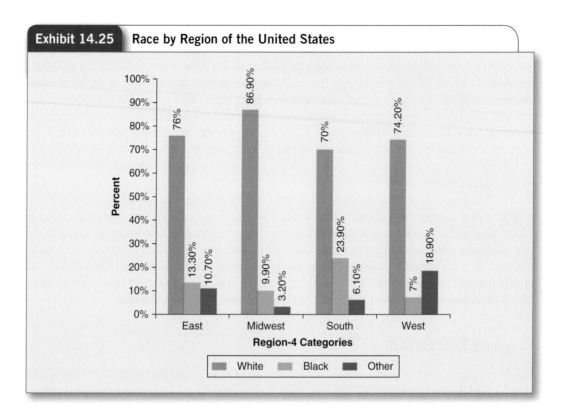

Exhibit 14.26 Violent Crime Rates, 1973–1995[a]

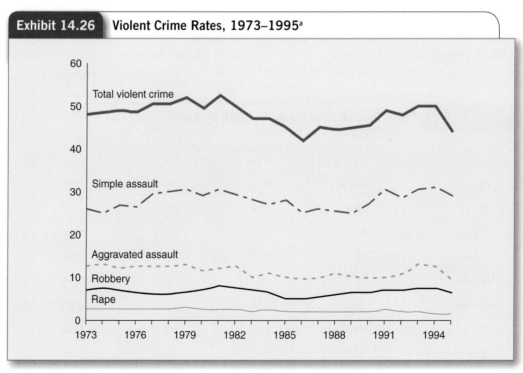

a. Victimization rate per 1,000 persons age 12 or older.

Describing Association

A cross-tabulation table reveals four aspects of the association between two variables:

- *Existence.* Do the percentage distributions vary at all between categories of the independent variable?

- *Strength.* How much do the percentage distributions vary between categories of the independent variable? In most analyses, the analyst would not pay much attention to differences of less than 10 percentages points between categories of the independent variable.

- *Direction.* For quantitative variables, do values on the dependent variable tend to increase or decrease with an increase in value on the independent variable?

- *Pattern.* For quantitative variables, are changes in the percentage distribution of the dependent variable fairly regular (simply increasing or decreasing), or do they vary (perhaps increasing, then decreasing, or perhaps gradually increasing, then rapidly increasing)?

Looking back at Exhibit 14.23, an association exists; it is moderately strong (the difference in percentages between those who voted in the first and last column is 20.5 percentage points); and the direction of association between likelihood of voting and family income is positive. The pattern in this table is close to what is termed **monotonic**. In a monotonic relationship, the value of cases consistently increases (or decreases) on one variable as the value of cases increases on the other variable. The relationship in the table that we will examine in Exhibit 14.33, involving income and education, is also monotonic.

Monotonic A pattern of association in which the value of cases on one variable increases or decreases fairly regularly across the categories of another variable.

Curvilinear Any pattern of association between two quantitative variables that does not involve a regular increase or decrease.

Monotonic is often defined a bit less strictly, with the idea being that as the values of cases on one variable increase (or decrease), the values of cases on the other variable tends to increase (or decrease), and at least do not change direction. This describes the relationship between voting and income: The likelihood of voting increases as family income increases, although the increase levels off in the middle two categories, with the result that the association is not strictly monotonic. There is also a moderately strong positive association between age and voting in Exhibit 14.24, with likelihood of voting rising 30 percentage points between the ages of 20 and 89. However, the pattern of this relationship is **curvilinear** rather than monotonic: The increase in voting with age occurs largely between the ages of 20 and 40, before levelling off and then rising a bit for people 60 or older.

The relationship between the measure of trust and voting appears in Exhibit 14.27. There is an association, and in the direction I hypothesized: 83.4% of those who believe that people can be trusted voted, compared

Exhibit 14.27	**Voting in 2008 by Interpersonal Trust**	
	People Can Be Trusted	
Voting	**Can Trust or Depends**	**Cannot Trust**
Voted	83.4%	64.7%
Did not vote	16.6	35.3
Total	100.0%	100.0%
(n)	(501)	(782)

with 64.7% of those who believe that people cannot be trusted. Because both variables are dichotomies, there can be no pattern to the association beyond the difference between the two percentages. (Comparing the column percentages in either the first or the second row gives the same picture.)

Exhibit 14.28, by contrast, gives less evidence of an association between gender and voting. The difference between the percentage of men and women who voted is 6.7 percentage points.

Evaluating Association

You will find when you read research reports and journal articles that social scientists usually make decisions about the existence and strength of association on the basis of more statistics than just a cross-tabulation table.

A **measure of association** is a type of descriptive statistic used to summarize the strength of an association. There are many measures of association, some of which are appropriate for variables measured at particular levels. One popular measure of association in cross-tabular analyses with variables measured at the ordinal level is **gamma**. As with many measures of association, the possible values of gamma vary from –1, meaning the variables are perfectly associated in an inverse direction; to 0, meaning there is no association of the type that gamma measures; to +1, meaning there is a perfect positive association of the type that gamma measures.

Exhibit 14.29 provides a rough guide to interpreting the value of a measure of association like gamma that can range from 0 to –1 and +1. For example, if the value

> **Measure of association** A type of descriptive statistic that summarizes the strength of an association.
>
> **Gamma** A measure of association that is sometimes used in cross-tabular analysis.

Exhibit 14.28 **Voting in 2008 by Gender**

Voting	Gender	
	Male	Female
Voted	68.6%	75.3%
Did not vote	31.4	24.7
Total	100%	100%
(n)	(829)	(1,091)

Exhibit 14.29 **A Guide to Interpreting Strong and Weak Relationships**

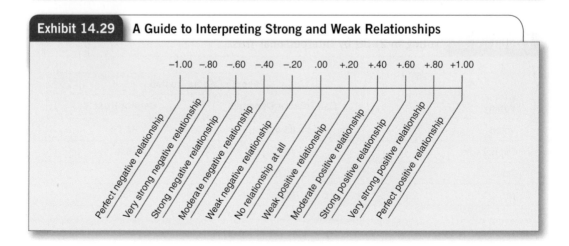

of gamma is –.23, we could say that there is a *weak negative relationship* between the two variables. If the value of gamma is +.61, we could say that there is a *strong positive relationship* between the two variables. A value of 0 always means that there is no relationship (although this really means there is no relationship that this particular statistic can identify). This "rough guide" to interpretation must be modified for some particular measures of association and your interpretations must also take into account the results of previous research and the particular methods you used to collect your data. For now, however, this rough guide will get you further along in your statistical interpretations.

Inferential statistics are used in deciding whether it is likely that an association exists in the larger population from which the sample was drawn. Even when the association between two variables is consistent with the researcher's hypothesis, it is possible that the association was just due to the vagaries of sampling on a random basis (of course, the problem is even worse if the sample is not random). It is conventional in statistics to avoid concluding that an association exists in the population from which the sample was drawn unless the probability that the association was due to chance is less than 5%. In other words, a statistician normally will not conclude that an association exists between two variables unless he or she can be at least 95% confident that the association was not due to chance. This is the same type of logic that you learned about earlier in this chapter, which introduced the concept of 95% confidence limits for the mean. Estimation of the probability that an association is not due to chance will be based on one of several inferential statistics, **chi-square** being the one used in most cross-tabular analyses. The probability is customarily reported in a summary form such as $p < .05$, which can be translated as "The probability that the association was due to chance is less than 5 out of 100 (5%)."

> **Chi-square** An inferential statistic used to test hypotheses about relationships between two or more variables in a cross-tabulation.

The tables in exhibit 14.30 and 14.31 will help you to understand the meaning of chi-square, without getting into the details of how it is calculated. Let's propose, as our "null hypothesis," that trust in other people has no association with family income. In that case, trust in people would be the same percent for all columns in the table—the same as for the overall sample. So if there were no association, we would expect on the basis of chance that 40% of the 293 people with family income below $20,000 (the first column) will vote. Since 40% of 293 equals 117.2, that is the number of people we would "expect" to be trusting and of low income if only chance factors were at work. This is the *expected count* and it differs from the *actual count* of 84, leaving a *residual* of –33.2. You can see that this process is repeated with respect to each cell of the table. The larger the deviations of the expected from the observed counts in the various table cells, the less likely it is that the association is due only to chance. Chi-square is calculated with a formula that combines the residuals in each cell. SPSS then compares the value of chi-square to a table that indicates how likely it is in a table of the given size that this value could have been obtained on the basis of chance. In the crosstab of family income and trust, the value of chi-square was 52.9 and the probability that a chi-square value of this magnitude was obtained on the basis of chance was less than

---

Exhibit 14.30 **Determining the Value of Chi-Square (Actual/Expected Counts)**

People Can Be Trusted	Family Income				Total
	< $20,000	$20–$39,000	$40–74,999	$75,000+	
Can trust or depends	84/117.2	88/110.4	138/129.2	174/127.2	484/484.0
Cannot trust	209/175.8	188/165.6	185/193.8	144/190.8	726/726.0
Total	100%	100%	100%	100%	
(n)	(293)	(276)	(323)	(318)	(1210)

1 in 1000 ($p < .001$). We could therefore feel confident that an association between these two variables exists in the U.S. adult population as a whole.

When the analyst feels reasonably confident (at least 95% confident) that an association was not due to chance, it is said that the association is statistically significant. **Statistical significance** means that an association is not likely to be due to chance, according to some criterion set by the analyst. Convention (and the desire to avoid concluding that an association exists in the population when it doesn't) dictates that the criterion be a probability less than 5%.

> **Statistical significance** The mathematical likelihood that an association is due to chance, judged by a criterion set by the analyst (often that the probability is less than 5 out of 100 or $p < .05$).

But statistical significance is not everything. You may remember from Chapter 5 that sampling error decreases as sample size increases. For this same reason, an association is less likely to appear on the basis of chance in a larger sample than in a smaller sample. In a table with more than 1,000 cases, such as those involving the full 2010 GSS sample, the odds of a chance association are often very low indeed. For example, with our table based on 1,708 cases, the probability that the association between income and voting (Exhibit 14.23) was due to chance was less than 1 in 1,000 ($p < .001$)! The association in that table was only moderate, as indicated by a gamma of .43. Even weak associations can be statistically significant with such a large random sample, which means that the analyst must be careful not to assume that just because a statistically significant association exists, it is therefore important. In a large sample, an association may be statistically significant but still be too weak to be substantively significant. All this boils down to another reason for evaluating carefully both the existence and the strength of an association.

Controlling for a Third Variable

Cross-tabulation can also be used to study the relationship between two variables while controlling for other variables. We will focus our attention on controlling for a third variable in this section, but I will say a bit about controlling for more variables at the section's end. We will examine three different uses for three-variable cross-tabulation: (1) identifying an intervening variable, (2) testing a relationship for spuriousness, and (3) specifying the conditions for a relationship. Each type of three-variable crosstab helps strengthen our understanding of the "focal relationship" involving our dependent and independent variables (Aneshensel 2002). Testing a relationship for possible spuriousness helps to meet the nonspuriousness criterion for causality; identifying an intervening variable can help chart the causal mechanism by which variation in the independent variable influences variation in the dependent variable; and specifying the conditions when a relationship occurs can help improve our understanding of the nature of that relationship.

> **Elaboration analysis** The process of introducing a third variable into an analysis to better understand—to elaborate—the bivariate (two-variable) relationship under consideration. Additional control variables also can be introduced.

All three uses for three-variable cross-tabulation are aspects of **elaboration analysis**: the process of introducing control variables into a bivariate relationship to better understand the relationship (Davis 1985; Rosenberg 1968). We will examine the gamma and chi-square statistics for each table in this analysis.

Intervening Variables

We will first complete our test of one of the implications of the causal model of voting in Exhibit 14.1: that trust (or efficacy) intervenes in the relationship between social status and voting. You already have seen that both income (one of our social status indicators) and trust in people are associated with the likelihood of voting. Both relationships are predicted by the model: so far, so good. You can also see in Exhibit 14.31 that trust is related to income: Higher income is associated with the belief that people can be trusted (gamma = .31; $p < .001$). Another prediction of the model is confirmed. But to determine whether the trust variable is an intervening variable in this relationship, we must determine whether it explains (transmits) the influence of income on trust. We therefore examine the relationship between income and voting while controlling for the respondent's belief that people can be trusted.

Exhibit 14.31　Cross-Tabulation of Interpersonal Trust by Income

	Family Income			
People Can Be Trusted	**<$20,000**	**$20,000–$39,999**	**$40,000–$74,999**	**$75,000+**
Can trust or depends	28.7%	31.9%	42.7%	54.7%
Cannot trust	71.3	68.1	57.3	45.3
Total	100%	100%	100%	100%
(n)	(293)	(276)	(323)	(318)

According to the causal model, income (social status) influences voting (political participation) by influencing trust in people (our substitute for efficacy), which, in turn, influences voting. We can evaluate this possibility by reading the two subtables in Exhibit 14.32. **Subtables** such as those in Exhibit 14.32 describe the relationship between two variables within the discrete categories of one or more control variables. The control variable in Exhibit 14.32 is trust in people, and the first subtable is the income-voting crosstab for only those respondents who believe that people can be trusted. The second subtable is for those respondents who believe that people can't be trusted. They are called subtables because together they make up the table in Exhibit 14.23. If trust in ordinary people intervened in the income-voting relationship, the effect of controlling for this third variable would be to

Subtables Tables describing the relationship between two variables within the discrete categories of one or more other control variables.

Exhibit 14.32　Voting in 2008 by Family Income by Interpersonal Trust

	Family Income			
Voting	**<$20,000**	**$20,000–$39,999**	**$40,000–$74,999**	**$75,000+**
People Can Be Trusted or It Depends				
Voted	72.4%	81.3%	80.2%	93.4%
Did not vote	27.6	18.8	19.8	6.6
Total	100%	100%	100%	100%
(n)	(76)	(80)	(131)	(166)
People Cannot Be Trusted				
Voted	50.8%	58.2%	73.9%	77.0%
Did not vote	49.2	41.8	26.1	23.0
Total	100%	100%	100%	100%
(n)	(193)	(177)	(176)	(139)

eliminate, or at least substantially reduce, this relationship—the distribution of voting would be the same for every income category in both subtables in Exhibit 14.32.

A quick inspection of the subtables in Exhibit 14.32 reveals that trust in people does not intervene in the relationship between income and voting. There is only a modest difference in the strength of the income-voting association in the subtables (gamma is −.39 in the first subtable and −.34 in the second). In both subtables, the likelihood that respondents voted rose with their incomes. Of course, this finding does not necessarily mean that the causal model was wrong. This one measure is a measure of trust in people, which is not the same as the widely studied concept of political efficacy; a better measure, from a different survey, might function as an intervening variable. But for now we should be less confident in the model.

Extraneous Variables

Another reason for introducing a third variable into a bivariate relationship is to see whether that relationship is spurious due to the influence of an extraneous variable (see Chapter 6)—a variable that influences both the independent and dependent variables, creating an association between them that disappears when the extraneous variable is controlled. Ruling out possible extraneous variables will help strengthen considerably the conclusion that the relationship between the independent and dependent variables is causal, particularly if all the variables that seem to have the potential for creating a spurious relationship can be controlled.

One variable that might create a spurious relationship between income and voting is education. You have already seen that the likelihood of voting increases with income. Is it not possible, though, that this association is spurious due to the effect of education? Education, after all, is associated with both income and voting, and we might surmise that it is what students learn in school about civic responsibility that increases voting, not income itself. Exhibit 14.33 diagrams this possibility, and Exhibit 14.34 shows the bivariate associations among education and voting, and education and income. As the model in Exhibit 14.33 predicts, education is associated with both income and voting. So far, so good. If education actually does create a spurious relationship between income and voting, there should be no association between income and voting after controlling for education. Because we are using crosstabs, this means there should be no association in any of the income-voting subtables for any value of education.

The trivariate cross-tabulation in Exhibit 14.35 shows that the relationship between voting and income is not spurious due to the effect of education; if it were, an association between voting and family income wouldn't appear in any of the subtables—somewhat like the first subtable, in which gamma is only −.17.

The association between family income and voting is higher in the other three subtables in Exhibit 14.35, for respondents with a high school, some college, or a college education. The strength of that association as measured by gamma is −.24 for those with a high school education and −.22 for those with at least some

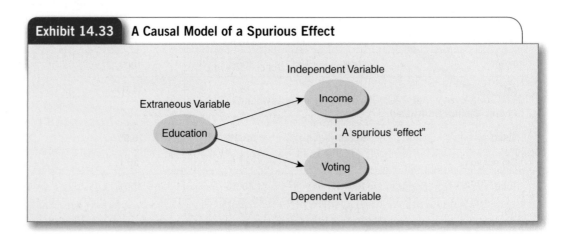

Exhibit 14.33 **A Causal Model of a Spurious Effect**

Independent Variable

Income

Extraneous Variable

Education

A spurious "effect"

Voting

Dependent Variable

Exhibit 14.34 Voting in 2008 by Education and Income by Education

Voting by Education

Voting	Education			
	Grade School	High School	Some College	College Graduate
Voted	45.0%	68.2%	76.0%	87.6%
Did not vote	55.0	31.8	24.0	12.4
Total	100%	100%	100%	100%
(n)	(307)	(529)	(499)	(582)

Family Income by Education

Family Income	Education			
	Grade School	High School	Some College	College Graduate
<$20,000	52.7%	26.6%	23.6%	10.1%
$20,000–$39,999	30.7	30.1	22.3	12.8
$40,000–$74,999	12.4	27.8	30.1	26.8
$75,000+	4.2	15.5	24.0	50.3
Total	100%	100%	100%	100%
(n)	(283)	(489)	(475)	(555)

college, and it is −.36 among college graduates. So our hypothesis—that income as a social status indicator leads to higher rates of voting—does not appear to be spurious due to the effect of education. The next section elaborates on the more complex pattern that we found.

Specification

By adding a third variable to an evaluation of a bivariate relationship, the data analyst can also specify the conditions under which the bivariate relationship occurs. A **specification** occurs when the association between the independent and dependent variables varies across the categories of one or more other control variables. This is what we just found in Exhibit 14.35.

The subtables in Exhibit 14.36 allow an evaluation of whether race specifies the effect of income on voting, as suggested by previous research. The percentages who voted in each of the family income categories vary less among African Americans (gamma = −.21) than among whites (gamma = −.47) or respondents who identify themselves as members of other minority groups (gamma = −.43). Race, therefore, does appear to specify somewhat the association between income and voting: The likelihood of African American respondents having voted varies less with their family income than it does among whites or other minority respondents. The lower rate of voting among members of other minority groups is itself of interest; investigation of the reason for this would make for an

> **Specification** A type of relationship involving three or more variables in which the association between the independent and dependent variables varies across the categories of one or more other control variables.

Exhibit 14.35 Voting in 2008 by Income and Education

Voting	Family Income			
	<$20,000	$20,000–$39,999	$40,000–$74,999	$75,000+
Education = Grade school				
Voted	38.0%	40.5%	55.2%	55.6%
Did not vote	62.0	59.5	44.8	44.4
Total	100%	100%	100%	100%
(n)	(137)	(79)	(29)	(9)
Education = High school				
Voted	58.4%	67.4%	65.6%	85.1%
Did not vote	41.6	32.6	34.4	14.9
Total	100%	100%	100%	100%
(n)	(125)	(138)	(131)	(74)
Education = Some college				
Voted	63.5%	80.6%	78.3%	81.5%
Did not vote	36.5	19.4	21.7	18.5
Total	100%	100%	100%	100%
(n)	(104)	(103)	(138)	(108)
Education = College graduate				
Voted	75.5%	82.6%	86.1%	92.2%
Did not vote	24.5	17.4	13.9	7.8
Total	100%	100%	100%	100%
(n)	(49)	(69)	(144)	(269)

interesting contribution to the literature. Is it because members of other minority groups are more likely to be recent immigrants and so less engaged in the U.S. political system than whites or African Americans? Can you think of other possibilities?

I should add one important caution about constructing tables involving three or more variables. Because the total number of cells in the subtables becomes large as the number of categories of the control (third) variable increases, the number of cases that each cell percentage is based on will become correspondingly small. This effect has two important consequences. First, the number of comparisons that must be made to identify the patterns in the table as a whole becomes substantial—and the patterns may

| Exhibit 14.36 | Voting in 2008 by Income and Race |

Voting	Family Income			
	<$20,000	$20,000–$39,999	$40,000–$74,999	$75,000+
Race = White				
Voted	51.7%	70.2%	76.1%	88.7%
Did not vote	48.3	29.8	23.9	11.3
Total	100%	100%	100%	100%
(n)	(261)	(292)	(360)	(399)
Race = African American				
Voted	70.1%	73.1%	78.4%	85.7%
Did not vote	29.9	26.9	21.6	14.3
Total	100%	100%	100%	100%
(n)	(107)	(67)	(51)	(35)
Race = Other minority				
Voted	38.3%	38.7%	62.5%	76.9%
Did not vote	61.7	61.3	37.5	23.1
Total	100%	100%	100%	100%
(n)	(47)	(31)	(32)	(26)

become too complex to make much sense of them. Second, as the number of cases per category decreases, the odds that the distributions within each category vary due to chance become greater. This problem of having too many cells and too few cases can be lessened by making sure that the control variable has only a few categories and by drawing a large sample, but often neither of these steps will be sufficient to resolve the problem completely.

Regression Analysis

My goal in introducing you to cross-tabulation has been to help you think about the association among variables and to give you a relatively easy tool for describing association. In order to read most statistical reports and to conduct more sophisticated analyses of social data, you will have to extend your statistical knowledge.

Many statistical reports and articles published in social science journals use a statistical technique called **regression analysis** or **correlational analysis** to describe the association between two or more quantitative variables. The terms actually refer to different aspects of the same technique. Statistics based on regression and correlation are used very often in social science and have many advantages over cross-tabulation—as well as some disadvantages.

> **Regression analysis** A statistical technique for characterizing the pattern of a relationship between two quantitative variables in terms of a linear equation and for summarizing the strength of this relationship in terms of its deviation from that linear pattern.
>
> **Correlational analysis** A statistical technique that summarizes the strength of a relationship between two quantitative variables in terms of its adherence to a linear pattern.

I give you only an overview here of this approach. Take a look at Exhibit 14.37. It's a plot, termed a *scatterplot,* of the relationship in the GSS 2010 sample between years of education and occupational prestige (a score that ranges from 0 to 100, reflecting the prestige accorded to the respondent's occupation by people in America). You can see that I didn't collapse the values of either of these variables into categories, as I had to do in order to use them in the preceding cross-tabular analysis. Instead, the scatterplot shows the location of each case in the data in terms of years of education (the horizontal axis) and occupational prestige level (the vertical axis).

You can see that the data points in the scatterplot tend to run from the lower left to the upper right of the chart, indicating a positive relationship: The more the years of education, the higher the occupational prestige. The line drawn through the points is the regression line. The regression line summarizes this positive relationship between years of education, which is the independent variable (often simply termed X in regression analysis), and occupational prestige, the dependent variable (often simply termed Y in regression analysis). This regression line is the "best fitting" straight line for this relationship—it is the line that lies closest to all the points in the chart, according to certain criteria.

How well does the regression line fit the points? In other words, how close does the regression line come to the points? (Actually, it's the square of the vertical distance, on the *y*-axis, between the points and the regression

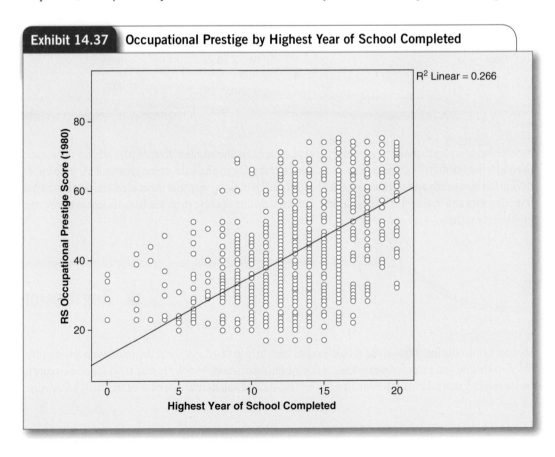

Exhibit 14.37   Occupational Prestige by Highest Year of School Completed

R^2 Linear = 0.266

line that is used as the criterion.) The **correlation coefficient**, also called *Pearson's r*, or just *r*, gives one answer to that question. The value of *r* for this relationship is .52, which indicates a moderately strong positive linear relationship (if it were a negative relationship, *r* would have a negative sign). The value of *r* is 0 when there is absolutely no linear relationship between the two variables, and it is 1 when all the points representing all the cases lie exactly on the regression line (which would mean that the regression line describes the relationship perfectly).

So the correlation coefficient does for a scatterplot such as this what gamma does for a cross-tabulation table: It is a summary statistic that tells us about the strength of the association between the two variables. Values of *r* close to 0 indicate that the relationship is weak; values of *r* close to ±1 indicate the relationship is strong—in between there is a lot of room for judgment. You will learn in a statistics course that r^2 is often used instead of *r*.

You can also use correlation coefficients and regression analysis to study simultaneously the association between three or more variables. In such a *multiple regression analysis,* you could test to see whether several other variables in addition to education are associated simultaneously with occupational prestige scores—that is, whether the variables have independent effects on occupational prestige. As an example, Exhibit 14.38 presents the key statistics obtained in a multiple regression analysis I conducted with the GSS 2010 data to do just that: I regressed occupational prestige score on years of schooling, age, sex, and race (dichotomized).

First look at the numbers under the Beta Coefficient heading. Beta coefficients are standardized statistics that indicate how strong the linear association is between the dependent variable (occupational prestige, in this case) and each independent variable, while the other independent variables are controlled. Like the correlation coefficient (*r*), values of beta range from 0, when there is no linear association, to ±1.0, when the association falls exactly on a straight line. You can see in the beta column that education has a moderate positive independent association (or with occupational prestige, while age has a weak association and race has a marginal one). In the Significance Level column, we can see that each of these three effects is statistically significant at the .05 level or the .001 level. Sex does not have any independent association with occupational prestige. We learn from the summary statistic, R^2 (*r*-squared), that the four independent variables together explain, or account for, 28.5% of the total variation in occupational prestige scores.

You will need to learn more about when correlation coefficients and regression analysis are appropriate (e.g., both variables have to be quantitative, and the relationship has to be linear [not curvilinear]). But that's for another time and place. To learn more about correlation coefficients and regression analysis, you should

> **Correlation coefficient** A summary statistic that varies from 0 to 1 or –1, with 0 indicating the absence of a linear relationship between two quantitative variables and 1 or –1 indicating that the relationship is completely described by the line representing the regression of the dependent variable on the independent variable.

Exhibit 14.38   **Multiple Regression of Determinants of Voting**

Variable	Beta Coefficient	Significance Level
Education	.52	$p < .001$
Age	.12	$p < .001$
Sex	−.02	NS
Race (dichotomized)	−.05	$p < .05$
R^2	.285	
N	1885	

Source: 2010 General Social Survey.

take an entire statistics course. For now, this short introduction will enable you to make sense of more of the statistical analyses you find in research articles. You can learn more about these techniques with the tutorials on the text's study site.

Analyzing Data Ethically: How Not to Lie About Relationships

When the data analyst begins to examine relationships among variables in some real data, social science research becomes most exciting. The moment of truth, it would seem, has arrived. Either the hypotheses are supported or they are not. But, in fact, this is also a time to proceed with caution and to evaluate the analyses of others with even more caution. Once large data sets are entered into a computer, it becomes very easy to check out a great many relationships; when relationships are examined among three or more variables at a time, the possibilities become almost endless. In fact, regression analysis (extended to what is termed *multiple regression analysis*), allows a researcher to test easily for many relationships at the same time.

This range of possibilities presents a great hazard for data analysis. It becomes tempting to search around in the data until something interesting emerges. Rejected hypotheses are forgotten in favor of highlighting what's going on in the data. It's not wrong to examine data for unanticipated relationships; the problem is that inevitably some relationships among variables will appear just on the basis of chance association alone. If you search hard and long enough, it will be possible to come up with something that really means nothing.

A reasonable balance must be struck between deductive data analysis to test hypotheses and inductive analysis to explore patterns in a data set. Hypotheses formulated in advance of data collection must be tested as they were originally stated; any further analyses of these hypotheses that involve a more exploratory strategy must be labeled in research reports as such. Serendipitous findings do not need to be ignored, but they must be reported as such. Subsequent researchers can try to test deductively the ideas generated by our explorations.

We also have to be honest about the limitations of using survey data to test causal hypotheses. The usual practice for those who seek to test a causal hypothesis with nonexperimental survey data is to test for the relationship between the independent and dependent variables, controlling for other variables that might possibly create a spurious relationship. This is what we did by examining the relationship between income and voting while controlling for education (Exhibit 14.35). (These subtables actually show that education specifies the relationship between family income and voting—there is no relationship for those with only a grade-school education, but the relationship exists for those who finished high school and those who attended college. Education does not actually explain the income–voting relationship.)

But finding that a hypothesized relationship is not altered by controlling for just one variable does not establish that the relationship is causal—nor does controlling for two, three, or many more variables. There always is a possibility that some other variable that we did not think to control, or that was not even measured in the survey, has produced a spurious relationship between the independent and dependent variables in our hypothesis (Lieberson 1985). We have to think about the possibilities and be cautious in our causal conclusions.

It is also important to understand the statistical techniques we are using and to use them appropriately. In particular, the analyst who uses regression analysis has to make a number of assumptions about the variables in the analysis; when these assumptions are violated, regression results can be very misleading. (You just might want to rush right out and buy this statistics text at this point: Frankfort-Nachmias, Chava and Anna Leon-Guerrero. 2008. *Social Statistics for a Diverse Society,* 5th ed. Thousand Oaks, CA: Sage Publications.)

🔲 **Conclusions**

This chapter has demonstrated how a researcher can describe social phenomena, identify relationships among them, explore the reasons for these relationships, and test hypotheses about them. Statistics provide a remarkably useful tool for developing our understanding of the social world, a tool that we can use both to test our ideas and to generate new ones.

Unfortunately, to the uninitiated, the use of statistics can seem to end debate right there—you can't argue with the numbers. But you now know better than that. The numbers will be worthless if the methods used to generate the data are not valid; and the numbers will be misleading if they are not used appropriately, taking into account the type of data to which they are applied. And even assuming valid methods and proper use of statistics, there's one more critical step, because the numbers do not speak for themselves. Ultimately, it is how we interpret and report the statistics that determines their usefulness.

Key Terms

Bar chart   452
Base number (N)   455
Bimodal   461
Central tendency   451
Chi-square   477
Combined frequency display   458
Compressed frequency display   458
Contingency table   470
Correlation coefficient   485
Correlational analysis   484
Cross-tabulation (crosstab)   470
Curvilinear   475
Data cleaning   449
Descriptive statistics   446

Elaboration analysis   478
Extraneous variable   480
Frequency distribution   454
Frequency polygon   453
Gamma   476
Histogram   452
Interquartile range   466
Marginal distribution   471
Mean   462
Measure of association   476
Median   461
Mode   461
Monotonic   475
Normal distribution   467

Outlier   466
Percentages   471
Probability average   461
Quartiles   466
Range   466
Regression analysis   484
Skewness   451
Specification   481
Standard deviation   467
Statistical significance   478
Subtables   479
Unimodal   461
Variability   451
Variance   466

Highlights

- Data entry options include direct collection of data through a computer, use of scannable data entry forms, and use of data entry software. All data should be cleaned during the data entry process.

- Use of secondary data can save considerable time and resources but may limit data analysis possibilities.

- Bar charts, histograms, and frequency polygons are useful for describing the shape of distributions. Care must be taken with graphic displays to avoid distorting a distribution's apparent shape.

- Frequency distributions display variation in a form that can be easily inspected and described. Values should be grouped in frequency distributions in a way that does not alter the shape of the distribution. Following several guidelines can reduce the risk of problems.

- Summary statistics often are used to describe the central tendency and variability of distributions. The appropriateness of the mode, mean, and median vary with a variable's level of measurement, the distribution's shape, and the purpose of the summary.

- The variance and standard deviation summarize variability around the mean. The interquartile range is usually preferable to the range to indicate the interval spanned by cases, due to the effect of outliers on the range. The degree of skewness of a distribution is usually described in words rather than with a summary statistic.

- Some of the data in many reports can be displayed more efficiently by using combined and compressed statistical displays.

- Honesty and openness are the key ethical principles that should guide data summaries.

- Cross-tabulations should normally be percentaged within the categories of the independent variable. A cross-tabulation can be used to determine the existence, strength, direction, and pattern of an association.

- Elaboration analysis can be used in cross-tabular analysis to test for spurious and mediating relationships and to specify the conditions under which relationships occur.

- Inferential statistics are used with sample-based data to estimate the confidence that can be placed in a statistical estimate of a population parameter. Estimates of the probability that an association between variables may have occurred on the basis of chance are also based on inferential statistics.

- Regression analysis is a statistical method for characterizing the relationship between two or more quantitative variables with a linear equation and for summarizing the extent to which the linear equation represents that relationship. Correlation coefficients summarize the fit of the relationship to the regression line.

STUDENT STUDY SITE

To assist in completing the web exercises, please access the study site at **www.sagepub.com/schuttisw7e**, where you will find the web exercises with accompanying links. You'll find other useful study materials such as self-quizzes and e-flashcards for each chapter, along with a group of carefully selected articles from research journals that illustrate the major concepts and techniques presented in the book.

Discussion Questions

1. I presented in this chapter several examples of bivariate and trivariate cross-tabulations involving voting in the 2008 presidential election. What additional influences would you recommend examining to explain voting in elections? Suggest some additional independent variables for bivariate analyses with voting as well as several additional control variables to be used in three-variable crosstabs.

2. When should we control . . . just to be honest? In the evaluation project that I described in Chapter 11, I analyzed with some colleagues the effect on cognitive functioning of living in group homes rather than individual apartments. I found that living in group homes resulted in gains in cognitive functioning, compared with living in individual apartments. However, this benefit of group homes occurred only for residents who were not substance abusers; substance abusers did not gain cognitively from living in group (or individual) homes (Caplan et al. 2006). Would it have been alright if we had just reported the bivariate association between housing type and change in cognitive functioning? Should social researchers be expected to investigate alternative explanations for their findings? Should they be expected to check to see if the associations they find occur for different subgroups in their samples?

Practice Exercises

1. Exhibit 14.39 shows a frequency distribution of "trust in people" as produced by the Statistical Package for the Social Sciences (SPSS) with the 2010 GSS data. As you can see, the table includes abbreviated labels for the variable and its response choices, as well as the raw frequencies and three percentage columns. The first percentage column (Percent) shows the percentage in each category of trust; the next percentage column (Valid Percent) is based on the total number of respondents who gave valid answers (1,372 in this instance). It is the Valid Percent column that normally should be used to construct a

frequency distribution for presentation. The last percentage column is Cumulative Percent, adding up the valid percents from top to bottom.

Redo the table for presentation, using the format of the frequency distributions presented in the text.

2. Try your hand at recoding. Start with the distribution of the political ideology variable from Exhibit 14.10. It is named POLVIEWS in the GSS. Recode it to just three categories. What decision did you make about grouping? What was the consequence of this decision for the shape of the distribution? For the size of the middle category?

3. Cross-tabulations produced by most statistical packages are not in the proper format for inclusion in a report, and so they have to be reformatted. Referring to Exhibit 14.40, rewrite the table in presentational format, using one of the other tables as your guide. Describe the association in the table in terms of each of the four aspects of association. A chi-square test of statistical significance resulted in a *p* value of .000, meaning that the actual value was less than .001. State the level of confidence that you can have that the association in the table is not due to chance.

4. What if you had to answer this question: What was the income distribution of voters in the 2008 presidential election, and how did it compare with the income distribution for those who didn't vote? Can you answer this question exactly with Exhibit 14.23? If not, change the column percentages in the table to row percentages. To do this, you will first have to convert the column percentages back to cell frequencies (although the frequencies are included in the table, so you can check your work). You can do this by multiplying the column percentage by the number of cases in the column, and then dividing by 100 (you will probably have fractional values because of rounding error). Then, compute the row percentage from these frequencies and the row totals.

5. Exhibit 14.40 contains a cross-tabulation of voting by education (recoded) directly as output by SPSS from the 2010 GSS data set. Describe the row and column marginal distributions. Try to calculate a cell percentage using the frequency (count) in that cell and the appropriate base number of cases.

6. Now, review the data analysis presented in one of the articles on the book's website, www.sagepub.com/schuttisw7e, in which statistics were used. What do you learn from the data analysis? Which questions do you have about the meaning of the statistics used?

Ethics Questions

1. Review the frequency distributions and graphs in this chapter. Change one of these data displays so that you are "lying with statistics." (You might consider using the graphic technique discussed by Orcutt & Turner 1993.)

2. Consider the relationship between voting and income that is presented in Exhibit 14.23. What third variables do you think should be controlled in the analysis to better understand the basis for this relationship? How might social policies be affected by finding out that this relationship was due to differences in neighborhood of residence rather than to income itself?

Exhibit 14.39 **Distribution of "Can People Be Trusted?"**

		Frequency	Percent	Valid Percent	Cumulative Percent
Valid	CAN TRUST	458	22.4	33.4	33.4
	CANNOT TRUST	839	41.0	61.2	94.5
	DEPENDS	75	3.7	5.5	100.0
	Total	1372	67.1	100.0	

Exhibit 14.40 Vote in 2008 Election by Education (Trichotomized)

			Education Trichotomized			
			Grade School	High School	Some College	Total
Voting in 2008 election	Voted	Count	138	361	889	1388
		% within education	45.0%	68.2%	82.2%	72.4%
	Did not vote	Count	169	168	192	529
		% within education	55.0%	31.8%	17.8%	27.6%
	Total	Count	307	529	1081	1917
		% within education	100.0%	100.0%	100.0%	100.0%

Web Exercises

1. Search the web for a social science example of statistics. Using the key terms from this chapter, describe the set of statistics you have identified. Which social phenomena does this set of statistics describe? What relationships, if any, do the statistics identify?

2. Go to the Roper Center for Public Opinion Research website at www.ropercenter.uconn.edu. Now, pick the presidential approval ratings data, at http://webapps.ropercenter.uconn .edu/CFIDE/roper/presidential/webroot/presidential_rating .cfm. Choose any two U.S. presidents from Franklin D. Roosevelt to the present. By using the website links, locate the presidential job performance poll data for the two presidents you have chosen.

    Based on poll data on presidential job performance, create a brief report that includes the following for each president you chose: the presidents you chose and their years in office; the question asked in the polls; and bar charts showing years when polls were taken, average of total percentage approving of job performance, average of total percentage disapproving of job performance, and average of total percentage with no opinion on job performance. Write a brief summary comparing and contrasting your two bar charts.

3. Do a web search for information on a social science subject you are interested in. How much of the information you find relies on statistics as a tool for understanding the subject? How do statistics allow researchers to test their ideas about the subject and generate new ideas? Write your findings in a brief report, referring to the websites upon which you relied.

SPSS Exercises

1. Develop a description of the basic social and demographic characteristics of the U.S. population in 2010. Examine each characteristic with three statistical techniques: a graph, a frequency distribution, and a measure of central tendency (and a measure of variation, if appropriate).

    a. From the menu, select Graphs and then Legacy Dialogs and Bar. Select Simple Define [Marital—Category Axis]. Bars represent % of cases. Select Options (do not display groups defined by missing values). Finally, select Histogram for each of the variables [EDUC, EARNRS, TVHOURS, ATTEND].

b. Describe the distribution of each variable.

c. Generate frequency distributions and descriptive statistics for these variables. From the menu, select Analyze/Descriptive Statistics/Frequencies. From the Frequencies window, set MARITAL, EDUC, EARNRS, TVHOURS, ATTEND. For the Statistics, choose the mean, median, range, and standard deviation.

d. Collapse the categories for each distribution. Be sure to adhere to the guidelines given in the section "Grouped Data." Does the general shape of any of the distributions change as a result of changing the categories?

e. Which statistics are appropriate to summarize the central tendency and variation of each variable? Do the values of any of these statistics surprise you?

2. Try describing relationships with support for capital punishment by using graphs. Select two relationships you identified in previous exercises and represent them in graphic form. Try drawing the graphs on lined paper (graph paper is preferable).

3. On the study site, you will find a file containing a subset of the 2010 General Social Survey, allowing you to easily replicate the tables in this chapter. It would be a good exercise for you to do, but remember that if you are using the GSS2010x file, your tables will contain only a sample of cases from the complete 2010 GSS file that I used. The computer output you get will probably not look like the tables shown here, because I reformatted the tables for presentation, as you should do before preparing a final report. At this point, I'll let you figure out the menu commands required to generate these graphs, frequency distributions, and cross-tabulations. If you get flummoxed, go to the book study site and review my tutorial on cross-tabulation with SPSS.

4. Propose a variable that might have created a spurious relationship between income and voting. Explain your thinking. Propose a variable that might result in a conditional effect of income on voting, so that the relationship between income and voting would vary across the categories of the other variable. Test these propositions with three-variable cross-tabulations. Were any supported? How would you explain your findings?

Developing a Research Proposal

Use the GSS data to add a pilot study to your proposal (Exhibit 2.14, #18). A pilot study is a preliminary effort to test out the procedures and concepts that you have proposed to research.

1. In SPSS, review the GSS2010x variable list and identify some variables that have at least some connection to your research problem. If possible, identify one variable that might be treated as independent in your proposed research and one that might be treated as dependent.

2. Request frequencies for these variables.

3. Request a cross-tabulation of one dependent variable by a variable you are treating as independent. If necessary, recode the independent variable to five or fewer categories.

4. Write a brief description of your findings and comment on their implications for your proposed research. Did you learn any lessons from this exercise for your proposal?

Summarizing and Reporting Research

Y ou learned in Chapter 2 that research is a circular process, so it is appropriate that we end this book where we began. The stage of reporting research results is also the point at which the need for new research is identified. It is the time when, so to speak, "the rubber hits the road"—when we have to make our research make sense to others. To whom will our research be addressed? How should we present our results to them? Should we seek to influence how our research report is used?

      The primary goals of this chapter are to guide you in comparing completed research studies, writing worthwhile reports of your own, and communicating with the public about research. I will introduce one new research technique—**meta-analysis**, which is a quantitative method for statistically evaluating the results of a large body of prior research on a specific topic. This chapter also gives

Meta-analysis The quantitative analysis of findings from multiple studies.

particular attention to the writing process itself and points out how that process can differ when writing up qualitative versus quantitative research. I will conclude by considering some of the ethical issues unique to the reporting process, with special attention to the problem of plagiarism.

▣ Comparing Research Designs

The central features of experiments, surveys, qualitative methods, and comparative historical methods provide distinct perspectives even when used to study the same social processes. Comparing subjects randomly assigned to a treatment and to a comparison group, asking standard questions of the members of a random sample, observing while participating in a natural social setting, recording published statistics on national characteristics, and reading historical documents involve markedly different decisions about measurement, causality, and generalizability. As you can see in Exhibit 15.1, not one of these methods can reasonably be graded as superior to the others in all respects, and each varies in its suitability to different research questions and goals. Choosing among them for a particular investigation requires consideration of the research problem, opportunities and resources, prior research, philosophical commitments, and research goals.

Experimental designs are strongest for testing nomothetic causal hypotheses and are most appropriate for studies of treatment effects (see Chapter 7). Research questions that are believed to involve basic social psychological processes are most appealing for laboratory studies because the problem of generalizability is reduced. Random assignment reduces the possibility of preexisting differences between treatment and comparison groups to small, specifiable, chance levels; therefore, many of the variables that might create a spurious association are controlled. But in spite of this clear advantage, an experimental design requires a degree of control that cannot always be achieved outside the laboratory. It can be difficult to ensure in real-world settings that a treatment is delivered as intended and that other influences do not intrude. As a result, what appears to be a

| Exhibit 15.1 | Comparison of Research Methods |

Design	Measurement Validity	Generalizability	Type of Causal Assertions	Causal Validity
Experiments	+	−	Nomothetic	+
Surveys	+	+	Nomothetic	−/+[a]
Qualitative Methods	−/+[b]	−	Idiographic	−
Comparative[c]	−	−/+	Idiographic or Nomothetic	−

a. Surveys are a weaker design for identifying casual effects than true experiments, but use of statistical controls can strengthen causal arguments.

b. Reliability is low compared with surveys, and systematic evaluation of measurement validity is often not possible. However, direct observations may lead to great confidence in the validity of measures.

c. All conclusions about this type of design vary with the specific approach used. See Chapter 12.

treatment effect or noneffect may be something else altogether. Field experiments thus require careful monitoring of the treatment process. Unfortunately, most field experiments also require more access arrangements and financial resources than can often be obtained.

Laboratory experiments permit much more control over conditions, but at the cost of less generalizable findings. People must volunteer for most laboratory experiments, and so there is a good possibility that experimental subjects differ from those who do not volunteer. Ethical and practical constraints limit the types of treatments that can be studied experimentally (you can't assign social class or race experimentally). The problem of generalizability in an experiment using volunteers lessens when the object of investigation is an orientation, behavior, or social process that is relatively invariant among people, but it is difficult to know which orientations, behaviors, or processes are so invariant. If a search of the research literature on the topic identifies many prior experimental studies, the results of these experiments will suggest the extent of variability in experimental effects and point to the unanswered questions about these effects.

Both surveys and experiments typically use standardized, quantitative measures of attitudes, behaviors, or social processes. Closed-ended questions are most common and are well suited for the reliable measurement of variables that have been studied in the past and whose meanings are well understood (see Chapter 4). Of course, surveys often include measures of many more variables than are included in an experiment (Chapter 8), but this feature is not inherent in either design. Phone surveys may be quite short, whereas some experiments can involve very lengthy sets of measures (see Chapter 7). The set of interview questions we used at baseline in the Boston housing study (Chapter 11), for example, required more than 10 hours to complete. The level of funding for a survey will often determine which type of survey is conducted and thus how long the questionnaire is.

Most social science surveys rely on random sampling for their selection of cases from some larger population, and it is this feature that makes them preferable for descriptive research that seeks to develop generalizable findings (see Chapter 5). However, survey questionnaires can only measure what respondents are willing to report verbally; they may not be adequate for studying behaviors or attitudes that are regarded as socially unacceptable. Surveys are also often used to test hypothesized causal relationships. When variables that might create spurious relationships are included in the survey, they can be controlled statistically in the analysis and thus eliminated as rival causal influences.

Research in the News

JULIET, JULIET, WHEREFORE ART THOU?

Research has shown marital discord has an equally negative effect on both spouses, but the same is not true among unmarried young adults. It is young men who benefit more than women from support and are more harmed than women by strain in romantic relationships—although women are more affected by whether they're in a relationship at all. Could it be that women benefit from other ongoing intimate friendships, while men have only their female companion for ongoing care?

Source: Paul, Pamela. 2010. "A Young Man's Lament: Love Hurts!" *The New York Times,* July 25:L6.

Qualitative methods presume an intensive measurement approach in which indicators of concepts are drawn from direct observation or in-depth commentary (see Chapter 9). This approach is most appropriate when it is not clear what meaning people attach to a concept or what sense they might make of particular questions about it. Qualitative methods are also admirably suited to the exploration of new or poorly understood social settings, when it is not even clear what concepts would help one understand the situation. They may also be used instead of survey methods when the population of interest is not easily identifiable or seeks to remain hidden. For these reasons, qualitative methods tend to be preferred when exploratory research questions are posed or when new groups are investigated. But, of course, intensive measurement necessarily makes the study of large numbers of cases or situations difficult, resulting in the limitation of many field research efforts to small numbers of people or unique social settings. The individual field researcher may not require many financial resources, but the amount of time required for many field research projects serves as a barrier to many would-be field researchers.

When qualitative methods can be used to study several individuals or settings that provide marked contrasts in terms of a presumed independent variable, it becomes possible to evaluate nomothetic causal hypotheses with these methods. However, the impossibility of taking into account many possible extraneous influences in such limited comparisons makes qualitative methods a weak approach to hypothesis testing. Qualitative methods are more suited to the elucidation of causal mechanisms. In addition, qualitative methods can be used to identify the multiple successive events that might have led to some outcome, thus identifying idiographic causal processes.

Historical and comparative methods range from cross-national quantitative surveys to qualitative comparison of social features and political events (see Chapter 12). Their suitability for exploration, description, explanation, and evaluation varies in relation to the particular method used, but they are essential for research on historical processes and national differences. If the same methods are used to study multiple eras or nations rather than just one nation at one time, both the generalizability and the causal validity of conclusions can be increased.

Performing Meta-Analyses

A meta-analysis is a quantitative method for identifying patterns in findings across multiple studies of the same research question (Cooper & Hedges 1994). Unlike a traditional literature review, which describes previous research studies verbally, meta-analyses treat previous studies as cases, whose features are measured as variables and then analyzed statistically. It is like conducting a survey in which the "respondents" are previous studies. Meta-analysis shows how evidence about social processes varies across research studies. If the methods used in these studies varied, then meta-analysis can describe how this variation affected the study findings. If social contexts varied across the studies, then meta-analysis will indicate how social context affected the study findings.

Meta-analysis can be used when a number of studies have attempted to answer the same research question with similar quantitative methods, most often experiments. Meta-analysis is not appropriate for evaluating results from qualitative studies or from multiple studies that used different methods or measured different dependent variables. It is also not very sensible to use meta-analysis to combine study results when the original case data from these studies are available and can actually be combined and analyzed together (Lipsey & Wilson 2001). Meta-analysis is a technique for combination and statistical analysis of published research reports.

After a research problem is formulated based on the findings of prior research, the literature must be searched systematically to identify the entire population of relevant studies. Typically, multiple bibliographic databases are used; some researchers also search for relevant dissertations and conference papers. Once the studies are identified, their findings, methods, and other features are coded (e.g., sample size, location of sample, and strength of the association between the independent and dependent variables). Eligibility criteria must be specified carefully to determine which studies to include and which to omit as too different. Mark Lipsey and David Wilson (2001:16–21) suggested that eligibility criteria include the following:

- *Distinguishing features:* This includes the specific intervention tested and perhaps the groups compared.

- *Research respondents:* The pertinent characteristics of the research respondents (subject sample) who provided study data must be similar to those of the population about which generalization is sought.

- *Key variables:* These must be sufficient to allow tests of the hypotheses of concern and controls for likely additional influences.

- *Research methods:* Apples and oranges cannot be directly compared, but some trade-off must be made between including the range of studies about a research question and excluding those that are so different in their methods as not to yield comparable data.

- *Cultural and linguistic range:* If the study population is going to be limited to English-language publications, or limited in some other way, this must be acknowledged, and the size of the population of relevant studies in other languages should be estimated.

- *Time frame:* Social processes relevant to the research question may have changed for reasons such as historical events or the advent of new technologies, so temporal boundaries around the study population must be considered.

- *Publication type:* It must be determined whether the analysis will focus only on published reports in professional journals, or include dissertations and/or unpublished reports.

Statistics are then calculated to identify the average effect of the independent variable on the dependent variable, as well as the effect of methodological and other features of the studies (Cooper & Hedges 1994). The **effect size** statistic is the key to capturing the association between the independent and dependent variables across multiple studies. The effect size statistic is a standardized measure of association—often the difference between the mean of the experimental group and the mean of the control group on the dependent variable, adjusted for the average variability in the two groups (Lipsey & Wilson 2001).

> **Effect size** A standardized measure of association—often the difference between the mean of the experimental group and the mean of the control group on the dependent variable, adjusted for the average variability in the two groups.

The meta-analytic approach to synthesizing research findings can result in much more generalizable findings than those obtained with just one study. Methodological weaknesses in the studies included in the meta-analysis are still a problem, however; it is only when other studies without particular methodological weaknesses are included that we can estimate effects with some confidence. In addition, before we can place any confidence in the results of a meta-analysis, we must be confident that all (or almost all) relevant studies were included and that the information we need to analyze was included in all (or most) of the studies (Matt & Cook 1994).

Case Study: Patient–Provider Race-Concordance and Minority Health Outcomes

Do minority patients have better health outcomes when they receive treatment from a provider of the same race or ethnicity? Salimah H. Meghani and other researchers in nursing at the University of Pennsylvania and other Pennsylvania institutions sought to answer this question with a meta-analysis of published research. Their research report illustrates the key steps in a meta-analysis (Meghani et al. 2009).

They began their analysis with a comprehensive review of published research that could be located in three health-related bibliographic databases with searches for English language research articles linked to the key words *race, ethnicity, concordance,* or *race-concordance.* This search identified 159 articles; after reading the abstracts of these articles, 27 were identified that had investigated a research question about the effect of patient–provider race-concordance on minority patients' health outcomes (see Exhibit 15.2).

Meghani and her coauthors then summarized the characteristics and major findings of the selected studies (see Exhibit 15.3). Finally, each study was classified according to the health outcome(s) examined and its findings about the effect of race-concordance on each outcome (see Exhibit 15.4). Because only 9 of the 27 studied

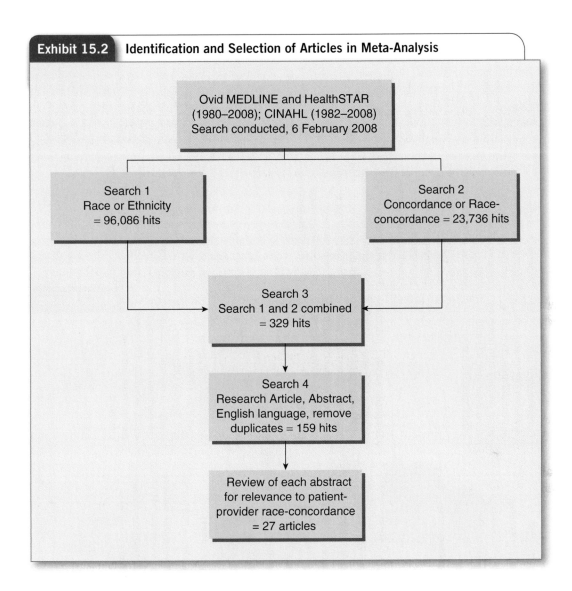

Exhibit 15.2 Identification and Selection of Articles in Meta-Analysis

provided support for a positive effect of race-concordance on outcomes, and in many of these studies the effects were modest, Meghani and her coauthors concluded that patient–provider racial concordance had little relevance to health care outcomes.

Case Study: Broken Homes and Delinquency

Many studies have tested the hypothesis that juveniles from broken homes have higher rates of delinquency than those from homes with intact families, but findings have been inconclusive. L. Edward Wells and Joseph Rankin (1991) were able to find 50 studies that tested this hypothesis, with estimates of the increase in delinquency among juveniles from broken homes ranging from 1% to 50%. To explain this variation, Wells and Rankin coded key characteristics of the research studies, such as the population sampled—the general population? a specific age range?—and the measures used: Did researchers take account of stepparents? Did they measure juveniles' relations with the absent parent? Was delinquency measured with official records or

Exhibit 15.3 Summary of Studies in Meta-Analysis

Citation	Aims	Sample Characteristics	Design/Setting	Major Findings	Limitations
Studies supporting race-concordance hypothesis (n = 9)					
Cooper et al. (2003)	Does race-concordance affect patient-physician communication and patients' ratings of physicians' participatory decision making?	Patients (N = 252) • Whites = 110 (44%) • African Americans = 142 (56%) Physicians (N = 31) • Whites = 13 (42%) • African Americans = 18 (58%)	Cohort study of pre and post-visit follow-up surveys and audiotaped analysis from 16 urban primary care practices in the Baltimore, MD and Washington DC area.	Patients in ethnic-concordant encounters had longer and more meaningful visits, had higher coder rating for positive affect, and had higher patient ratings for satisfaction and positive judgments of their physician's participatory decision-making style.	– Small physician sample. – Limited statistical power to detect differences in speech speed, patient centered interviewing and physicians' positive affect.
Cooper-Patrick et al. (1999)	What is the association between race and gender concordance or discordance in the patient-physician relationship and participatory decision making?	Patients (N = 1816) • Whites = 784 (43%) • African Americans = 814 (45%) • Hispanics = 218 (12%) Physicians (N = 64) • Whites = 36 (56%) • African Americans = 16 (25%) • Hispanics = 2 (3%) • Asians = 10 (15%)	Telephone survey conducted between 1996 and 1998 of adults who had attended one of the 32 primary care practices in an urban, primary care setting in Washington, DC area.	Patients in race-concordant relationships with their physicians rated their visits as significantly more participatory than patients in race-discordant relationships.	– Small sample of minority physicians. – Data based on patient self-report. – Data may be confounded by physician or practice-related variables not included in the study.
King et al. (2004)	Does race-concordance explain why African Americans are less likely than Whites to receive antiretroviral treatment?	Patients (N = 1241) • White-White = 803 (61%) • African American patients-White providers = 341 (32%) • African American-African American = 86 (6%)	Secondary analysis of data from 1996 HIV Cost and Service Utilization Study – a national probability, prospective cohort study of adults receiving HIV-related medical care and their providers.	African-American patients with White providers received protease inhibitors significantly later than African Americans with African American providers and White patients with White providers.	– Relied on self-reported dates for the receipt of protease inhibitor. – Not enough patients to allow both races cared for by same provider. – Small sample for African American providers.

Exhibit 15.4 **Identification and Selection of Articles in Meta-Analysis**

Category	Specific race-concordance outcome studied	Support for race-concordance?
Provision of health care (*n* = 8)		
King et al. (2004)	Time to receipt of protease inhibitor in HIV-positive patients	+
Malat (2001)[a]	Rating of time spent during last medical visit	−
McKinlay et al. (2002)	Diagnosis of depression and polymyalgia rheumatica, level of certainty, and test ordering	−
Modi et al. (2007)	Recommendations for percutaneous endoscopic gastrostomy in patients with advanced dementia	+
Stevens et al. (2003)	Parents' report of primary care experiences of their children	−
Stevens et al. (2005)	Receipt of basic preventative services or family centered care	−
Tai-Seale et al. (2005)	Assessment of depression in elderly	−
Zayas et al. (2005)	Diagnoses of psychiatric illness	−
Utilization of health care (*n* = 7)		
Konrad et al. (2005)	Use of antihypertensive medications	+/−
Lasser et al. (2005)	Missed appointment rates in primary care	+
LaVeist et al. (2003)	Failure to use needed care and delay in using needed care	+/−
Murray-Garcia et al. (2001)	Visits made to race-concordant residents	+
Saha et al. (1999)	Use of preventive care and needed health services	+/−
Saha et al. (2003)[a]	Use of basic health-care services	−
Sterling et al. (2001)	Retention in outpatient substance abuse treatment	−
Patient — provider communication (*n* = 5)		
Brown et al. (2007)	Pediatrician-parent communication patterns in medical encounters	+/−
Clark et al. (2004)	Physician-patient agreement on change in behavior (diet, exercise, medication, smoking, stress and weight)	−
Cooper-Patrick et al. (1999)	Patient-provider participatory decision-making	+
Cooper et al. (2003)[a]	Patient-centered communication	+
Gordon et al. (2006)	Doctors' information-giving in lung cancer consultations	+/−
Patient satisfaction (*n* = 5)		
Cooper et al. (2003)[a]	Satisfaction and rating of care	+
LaVeist & Nuru-Jeter (2002)	Patient satisfaction with provider of same race	+
LaVeist & Carroll (2002)	Patient satisfaction with provider of same race	+
Saha et al. (1999)	Patient satisfaction with provider of same race	+/−
Saha et al. (2003)[a]	Patient satisfaction with health care	−

by self-report? What types of delinquency were measured? Unlike Meghani et al.'s (2009) meta-analysis, Wells and Rankin conducted a statistical analysis of effects across the studies.

The average effect of broken homes across the studies was an increase in the likelihood of delinquency by about 10% to 15% (see Exhibit 15.5). Effects varied with the studies' substantive features and their methods, however. Juveniles from broken homes were more likely to be involved in status offenses (such as truancy and running away) and drug offenses but were no more likely to commit crimes involving theft or violence than were juveniles from intact homes. Juveniles' race, sex, and age and whether a stepparent was present did not have consistent effects. On the other hand, differences in methods accounted for much of the variation among the studies in the estimated effect of broken homes. The effect of broken homes on delinquency tended to be greater in studies using official records rather than surveys and in studies of smaller special populations rather than of the general population. In general, the differences in estimates of the association between broken homes and delinquency were due primarily to differences in study methods and only secondarily to differences in the social characteristics of the people studied.

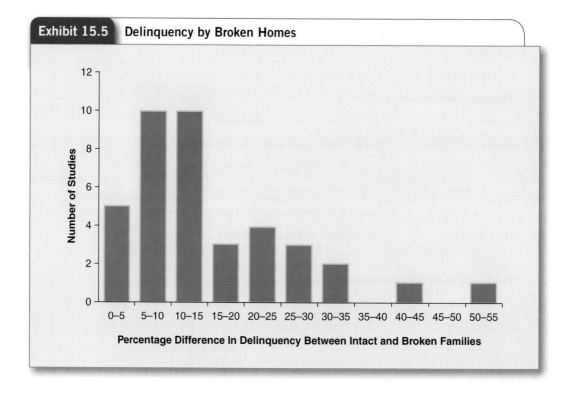

Exhibit 15.5 **Delinquency by Broken Homes**

Meta-analyses such as the Meghani et al. (2009) and Wells and Rankin (1991) studies make us aware of how hazardous it is to base understanding of social processes on single studies that are limited in time, location, and measurement. Although one study may not support the hypothesis that we deduced from what seemed to be a compelling theory, this is not a sufficient basis for discarding the theory itself or even for assuming that the hypothesis is no longer worthy of consideration in future research. You can see that a meta-analysis combining the results of many studies may identify the conditions for which the hypothesis is supported and others for which it is not.

Of course, we need to have our wits about us when we read reports of meta-analytic studies. It is not a good idea to assume that a meta-analysis is the definitive word on a research question just because it cumulates the results of multiple studies. Fink (2005:202–215) suggests evaluating meta-analytic studies in terms of the following seven criteria:

1. *A clear statement of the analytic objectives:* The study's methods cannot be evaluated without knowledge of the objectives they were intended to achieve. Meta-analyses are most appropriate for summarizing research conducted to identify the effect of some type of treatment or some other readily identifiable individual characteristic.

2. *Explicit inclusion and exclusion criteria:* On what basis were research reports included in the analysis? Were high-quality studies—those that used a rigorous research design—distinguished from low-quality studies? When high-quality and low-quality studies are distinguished, their results should be summarized separately so that it is clear if better research designs led to different results.

3. *Satisfactory search strategies:* Both electronic and written reference sources should be searched. Was some method used to find studies that were conducted but not published? It may be necessary to write directly to researchers in the field and to consult lists of papers presented at conferences.

4. *A standardized protocol for screening the literature:* Screening involves rating the quality of the study and its relevance to the research question. This screening should be carried out with a simple rating form.

5. *A standardized protocol for collecting data:* It is best to have two reviewers use a standard form for coding the characteristics of the reported research. The level of agreement between these reviewers should be assessed.

6. *Complete explanation of the method of combining results:* Some checks should be conducted to determine where particular study features influenced the size of the treatment effect.

7. *Report of results, conclusions, and limitations:* This seems obvious, but it's easy for a researcher to skirt over study limitations or some aspects of the findings.

🖳 Writing Research

The goal of research is not just to discover something but also to communicate that discovery to a larger audience: other social scientists, government officials, your teachers, the general public—perhaps several of these audiences. Whatever the study's particular outcome, if the intended audience for the research comprehends the results and learns from them, the research can be judged a success. If the intended audience does not learn about the study's results, the research should be judged a failure—no matter how expensive the research, how sophisticated its design, or how much you (or others) invested in it.

Successful research reporting requires both good writing and a proper publication outlet. We will first review guidelines for successful writing before we look at particular types of research publications.

Consider the following principles formulated by experienced writers (Booth, Colomb, & Williams 1995:150–151):

- Respect the complexity of the task and don't expect to write a polished draft in a linear fashion. Your thinking will develop as you write, causing you to reorganize and rewrite.

- Leave enough time for dead ends, restarts, revisions, and so on, and accept the fact that you will discard much of what you write.

- Write as fast as you comfortably can. Don't worry about spelling, grammar, and so on until you are polishing things up.

- Ask anyone whom you trust for reactions to what you have written.

- Write as you go along, so you have notes and report segments drafted even before you focus on writing the report.

It is important to outline a report before writing it, but neither the report's organization nor the first written draft should be considered fixed. As you write, you will get new ideas about how to organize the report. Try them out. As you review the first draft, you will see many ways to improve your writing. Focus particularly on how to shorten and clarify your statements. Make sure that each paragraph concerns only one topic. Remember the golden rule of good writing: Writing is revising!

You can ease the burden of report writing in several ways:

- Draw on the research proposal and on project notes.

- Use a word processing program on a computer to facilitate reorganizing and editing.

- Seek criticism from friends, teachers, and other research consumers before turning in the final product.

I often find it helpful to use what I call **reverse outlining**: After you have written a first complete draft, outline it on a paragraph-by-paragraph basis, ignoring the actual section headings you used. See if the paper you wrote actually fits the outline you planned. How could the organization be improved?

Reverse outlining Outlining the sections in an already written draft of a paper or report to improve its organization in the next draft.

Most important, leave yourself enough time so that you can revise, several times if possible, before turning in the final draft. Here are one student's reflections on writing and revising:

I found the process of writing and revising my paper longer than I expected. I think it was something I was doing for the first time—working within a committee—that made the process not easy. The overall experience was very good, since I found that I have learned so much. My personal computer also did help greatly.

Revision is essential until complete clarity is achieved. This took most of my time. Because I was so close to the subject matter, it became essential for me to ask for assistance in achieving clarity. My committee members, English editor, and fellow students were extremely helpful. Putting it on disk was also, without question, a timesaver. Time was the major problem.

The process was long, hard, and time-consuming, but it was a great learning experience. I work full time so I learned how to budget my time. I still use my time productively and am very careful of not wasting it. (Graduate Program in Applied Sociology 1990:13)

For more suggestions about writing, see Becker (1986), Booth et al. (1995), Cuba (2002), Strunk and White (2000), and Turabian (1996).

▣ Reporting Research

You begin writing your research report when you are working on the research proposal and writing your literature review (see Chapter 2). You will find that the final report is much easier to write, and more adequate, if you write more material for it as you work out issues during the project. It is very disappointing to discover that something important was left out when it is too late to do anything about it. And, I don't need to point out that students (and professional researchers) often leave final papers (and reports) until the last possible minute (often for understandable reasons, including other coursework and job or family responsibilities). But be forewarned: *The last-minute approach does not work for research reports.*

The organization of your research report will depend to some extent on the audience for which you are writing and the type of research you have conducted. Articles that will be submitted to an academic journal will differ from research reports written for a funding agency or for the general public. Research reports based on qualitative research will differ in some ways from those based on quantitative research. Students writing course papers are often required to structure their research report using the journal article format, and may be asked to present their results differently if they have used qualitative (or mixed) methods. The following sections outline the major issues to consider.

Journal Articles

Writing for academic journals is perhaps the toughest form of writing, because articles are submitted to several experts in your field for careful review—anonymously, with most journals—prior to acceptance for publication. Perhaps it wouldn't be such an arduous process if so many academic journals did not have rejection rates in excess of 90% and turnaround times for reviews that are usually several months. Even the best articles, in the judgment of the reviewers, are given a "revise and resubmit" after the first review and then are evaluated all over again after the revisions are concluded.

But journal article review procedures have some important benefits. First and foremost is the identification of areas in need of improvement, as the eyes of the author(s) are replaced by those of previously uninvolved, subject-matter experts and methodologists. A good journal editor makes sure that he or she has a list of many different types of experts available for reviewing whatever types of articles the journal is likely to receive. There is a parallel benefit for the author(s): It is always beneficial to review criticisms of your own work by people who know the field well. It can be a painful and time-consuming process, but the entire field moves forward as researchers continually critique and suggest improvements in each other's research reports.

Exhibit 15.6 presents an outline of the sections in an academic journal article, with some illustrative quotes. It is essential to begin with a clear abstract of the article, which summarizes in one paragraph the research question, prior research, study methods and major findings, and key conclusions. Many others who search the literature about the topic of your article will never read the entire article unless they are persuaded by the abstract that the article provides worthwhile and relevant information. The article's introduction highlights the importance of the problem selected—the relationship between marital disruption (divorce) and depression. The introduction, which in this article includes the literature review, also identifies clearly the gap in the research literature that the article is meant to fill: the untested possibility that depression might cause marital disruption rather than, or in addition to, marital disruption causing depression. Literature reviews in journal articles should be integrated reviews that highlight the major relevant findings and identify key methodological lessons from the prior research as a whole, rather than presenting separate summaries of prior research studies (see Chapter 2). The findings section (titled "Results") begins by presenting the basic

---

Exhibit 15.6 **Sections in a Journal Article**

Aseltine, Robert H. Jr. and Ronald C. Kessler. 1993. "Marital Disruption and Depression in a Community Sample." Journal of Health and Social Behavior *34(September):237–251.*

INTRODUCTION
    Despite 20 years of empirical research, the extent to which marital disruption causes poor mental health remains uncertain. The reason for this uncertainty is that previous research has consistently overlooked the potentially important problems of selection into and out of marriage on the basis of prior mental health. (p. 237)

SAMPLE AND MEASURES
Sample
Measures

RESULTS
The Basic Association Between Marital Disruption and Depression
Sex Differences
The Impact of Prior Marital Quality
The Mediating Effects of Secondary Changes
The Modifying Effects of Transitions to Secondary Roles

DISCUSSION [includes conclusions]
. . . According to the results, marital disruption does in fact cause a significant increase in depression compared to pre-divorce levels within a period of three years after the divorce. (p. 245)

---

association between marital disruption and depression. Then it elaborates on this association by examining sex differences, the impact of prior marital quality, and various mediating and modifying effects. Tables and perhaps graphs are used to present the data corresponding to each of the major findings in an easily accessible format. As indicated in the combined discussion and conclusions section, the analysis shows that marital disruption does indeed increase depression and specifies the time frame (3 years) during which this effect occurs.

These basic article sections present research results well, but many research articles include subsections tailored to the issues and stages in the specific study being reported. Most journals require a short abstract at the beginning, which summarizes the research question and findings. Most research articles include a general Methodology section that will include subsections on measurement and sampling. A Conclusions section is often used to present the most general conclusions, reflections, and limitations, rather than including these in the Discussion section.

Applied Research Reports

Applied research reports are written for a different audience from the professional social scientists and students who read academic journals. Typically, an applied report is written with a wide audience of potential users in mind and to serve multiple purposes. Often, both the audience and the purpose are established by the agency or other organization that funded the research project on which the report is based. Sometimes, the researcher may use the report to provide a broad descriptive overview of the study findings, which will be presented more succinctly in a subsequent journal article. In either case, an applied report typically provides much more information about a research project than does a journal article and relies primarily on descriptive statistics rather than only those statistics useful for the specific hypothesis tests that are likely to be the primary focus of a journal article.

Exhibit 15.7 outlines the sections in an applied research report. This particular report was mandated by the California state legislature to review a state-funded program for the homeless mentally disabled. The goals of the report are described as both description and evaluation. Applied reports begin with an Executive Summary that presents the highlights from each section of the report, including major findings, in a brief

---

Exhibit 15.7 Sections in an Applied Report

Vernez, Georges, M. Audrey Burnam, Elizabeth A. McGlynn, Sally Trude, and Brian S. Mittman. 1988. Review of California's Program for the Homeless Mentally Disabled. *Santa Monica, CA: The RAND Corporation.*

SUMMARY
   In 1986, the California State Legislature mandated an independent review of the HMD programs that the counties had established with the state funds. The review was to determine the accountability of funds; describe the demographic and mental disorder characteristics of persons served; and assess the effectiveness of the program. This report describes the results of that review. (p. v)

INTRODUCTION
   Background
   California's Mental Health Services Act of 1985...allocated $20 million annually to the state's 58 counties to support a wide range of services, from basic needs to rehabilitation. (pp. 1–2)
   Study Objectives
   Organization of the Report

HMD PROGRAM DESCRIPTION AND STUDY METHODOLOGY
The HMD Program
Study Design and Methods
Study Limitations

COUNTING AND CHARACTERIZING THE HOMELESS
   Estimating the Number of Homeless People
   Characteristics of the Homeless Population

THE HMD PROGRAM IN 17 COUNTIES
   Service Priorities
   Delivery of Services
   Implementation Progress
   Selected Outcomes
   Effects on the Community and on County Service Agencies
   Service Gaps

DISCUSSION
   Underserved Groups of HMD
   Gaps in Continuity of Care
   A particularly large gap in the continuum of care is the lack of specialized housing alternatives for the mentally disabled. The nature of chronic mental illness limits the ability of these individuals to live completely independently. But their housing needs may change, and board-and-care facilities that are acceptable during some periods of their lives may become unacceptable at other times. (p. 57)

(Continued)

Exhibit 15.7 (Continued)

Improved Service Delivery
Issues for Further Research

Appendix

A. SELECTION OF 17 SAMPLED COUNTIES
B. QUESTIONNAIRE FOR SURVEY OF THE HOMELESS
C. GUIDELINES FOR CASE STUDIES
D. INTERVIEW INSTRUMENTS FOR TELEPHONE SURVEY
E. HOMELESS STUDY SAMPLING DESIGN, ENUMERATION, AND SURVEY WEIGHTS
F. HOMELESS SURVEY FIELD PROCEDURES
G. SHORT SCREENER FOR MENTAL AND SUBSTANCE USE DISORDERS
H. CHARACTERISTICS OF THE COUNTIES AND THEIR HMD-FUNDED PROGRAMS
I. CASE STUDIES FOR FOUR COUNTIES' HMD PROGRAMS

---

format, often using bullets. The body of the report presents findings on the number and characteristics of homeless persons and on the operations of the state-funded program in each of 17 counties. This is followed by the discussion section, which highlights the service needs that are not being met, as well as a recommendations section, which focuses attention on policy implications of the research findings. Nine appendixes provide details on the study methodology and the counties studied.

One of the major differences between an applied research report and a journal article is that a journal article must focus on answering a particular research question, whereas an applied report is likely to have the broader purpose of describing a wide range of study findings and attempting to meet the diverse needs of multiple audiences for the research. But a research report that simply describes "findings" without some larger purpose in mind is unlikely to be effective in reaching any audience. Anticipating the needs of the intended audience(s) and identifying the ways in which the report can be useful to them will result in a product that is less likely to be ignored.

Findings From Welfare Reform: What About the Children?

A good example of applied research reporting comes from the Robert Wood Johnson Foundation, which presented an online research report to make widely available the findings from an investigation of the impact of welfare reform (Sunderland 2005). With funding from the Robert Wood Johnson Foundation and 14 other private foundations, the National Institute of Child Health and Human Development, the Social Security Administration, and the National Institute of Mental Health, social scientist P. Lindsay Chase-Lansdale and colleagues (2003) examined the impact of the 1996 federal act mandating changes in welfare requirements—the Personal Responsibility and Work Opportunity Reconciliation Act (PRWORA). Their three-city (Boston, Chicago, San Antonio) research design sought to test the alternative arguments of reform proponents and opponents about the consequences of the reforms for families and children:

> Proponents of welfare reform argued that . . . moving mothers from welfare to work would benefit children because it would increase their families' income, model disciplined work behavior, and better structure their family routines. Opponents of PRWORA countered that . . . the reforms [would] reduce the time mothers and children spend together, . . . increase parental stress and decrease responsive parenting, and . . . move children into low-quality childcare or unsupervised settings while their parents worked. (p. 1548)

The online report describes the three different components of the research design and refers leaders to sources for details about each (Sunderland 2005:4):

1. A longitudinal survey of adults and children designed to provide information on the health, cognitive, behavioral, and emotional development of the children and on their primary caregivers' work-related behavior, welfare experiences, health, well-being, family lives, and use of social services

2. An "embedded" developmental study of a subsample of children to improve the breadth and depth of the child evaluations

3. Ethnographic studies of low-income families in each city to describe how changes in welfare policy affected their daily lives and influenced neighborhood resources over time

The online report summarizes findings and recommendations that have been reported in various publications and other reports. It also highlights several methodological limitations of the study:

Findings from the March 7, 2003 issue of *Science* [Chase-Lansdale et al. 2003]. Investigators reported findings from the longitudinal survey and embedded developmental study. They looked at how the mothers' employment and welfare transitions were related to outcomes in three areas of the child's development—cognitive achievement, problem behaviors, and psychological well-being. They also analyzed whether employment and welfare transitions meant changes in family income and mothers' time apart from their children, to test whether time and money could explain changes in child outcomes.

- *For preschoolers, neither mothers' employment transitions nor their welfare transitions appear to be problematic or beneficial for cognitive achievement or behavioral problems. . . .* Whether or not a mother left welfare, entered welfare, took a job or left a job between the interviews had no discernable link with preschoolers' development.

- *For adolescents, the dominant pattern was also one of few associations. But where findings did occur, the most consistent pattern was that mothers' transitions into employment were related to improvements in adolescents' mental health. Adolescents whose mothers began working—* whether for one or more hours or 40 hours and whether short- or long-term—reported statistically significant declines in psychological distress. This pattern was strongest for their symptoms of anxiety.

- *Increased income may explain why adolescents' mental health improved when mothers went to work . . .* but the fact that improvements were not seen in other preschool and adolescent outcomes implies that higher family income from maternal employment may not benefit children uniformly and is not the only explanatory factor. In families with either preschoolers or adolescents, mothers' entry into employment was related to a significant increase in family income. . . . For example, teenagers' mothers who went to work had household income-to-needs ratios [i.e., the ratio of their income to the cost of household necessities] that rose from 0.65 to 1.26, on average, bringing the majority of families above the poverty line. In contrast, exits from employment were generally related to decreases in income. However, income did not change when mothers went onto or left the welfare rolls.

- *Adolescents in our study did not experience much additional separation from their mothers due to employment (about 45 minutes each day). . . .* When mothers of adolescents entered the labor force, they compensated for time away from the young teenagers by cutting down on time apart when they were not on the job.

- *Preschoolers . . . experienced a significant decline in time spent with their mothers.* When mothers moved into employment, they decreased total time with their preschoolers by an average of 2.1 hours per day.

- *Time-use data suggest that when mothers went to work, they cut back on personal, social, and educational activities that did not involve their children. . . .* The quality of mothers' parenting (e.g., structured family routines, cognitive stimulation) rarely changed with employment at a statistically significant rate, so parenting may not be an explanatory mechanism. (Sunderland 2005:5–6)

Limitations

The researchers noted two limitations.

1. Researchers conducted the first two waves of the study during an economic boom that lowered unemployment and increased wages for less skilled workers. Their findings might not be replicated during an extended period with different economic conditions.

2. Investigators' findings only pertain to children's development within a 16-month interval; harmful or beneficial effects could arise after a longer interval.

Conclusions

Researchers concluded:

This study suggests that mothers' welfare and employment transitions during this unprecedented era of welfare reform are not associated with negative outcomes for preschoolers or young adolescents. A few positive associations were tenuously indicated for adolescents. . . . The well-being of preschoolers appeared to be unrelated to their mothers' leaving welfare or entering employment, at least as indexed in measures of cognitive achievement and behavior problems. (Sunderland 2005:6)

The Robert Wood Johnson online report also summarized findings about the characteristics of children at the start of the study (Sunderland 2005):

Within a sample of 1,885 low-income children and their families, preschoolers and adolescents show patterns of cognitive achievement and problem behavior that should be of concern to policy makers. The preschoolers and adolescents in [the] sample are more developmentally at risk compared to middle-class children in national samples. In addition, adolescents whose mothers were on welfare in 1999 have lower levels of cognitive achievement and higher levels of behavioral and emotional problems than do adolescents whose mothers had left welfare or whose mothers had never been on welfare. For preschoolers, mothers' current or recent welfare participation is linked with poor cognitive achievement; preschoolers of recent welfare leavers have the most elevated levels of problem behavior. Preschoolers and adolescents in sanctioned families also show problematic cognitive and behavioral outcomes (sanctioning is the withholding of all or part of a family's TANF benefits for noncompliance with work requirements or other rules). Mothers' marital, educational, mental, and physical health status, as well as their parenting practices, seems to account for most of the welfare group differences. (p. 7)

Because so few families in the sample had reached their time limits, investigators could not tell if families that leave the welfare system after reaching their time limits would show patterns similar to those that leave after sanctions.

Recommendations

The policy brief recommended:

- The intense focus on welfare reform in our country should not impede a general concern and plan of action for all children in poverty, whether on welfare or not. In order to lessen developmental risks and improve the developmental trajectories of these children, numerous avenues should be pursued for the provision of supportive mental health and educational services.

- State and federal governments should explore options for identifying and reaching out to the most disadvantaged and high-risk families involved in the welfare system—families experiencing welfare sanctions. Sanctioned families have a number of characteristics that serve as markers of concern for the healthy development of children and youth. . . . Possible policy options include assistance to bring families into compliance with rules before they are sanctioned, closer monitoring of sanctioned families and the provision of additional supports, such as mental health services, academic enrichment, after-school programs, and other family support services.

Framing an Applied Report

What can be termed the **front matter** and the **back matter** of an applied report also are important. Applied reports usually begin with an executive summary: a summary list of the study's main findings, often in bullet fashion. Appendixes, the back matter, may present tables containing supporting data that were not discussed in the body of the report. Applied research reports also often append a copy of the research instrument(s).

> **Front matter** The section of an applied research report that includes an executive summary, abstract, and table of contents.
>
> **Back matter** The section of an applied research report that may include appendixes, tables, and the research instrument(s).

An important principle for the researcher writing for a nonacademic audience is that the findings and conclusions should be engaging and clear. You can see how I did this in a report from a class research project I designed with my graduate methods students (and in collaboration with several faculty knowledgeable about substance abuse) (Exhibit 15.8). These report excerpts indicate how I summarized key findings in an executive summary (Schutt et al. 1996:iv), emphasized the importance of the research in the introduction (Schutt et al. 1996:1), used formatting and graphing to draw attention to particular findings in the body of the text (Schutt et al. 1996:5), and tailored recommendations to our own university context (Schutt et al. 1996:26).

Reporting Quantitative and Qualitative Research

The requirements for good research reports are similar in many respects for quantitative and qualitative research projects. Every research report should include good writing, a clear statement of the research question, an integrated literature and presentation of key findings with related discussion, conclusions, and limitations. The outline used in Aseltine and Kessler's (1993) report of a quantitative project may also be used by some authors of qualitative research reports. The Robert Wood Johnson research report also provides an example of how a research report of a mixed methods study can integrate the results of analyses of both types of data. However, the differences between qualitative and quantitative research approaches mean that it is often desirable for research reports based on qualitative research to diverge in some respects from those reflecting quantitative research.

Reports based on qualitative research should be enriched in each section with elements that reflect the more holistic and reflexive approach of qualitative projects. The introduction should include background about the development of the researcher's interest in the topic, while the literature review should include some attention to the types of particular qualitative methods used in prior research. The methodology section should

Exhibit 15.8 Student Substance Abuse, Report Excerpts

Executive Summary

- Rates of substance abuse were somewhat lower at UMass–Boston than among nationally selected samples of college students.
- Two-thirds of the respondents reported at least one close family member whose drinking or drug use had ever been of concern to them—one-third reported a high level of concern.
- Most students perceived substantial risk of harm due to illicit drug use, but just one-quarter thought alcohol use posed a great risk of harm.

Introduction

Binge drinking, other forms of alcohol abuse, and illicit drug use create numerous problems on college campuses. Deaths from binge drinking are too common and substance abuse is a factor in as many as two-thirds of on-campus sexual assaults (Finn 1997; National Institute of Alcohol Abuse and Alcoholism 1995). College presidents now rate alcohol abuse as the number one campus problem (Wechsler, Davenport, Dowdall, Moeykens, & Castillo 1994) and many schools have been devising new substance abuse prevention policies and programs. However, in spite of increasing recognition of and knowledge about substance abuse problems at colleges as a whole, little attention has been focused on substance abuse at commuter schools.

Findings

The composite index identifies 27% of respondents as at risk of substance abuse (an index score of 2 or higher). One-quarter reported having smoked or used smokeless tobacco in the past two weeks.

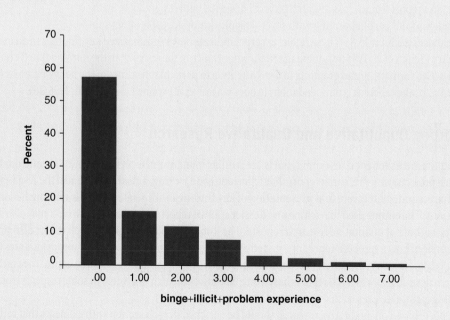

> 27% of respondents were identified as at risk of substance abuse.

describe how the researcher gained access to the setting or individuals studied and the approach used to managing relations with research participants. The presentation of findings in qualitative studies may be organized into sections reflecting different themes identified in interviews or observational sessions. Quotes from participants or from observational notes should be selected to illustrate these themes, although qualitative research reports differ in the extent to which the researcher presents findings in summary form or uses direct quotes to identify key issues. The findings sections in a qualitative report may alternate between presentations of quotes or observations about the research participants, the researcher's interpretations of this material, and some commentary on how the researcher reacted in the setting, although some qualitative researchers will limit their discussion of their reactions to the discussion section.

Reports on mixed methods projects should include subsections in the methods section that introduce each method, and then distinguish findings from qualitative and quantitative analyses in the findings section. Some mixed-methods research reports may present analyses that use both qualitative and quantitative data in yet another subsection, while others may just discuss implications of analyses of each type for the overall conclusions in the discussions and conclusions sections (Dahlberg, Wittink, & Gallo 2010:785–791). However findings based on each method are presented, it is important to consider explicitly both the ways in which the specific methods affected findings obtained with those methods and also to discuss the implications of findings obtained using both methods for the overall study conclusions.

🔲 Ethics, Politics, and Research Reports

It is at the time of reporting research results that the researcher's ethical duty to be honest and open becomes paramount. Here are some guidelines:

- *Provide an honest accounting of how the research was carried out and where the initial research design had to be changed.* Readers do not have to know about every change you made in your plans and each new idea you had, but they should be informed about major changes in hypotheses or research design. If important hypotheses were not supported, acknowledge this, rather than conveniently forgetting to mention them (Brown & Hedges 2009:383). If a different approach to collecting or analyzing the data could have led to different conclusions, this should be acknowledged in the limitations section (Bergman 2008:588–590).

- *Evaluate honestly the strengths and weaknesses of your research design.* Systematic evaluations suggest that the stronger the research design from the standpoint of establishing internal (causal) validity, the weaker the empirical support that is likely to be found for hypothesized effects (cf., Weisburd, Lum, & Petrosino 2001:64). Finding support for a hypothesis tested with a randomized experimental design is stronger evidence than support based on correlations among variables measured in a cross-sectional survey.

- *Refer to prior research and interpret your findings within the body of literature resulting from that prior research.* Your results are likely to be only the latest research conducted to investigate a research question that others have studied. It borders on unethical practice to present your findings as if they are the only empirical information with which to answer your research question, yet many researchers commit this fundamental mistake (Bergman 2008:599). For example, a systematic evaluation of citation frequency in articles reporting clinical trial results in medical journals found that on average, just 21% of the available prior research was cited (for trials with at least three prior articles that could have been cited) (Robinson & Goodman 2011:53).

The result of such omission is that readers may have no idea whether your own research supports a larger body of evidence or differs from it—and so should be subject to even greater scrutiny.

- *Maintain a full record of the research project so that questions can be answered if they arise.* Many details will have to be omitted from all but the most comprehensive reports, but these omissions should not make it impossible to track down answers to specific questions about research procedures that may arise in the course of data analysis or presentation. Tests of relationships that were conducted but not included in the report should be acknowledged.

- *Avoid "lying with statistics" or using graphs to mislead.* (See Chapter 14 for more on this topic.) There is a more subtle problem to be avoided, which is "cherry-picking" results to present. Although some studies are designed to test only one hypothesis involving variables that each are measured in only one way, many studies collect data that can be used to test many hypotheses, often with alternative measures. If many possible relationships have been examined with the data collected and only those found to yield a statistically significant result are reported, the odds of capitalizing on chance findings are multiplied. This is a major temptation in research practice and has the unfortunate result that most published findings are not replicated or do not stand up to repeated tests over time (Lehrer 2010:57). Every statistical test presented can only be adequately understood in light of the entire body of statistical analyses that led to that particular result.

- *Acknowledge the sponsors of the research.* This is important, in part, so that others can consider whether this sponsorship may have tempted you to bias your results in some way. Whether you conducted your research for a sponsor, or together with members of an underserved community, give research participants an opportunity to comment on your main findings before you release them to the public. Consider revising your report based on their suggestions or, if you disagree with their suggestions, include their comments in footnotes at relevant points in your report or as an appendix to it (Bledsoe & Hopson 2009:392).

- *Thank staff who made major contributions.* This is an ethical as well as a political necessity. Let's maintain our social relations!

- *Be sure that the order of authorship for coauthored reports is discussed in advance and reflects agreed-on principles.* Be sensitive to coauthors' needs and concerns.

Ethical research reporting should not mean ineffective reporting. You need to tell a coherent story in the report and avoid losing track of the story in a thicket of minuscule details. You do not need to report every twist and turn in the conceptualization of the research problem or the conduct of the research. But be suspicious of reports that don't seem to admit to the possibility of any room for improvement. Social science is an ongoing enterprise in which one research report makes its most valuable contribution by laying the groundwork for another, more sophisticated, research project. Highlight important findings in the research report but also use the research report to point out what are likely to be the most productive directions for future researchers.

Communicating With the Public

Even following appropriate guidelines such as these, however, will not prevent controversy and conflict over research on sensitive issues. The sociologist Peter Rossi (1999) recounts the controversy that arose when he released a summary of findings conducted in his 1989 study of homeless persons in Chicago (see Chapter 5). In spite of important findings about the causes and effects of homelessness, media attention focused on Rossi's markedly smaller estimate of the numbers of homeless persons in Chicago compared with the "guesstimates" that had been publicized by local advocacy groups. "Moral of the story: Controversy is news, to which careful empirical findings cannot be compared" (Rossi 1999:2).

Does this mean that ethical researchers should avoid political controversy by sidestepping media outlets for their work? Many social scientists argue that the media offers one of the best ways to communicate the practical application of sociological knowledge and that when we avoid these opportunities, "some of the best sociological insights never reach policy makers because sociologists seldom take advantage of useful mechanisms to get their ideas out" (Wilson 1998:435).

The sociologist William Julius Wilson (1998:438) urges the following principles for engaging the public through the media:

1. Focus on issues of national concern, issues that are high on the public agenda.

2. Develop creative and thoughtful arguments that are clearly presented and devoid of technical language.

3. Present the big picture whereby the arguments are organized and presented so that the readers can see how the various parts are interrelated.

Ultimately each researcher must make a decision about the most appropriate and important outlets for his or her work.

Plagiarism

It may seem depressing to end a book on research methods with a section on plagiarism, but it would be irresponsible to avoid the topic. Of course, you may have a course syllabus detailing instructor or university policies about plagiarism and specifying the penalties for violating that policy, so I'm not simply going to repeat that kind of warning. You probably realize that the practice of selling term papers is revoltingly widespread (my search of "term papers" on Google returned 17,600,000 websites on May 30, 2011); so I'm not going to just repeat that academic dishonesty is widespread. Instead, I will use this section to review the concept of plagiarism and to show how that problem connects to the larger issue of the integrity of social research. When you understand the dimensions of the problem and the way it affects research, you should be better able to detect plagiarism in other work and avoid it in your own.

You learned in Chapter 3 that maintaining professional integrity—honesty and openness in research procedures and results—is the foundation for ethical research practice. When it comes to research publications and reports, being honest and open means avoiding plagiarism—that is, presenting as one's own the ideas or words of another person or persons for academic evaluation without proper acknowledgment (Hard, Conway, & Moran 2006:1059).

An increasing body of research suggests that plagiarism is a growing problem on college campuses. Jason Stephen, Michael Young, and Thomas Calabrese (2007:243) found in a web-based survey of self-selected students at two universities that one-quarter acknowledged having plagiarized a few sentences (24.7%) or a complete paper (.3%) in coursework within the past year (many others admitted to other forms of academic dishonesty, such as copying homework). Hard et al. (2006) conducted an anonymous survey in selected classes in one university, with almost all students participating, and found much higher plagiarism rates: 60.6% reported that they had copied "sentences, phrases, paragraphs, tables, figures, or data directly or in slightly modified form from a book, article, or other academic source without using quotation marks or giving proper acknowledgment to the original author or source" (p. 1069) and 39.4% reported that they had "copied information from Internet web sites and submitted it as [their] work" (p. 1069).

So the plagiarism problem is not just about purchasing term papers—although that is really about as bad as it gets (Broskoske 2005:1); plagiarism is also about what you do with the information you obtain from a literature review or inspection of research reports. And rest assured that this is not only about student papers; it also is about the work of established scholars and social researchers who publish reports that you want to rely on for accurate information. Several noted historians have been accused of plagiarizing passages that they used in popular books; some have admitted to not checking the work of their research assistants, to not keeping track of their sources, or to being unable to retrieve the data they claimed they had analyzed. Whether the cause is cutting corners to meet deadlines or consciously fudging facts, the effect is to undermine the trustworthiness of social research.

Now that you are completing this course in research methods, it's time to think about how to do your part to reduce the prevalence of plagiarism. Of course, the first step is to maintain careful procedures for documenting the sources that you rely on for your own research and papers, but you should also think about how best to reduce temptations among others. After all, what people believe about what others do is a strong influence on their own behavior (Hard et al. 2006:1058).

Reviewing the definition of plagiarism and how your discipline's professional association enforces it is an important first step. These definitions and procedures reflect a collective effort to help social scientists maintain standards throughout the discipline. Awareness is the first step (American Sociological Association [ASA] 1999:19).

Sociologists have an obligation to be familiar with their Code of Ethics, other applicable ethics codes, and their application to sociologists' work. Lack of awareness or misunderstanding of an ethical standard is not, in itself, a defense to a charge of unethical conduct.

The ASA's (1999) *Code of Ethics* includes an explicit prohibition of plagiarism:

14. Plagiarism

(a) In publications, presentations, teaching, practice, and service, sociologists explicitly identify, credit, and reference the author when they take data or material verbatim from another person's written work, whether it is published, unpublished, or electronically available.

(b) In their publications, presentations, teaching, practice, and service, sociologists provide acknowledgment of and reference to the use of others' work, even if the work is not quoted verbatim or paraphrased, and they do not present others' work as their own whether it is published, unpublished, or electronically available. (p. 16)

The next step toward combating the problem and temptation of plagiarism is to keep focused on the goal of social research methods: investigating the social world. If researchers are motivated by a desire to learn about social relations, to understand how people understand society, and to discover why conflicts arise and how they can be prevented, they will be as concerned with the integrity of their research methods as are those, like yourself, who read and use the results of their research. Throughout *Investigating the Social World,* you have been learning how to use research processes and practices that yield valid findings and trustworthy conclusions. Failing to report honestly and openly on the methods used or sources consulted derails progress toward that goal.

It works the same as with cheating in school. When students are motivated only by the desire to "ace" their tests and receive better grades than others, they are more likely to plagiarize and use other illicit means to achieve that goal. Students who seek first to improve their understanding of the subject matter and to engage in the process of learning are less likely to plagiarize sources or cheat on exams (Kohn 2008:6–7). They are also building the foundation for becoming successful social researchers who help others understand our world.

🔲 Conclusions

A well-written research article or report requires (to be just a bit melodramatic) blood, sweat, and tears and more time than you will, at first, anticipate. But the process of writing one will help you write the next. And the issues you consider, if you approach your writing critically, will be sure to improve your subsequent research projects and sharpen your evaluations of other investigators' research projects.

Good critical skills are essential when evaluating research reports, whether your own or those produced by others. There are *always* weak points in any research, even published research. It is an indication of strength, not weakness, to recognize areas where one's own research needs to be, or could have been, improved. And it is really not just a question of sharpening your knives and going for the jugular. You need to be able to weigh the strengths and weaknesses of particular research results and to evaluate a study in terms of its contribution to understanding the social world—not in terms of whether it gives a definitive answer for all time.

But this is not to say that anything goes. Much research lacks one or more of the three legs of validity—measurement validity, causal validity, or generalizability—and contributes more confusion than understanding about the social world. Top journals generally maintain very high standards, partly because they have good critics in the review process and distinguished editors who make the final acceptance decisions. But some daily newspapers do a poor job of screening, and research-reporting standards in many popular magazines, TV shows, and books are often abysmally poor. Keep your standards high and your views critical when reading research reports, but not so high or so critical that you turn away from studies that make tangible contributions to understanding the social world—even if they don't provide definitive answers. And don't be so intimidated by the need to maintain high standards that you shrink from taking advantage of opportunities to conduct research yourself.

The growth of social science methods from its infancy to adolescence, perhaps to young adulthood, ranks as a key intellectual accomplishment of the 20th century. Opinions about the causes and consequences of homelessness no longer need to depend on the scattered impressions of individuals; criminal justice policies can be shaped by systematic evidence of their effectiveness; and changes in the distribution of poverty and wealth in populations can be identified and charted. Employee productivity, neighborhood cohesion, and societal conflict may each be linked to individual psychological processes and to international economic strains.

Of course, social research methods are no more useful than the commitment of researchers to their proper application. Research methods, like all knowledge, can be used poorly or well, for good purposes or bad, when appropriate or not. A claim that a belief is based on social science research in itself provides no extra credibility. As you have learned throughout this book, we must first learn which methods were used, how they were applied, and whether interpretations square with the evidence. To investigate the social world, we must keep in mind the lessons of research methods.

Key Terms

Highlights

- Research reports should be evaluated systematically, using the review guide in Appendix A and also taking account of the inter-relations among the design elements.

- Proposal writing should be a time for clarifying the research problem, reviewing the literature, and thinking ahead about the report that will be required. Trade-offs between different design elements should be considered and the potential for mixing methods evaluated.

- Different types of reports typically pose different problems. Authors of student papers must be guided in part by the expectations of their professor. Thesis writers have to meet the requirements of different committee members but can benefit greatly from the areas of expertise represented on a typical thesis committee. Applied research reports are constrained by the expectations of the research sponsor; an advisory committee from the applied setting can help avoid problems. Journal articles must pass a peer review by other social scientists and are often much improved in the process.

- Research reports should include an introductory statement of the research problem, a literature review, a methodology section, a findings section with pertinent data displays, and a conclusions section that identifies any weaknesses in the research design and points out implications for future research and theorizing. This basic report format should be modified according to the needs of a particular audience.

- All reports should be revised several times and critiqued by others before being presented in final form.

- The central ethical concern in research reporting is to be honest. This honesty should include providing a truthful accounting of how the research was carried out, maintaining a full record about the project, using appropriate statistics and graphs, acknowledging the research sponsors, and being sensitive to the perspectives of coauthors.

- Plagiarism is a grievous violation of scholarly ethics. All direct quotes or paraphrased material from another author's work must be appropriately cited.

- Credit must be given where credit is due. The contributions of persons and organizations to a research project must be acknowledged in research reports.

- Social scientists are obligated to evaluate the credibility of information obtained from any source before using it in their research reports. Special attention should be given to the credibility of information acquired through the Internet.

STUDENT STUDY SITE

To assist in completing the web exercises, please access the study site at **www.sagepub.com/schuttisw7e**, where you will find the web exercises with accompanying links. You'll find other useful study materials, such as self-quizzes and e-flashcards, for each chapter, along with a group of carefully selected articles from research journals that illustrate the major concepts and techniques presented in the book.

Discussion Questions

1. A good place to start developing your critical skills would be with one of the articles reviewed in this chapter. Try reading one, and fill in the answers to the article review questions that I did not cover (Appendix A). Do you agree with my answers to the other questions? Could you add some points to my critique or to the lessons on research design that I drew from these critiques?

2. Read the journal article "Marital Disruption and Depression in a Community Sample," by Aseltine and Kessler, in the September 1993 issue of *Journal of Health and Social Behavior*.

How effective is the article in conveying the design and findings of the research? Could the article's organization be improved at all? Are there bases for disagreement about the interpretation of the findings? Did reading the full article increase your opinion of its value?

3. Rate four journal articles for the overall quality of the research and for the effectiveness of the writing and data displays. Discuss how each could have been improved.

Practice Exercises

1. Call a local social or health service administrator or a criminal justice official, and arrange for an interview. Ask the official about his or her experience with applied research reports and conclusions about the value of social research and the best techniques for reporting to practitioners.

2. Interview a student who has written an independent paper or thesis based on original data. Ask your subject to describe his or her experiences while writing the thesis. Review the decisions made in designing the research, and ask about the stages of research design, data collection and analysis, and report writing that proved to be difficult.

Ethics Questions

1. Plagiarism is no joke. What are the regulations on plagiarism in class papers at your school? What do you think the ideal policy would be? Should this policy take into account cultural differences in teaching practices and learning styles? Do you think this ideal policy is likely to be implemented? Why or why not? Based on your experiences, do you believe that most student plagiarism is the result of misunderstanding about proper citation practices, or is it the result of dishonesty? Do you think that students who plagiarize while in school are less likely to be honest as social researchers?

2. Full disclosure of funding sources and paid consulting and other business relationships is now required by most journals. Should researchers publishing in social science journals also be required to fully disclose all sources of funding, including receipt of payment for research done as a consultant? Should full disclosure of all previous funding sources be required in each published article? Write a short justification of the regulations you propose.

Web Exercises

1. Go to the National Science Foundation's (NSF) Sociology Program website at www.nsf.gov/funding/pgm_summ.jsp?pims_id=5369. What are the components that the NSF's Sociology Program looks for in a proposed piece of research? Examine the Table of Contents (p. 6) for an NSF proposal at www.nsf.gov/pubs/gpg/nsf04_23/nsf04_23.pdf. Now outline a research proposal to the NSF to investigate a research question of your choice.

2. The National Academy of Sciences wrote a lengthy report on ethics issues in scientific research. Visit the site and read the free executive summary. Go to www.nap.edu/catalog.php?record_id=10430 and click on "Download Free" at the bottom of the page. Summarize the information and guidelines in the report.

3. Using the web, find five different examples of social science research projects that have been completed. Briefly describe each. How does each differ in its approach to reporting the research results? To whom do you think the author(s) of each are "reporting" (i.e., who is the audience)? How do you think the predicted audience has helped shape the author's approach to reporting the results? Be sure to note the websites at which you located each of your five examples.

SPSS Exercises

1. Review the output you have generated in previous SPSS exercises. Select the distributions, statistics, and crosstabs that you believe provide a coherent and interesting picture of support for capital punishment in America.

Prepare these data displays in presentation format and number them (Table 1, Table 2, etc.).

2. Write a short report based on the analyses you conducted for the SPSS exercises throughout this book, including the data displays you have just prepared. Include in your report a brief introduction and literature review (you might use the articles I referred to in the SPSS exercises for Chapters 1 and 3). In a short methods section, review the basic methods used in the GSS 2010, and list the variables you have used for the analysis.

In your conclusions section, include some suggestions for additional research on support for capital punishment.

Developing a Research Proposal

Now it's time to bring all the elements of your proposal together (Exhibit 2.14, #19 to #23).

1. Organize the proposal material you wrote for previous chapters in a logical order. Select what you feel is the strongest research method (Chapters 7–13) as your primary method.

2. Add a multiple-method component to your research design with one of the other methods sections you prepared in Chapters 7 to 13.

3. Rewrite the entire proposal, adding an introduction. Also add sections that outline a budget, and state the limitations of your study.

4. Review the proposal with the "Decisions in Research" checklist (Exhibit 2.14). Answer each question (or edit your previous answers), and justify your decision at each checkpoint.

Glossary

Alternate-forms reliability   A procedure for testing the reliability of responses to survey questions in which subjects' answers are compared after the subjects have been asked slightly different versions of the questions or when randomly selected halves of the sample have been administered slightly different versions of the questions.

Anomalous findings   Unexpected findings in data analysis that are inconsistent with most other findings with those data.

Anonymity   Provided by research in which no identifying information is recorded that could be used to link respondents to their responses.

Assignment variable   The variable used to specify a cutoff score for eligibility in some treatment in a regression-discontinuity design.

Association   A criterion for establishing a nomothetic causal relationship between two variables: Variation in one variable is related to variation in another variable.

Authenticity   When the understanding of a social process or social setting is one that reflects fairly the various perspectives of participants in that setting.

Availability sampling   Sampling in which elements are selected on the basis of convenience.

Back matter   The section of an applied research report that may include appendixes, tables, and the research instrument(s).

Balanced response choices   An equal number of responses to a fixed-choice survey question express positive and negative choices in comparable language.

Bar chart   A graphic for qualitative variables in which the variable's distribution is displayed with solid bars separated by spaces.

Base number *(N)*   The total number of cases in a distribution.

Before-and-after design   A quasi-experimental design consisting of several before-after comparisons involving the same variables but no comparison group.

Behavior coding   Observation in which the research categorizes, according to strict rules, the number of times certain behaviors occur.

Bimodal   A distribution that has two nonadjacent categories with about the same number of cases, and these categories have more cases than any others.

Black box evaluation   This type of evaluation occurs when an evaluation of program outcomes ignores, and does not identify, the process by which the program produced the effect.

Case-oriented research   Research that focuses attention on the nation or other unit as a whole.

Case-oriented understanding   An understanding of social processes in a group, formal organization, community, or other collectivity that reflects accurately the standpoint of participants.

Case study A setting or group that the analyst treats as an integrated social unit that must be studied holistically and in its particularity.

Causal effect (idiographic perspective)   When a series of concrete events, thoughts, or actions result in a particular event or individual outcome.

Causal effect (nomothetic perspective)   When variation in one phenomenon, an independent variable, leads to or results, on average, in variation in another phenomenon, the dependent variable.

Causal validity (internal validity)   Exists when a conclusion that A leads to or results in B is correct.

Census   Research in which information is obtained through responses from or information about all available members of an entire population.

Central tendency   The most common value (for variables measured at the nominal level) or the value around which cases tend to center (for a quantitative variable).

Certificate of Confidentiality   A certificate issued to a researcher by the National Institutes of Health that ensures the right to protect information obtained about high-risk populations or behaviors—except child abuse or neglect—from legal subpoenas.

Ceteris paribus   Latin phrase meaning "other things being equal."

Chi-square   An inferential statistic used to test hypotheses about relationships between two or more variables in a cross-tabulation.

Closed-ended (fixed-choice) question   A survey question that provides preformatted response choices for the respondent to circle or check.

Cluster   A naturally occurring, mixed aggregate of elements of the population.

Cluster sampling   Sampling in which elements are selected in two or more stages, with the first stage being the random selection of naturally occurring clusters and the last stage being the random selection of elements within clusters.

Cognitive interview   A technique for evaluating questions in which researchers ask people test questions and then probe with follow-up questions to learn how they understood the question and what their answers mean.

Cohort   Individuals or groups with a common starting point. Examples include college class of 1997, people who graduated from high school in the 1980s, General Motors employees who started work between the years 1990 and 2000, and people who were born in the late 1940s or the 1950s (the baby boom generation).

Combined frequency display   A table that presents together the distributions for a set of conceptually similar variables having the same response categories; common headings are used for the responses.

Comparative historical research   Research comparing data from more than one time period in more than one nation.

Comparison group   In an experiment, a group that has been exposed to a different treatment (or value of the independent variable) than the experimental group.

Compensatory rivalry (John Henry effect)   A type of contamination in experimental and quasi-experimental designs that occurs when control group members are aware that they are being denied some advantage and increase their efforts by way of compensation.

Complete observer   A role in participant observation in which the researcher does not participate in group activities and is publicly defined as a researcher.

Compressed frequency display   A table that presents cross-classification data efficiently by eliminating unnecessary percentages, such as the percentage corresponding to the second value of a dichotomous variable.

Computer-assisted personal interview (CAPI)   A personal interview in which the laptop computer is used to display interview questions and to process responses that the interviewer types in, as well as to check that these responses fall within allowed ranges.

Computer-assisted qualitative data analysis   Uses special computer software to assist qualitative analyses through creating, applying, and refining categories; tracing linkages between concepts; and making comparisons between cases and events.

Computer-assisted telephone interview (CATI)   A telephone interview in which a questionnaire is programmed into a computer, along with relevant skip patterns, and only valid entries are allowed; incorporates the tasks of interviewing, data entry, and some data cleaning.

Concept   A mental image that summarizes a set of similar observations, feelings, or ideas.

Conceptualization   The process of specifying what we mean by a term. In deductive research, conceptualization helps to translate portions of an abstract theory into specific variables that can be used in testable hypotheses. In inductive research, conceptualization is an important part of the process used to make sense of related observations.

Concurrent validity   The type of validity that exists when scores on a measure are closely related to scores on a criterion measured at the same time.

Confidentiality   Provided by research in which identifying information that could be used to link respondents to their responses is available only to designated research personnel for specific research needs.

Conflict theory   Identifies conflict between social groups as the primary force in society. Understanding the bases and consequences of the conflict is key to understanding social processes.

Conflicts of interest   When a researcher has a significant stake in the design or outcome of his or her own research.

Constant   A number that has a fixed value in a given situation; a characteristic or value that does not change.

Constructivism   Methodology based on questioning belief in an external reality. Emphasizes the importance of exploring the way in which different stakeholders in a social setting construct their beliefs.

Construct validity   The type of validity that is established by showing that a measure is related to other measures as specified in a theory.

Contamination   A source of causal invalidity that occurs when either the experimental and/or the comparison group is aware of the other group and is influenced in the posttest as a result.

Content analysis   A research method for systematically analyzing and making inferences from text.

Content validity   The type of validity that exists when the full range of a concept's meaning is covered by the measure.

Context   A focus of idiographic causal explanation; a particular outcome is understood as part of a larger set of interrelated circumstances.

Context effects   Occur in a survey when one or more questions influence how subsequent questions are interpreted.

Contextual effects   Relationships among variables that vary among geographic units or other social settings.

Contingent question   A question that is asked of only a subset of survey respondents.

Continuous measure   A measure with numbers indicating the values of variables that are points on a continuum.

Control group   A comparison group that receives no treatment.

Convergent validity   The type of validity achieved when one measure of a concept is associated with different types of measures of the same concept.

Correlational analysis   A statistical technique that summarizes the strength of a relationship between two quantitative variables in terms of its adherence to a linear pattern.

Correlation coefficient   A summary statistic that varies from 0 to 1 or −1, with 0 indicating the absence of a linear relationship between two quantitative variables and 1 or −1 indicating that the relationship is completely described by the line representing the regression of the dependent variable on the independent variable.

Cost-benefit analysis   A type of evaluation research that compares program costs with the economic value of program benefits.

Cost-effectiveness analysis   A type of evaluation research that compares program costs with actual program outcomes.

Counterfactual   The situation that would have occurred if the subjects who were exposed to the treatment actually were not exposed, but otherwise had had identical experiences to those they underwent during the experiment.

Cover letter   The letter sent with a mailed questionnaire. It explains the survey's purpose and auspices and encourages the respondent to participate.

Covert observer   A role in participant observation in which the researcher does not participate in group activities and is not publicly defined as a researcher.

Covert participant   A role in field research in which the researcher does not reveal his or her identity as a researcher to those who are observed, while participating.

Criterion validity   The type of validity that is established by comparing the scores obtained on the measure being validated to those obtained with a more direct or already validated measure of the same phenomenon (the criterion).

Cronbach's alpha   A statistic commonly used to measure interitem reliability.

Cross-population generalizability (external validity)   Exists when findings about one group, population, or setting hold true for other groups, populations, or settings.

Cross-sectional comparative research   Research comparing data from one time period between two or more nations.

Cross-sectional research design   A study in which data are collected at only one point in time.

Cross-tabulation (crosstab)   In the simplest case, a bivariate (two-variable) distribution, showing the distribution of one variable for each category of another variable; can be elaborated using three or more variables.

Curvilinear   Any pattern of association between two quantitative variables that does not involve a regular increase or decrease.

Data cleaning   The process of checking data for errors after the data have been entered in a computer file.

Debriefing   A researcher's informing subjects after an experiment about the experiment's purposes and methods and evaluating subjects' personal reactions to the experiment.

Deductive research   The type of research in which a specific expectation is deduced from a general premise and is then tested.

Demography   The statistical and mathematical study of the size, composition, and spatial distribution of human populations and how these features change over time.

Demoralization   A type of contamination in experimental and quasi-experimental designs that occurs when control group members feel they have been left out of some valuable treatment and perform worse as a result.

Dependent variable   A variable that is hypothesized to vary depending on, or under the influence of, another variable.

Descriptive research   Research in which social phenomena are defined and described.

Descriptive statistics   Statistics used to describe the distribution of and relationship among variables.

Dichotomy   Variable having only two values.

Differential attrition (mortality)   A problem that occurs in experiments when comparison groups become different because subjects are more likely to drop out of one of the groups for various reasons.

Direction of association   A pattern in a relationship between two variables—the values of variables tend to change consistently in relation to change on the other variable. The direction of association can be either positive or negative.

Discrete measure   A measure that classifies cases in distinct categories.

Discriminant validity   An approach to construct validation; the scores on the measure to be validated are compared with scores on another measure of the same variable and to scores on variables that measure different but related concepts. Discriminant validity is achieved if the measure to be validated is related most strongly to its comparison measure and less so to the measures of other concepts.

Disproportionate stratified sampling   Sampling in which elements are selected from strata in different proportions from those that appear in the population.

Double-barreled question   A single survey question that actually asks two questions but allows only one answer.

Double-blind procedure   An experimental method in which neither the subjects nor the staff delivering experimental treatments know which subjects are getting the treatment and which are receiving a placebo.

Double negative   A question or statement that contains two negatives, which can muddy the meaning of the question.

Ecological fallacy   An error in reasoning in which incorrect conclusions about individual-level processes are drawn from group-level data.

Effect size   A standardized measure of association—often the difference between the mean of the experimental group and the mean of the control group on the dependent variable, adjusted for the average variability in the two groups.

Efficiency analysis   A type of evaluation research that compares program costs with program effects. It can be either a cost-benefit analysis or a cost-effectiveness analysis.

Elaboration analysis   The process of introducing a third variable into an analysis in order to better understand—to elaborate—the bivariate (two-variable) relationship under consideration. Additional control variables also can be introduced.

Electronic survey   A survey that is sent and answered by computer, either through e-mail or on the web.

Elements   The individual members of the population whose characteristics are to be measured.

Emic focus   Representing a setting with the participants' terms.

Empirical generalization   A statement that describes patterns found in data.

Endogenous change   A source of causal invalidity that occurs when natural developments or changes in the subjects (independent of the experimental treatment itself) account for some or all of the observed change from the pretest to the posttest.

Enumeration units   Units that contain one or more elements and that are listed in a sampling frame.

Ethnography   The study of a culture or cultures that some group of people shares, using participant observation over an extended period of time.

Ethnomethodology   A qualitative research method focused on the way that participants in a social setting create and sustain a sense of reality.

Etic focus   Representing a setting with the researchers' terms.

Evaluability assessment   A type of evaluation research conducted to determine whether it is feasible to evaluate a program's effects within the available time and resources.

Evaluation research   Research that describes or identifies the impact of social policies and programs.

Event-based design (cohort study)   A type of longitudinal study in which data are collected at two or more points in time from individuals in a cohort.

Event-structure analysis   A systematic method of developing a causal diagram showing the structure of action underlying some chronology of events; the end result is an idiographic causal explanation.

Exhaustive   Every case can be classified as having at least one attribute (or value) for the variable.

Expectancies of experimental staff   A source of treatment misidentification in experiments and quasi-experiments that occurs when change among experimental subjects is due to the positive expectancies of the staff who are delivering the treatment rather than to the treatment itself; also called a self-fulfilling prophecy.

Experience sampling method (ESM)   A technique for drawing a representative sample of everyday activities, thoughts, and experiences. Participants carry a pager and are beeped at random times over several days or weeks; on hearing the beep, participants complete a report designed by the researcher.

Experimental group   In an experiment, the group of subjects that receives the treatment or experimental manipulation.

Explanatory research   Seeks to identify causes and effects of social phenomena and to predict how one phenomenon will change or vary in response to variation in some other phenomenon.

Exploratory research   Seeks to find out how people get along in the setting under question, what meanings they give to their actions, and what issues concern them.

Ex post facto control group design   A nonexperimental design in which comparison groups are selected after the treatment, program, or other variation in the independent variable has occurred.

External events   A source of causal invalidity that occurs when events external to the study influence posttest scores. Also called an effect of history.

Extraneous variable   A variable that influences both the independent and dependent variables so as to create a spurious association between them that disappears when the extraneous variable is controlled.

Face validity   The type of validity that exists when an inspection of items used to measure a concept suggests that they are appropriate "on their face."

Factorial survey   A survey in which randomly selected subsets of respondents are asked different questions, or are asked to respond to different vignettes, in order to determine the causal effect of the variables represented by these differences.

Feedback   Information about service delivery system outputs, outcomes, or operations that is available to any program input.

Feminist research   Research with a focus on women's lives and often including an orientation to personal experience, subjective orientations, the researcher's standpoint, and emotions.

Fence-sitters   Survey respondents who see themselves as being neutral on an issue and choose a middle (neutral) response that is offered.

Field experiment   A study conducted in a real-world setting.

Field notes   Notes that describe what has been observed, heard, or otherwise experienced in a participant observation study. These notes usually are written after the observational session.

Field research   Research in which natural social processes are studied as they happen and left relatively undisturbed.

Field researcher   A researcher who uses qualitative methods to conduct research in the field.

Filter question   A survey question used to identify a subset of respondents who then are asked other questions.

Fixed-sample panel design (panel study)   A type of longitudinal study in which data are collected from the same individuals—the panel—at two or more points in time. In another type of panel design, panel members who leave are replaced with new members.

Floaters   Survey respondents who provide an opinion on a topic in response to a closed-ended question that does not include a "Don't know" option, but who will choose "Don't know" if it is available.

Focus groups   A qualitative method that involves unstructured group interviews in which the focus group leader actively encourages discussion among participants on the topics of interest.

Forced-choice questions   Closed-ended survey questions that do not include "Don't know" as an explicit response choice.

Formative evaluation   Process evaluation that is used to shape and refine program operations.

Frequency distribution   Numerical display showing the number of cases, and usually the percentage of cases (the relative frequencies), corresponding to each value or group of values of a variable.

Frequency polygon   A graphic for quantitative variables in which a continuous line connects data points representing the variable's distribution.

Front matter   The section of an applied research report that includes an executive summary, abstract, and table of contents.

Gamma   A measure of association that is sometimes used in cross-tabular analysis.

Gatekeeper   A person in a field setting who can grant researchers access to the setting.

Generalizability   Exists when a conclusion holds true for the population, group, setting, or event that we say it does, given the conditions that we specify.

Grand tour question   A broad question at the start of an interview that seeks to engage the respondent in the topic of interest.

Grounded theory   Systematic theory developed inductively, based on observations that are summarized into conceptual categories, reevaluated in the research setting, and gradually refined and linked to other conceptual categories.

Group-administered survey   A survey that is completed by individual respondents who are assembled in a group.

Hawthorne effect   A type of contamination in research designs that occurs when members of the treatment group change in terms of the dependent variable because their participation in the study makes them feel special.

Hermeneutic circle   Represents the dialectical process in which the researcher obtains information from multiple stakeholders in a setting, refines his or her understanding of the setting, and then tests that understanding with successive respondents.

Histogram   A graphic for quantitative variables in which the variable's distribution is displayed with adjacent bars.

Historical events research   Research in which social events are studied at one past time period.

Historical process research   Research in which historical processes are studied over a long period of time.

History effect   A source of causal invalidity that occurs when an outside event other than the treatment influences outcome scores; also called an effect of external events.

Hypothesis   A tentative statement about empirical reality, involving a relationship between two or more variables.

Idiographic causal explanation   An explanation that identifies the concrete, individual sequence of events, thoughts, or actions that resulted in a particular outcome for a particular individual or that led to a particular event; may be termed an *individualist* or *historicist* explanation.

Idiosyncratic errors   Errors that affect a relatively small number of individuals in unique ways that are unlikely to be repeated in just the same way.

Idiosyncratic variation   Variation in responses to questions that is caused by individuals' reactions to particular words or ideas in the question instead of by variation in the concept that the question is intended to measure.

Illogical reasoning   When we prematurely jump to conclusions or argue on the basis of invalid assumptions.

Impact evaluation (or analysis)   Analysis of the extent to which a treatment or other service has an effect. Also known as summative evaluation.

Inaccurate observation   An observation based on faulty perceptions of empirical reality.

Independent variable   A variable that is hypothesized to cause, or lead to, variation in another variable.

Index   The sum or average of responses to a set of questions about a concept.

Indicator   The question or other operation used to indicate the value of cases on a variable.

Inductive research   The type of research in which general conclusions are drawn from specific data.

Inferential statistics   A mathematical tool for estimating how likely it is that a statistical result based on data from a random sample is representative of the population from which the sample is assumed to have been selected.

In-person interview   A survey in which an interviewer questions respondents face-to-face and records their answers.

Inputs   The resources, raw materials, clients, and staff that go into a program.

Institutional Review Board (IRB)   A group of organizational and community representatives required by federal law to review the ethical issues in all proposed research that is federally funded, involves human subjects, or has any potential for harm to subjects.

Integrative approach   An orientation to evaluation research that expects researchers to respond to the concerns of people involved with the program—stakeholders—as well as to the standards and goals of the social scientific community.

Intensive (depth) interviewing   A qualitative method that involves open-ended, relatively unstructured questioning in which the interviewer seeks in-depth information on the interviewee's feelings, experiences, and perceptions.

Intent-to-treat analysis   When analysis of the effect of a treatment on outcomes in an experimental design compares outcomes for all those who were assigned to the treatment group to outcomes for all those who were assigned to the control group, whether or not participants remained in the treatment group.

Interactive voice response (IVR)   A survey in which respondents receive automated calls and answer questions by pressing numbers on their touch-tone phones or speaking numbers that are interpreted by computerized voice recognition software.

Interitem reliability   An approach that calculates reliability based on the correlation among multiple items used to measure a single concept. Also known as internal consistency.

Interobserver reliability   When similar measurements are obtained by different observers rating the same persons, events, or places.

Interpretive questions   Questions included in a questionnaire or interview schedule to help explain answers to other important questions.

Interpretivism   Methodology based on the belief that reality is socially constructed and that the goal of social scientists is to understand what meanings people give to that reality.

Interquartile range   The range in a distribution between the end of the first quartile and the beginning of the third quartile.

Intersubjective agreement   Agreement between scientists about the nature of reality; often upheld as a more reasonable goal for science than certainty about an objective reality.

Interval level of measurement   A measurement of a variable in which the numbers indicating a variable's values represent fixed measurement units but have no absolute, or fixed, zero point.

Interval-ratio level of measurement   A measurement of a variable in which the numbers indicating a variable's values represent fixed measurement units but may not have an absolute, or fixed, zero point.

Interview schedule   The survey instrument containing the questions asked by the interviewer in an in-person or phone survey.

Intrarater (or intraobserver) reliability   Consistency of ratings by an observer of an unchanging phenomenon at two or more points in time.

Jottings   Brief notes written in the field about highlights of an observation period.

Key informant   An insider who is willing and able to provide a field researcher with superior access and information, including answers to questions that arise in the course of the research.

Level of measurement   The mathematical precision with which the values of a variable can be expressed. The nominal level of measurement, which is qualitative, has no mathematical interpretation; the quantitative levels of measurement—ordinal, interval, and ratio—are progressively more precise mathematically.

Longitudinal research design   A study in which data are collected that can be ordered in time; also defined as research in which data are collected at two or more points in time.

Mailed survey   A survey involving a mailed questionnaire to be completed by the respondent.

Marginal distribution   The summary distributions in the margins of a cross-tabulation that correspond to the frequency distribution of the row variable and of the column variable.

Matching   A procedure for equating the characteristics of individuals in different comparison groups in an experiment. Matching can be done on either an individual or an aggregate basis. For individual matching, individuals who are similar in terms of key characteristics are paired prior to assignment, and then the two members of each pair are assigned to the two groups. For aggregate matching, groups are chosen for comparison that are similar in terms of the distribution of key characteristics.

Matrix   A form on which can be recorded systematically particular features of multiple cases or instances that a qualitative data analyst needs to examine.

Mean   The arithmetic, or weighted, average, computed by adding up the value of all the cases and dividing by the total number of cases.

Measurement   The process of linking abstract concepts to empirical indicants.

Measurement validity   Exists when a measure measures what we think it measures.

Measure of association   A type of descriptive statistic that summarizes the strength of an association.

Mechanism   A discernible process that creates a causal connection between two variables.

Median   The position average, or the point that divides a distribution in half (the 50th percentile).

Meta-analysis   The quantitative analysis of findings from multiple studies.

Method of agreement   A method proposed by John Stuart Mill for establishing a causal relation, in which the values of cases that agree on an outcome variable also agree on the value of the variable hypothesized to have a causal effect, while they differ in terms of other variables.

Method of difference   A method proposed by John Stuart Mill for establishing a causal relation, in which the values of cases that differ on an outcome variable also differ on the value of the variable hypothesized to have a causal effect, while they agree in terms of other variables.

Mixed-mode survey   A survey that is conducted by more than one method, allowing the strengths of one survey design to compensate for the weaknesses of another and maximizing the likelihood of securing data from different types of respondents; for example, nonrespondents in a mailed survey may be interviewed in person or over the phone.

Mode   The most frequent value in a distribution; also termed the probability average.

Monotonic   A pattern of association in which the value of cases on one variable increases or decreases fairly regularly across the categories of another variable.

Multiple group before-and-after design   A type of quasi-experimental design in which several before-and-after comparisons are made involving the same independent and dependent variables but different groups.

Mutually exclusive   A variable's attributes (or values) are mutually exclusive when every case can be classified as having only one attribute (or value).

Narrative analysis   A form of qualitative analysis in which the analyst focuses on how respondents impose order on the flow of experience in their lives and thus make sense of events and actions in which they have participated.

Narrative explanations   An idiographic causal explanation that involves developing a narrative of events and processes that indicate a chain of causes and effects.

Needs assessment   A type of evaluation research that attempts to determine the needs of some population that might be met with a social program.

Netnography   The use of ethnographic methods to study online communities.

Nominal level of measurement   Variables whose values have no mathematical interpretation; they vary in kind or quality, but not in amount.

Nomothetic causal explanation   An explanation that identifies common influences on a number of cases or events.

Nonequivalent control group design   A quasi-experimental design in which there are experimental and comparison groups that are designated before the treatment occurs but are not created by random assignment.

Nonprobability sampling method   Sampling method in which the probability of selection of population elements is unknown.

Nonrespondents   People or other entities who do not participate in a study although they are selected for the sample.

Nonspuriousness   A criterion for establishing a causal relation between two variables; when a relationship between two variables is not due to variation in a third variable.

Normal distribution   A symmetric, bell-shaped distribution that results from chance variation around a central value.

Normal science   The gradual, incremental research conducted by scientists within the prevailing scientific paradigm.

Omnibus survey   A survey that covers a range of topics of interest to different social scientists.

Open-ended question   A survey question to which the respondent replies in his or her own words, either by writing or by talking.

Operations   A procedure for identifying or indicating the value of cases on a variable.

Operationalization   The process of specifying the operations that will indicate the value of cases on a variable.

Oral history   Data collected through intensive interviews with participants in past events.

Ordinal level of measurement   A measurement of a variable in which the numbers indicating a variable's values specify only the order of the cases, permitting *greater than* and *less than* distinctions.

Outcomes   The impact of the program process on the cases processed.

Outlier   An exceptionally high or low value in a distribution.

Outputs   The services delivered or new products produced by the program process.

Overgeneralization   Occurs when we unjustifiably conclude that what is true for some cases is true for all cases.

Paradigm wars   The intense debate from the 1970s to the 1990s between social scientists over the value of positivist and interpretivist research philosophies.

Participant observation   A qualitative method for gathering data that involves developing a sustained relationship with people while they go about their normal activities.

Participant observer   A researcher who gathers data through participating and observing in a setting where he or she develops a sustained relationship with people while they go about their normal activities. The term *participant observer* is often used to refer to a continuum of possible roles, from complete observation, in which the researcher does not participate along with others in group activities, to complete participation, in which the researcher participates without publicly acknowledging being an observer. Also termed *overt participant.*

Participatory action research (PAR)   A type of research in which the researcher involves some organizational members as active participants throughout the process of studying an organization; the goal is to make changes in the organization. Also termed *community-based participatory research.*

Part-whole question effects   These occur when responses to a general or summary question about a topic are influenced by responses to an earlier, more specific question about that topic.

Percentages   Relative frequencies, computed by dividing the frequency of cases in a particular category by the total number of cases and then multiplying by 100.

Periodicity   A sequence of elements (in a list to be sampled) that varies in some regular, periodic pattern.

Phone survey   A survey in which interviewers question respondents over the phone and then record their answers.

Photo voice   A method in which research participants take pictures of their everyday surroundings with cameras the researcher distributes, and then meet in a group with the researcher to discuss the pictures' meaning.

Placebo effect   A source of treatment misidentification that can occur when subjects receive a fake "treatment" that they consider likely to be beneficial and improve because of that expectation even though they did not receive the actual treatment.

Population parameter   The value of a statistic, such as a mean, computed using the data for the entire population; a sample statistic is an estimate of a population parameter.

Positivism   The belief, shared by most scientists, that there is a reality that exists quite apart from our own perception of it, that it can be understood through observation, and that it follows general laws.

Postpositivism   A philosophical view that modifies the positivist premise of an external, objective reality by recognizing its complexity, the limitations of human observers, and therefore the impossibility of developing more than a partial understanding of reality.

Posttest   In experimental research, the measurement of an outcome (dependent) variable after an experimental intervention or after a presumed independent variable has changed for some other reason.

Predictive validity   The type of validity that exists when a measure predicts scores on a criterion measured in the future.

Pretest   In experimental research, the measurement of an outcome (dependent) variable prior to an experimental intervention or change in a presumed independent variable for some other reason. The pretest is exactly the same "test" as the posttest, but it is administered at a different time.

Probability average   The most frequent value in a distribution.

Probability of selection   The likelihood that an element will be selected from the population for inclusion in the sample. In a census of all the elements of a population, the probability that any particular element will be selected is 1.0. If half the elements in the population are sampled on the basis of chance (say, by tossing a coin), the probability of selection for each element is one-half, or .5. As the size of the sample as a proportion of the population decreases, so does the probability of selection.

Probability sampling method   A sampling method that relies on a random, or chance, selection method so that the probability of selection of population elements is known.

Process evaluation   Evaluation research that investigates the process of service delivery.

Program process   The complete treatment or service delivered by the program.

Program theory   A descriptive or prescriptive model of how a program operates and produces effects.

Progressive focusing   The process by which a qualitative analyst interacts with the data and gradually refines his or her focus.

Proportionate stratified sampling   Sampling method in which elements are selected from strata in exact proportion to their representation in the population.

Pseudoscience   Claims presented so that they appear scientific even though they lack supporting evidence and plausibility.

Purposive sampling   A nonprobability sampling method in which elements are selected for a purpose, usually because of their unique position.

Qualitative comparative analysis (QCA)   A systematic type of qualitative analysis that identifies the combination of factors that had to be present across multiple cases to produce a particular outcome.

Qualitative methods   Methods such as participant observation, intensive interviewing, and focus groups that are designed to capture social life as participants experience it rather than in categories predetermined by the researcher. These methods rely on written or spoken words or observations that do not have a direct numerical interpretation and typically involve exploratory research questions, inductive reasoning, an orientation to social context and human objectivity, and the meanings attached by participants to events and to their lives.

Quantitative methods   Methods such as surveys and experiments that record variation in social life in terms of categories that vary in amount. Data that are treated as quantitative are either numbers or attributes that can be ordered in terms of magnitude.

Quartiles   The points in a distribution corresponding to the first 25% of the cases, the first 50% of the cases, and the first 75% of the cases.

Quasi-experimental design   A research design in which there is a comparison group that is comparable with the experimental group in critical ways, but subjects are not randomly assigned to the comparison and experimental groups.

Questionnaire   The survey instrument containing the questions in a self-administered survey.

Quota sampling   A nonprobability sampling method in which elements are selected to ensure that the sample represents certain characteristics in proportion to their prevalence in the population.

Random assignment   A procedure by which each experimental subject is placed in a group randomly.

Random digit dialing   The random dialing by a machine of numbers within designated phone prefixes, which creates a random sample for phone surveys.

Randomization   The random assignment of cases, as by the toss of a coin.

Randomized comparative change design   The classic true experimental design in which subjects are assigned randomly to two groups; both these groups receive a pretest, and then one group receives the experimental intervention, and then both groups receive a posttest. Also known as a *pretest-posttest control group design*.

Randomized comparative posttest design   A true experimental design in which subjects are assigned randomly to two groups—one group then receives the experimental intervention and both groups receive a posttest; there is no pretest. Also known as *posttest-only control group design*.

Random number table   A table containing lists of numbers that are ordered solely on the basis of chance; it is used for drawing a random sample.

Random sampling   A method of sampling that relies on a random, or chance, selection method so that every element of the sampling frame has a known probability of being selected.

Random sampling error (chance sampling error)   Differences between the population and the sample that are due only to chance factors (random error), not to systematic sampling error. Random sampling error may or may not result in an unrepresentative sample. The magnitude of sampling error due to chance factors can be estimated statistically.

Range   The true upper limit in a distribution minus the true lower limit (or the highest rounded value minus the lowest rounded value, plus one).

Ratio level of measurement   A measurement of a variable in which the numbers indicating a variable's values represent fixed measuring units and an absolute zero point.

Rational choice theory   A social theory that explains individual action with the principle that actors choose actions that maximize their gains from taking that action.

Reactive effects   The changes in individual or group behavior that are due to being observed or otherwise studied.

Reductionist fallacy (reductionism)   An error in reasoning that occurs when incorrect conclusions about group-level processes are based on individual-level data. Also known as individualist fallacy.

Regression analysis   A statistical technique for characterizing the pattern of a relationship between two quantitative variables in terms of a linear equation and for summarizing the strength of this relationship in terms of its deviation from that linear pattern.

Regression effect   A source of causal invalidity that occurs when subjects who are chosen for a study because of their extreme scores on the dependent variable become less extreme on the posttest due to natural cyclical or episodic change in the variable.

Regression-discontinuity design   A quasi-experimental design in which individuals are assigned to a treatment and a comparison group solely on the basis of a cutoff score on some assignment variable and then treatment effects are identified by a discontinuity in the regression line that displays the relation between the outcome and the assignment variable at the cutoff score.

Reliability   A measurement procedure yields consistent scores when the phenomenon being measured is not changing.

Reliability measure   Statistics that summarize the consistency among a set of measures. Cronbach's alpha is the most common measure of the reliability of a set of items included in an index.

Repeated cross-sectional design (trend study)   A longitudinal study in which data are collected at two or more points in time from different samples of the same population.

Repeated measures panel design   A quasi-experimental design consisting of several pretest and posttest observations of the same group.

Replacement sampling   A method of sampling in which sample elements are returned to the sampling frame after being selected, so they may be sampled again. Random samples may be selected with or without replacement.

Replications   Repetitions of a study using the same research methods to answer the same research question.

Representative sample   A sample that "looks like" the population from which it was selected in all respects that are potentially relevant to the study. The distribution of characteristics among the elements of a representative sample is the same as the distribution of those characteristics among the total population. In an unrepresentative sample, some characteristics are overrepresented or underrepresented.

Research circle   A diagram of the elements of the research process, including theories, hypotheses, data collection, and data analysis.

Resistance to change   The reluctance to change our ideas in light of new information.

Reverse outlining   Outlining the sections in an already written draft of a paper or report to improve its organization in the next draft.

Sample generalizability   Exists when a conclusion based on a sample, or subset, of a larger population holds true for that population.

Sample statistic   The value of a statistic, such as a mean, computed from sample data.

Sampling error   Any difference between the characteristics of a sample and the characteristics of a population. The larger the sampling error, the less representative the sample.

Sampling frame   A list of all elements or other units containing the elements in a population.

Sampling interval   The number of cases from one sampled case to another in a systematic random sample.

Sampling units   Units listed at each stage of a multistage sampling design.

Saturation point   The point at which subject selection is ended in intensive interviewing, when new interviews seem to yield little additional information.

Science   A set of logical, systematic, documented methods for investigating nature and natural processes; the knowledge produced by these investigations.

Scientific paradigm   A set of beliefs that guide scientific work in an area, including unquestioned presuppositions, accepted theories, and exemplary research findings.

Scientific revolution   The abrupt shift from one dominant scientific paradigm to an alternative paradigm that may be developed after accumulation of a large body of evidence that contradicts the prevailing paradigm.

Secondary data   Previously collected data that are used in a new analysis.

Secondary data analysis   The method of using preexisting data in a different way or to answer a different research question than intended by those who collected the data.

Selection bias   A source of internal (causal) invalidity that occurs when characteristics of experimental and comparison group subjects differ in any way that influences the outcome.

Selective distribution of benefits   An ethical issue about how much researchers can influence the benefits subjects receive as part of the treatment being studied in a field experiment.

Selective observation   Choosing to look only at things that are in line with our preferences or beliefs.

Serendipitous findings   Unexpected patterns in data, which stimulate new ideas or theoretical approaches. Also known as anomalous findings.

Simple random sampling   A method of sampling in which every sample element is selected only on the basis of chance, through a random process.

Skewness   The extent to which cases are clustered more at one or the other end of the distribution of a quantitative variable rather than in a symmetric pattern around its center. Skew can be positive (a right skew), with the number of cases tapering off in the positive direction, or negative (a left skew), with the number of cases tapering off in the negative direction.

Skip pattern   The unique combination of questions created in a survey by filter questions and contingent questions.

Snowball sampling   A method of sampling in which sample elements are selected as they are identified by successive informants or interviewees.

Social research question   A question about the social world that is answered through the collection and analysis of firsthand, verifiable, empirical data.

Social science   The use of scientific methods to investigate individuals, societies, and social processes; the knowledge produced by these investigations.

Social science approach   An orientation to evaluation research that expects researchers to emphasize the importance of researcher expertise and maintenance of autonomy from program stakeholders.

Solomon four-group design   A type of experimental design that combines a randomized pretest-posttest control group design with a randomized posttest-only design, resulting in two experimental groups and two comparison groups.

Specification   A type of relationship involving three or more variables in which the association between the independent and dependent variables varies across the categories of one or more other control variables.

Split-ballot design   Unique questions or other modifications in a survey administered to randomly selected subsets of the total survey sample, so that more questions can be included in the entire survey or so that responses to different question versions can be compared.

Split-halves reliability   Reliability achieved when responses to the same questions by two randomly selected halves of a sample are about the same.

Spurious relationship   A relationship between two variables that is due to variation in a third variable.

Stakeholder approach   An orientation to evaluation research that expects researchers to be responsive primarily to the people involved with the program. Also termed *responsive evaluation.*

Stakeholders   Individuals and groups who have some basis of concern with the program.

Standard deviation   The square root of the average squared deviation of each case from the mean.

Statistical control   A method in which one variable is held constant so that the relationship between two (or more) other variables can be assessed without the influence of variation in the control variable.

Statistical significance   The mathematical likelihood that an association is not due to chance, judged by a criterion set by the analyst.

Stratified random sampling   A method of sampling in which sample elements are selected separately from population strata that are identified in advance by the researcher.

Street-level bureaucrats   Officials who serve clients and have a high degree of discretion.

Subject fatigue   Problems caused by panel members growing weary of repeated interviews and dropping out of a study or becoming so used to answering the standard questions in the survey that they start giving stock or thoughtless answers.

Subtables   Tables describing the relationship between two variables within the discrete categories of one or more other control variables.

Survey pretest   A method of evaluating survey questions and procedures by testing them out on a small sample of individuals like those to be included in the actual survey, and then reviewing responses to the questions and reactions to the survey procedures.

Survey research   Research in which information is obtained from a sample of individuals through their responses to questions about themselves or others.

Symbolic interaction theory   Focuses on the symbolic nature of social interaction—how social interaction conveys meaning and promotes socialization.

Systematic bias   Overrepresentation or underrepresentation of some population characteristics in a sample due to the method used to select the sample. A sample shaped by systematic sampling error is a biased sample.

Systematic observation   A strategy that increases the reliability of observational data by using explicit rules that standardize coding practices across observers.

Systematic random sampling   A method of sampling in which sample elements are selected from a list or from sequential files, with every nth element being selected after the first element is selected randomly within the first interval.

Tacit knowledge   In field research, a credible sense of understanding of social processes that reflects the researcher's awareness of participants' actions as well as their words, and of what they fail to state, feel deeply, and take for granted.

Target population   A set of elements larger than or different from the population sampled and to which the researcher would like to generalize study findings.

Test-retest reliability   A measurement showing that measures of a phenomenon at two points in time are highly correlated, if the phenomenon has not changed, or have changed only as much as the phenomenon itself.

Theoretical sampling   A sampling method recommended for field researchers by Glaser and Strauss (1967). A theoretical sample is drawn in a sequential fashion, with settings or individuals selected for study as earlier observations or interviews indicate that these settings or individuals are influential.

Theory   A logically interrelated set of propositions about empirical reality.

Theory-driven evaluation   A program evaluation that is guided by a theory that specifies the process by which the program has an effect.

Thick description   A rich description that conveys a sense of what it is like from the standpoint of the natural actors in that setting.

Time order   A criterion for establishing a causal relation between two variables. The variation in the presumed cause (the independent variable) must occur before the variation in the presumed effect (the dependent variable).

Time series design   A quasi-experimental design consisting of many pretest and posttest observations of the same group.

Treatment misidentification   A problem that occurs in an experiment when the treatment itself is not what causes the outcome, but rather the outcome is caused by some intervening process that the researcher has not identified and is not aware of.

Triangulation   The use of multiple methods to study one research question. Also used to mean the use of two or more different measures of the same variable.

True experiment   Experiment in which subjects are assigned randomly to an experimental group that receives a treatment or other manipulation of the independent variable and a comparison group that does not receive the treatment or receives some other manipulation. Outcomes are measured in a posttest.

Unbalanced response choices   A fixed-choice survey question has a different number of positive and negative response choices.

Unimodal   A distribution of a variable in which there is only one value that is the most frequent.

Units of analysis   The level of social life on which a research question is focused, such as individuals, groups, towns, or nations.

Units of observation   The cases about which measures actually are obtained in a sample.

Unobtrusive measures   A measurement based on physical traces or other data that are collected without the knowledge or participation of the individuals or groups that generated the data.

Validity   The state that exists when statements or conclusions about empirical reality are correct.

Variability   The extent to which cases are spread out through the distribution or clustered in just one location.

Variable   The extent to which cases are spread out through the distribution or clustered in just one location.

Variable-oriented research   Research that focuses attention on variables representing particular aspects of the units studied and then examines the relations among these variables across sets of cases.

Variance   A statistic that measures the variability of a distribution as the average squared deviation of each case from the mean.

Web survey   A survey that is accessed and responded to on the World Wide Web.

Appendix A

Questions to Ask About a Research Article

1. What is the basic research question, or problem? Try to state it in just one sentence. (Chapter 2)

2. Is the purpose of the study explanatory, evaluative, exploratory, or descriptive? Did the study have more than one purpose? (Chapter 1)

3. Was a theoretical framework presented? What was it? Did it seem appropriate for the research question addressed? Can you think of a different theoretical perspective that might have been used? What philosophy guides the research? Is this philosophy appropriate to the research question? (Chapters 2, 3)

4. What prior literature was reviewed? Was it relevant to the research problem? To the theoretical framework? Does the literature review appear to be adequate? Are you aware of (or can you locate) any important studies that have been omitted? (Chapter 2)

5. How well did the study live up to the guidelines for science? Do you need additional information in any areas to evaluate the study? To replicate it? (Chapter 3)

6. Did the study seem consistent with current ethical standards? Were any trade-offs made between different ethical guidelines? Was an appropriate balance struck between adherence to ethical standards and use of the most rigorous scientific practices? (Chapter 3)

7. Were any hypotheses stated? Were these hypotheses justified adequately in terms of the theoretical framework? In terms of prior research? (Chapter 2)

8. What were the independent and dependent variables in the hypothesis or hypotheses? Did these variables reflect the theoretical concepts as intended? What direction of association was hypothesized? Were any other variables identified as potentially important? (Chapter 2)

9. What were the major concepts in the research? How, and how clearly, were they defined? Were some concepts treated as unidimensional that you think might best be thought of as multidimensional? (Chapter 4)

10. Did the instruments used, the measures of the variables, seem valid and reliable? How did the authors attempt to establish this? Could any more have been done in the study to establish measurement validity? (Chapter 4)

11. Was a sample or the entire population of elements used in the study? What type of sample was selected? Was a probability sampling method used? Did the authors think the sample was generally representative of the population from which it was drawn? Do you? How would you evaluate the likely generalizability of the findings to other populations? (Chapter 5)

12. Was the response rate or participation rate reported? Does it appear likely that those who did not respond or participate were markedly different from those who did participate? Why or why not? Did the author(s) adequately discuss this issue? (Chapters 5, 8)

13. What were the units of analysis? Were they appropriate for the research question? If some groups were the units of analysis, were any statements made at any point that are open to the ecological fallacy? If individuals were the units of analysis, were any statements made at any point that suggest reductionist reasoning? (Chapter 6)

14. Was the study design cross-sectional or longitudinal, or did it use both types of data? If the design was longitudinal, what type of longitudinal design was it? Could the longitudinal design have been improved in any way, as by collecting panel data rather than trend data, or by decreasing the dropout rate in a panel design? If cross-sectional data were used, could the research question have been addressed more effectively with longitudinal data? (Chapter 6)

15. Were any causal assertions made or implied in the hypotheses or in subsequent discussions? What approach was used to demonstrate the existence of causal effects? Were all five issues in establishing causal relationships addressed? What, if any, variables were controlled in the analysis to reduce the risk of spurious relationships? Should any other variables have been measured and controlled? How satisfied are you with the internal validity of the conclusions? (Chapter 6)

16. Was an experimental survey, participant observation, historical comparative, or some other research design used? How well was this design suited to the research question posed and the specific hypotheses tested, if any? Why do you suppose the author(s) chose this particular design? How was the design modified in response to research constraints? How was it modified in order to take advantage of research opportunities? (Chapters 7–13)

17. Was this an evaluation research project? If so, which type of evaluation was it? Which design alternatives did it use? (Chapter 11)

18. Was a historical comparative design used? Which type was it? Were problems due to using historical and/or cross-national data addressed? (Chapter 12)

19. Was any attention given to social context and subjective meanings? If so, what did this add? If not, would it have improved the study? Explain. (Chapter 1)

20. Summarize the findings. How clearly were statistical and/or qualitative data presented and discussed? Were the results substantively important? (Chapters 10, 14)

21. Did the author(s) adequately represent the findings in the discussion and/or conclusions sections? Were conclusions well grounded in the findings? Are any other interpretations possible? (Chapter 15)

22. Compare the study to others addressing the same research question. Did the study yield additional insights? In what ways was the study design more or less adequate than the design of previous research? (Chapter 15)

23. What additional research questions and hypotheses are suggested by the study's results? What light did the study shed on the theoretical framework used? On social policy questions? (Chapters 2, 15)

Appendix B

How to Read a Research Article

The discussions of research articles throughout the text may provide all the guidance you need to read and critique research on your own. But reading about an article in bits and pieces in order to learn about particular methodologies is not quite the same as reading an article in its entirety in order to learn what the researcher found out. The goal of this appendix is to walk you through an entire research article, answering the review questions introduced in Appendix B. Of course, this is only one article and our "walk" will take different turns than would a review of other articles, but after this review you should feel more confident when reading other research articles on your own.

We will use for this example an article by South and Spitze (1994) on housework in marital and nonmarital households, reprinted on pages C-8 to C-28 of this appendix. It focuses on a topic related to everyone's life experiences as well as to important questions in social theory. Moreover, it is a solid piece of research published in a top journal, the American Sociological Association's *American Sociological Review*.

I have reproduced below each of the article review questions from Appendix B, followed by my answers to them. After each question, I indicate the chapter where the question was discussed and after each answer I cite the article page or pages that I am referring to. You can also follow my review by reading through the article itself and noting my comments.

1. *What is the basic research question, or problem? Try to state it in just one sentence.* (Chapter 2)

The clearest statement of the research question—actually three questions—is that "we seek to determine how men and women in these [six] different situations [defined by marital status and living arrangement] compare in the amounts of time they spend doing

housework, whether these differences can be attributed to differences in other social and economic characteristics, and which household tasks account for these differences" (p. C-9). Prior to this point, the authors focus in on this research question, distinguishing it from the more general issue of how housework is distributed within marriages and explaining why it is an important research question.

2. *Is the purpose of the study explanatory, evaluative, exploratory, or descriptive? Did the study have more than one purpose?* (Chapter 1)

The problem statement indicates that the study will have both descriptive and explanatory purposes: it will "determine how men and women . . . compare" and then try to explain the differences in housework between them. The literature review that begins on page C-9 also makes it clear that the primary purpose of the research was explanatory, since the authors review previous explanations for gender differences in housework and propose a new perspective (pp. C-9–C-14).

3. *What prior literature was reviewed? Was it relevant to the research problem? To the theoretical framework? Does the literature review appear to be adequate? Are you aware of (or can you locate) any important studies that have been omitted?* (Chapter 2)

Literature is reviewed from the article's first page until the "Data and Methods" section (pp. C-8–C-14). It all seems relevant to the particular problem as well as to the general theoretical framework. In the first few paragraphs, several general studies are mentioned to help clarify the importance of the research problem (pp. C-8–C-9).

In the "Models of Household labor" section, alternative theoretical perspectives used in other studies are reviewed and the strength of the support for them is noted (pp. C-9–C-11). After identifying the theoretical perspective they will use, the authors then introduce findings from particular studies that are most relevant to their focus on how housework varies with marital status (pp. C-11–C-14). I leave it to you to find out whether any important studies were omitted.

4. *Was a theoretical framework presented? What was it? Did it seem appropriate for the research question addressed? Can you think of a different theoretical perspective that might have been used?* (Chapter 2)

The "gender perspective" is used as a framework for the research (p. C-10). This perspective seems very appropriate to the research question addressed because it highlights the importance of examining differences between married and other households. The authors themselves discuss three other theoretical perspectives on the division of household labor that might have been used as a theoretical framework, but identify weaknesses in each of them (pp. C-9–C-10).

5. *How well did the study live up to the guidelines for science? Do you need additional information in any areas to evaluate the study? To replicate it? (Chapter 3)*

It would be best to return to this question after reading the whole article. The study clearly involves a test of ideas against empirical reality as much as that reality could be measured; it was carried out systematically and disclosed, as far as we can tell, fully. Since the authors used an available dataset, others can easily obtain the complete documentation for the study and try to replicate the authors' findings. The authors explicitly note and challenge assumptions made in other theories of the division of housework (p. C-10), although they do not clarify their own assumptions as such. Two of their assumptions are that the appropriation of another's work is likely to occur "perhaps only" in heterosexual couple households (p. C-10) and that "a woman cannot display love for or subordination to a man through housework when no man is present" (p. C-11). The authors also assume that respondents' reports of the hours they have spent on various tasks are reasonably valid (p. C-15). These seem to me to be reasonable assumptions, but a moment's reflection should convince you that they are, after all, unproved assumptions that could be challenged. This is not in itself a criticism of the research, since some assumptions must be made in any study. The authors specified the meaning of key terms, as required in scientific research. They also searched for regularities in their data, thus living up to another guideline. A skeptical stance toward current knowledge is apparent in the literature review and in the authors' claim that they have found only "suggestive evidence" for their theoretical perspective (p. C-25). They aim clearly to build social theory and encourage others to build on their findings, "to further specify the conditions" (p. C-25). The study thus seems to exemplify adherence to basic scientific guidelines and to be very replicable.

6. *Did the study seem consistent with current ethical standards? Were any trade-offs made between different ethical guidelines? Was an appropriate balance struck between adherence to ethical standards and use of the most rigorous scientific practices?* (Chapter 3)

The authors use survey data collected by others and so encounter no ethical problems in their treatment of human subjects. The reporting seems honest and open. Although the research should help inform social policy, the authors' explicit focus is on how their research can inform social theory. This is quite appropriate for research reported in a scientific journal, so there are no particular ethical problems raised about the uses to which the research is put. The original survey used by the authors does not appear at all likely to have violated any ethical guidelines concerning the treatment of human subjects, although it would be necessary to inspect the original research report to evaluate this.

7. *Were any hypotheses stated? Were these hypotheses justified adequately in terms of the theoretical framework? In terms of prior research?* (Chapter 2)

Five primary hypotheses are stated, although they are labeled as "several important contrasts" that are suggested by the "doing gender" approach, rather than as hypotheses. For example, the first hypothesis is that "women in married-couple households [are expected] to spend more time doing housework than women in any other living situation" (p. C-11). A more general point is made about variation in housework across household types before these specific hypotheses are introduced. Several more specific hypotheses are then introduced about variations among specific types of households (pp. C-11–C-12). Some questions about patterns of housework in households that have not previously been studied are presented more as speculations than as definite hypotheses (pp. C-13–C-14). Three additional hypotheses are presented concerning the expected effects of the control variables (p. C-14).

8. *What were the independent and dependent variables in the hypothesis or hypotheses? Did these variables reflect the theoretical concepts as intended? What direction of association was hypothesized? Were any other variables identified as potentially important?* (Chapter 2)

The independent variable in the first hypothesis is marital status (married versus other); the dependent variable is time spent doing housework. The hypothesis states that more time will be spent by married women than by other women, and it is stated that this expectation is "net of other differences among the household types" (p. C-11). Can you identify the variables in the other hypotheses (the second and fourth hypotheses about men just restate the preceding hypotheses for women)? Another variable, gender differences in time spent on housework, is discussed throughout the article, but it is not in itself measured; rather, it is estimated by comparing the aggregate distribution of hours for men and women.

9. *What were the major concepts in the research? How, and how clearly, were they defined? Were some concepts treated as unidimensional that you think might best be thought of as multidimensional?* (Chapter 4)

The key concept in the research is that of "doing gender"; it is discussed at length and defined in a way that becomes reasonably clear when it is said that "housework 'produces' gender through the everyday enactment of dominance, submission, and other behaviors symbolically linked to gender" (p. C-10). The central concept of housework is introduced explicitly as "a major component of most people's lives" (p. C-11), but it is not defined conceptually—presumably because it refers to a widely understood phenomenon. A conceptual definition would have helped to justify the particular operationalization used, and the decision to exclude childcare from what is termed housework (p. C-15). (A good, practical reason for this exclusion is given in footnote 2.) The concept of housework is treated as multidimensional by distinguishing what are termed "male-typed," "female-typed," and "gender-neutral" tasks (p. C-23). Another key concept is that of marital status, which the authors define primarily by identifying its different categories (pp. C-11–C-14).

10. *Did the instruments used, the measures of the variables, seem valid and reliable? How did the authors attempt to establish this? Could any more have been done in the study to establish measurement validity?* (Chapter 4)

The measurement of the dependent variable was straightforward, but required respondents to estimate the number of hours per week they spent on various tasks. The authors report that some other researchers have used a presumably more accurate method—time diaries—to estimate time spent on household tasks, and that the results they obtain are very similar to those of the recall method used in their study. This increases confidence in the measurement approach used, although it does not in itself establish the validity or reliability of the self-report data. Measures of marital status and other variables involved relatively straightforward questions and do not raise particular concerns about validity. The researchers carefully explain in footnotes how they handled missing data.

11. *Was a sample or the entire population of elements used in the study? What type of sample was selected? Was a probability sampling method used? Did the authors think the sample was*

generally representative of the population from which it was drawn? Do you? How would you evaluate the likely generalizability of the findings to other populations? (Chapter 5)

The sample was a random (probability) sample of families and households. A disproportionate stratified sampling technique was used to ensure the representation of adequate numbers of single-parent families, cohabitors, and other smaller groups that are of theoretical interest (pp. C-14–C-15). The sample is weighted in the analysis to compensate for the disproportionate sampling method and is said to be representative of the U.S. population. The large size of the sample ($N = 11,016$ after cases with missing values were excluded) indicates that the confidence limits around sample statistics will be very small. Do you think the findings could be generalized to other countries with different cultural values about gender roles and housework?

12. *Was the response rate or participation rate reported? Does it appear likely that those who did not respond or participate were markedly different from those who did participate? Why or why not? Did the author(s) adequately discuss this issue?* (Chapters 5, 8)

The response rate was not mentioned—a major omission, although it could be found in the original research report. The authors omitted 2,001 respondents from the obtained sample due to missing data and adjusted values of variables having missing data for some other cases. In order to check the consequences of these adjustments, the authors conducted detailed analyses of the consequences of various adjustment procedures. They report that the procedures they used did not affect their conclusions (pp. C-14–C-15). This seems reasonable.

13. *What were the units of analysis? Were they appropriate for the research question? If some groups were the units of analysis, were any statements made at any point that are open to the ecological fallacy? If individuals were the units of analysis, were any statements made at any point that suggest reductionist reasoning?* (Chapter 6)

The survey sampled adults, although it was termed a survey of families and households; and it is data on individuals (and the households in which they live) that are analyzed. You can imagine this same study being conducted with households forming the units of analysis, and the dependent variable being the percentage of total time in the family spent on housework, rather than the hours spent by individuals on housework. The conclusions generally are appropriate to the use of individuals as the units of analysis, but there is some danger in reductionist misinterpretation of some of the interpretations, such as that "men and women must be 'doing gender' when they live together" (p. C-15). Conclusions like this would be on firmer ground if they were based on household-level data that revealed whether one person's approach to housework did, in fact, vary in relation to that of his or her partner.

14. *Was the study design cross-sectional or longitudinal, or did it use both types of data? If the design was longitudinal, what type of longitudinal design was it? Could the longitudinal design have been improved in any way, as by collecting panel data rather than trend data, or by decreasing the dropout rate in a panel design? If cross-sectional data were used, could the research question have been addressed more effectively with longitudinal data?* (Chapter 6)

The survey was cross-sectional. The research question certainly could have been addressed more effectively with longitudinal data that followed people over their adult lives, since many of the authors' interpretations reflect their interest in how individuals' past experiences with housework shape their approach when they enter a new marital status (pp. C-15–C-16).

15. *Were any causal assertions made or implied in the hypotheses or in subsequent discussions? What approach was used to demonstrate the existence of causal effects? Were all five issues in establishing causal relationships addressed? What, if any, variables were controlled in the analysis to reduce the risk of spurious relationships? Should any other variables have been measured and controlled? How satisfied are you with the internal validity of the conclusions?* (Chapter 6)

The explanatory hypotheses indicate that the authors were concerned with causality. Mention is made of a possible causal mechanism when it is pointed out that "doing gender"—the presumed causal influence—may operate at both unconscious and conscious levels (p. C-10). In order to reduce the risk of spuriousness in the presumed causal relationship (between marital status and housework time), variables such as age, education, earnings, and the presence of children are controlled (p. C-16). There are, of course, other variables that might have created a spurious relationship, but at least several of the most likely contenders have been controlled. For example, the use of cross-sectional data leaves us wondering whether some of the differences attributed to marital status might really be due to generational differences—the never-married group is likely to be younger and the widowed group older; controlling for age gives us more confidence that this is not the case. On the other hand, the lack of longitudinal data means that we do not know whether the differences in housework might have preceded marital status: perhaps women who got married also did more housework even before they were married than women who remained single.

16. *Was an experimental survey, participant observation, content analysis, or some other research design used? How well*

was this design suited to the research question posed and the specific hypotheses tested, if any? Why do you suppose the author(s) chose this particular design? How was the design modified in response to research constraints? How was it modified in order to take advantage of research opportunities? (Chapters 7–13)

Survey research was the method of choice, and probably was used for this article because the data set was already available for analysis. Survey research seems appropriate for the research questions posed, but the limitation of the survey to one point in time was a major constraint (p. C-14).

17. *Was this an evaluation research project? If so, which type of evaluation was it? Which design alternatives did it use?* (Chapter 11)

This study did not use an evaluation research design. The issues on which it focuses might profitably be studied in some program evaluations.

18. *Was a historical comparative design used? Which type was it? Was secondary data used? Were problems due to using historical and/or cross-national data addressed?* (Chapter 12)

This study did not use any type of historical or comparative design or secondary data. It is interesting to consider how the findings might have differed if comparisons to other cultures or to earlier times had been made.

19. *Was any attention given to social context? To biological processes? If so, what did this add? If not, would it have improved the study? Explain.* (Chapter 1)

In a sense, the independent variable in this study is social context: The combinations of marital status and living arrangements distinguish different social contexts in which gender roles are defined. However, no attention is given to the potential importance of larger social contexts, such as neighborhood, region, or nation. It is also possible to imagine future research that tests the influence of biological factors on the household division of labor, as in Udry's (1988) study of adolescents.

20. *Summarize the findings. How clearly were statistical and/ or qualitative data presented and discussed? Were the results substantively important?* (Chapters 10, 14)

Statistical data are presented clearly using descriptive statistics (multiple regression analysis, a multivariate statistical technique), and graphs that highlight the most central findings. In fact, the data displays are exemplary, because they effectively convey findings to a wide audience and also subject the hypotheses to rigorous statistical tests. No qualitative data are presented. The findings seem substantively important, since they identify large differences in the household roles of men and women and in how these roles vary in different types of households (pp. C-17–C-24).

21. *Did the author(s) adequately represent the findings in the discussion and/or conclusions sections? Were conclusions well grounded in the findings? Are any other interpretations possible?* (Chapter 15)

The findings are well represented in the discussion and conclusions section (pp. C-24–C-26). The authors point out in their literature review that a constant pattern of gender differences in housework across household types would "cast doubt on the validity of the gender perspective" (p. C-11), and the findings clearly rule this out. However, the conclusions give little consideration to the ways in which the specific findings might be interpreted as consistent or inconsistent with reasonable predictions from each of the three other theoretical perspectives reviewed. You might want to consider what other interpretations of the findings might be possible. Remember that other interpretations always are possible for particular findings—it is a question of the weight of the evidence, the persuasiveness of the theory used, and the consistency of the findings with other research.

22. *Compare the study to others addressing the same research question. Did the study yield additional insights? In what ways was the study design more or less adequate than the design of previous research?* (Chapter 15)

The study investigated an aspect of the question of gender differences in housework that had not previously received much attention (variation in gender differences across different types of households). This helped the authors to gain additional insights into gender and housework, although the use of cross-sectional data and a retrospective self-report measure of housework made their research in some ways less adequate than others.

23. *What additional research questions and hypotheses are suggested by the study's results? What light did the study shed on the theoretical framework used? On social policy questions?* (Chapters 2, 15)

The article suggests additional questions for study about "the conditions under which [the dynamics of doing gender] operate" and how equity theory might be used to explain the division of labor in households (p. C-25). The authors make a reasonable case for the value of their "gender perspective." Social policy questions are not addressed directly, but the article would be of great value to others concerned with social policy.

HOUSEWORK IN MARITAL AND NONMARITAL HOUSEHOLDS*

SCOTT J. SOUTH
State University of New York at Albany

GLENNA SPITZE
State University of New York at Albany

Although much recent research has explored the division of household labor between husbands and wives, few studies have examined housework patterns across marital statuses. This paper uses data from the National Survey of Families and Households to analyze differences in time spent on housework by men and women in six different living situations: never married and living with parents, never married and living independently, cohabiting, married, divorced, and widowed. In all situations, women spend more time than men doing housework, but the gender gap is widest among married persons. The time women spend doing housework is higher among cohabitants than among the never-married, is highest in marriage, and is lower among divorcees and widows. Men's housework time is very similar across both never-married living situations, in cohabitation, and in marriage. However, divorced and widowed men do substantially more housework than does any other group of men, and they are especially more likely than their married counterparts to spend more time cooking and cleaning. In addition to gender and marital status, housework time is affected significantly by several indicators of workload (e.g., number of children, home ownership) and time devoted to nonhousehold activities (e.g., paid employment, school enrollment)—most of these variables have greater effects on women's housework time than on men's. An adult son living at home increases women's housework, whereas an adult daughter at home reduces housework for women and men. These housework patterns are generally consistent with an emerging perspective that views housework as a symbolic enactment of gender relations. We discuss the implications of these findings for perceptions of marital equity.

Until 20 years ago, social science research on housework was largely nonexistent (Glazer-Malbin 1976; Huber and Spitze 1983), but since then, research on the topic has exploded. Patterns of housework and how housework is experienced by participants have been documented in both qualitative (e.g., Hochschild with Machung 1989; Oakley 1974) and quantitative studies (e.g., Berk 1985; Blair and Lichter 1991; Coverman and Sheley 1986; Goldscheider and Waite 1991; Rexroat and Shehan 1987; Ross 1987; Shelton 1990; Spitze 1986; Walker and Woods 1976). The vast majority of these studies have focused on married couples, but a few have examined cohabiting couples as well (e.g., Blumstein and Schwartz 1983; Shelton and John 1993; Stafford, Backman, and Dibona 1977). The rationale for fo-

cusing on couples is typically a research interest in equity (Benin and Agostinelli 1988; Blair and Johnson 1992; Ferree 1990; Peterson and Maynard 1981; Thompson 1991) and in how changes in women's employment and gender roles have changed, or failed to change, household production functions.

Very few studies have examined housework as performed in noncouple households composed of never-married, separated or divorced, or widowed persons (e.g., Grief 1985; Sanik and Mauldin 1986). Such studies are important for two reasons. First, people are spending increasing amounts of time in such households at various points in their lives due to postponed marriages, higher divorce rates, and a preference among adults in all age categories (including the later years) for independent living. For example, the proportion of households that includes married couples decreased from 76.3 percent to 60.9 percent between 1940 and 1980 (Sweet and Bumpass 1987), and the number of years adult women spend married has decreased by about seven years during the past several decades (Watkins, Menken, and Bon-

*Direct all correspondence to Scott J. South or Glenna Spitze, Department of Sociology, State University of New York at Albany, Albany, NY 12222. The authors contributed equally to this research and are listed alphabetically. We acknowledge with gratitude the helpful comments of several anonymous *ASR* reviewers.

gaarts 1987). It is important to learn how housework is experienced by this substantial segment of the population to understand the household production function in general and because performance of housework is related to decisions about paid work and leisure time for people in these categories.

Second, the housework experiences of single, divorced, and widowed persons go with them if they move into marriage or cohabitation—these experiences are part of the context in which they negotiate how to accomplish tasks jointly with a partner. People may use those prior experiences or assumptions about what they *would* do if the marriage or cohabiting relationship dissolved to set an alternative standard when assessing an equitable division of household labor, rather than simply comparing their own investment in housework to their partner's. Thus, by understanding factors affecting housework contributions by men and women not living in couple relationships, we can better understand what happens when they do form those relationships.

Our broadest objective in this paper is to analyze how time spent doing housework by men and women varies by marital status and to interpret this analysis in relation to the "gender perspective" on household labor. Focusing on six situations defined by marital status and living arrangement, we seek to determine how men and women in these different situations compare in the amounts of time they spend doing housework, whether these differences can be attributed to differences in other social and economic characteristics, and which household tasks account for these differences. We are particularly interested in those persons who are living independently and who are not married or cohabiting, since previous research has focused heavily on married persons and, to a lesser extent, on cohabiting couples (Shelton and John 1993; Stafford et al. 1977) and children still living at home (Benin and Edwards 1990; Berk 1985; Blair 1991; Goldscheider and Waite 1991; Hilton and Haldeman 1991).

MODELS OF HOUSEHOLD LABOR

Beginning with Blood and Wolfe's (1960) classic study, sociologists have attempted to explain the division of household labor between husbands and wives and to determine whether the division is changing over time. The *re-source-power perspective* originating in that work focuses on the economic and social contexts in which husbands and wives bring their individual resources (such as unequal earnings) to bear in bargaining over who will do which household chores. This resource-power theory has since been modified and elaborated upon in several ways, focusing on determining which resources are important and the conditions under which they are useful for bargaining. Rodman's (1967) theory of resources in cultural context and Blumberg and Coleman's (1989) theory of gender stratification (as applied to housework) suggest that there are limits on how effectively resources can be used, especially by women. Several observers suggest that wives' resources may be "discounted" by male dominance at the societal level (Aytac and Teachman 1992; Blumberg and Coleman 1989; Ferree 1991b; Gillespie 1971).

Two other perspectives are used frequently in the study of household labor. One focuses on *socialization and gender role attitudes*, suggesting that husbands and wives perform household labor in differing amounts depending upon what they have learned and have come to believe about appropriate behavior for men and women (see Goldscheider and Waite 1991). An alternative perspective, the *time availability hypothesis*, suggests that husbands and wives perform housework in amounts relative to the time left over after paid work time is subtracted. A variation on this, the demand response capability hypothesis (Coverman 1985), is somewhat broader and includes factors that increase the total amount of work to be done and spouses' availability to do it. The focus on time allocation as a rational process is akin to the economic perspective, most closely associated with Becker (1981; see also critique in Berk 1985). However, sociologists and economists differ in their views on this perspective: Economists assume that time allocation to housework and paid work is jointly determined and based on the relative efficiency of husbands and wives in both arenas; sociologists assume that decisions about paid work are causally prior (Godwin 1991; Spitze 1986).

The above three perspectives (power-resources, socialization-gender roles, and time availability) have guided much of the sociological research on household labor over the past 20 years (see reviews of these theories and their variations in Ferree 1991a; Godwin 1991; Shelton 1992; Spitze 1988). However, they have produced mixed results, and, as several reviewers have pointed out, much more variance is explained by gender per se than by any of the other factors in these models (Ferree 1991a; Thompson and Walker 1991). Moreover, studies show that women who earn more than their husbands often do a disproportionate share of the housework, perhaps in an attempt to prevent those earnings from threatening the husband's self-esteem (Thompson and Walker 1991). While both husbands' and wives' time in paid employment does affect the time they spend doing housework (Goldscheider and Waite 1991), it is argued that the basic distribution of household labor calls for an explanation of its gendered, asymmetrical nature (Thompson and Walker 1991).

A new direction in the explanation of household labor originates in West and Zimmerman's (1987) concept of "doing gender." They argue that gender can be understood as "a routine accomplishment embedded in everyday interaction" (1987:125). Berk (1985) applied their perspective to the division of household labor, observing that the current situation among husbands and wives is neither inherently rational (as the New Home Economics had argued; see Becker 1981) nor fair. Thus, Berk concludes that more than goods and services are "produced" through household labor. She describes the marital household as a "gender factory" where, in addition to accomplishing tasks, housework "produces" gender through the everyday enactment of dominance, submission, and other behaviors symbolically linked to gender (Berk 1985; see also Hartmann 1981; Shelton and John 1993).

Ferree (1991a) elaborates on the "gender perspective" and its application to household labor and argues that it challenges three assumptions of resource theory. First, as Berk pointed out in her critique of economic analyses of housework, housework is not allocated in the most efficient manner. Second, gender is more influential than individual resources in determining the division of household labor. And third, housework is not necessarily defined as "bad" and to be avoided. On the contrary, in addition to expressing subordination, housework can also express love and care, particularly for women (Ferree 1991a). Relatedly, DeVault (1989) describes in detail how the activities surrounding the planning and prepara-

tion of meals are viewed not only as labor but also as an expression of love. In support of the general argument that housework has important symbolic meanings, Ferree (1991a) points out that "housework-like chores are imposed in other institutions to instill discipline" (p. 113), such as KP in the army.

The process of "doing gender" is not assumed to operate at a conscious level; on the contrary, Berk (1985) points out that it goes on "without much notice being taken" (p. 207). Ferree (1991a) finds it "striking how little explicit conflict there is over housework in many families" (p. 113). Hochschild's (with Machung 1989) pathbreaking study shows how gender ideologies are enacted through the performance of housework and may operate in a contradictory manner at conscious and unconscious levels. She discovers through in-depth case studies that people's ideas about gender are often "fractured and incoherent" (p. 190) and that contradictions abound between what people say they believe, what they seem to feel, and how these beliefs and feelings are reflected in their household behavior.

This developing "doing gender" approach suggests several important contrasts between couple households (especially those of married couples) and other household types. Indeed, one could argue that *only* by examining a range of household types, including those *not* formed by couples, can one determine the usefulness of this explanation for the behavior of married or cohabiting persons. If gender is being "produced," one would expect this process to be more important in heterosexual couple households than in other household types—there would be less need or opportunity for either men or women to display dominance and subordination or other gender-linked behaviors when they are not involved in conjugal relations. Berk (1985) argues that "in households where the appropriation of *another's* work is possible, in practice the expression of work and the expression of gender become inseparable" (p. 204). Of course, we recognize that gender role socialization is likely to produce gender differentials, even among unmarried persons. However, this *appropriation* seems likely to occur mainly, or perhaps only, in heterosexual couple households, particularly when the couples are married. Berk observes a sharp contrast in the housework patterns of married couples versus same-sex roommate arrangements, the latter seeming "so uncomplicated" to respondents (1985:204).

If heterosexual couples indeed produce gender through performing housework, we would expect women in married-couple households to spend more time doing housework than women in any other living situation; we would expect men's time spent doing housework to be lower in married-couple households than in other household types. These expectations are net of other differences between the household types, such as the presence of children, that affect housework. We would expect women to display submission to and/or love for their husbands or male partners by performing a disproportionate share of the housework, whereas men would display their gender/dominance by avoiding housework that they might perform in other household settings—in particular female-typed housework that constitutes the vast majority of weekly housework time in households. Because a woman cannot display love for or subordination to a man through housework when no man is present, this avenue for displaying gender does not exist in one-adult households. Thus, we would predict smaller gender differences in noncouple than couple household settings once other relevant factors are controlled.

An alternative empirical outcome—one that would cast doubt on the validity of the gender perspective—would be a pattern across household type involving a more or less constant gender difference. We know that there is a gender gap in time spent doing housework between married men and women and between teenage boys and girls. We do not know, however, whether that gap is constant across other situations. If, for example, gender differences in childhood training produce standards or skill levels that vary with gender, one might argue that men and women would carry these attitudes or behaviors with them as they move among different household situations.

HOUSEWORK AND MARITAL STATUS

Housework is a major component of most people's lives, just as is paid work. It is first experienced in childhood as "chores" and continues into retirement. Yet, while housework is performed prior to marriage and after its dissolution, most studies of household labor focus exclusively on husbands and wives. This tends to create the false impression that housework occurs only within marital households.

Our analysis of housework is based on a categorization by marital status. We focus on men and women who have *never married*, or are currently *married, divorced*, or *widowed*. However, because a key aspect of our theoretical argument focuses on gender relations in heterosexual households, we add a "cohabiting" category, which includes persons who are currently cohabiting whether or not they have ever been married, divorced, or widowed. Further, the situation of never-married persons (who are not cohabiting) varies greatly depending upon whether they are *living independently* or *living in a parental household*; thus we divide never-married persons into two groups based on living situation. In the sections below, we review studies of housework performed by persons in each of these six categories.

Never-Married Persons Living in Their Parents' Homes

The performance of household chores is one of many gender-differentiated socialization experiences gained in families of origin. A number of studies have examined housework performed by boys and girls up to the age of 18 who are living with their parents. These studies have focused on three kinds of questions: how parents define the meaning of housework (White and Brinkerhoff 1981a), how children's contributions relate to or substitute for mothers' or fathers' work (Berk 1985; Goldscheider and Waite 1991), and how housework varies by the gender of the child, mother's employment, and number of parents in the household (e.g., Benin and Edwards 1990; Blair 1991; Hilton and Haldeman 1991).

Housework done by boys and by girls mirrors that of adults, with girls doing stereotypical "female" chores and spending more time doing housework than boys (Benin and Edwards 1990; Berk 1985; Blair 1991; Goldscheider and Waite 1991; Hilton and Haldeman 1991; Timmer, Eccles, and O'Brien 1985; White and Brinkerhoff 1981b). Patterns by gender and age suggest that, under certain conditions, children (particularly older girls) actually assist their parents. Gender differences increase with age, so that in the teenage years girls are spending about twice as much time per week as boys doing housework (Timmer et al.

1985), and the gender-stereotyping of tasks is at a peak. This pattern holds even in single-father families, where one might expect less traditional gender-typed behavior (Grief 1985). Adolescent girls' housework time has been shown to substitute for that of their mothers, while boys' housework time does not (Bergen 1991; Goldscheider and Waite 1991). Differences between single-parent and two-parent families also suggest more actual reliance on girls' work: Boys in single-parent households do less housework than do boys in two-parent households, while girls in single-parent households do more (Hilton and Haldeman 1991). Similar differences have been found between single- and dual-earner two-parent families. Again, girls do more when parents' time is constrained (dual earners) while boys do less, suggesting that parents actually rely on girls to substitute for their mothers' time doing housework (Benin and Edwards 1990).

One would expect parallel differences in the behavior of young adult men and women who still live with their parents. To our knowledge, only three studies have examined housework performed by adult children living in parental households. Ward, Logan, and Spitze (1992) find that adult children living with parents perform only a small proportion of total household tasks when compared to their parents, and parents whose adult children do not live at home actually perform fewer household tasks per month than do parents whose adult children live with them. There are also major differences between adult sons and adult daughters in the amount of housework they do, with daughters performing more tasks than sons when they live in a parent's home. This holds for all parent age groups, particularly those under 65. These gender differences are consistent with results on adult children's share of household tasks reported by Goldscheider and Waite (1991). Hartung and Moore (1992) report qualitative findings that are consistent with the conclusion that adult children, especially sons, contribute little to household chores and typically add to their mothers' burdens.

Never-Married Persons Living Independently

We know of no empirical research that focuses specifically on never-married persons living independently, so we will speculate briefly about

factors affecting them. One likely consequence of experiences with housework in the parental home is that girls acquire the skills required for independent living, including shopping, cooking, cleaning, and laundry. To the extent that they have already been doing significant amounts of housework at home, girls' transitions to independent living may not create a major change in the amount or types of housework they perform. The skills boys are more likely to learn in the parental home (e.g., yard work) may be less useful, particularly if their first independent living experience is in an apartment. They may reach adulthood enjoying housework less than women, feeling less competent at household tasks, holding lower standards of performance, embracing gender-stereotyped attitudes about appropriateness of tasks, and preferring to pay for substitutes (e.g., laundry, meals eaten out). On the other hand, single men living independently (and not cohabiting) are forced, to a certain extent, to do their own housework (Goldscheider and Waite 1991), because their living situations are unlikely to provide household services. Thus, the time spent by single men doing housework should increase when they move out of parental households.

Cohabiters

Cohabiting couples share some characteristics of both married and single persons (Shelton and John 1993; Stafford et al. 1977). As Rindfuss and VandenHeuvel (1992) point out, most discussions have used married persons as the comparison group, viewing cohabitation as an alternative kind of marriage or engagement. The division of household labor between cohabiters may be closer to that of married persons, but in other areas such as fertility plans, employment, school enrollment, and home ownership, cohabiters more closely resemble single persons (Rindfuss and VandenHeuvel 1992). Thus, we would expect cohabiters to fall at an intermediate position, between never-married living independently and married persons, in the allocation of time to housework.

A few empirical studies have examined housework by heterosexual cohabiting couples. One early study (Stafford et al. 1977) uses a relative contribution measure of housework and finds cohabiting couples to be fairly "traditional" in their division of household labor. A more recent study using an absolute measure of time expenditure in housework (Shelton and John 1993) sheds more light on the comparison between cohabiting and married couples. Adjusted means of time spent doing housework for cohabiting men are not significantly different from those for married men, but cohabiting women do less housework than do married women. These results are consistent with Blumstein and Schwartz's (1983) comparisons of married and cohabiting men and women. Blair and Lichter (1991) find no significant differences between married and cohabiting men's housework time, but find less task segregation by gender among cohabitants. As is true of comparisons on other dimensions (Rindfuss and VandenHeuvel 1992), studies of housework among cohabiting couples have used married persons as the comparison group, and there have been few comparisons of housework patterns in cohabiting relationships to patterns in other marital statuses.

Married Persons

Marriage often entails a number of changes that increase housework, including parenthood and home ownership, but it also might increase housework for less tangible reasons. Marriage and parenthood entail responsibility for the well-being of others, which is likely to be reflected in higher standards of cleanliness and nutrition, and thus require that more time be devoted to housework. However, the net result of this increase in total work is different for men and for women, and this gender division of household labor has been the subject of much research and theorizing in recent years. Averages tend to range widely depending on the definitions of housework used, but women generally report performing over 70 percent of total housework, even if they are employed (Bergen 1991; Ferree 1991a). One recent study reported married women (including nonemployed) doing 40 hours of housework per week and men 19 hours (Shelton and John 1993), and countless studies have documented that wives' employment has little effect on married men's housework load (see reviews in Spitze 1988; Thompson and Walker 1991). Clearly, wives are responsible for the vast bulk of household chores and for maintaining standards of cleanliness and health in the family. Married men have been described as doing less

housework than they create (Hartmann 1981). Further, when they do contribute to household chores, men are more likely to take on those jobs which are more pleasant, leaving women with those than can be described as "unrelenting, repetitive, and routine" (Thompson and Walker 1991:86). Thus, past empirical results for married persons are consistent with the gender perspective, but comparative analyses that include persons in other marital statuses are needed.

Divorced Persons

To our knowledge there have been no studies of the time divorced persons spend doing housework except those studies focusing on children's housework. Divorced persons (who are not cohabiting) have had the prior experience of living with a heterosexual partner. Women may experience a decrease in housework hours if in fact their partner was creating more housework than he was doing. Men's experience, on the other hand, may be similar to that of moving out of the parental household, that is, of having to do some household tasks for themselves that were previously performed by others. Those who never lived independently before may have to do some of these chores for the first time. Gove and Shin (1989) point out that both divorced and widowed men have more difficulty carrying out their daily household routines than do their female counterparts, who are more likely to experience economic strains.

Widowed Persons

In empirical studies, housework has been identified as an important source of strain for widowed men. Widowed men reduce the time they spend doing housework as the years since widowhood pass, and they are more likely than widows to have help doing it as time goes on (Umberson, Wortman, and Kessler 1992). Of course, today's widows and widowers came of age when the gendered division of labor in households was much more segregated than it is today and when living independently before or between marriages was much less common. While we expect widowed men today to have entered widowhood with relatively little experience in certain kinds of household chores, this may not be true in the future.

Widowed women may share some characteristics with divorced women; they may actually feel some relief from the strain of doing the bulk of household tasks for two (Umberson et al. 1992). Like widowed men, however, current cohorts of widowed women may have little experience in certain kinds of chores, in this case traditionally male chores such as yard work, car care, or financial management.

Other Factors Influencing Time Doing Housework

Men and women in different marital statuses are likely to differ on a variety of factors that can influence the performance of housework, such as their health, employment status, presence of children and other adults, and home ownership. We would expect the performance of housework to vary by marital status both because of these factors and because of the ways in which the marital status itself (or experience in a previous status) influences housework behavior. Here, we describe a model of time spent in housework that can be applied to persons in all marital situations. This model will then guide us in choosing control variables for the analysis of housework.

A person is expected to spend more time in housework as the *total amount to be done* increases. (Berk [1985] calls this the total "pie" in her study of married couple households.) We would expect the amount of housework to increase as the number of children increases, particularly when children are young, but to some extent for older children as well (Bergen 1991; Berk 1985; Ishii-Kuntz and Coltrane 1992; Rexroat and Shehan 1987). The amount of work will also increase with the addition of adults to the household, although of course they may perform housework as well. Work may also increase with the size of house and the responsibilities that go with home ownership, car ownership, and presence of a yard (Bergen 1991; Berk 1985).[1]

Note that the total housework to be done is to some extent a subjective concept. Two households with the same composition and type of home may accomplish different amounts of housework for several reasons. The standards held by the adults in the household will vary (Berk 1985) and may even vary systematically along dimensions such as education and age. Also, some households purchase more services than others, due to available income (Bergen 1991) and time constraints.

A second factor influencing the amount of housework a person does is the number of *other people* there are in the household with whom to share the work. Other people are most helpful if they are adults, and women are likely to contribute more than men. Teenagers and even grade-school-age children may be helpful, and their contribution may also vary by gender. The way that household labor is divided, and thus the amount performed by a particular man or woman, may also relate to gender-role attitudes that may vary with education, age, race, and other factors.

Third, persons with more *time and energy* will do more housework. Available time would be limited by hours spent in paid work, school enrollment status, health and disability status, and age (Coltrane and Ishii-Kuntz 1992; Ishii-Kuntz and Coltrane 1992; Rexroat and Shehan 1987). Concurrent roles, in addition to that of homemaker, detract from the time available to be devoted to housework.

DATA AND METHODS

Data for this study are drawn from the National Survey of Families and Households (NSFH), a national probability sample of 13,017 adults interviewed between March of 1987 and May of 1988 (Sweet, Bumpass, and Call 1988). The NSFH includes a wide variety of questions on sociodemographic background, household composition, labor force behavior, and marital and cohabitation experiences, as well as items describing respondents' allocation of time to household tasks. The NSFH oversamples single-parent families and cohabiters (as well as minorities and recently married persons), thus facilitating comparisons of household labor among persons in different—and relatively rare—household situations. Sample weights are used throughout the

---

[1] While owning appliances would be expected to decrease time spent doing housework, it has had much less clear-cut effects than expected, both over time and in cross-sectional studies (Gershuny and Robinson 1988).

analysis to achieve the proper representation of respondents in the U.S. population.

The dependent variable, hours devoted to housework in the typical week, is derived from a series of questions asking respondents how many hours household members spend on various tasks. Respondents were provided with a chart and instructed: "Write in the approximate number of hours per week that you, your spouse/partner, or others in the household normally spend doing the following things." Nine household tasks include "preparing meals," "washing dishes and cleaning up after meals," "cleaning house," "outdoor and other household maintenance tasks (lawn and yard work, household repair, painting, etc.)," "shopping for groceries and other household goods," "washing, ironing, mending," "paying bills and keeping financial records," "automobile maintenance and repair," and "driving other household members to work, school, or other activities." This analysis uses only the number of hours that the respondents report *themselves* as spending on these tasks. To construct the dependent variable, we sum the number of hours spent on each of the nine tasks.[2]

We make two adjustments to this dependent variable. First, because a few respondents reported spending inordinate numbers of hours on specific tasks, we recode values above the 95th percentile for each task to the value at that percentile. This adjustment reduces skewness in the individual items and therefore in the summed variable as well. Second, so we can include respondents who omit one or two of the nine questionnaire items, we impute values for the household tasks for these respondents.[3] In-

dividuals who failed to respond to more than two of the questions are excluded from the analysis. Omitting these respondents and excluding cases with missing values on the independent variables leaves 11,016 respondents available for analysis.

Given our focus on differences in housework between unmarried and married persons, it is essential that the dependent variable records the absolute number of hours devoted to housework rather than the proportional distribution of hours (or tasks) performed by various household members (e.g., Waite and Goldscheider 1992; Spitze 1986). Of course, estimates of time spent on household tasks made by respondents (as recorded in the NSFH) are likely to be less accurate than estimates from time diaries (for a review of validity studies dealing with time use, see Gershuny and Robinson 1988). Yet, estimates of the relative contribution of wives and husbands to household labor are generally comparable across different reporting methods (Warner 1986). Moreover, the effects of respondent characteristics on the time spent on housework shown here are quite similar to the effects observed in time diary studies. The size of the NSFH (approximately five times larger than the typical time-use survey), its oversampling of atypical marital statuses, and its breadth of coverage of respondent characteristics adequately compensate for the lack of time-diary data.

The key explanatory variable combines respondents' marital status' with aspects of their

---

[2] The research literature on housework is inconsistent regarding the inclusion of time spent in childcare. Many data sets commonly used to analyze household labor do not include childcare in their measure (e.g., Bergen 1991; Rexroat and Shehan 1987) or, as is the case here, childcare time is not included as a separate task (Coltrane and Ishii-Kuntz 1992), in part because respondents have difficulty separating time spent in childcare from leisure and from time spent in other tasks. Thus, we are not able to include childcare in our measure. This probably creates a downward bias in estimates of household labor time.

[3] The NSFH assigns four different codes to the household task items for respondents who did not give a numerical reply: some unspecified amount of time spent; inapplicable; don't know; and no answer. Our imputation procedure substitutes a value of 0 for

those who did not answer this question (but answered at least seven of the nine items) or who said the task was inapplicable. In the former case, skipping the item most likely indicates that the respondent spent no time on that task; in the latter case, the respondent most likely could not logically spend time on that task (e.g., persons without cars could not spend any time maintaining them). For respondents who indicated spending some unspecified amount of time on a task and for those who indicated they didn't know, our imputation procedure substitutes the mean value for that task. In both of these instances, respondents presumably spent at least some time on that task. Our explorations of alternative ways of handling missing data, including omitting respondents who failed to answer one or more of the questions, treating all nonnumerical responses as 0, and substituting all nonnumerical responses with the mean, showed quite clearly that our substantive conclusions are unaffected by the method used to handle missing data.

living arrangements. (For stylistic convenience, we refer to this variable simply as marital status.) We distinguish six mutually exclusive statuses: never married and living in the parental household, never married (not cohabiting) and living independently, cohabiting, currently married, divorced or separated (not cohabiting), and widowed (not cohabiting). Because we are interested in the impact of a spouse or partner on respondents' time doing housework, cohabiters include divorced, separated, and widowed cohabiters as well as never-married cohabiters.

The other explanatory variables measure respondents' demographic background, socioeconomic standing, household composition, concurrent roles, and disability status. As suggested above, several of these factors may help explain any differences that we observe in housework time by marital status and gender. *Age* is measured in years. Because housework demands are likely to peak during the middle adult years and to moderate at older ages, we also include *age squared* as an independent variable. *Education* is measured by years of school completed. *Household earnings* refers to the wage, salary, and self-employment income of all members of the household.[4] *Home ownership* is a dummy variable scored 1 for respondents who own their own home and 0 for those who do not.

Several variables reflect the presence in the household of persons who may create or perform housework. *Children* in the household are divided into the number of children younger than 5 years old, the number age 5 through 11, and the number age 12 through 18. Among the latter group, girls might be expected to create less (or perform more) housework than boys (Goldscheider and Waite 1991), and thus we include separate counts of male and female teenagers. We use several dummy variables to indicate the presence in the household of an *adult male* or *adult female* other than the respondent's spouse or cohabiting partner. Adult females are expected to reduce respondent's time devoted to housework, while adult males are expected to increase it. We further distinguish between adult household members who are the children of the respondent and those who are not.

Respondents who invest their time in activities outside the home are anticipated to devote less time to domestic labor. Employment status is measured by the usual number of *hours worked per week* in the labor force. And, whether the respondent is currently *attending school* is indicated by a dummy variable scored 1 for currently enrolled respondents and 0 for those not attending school.

Finally, *disability status* is measured by the response to the question, do you "have a physical or mental condition that limits your ability to do day-to-day household tasks?" Individuals reporting such a condition are scored 1 on this dummy variable; unimpaired respondents are scored 0.[5]

Our primary analytic strategy is to estimate OLS regression equations that examine the impact of gender, marital status, and the other explanatory variables on the time spent doing housework. Of particular importance for our theoretical model is whether marital status differences in housework time vary by gender—that is, do gender and marital status interact in affecting time spent doing housework? The "gender perspective" implies that marital status differences in housework will be more pronounced for women than for men and that the gender differences in housework will be greatest for married persons. The regression models are also used to determine the extent to which marital status differences in time doing

---

[4] So as not to lose an inordinate number of cases to missing data, we substituted the mean for missing values on household earnings, and we included a dummy variable for these respondents in the regression models (coefficients not shown). One potential difficulty with this procedure is that all respondents who were not the householder or the spouse of the householder receive the mean value, because respondents were not asked the earnings of other household members. Equations estimated only with repondents who are householders revealed effects almost identical to those reported in the text, although never-married respondents living in the parental household are necessarily excluded from these equations. Given that households with adult children include more adults than other households, the household earnings of these latter respondents are likely to be higher than average, but any bias in the effect of earnings is apt to be slight. With one exception (see footnote 5), the amount of missing data on the other explanatory variables is small.

[5] To retain the 5 percent of respondents who did not reply to the question on disability status, the regression equations also include a dummy variable for these respondents (coefficients not shown).

Table 1. Descriptive Statistics for Hours Spent in Housework per Week and for Explanatory Variables, by Gender: U.S. Men and Women, 1987 to 1988

Variable	Women		Men	
	Mean	Standard Deviation	Mean	Standard Deviation
Housework hours per week	32.62	18.18	18.14	12.88
Marital Status[a]				
Never married/living in parental home	.06	.23	.11	.32
Never married/living independently	.10	.30	.11	.32
Cohabiting	.04	.19	.04	.20
Married	.57	.50	.63	.48
Divorced/separated	.12	.33	.08	.26
Widowed	.12	.33	.03	.17
Number of children ages 0 to 4	.26	.59	.22	.55
Number of children ages 5 to 11	.33	.70	.29	.66
Number of girls ages 12 to 18	.16	.44	.15	.43
Number of boys ages 12 to 18	.17	.45	.15	.43
Adult male child present (0 = no; 1 = yes)	.10	.29	.07	.25
Adult male nonchild present (0 = no; 1 = yes)	.09	.29	.18	.38
Adult female child present (0 = no; 1 = yes)	.08	.27	.05	.22
Adult female nonchild present (0 = no; 1 = yes)	.14	.35	.17	.38
Home ownership (0 = no; 1 = yes)	.59	.49	.58	.49
Household earnings (in $1,000s)	28.72	37.69	31.64	36.51
Education 12.45	2.93	12.94	3.32	
Age	44.30	17.99	42.24	17.07
Age squared (/100)	22.86	17.81	20.75	16.38
Hours employed per week	18.43	20.01	31.81	22.55
School enrollment (0 = no; 1 = yes)	.06	.24	.07	.26
Disabled (0 = no; 1 = yes)	.06	.24	.05	.22
Number of cases	6,764		4,252	

[a]May not add to 1.00 because of rounding.

housework can be explained by other respondent characteristics and to assess whether the gender-specific impact of the explanatory variables holds for the general population (including unmarried people) in ways previously shown for married persons.

RESULTS

Table 1 presents descriptive statistics for all variables in the analysis. Immediately apparent is the sharp but unsurprising difference between men and women in the amount of time spent doing housework. In this sample, women report spending almost 33 hours per week on

household tasks, while men report spending slightly more than 18 hours. Both figures are roughly comparable to the findings of prior studies, although of course those studies did not include unmarried persons.

Gender differences in current marital status are relatively slight. Men are somewhat more likely than women to have never married, reflecting longstanding differences in age at marriage. And, among the never married, men are more likely than women to reside in the parental household. Women are more likely than men to be currently divorced or widowed, a probable consequence of their lower remarriage rates following divorce and men's higher mortality. Four percent of both sexes are co-habiters.

Differences between women and men on the other explanatory variables are also generally small. The sole exception is the number of hours worked outside the home, with women averaging approximately 18 hours per week and men 32 hours.

The regression analysis of time spent on housework is shown in Table 2. In our initial equations (not shown here), we pooled the male and female respondents and regressed housework hours on the explanatory variables, including dummy variables for gender and marital status. We then added to this equation product terms representing the interaction of gender and marital status. As predicted by the theoretical model, allowing marital status and gender to interact in their effects on housework significantly increases the variance explained ($F = 67.06$; $p < .001$). And specifically, the difference in housework hours between married women and married men is significantly larger than the housework hours differences between women and men in each of the other marital statuses. Product terms representing the interaction of gender with the other explanatory variables also revealed that several of the effects varied significantly by gender; thus, we estimate and present the equations separately for women and for men.[6]

The first equation in Table 2 is based only on the women respondents and regresses weekly housework hours on dummy variables representing five of the six marital statuses, with married respondents serving as the reference category. Persons in all five marital statuses work significantly fewer hours around the house than do the married respondents; at the extreme, married women spend over 17 hours more per week on housework than do never-married women who reside in the parental household. As anticipated, the amount of time spent on housework by women who are never

married and living independently, cohabiting, divorced (including separated), or widowed falls between that of women who have not married (and remain in the parental home) and those who have married.

The third column of Table 2 presents the parallel equation for men. As reflected in the constant term, married men report spending almost 18 hours per week in housework, compared to almost 37 hours for their female counterparts (the constant term in column 1). More importantly, marital status differences in housework hours among men are relatively small compared to the analogous differences among women. Married men do significantly more housework than never-married men who still live with their parents and significantly less than divorced and widowed men, but most of these differences are modest. Moreover, the pattern of time spent doing housework across marital statuses differs substantially between men and women; it is greatest for men during widowhood and greatest for women during marriage.

Equation 2 in Table 2 re-estimates marital status differences in housework hours for men and women, controlling for the other explanatory variables. As shown in column 2, differences among women in these additional variables account for some, though by no means all, of the marital status differences in housework. Controlling for these variables reduces the differences between married women and other women by between 17 percent (for widows) and 66 percent (for cohabiters). Further, the difference between married women and co-habiting women is no longer statistically significant once these variables are controlled. Thus, among women a moderate proportion of the marital status differences in time spent doing housework is attributable to compositional differences. Particularly important in accounting for these marital status differences in housework hours are the number of hours the respondent works outside the home and the presence of children in the household; both variables vary significantly by marital status and are at least moderately related to time spent doing housework. We discuss these and the other effects of the explanatory variables in detail below.

For men, in contrast, controlling for the other explanatory variables does somewhat less to explain marital status differences in house-

---

[6] The distribution of some of the factors that explain variation in housework hours differs by age group. For example, enrollment in school and the presence of children in the household are most prevalent for younger respondents, while disability and widowhood are more common among the aged. Yet, the correlation matrices showed little evidence of multicollinearity, and disaggregating the equations by age revealed patterns and determinants quite similar to those for the sample as a whole.

Table 2. OLS Coefficients for Regression of Hours Spent in Housework per Week on Marital Status and Other Explanatory Variables, by Gender: U.S. Men and Women, 1987 to 1988

Independent Variable	Women (1)	Women (2)	Men (1)	Men (2)
Marital Status				
Never married/living in parental home	−17.41***†	−9.73***†	−2.90***†	−.52†
	(.93)	(1.34)	(.63)	(1.18)
Never married/living independently	−11.62***†	−6.45***	1.09†	1.43†
	(.74)	(.84)	(.63)	(.80)
Cohabitating	−5.54***	−1.86†	1.34†	1.73†
	(1.14)	(1.14)	(.98)	(1.03)
Married	Reference		Reference	
Divorced/separated	−5.30***	−3.68***†	3.73***†	4.58***†
	(.66)	(.68)	(.75)	(.80)
Widowed	−9.08***	−7.51***	5.66***†	6.97***†
	(.67)	(.77)	(1.16)	(1.21)
Number of children ages 0 to 4	—	3.63***	—	.67†
		(.38)		(.39)
Number of children ages 5 to 11	—	3.77***†	—	.85***†
		(.31)		(.32)
Number of girls ages 12 to 18	—	1.62***†	—	−.64†
		(.46)		(.46)
Number of boys ages 12 to 18	—	1.88**	—	.74
		(.47)		(.47)
Adult male child parent (0 = no; 1 = yes)	—	1.79*	—	.91
		(.74)		(.82)
Adult male nonchild present (0 = no; 1 = yes)	—	−.10	—	−.37
		(.97)		(.72)
Adult female child present (0 = no; 1 = yes)	—	−2.46**	—	−2.93**
		(.80)		(.92)
Adult female nonchild present (0 = no; 1 = yes)	—	−1.18	—	−1.40
		(.85)		(.84)
Home ownership (0 = no; 1 = yes)	—	2.24**	—	−1.22*
		(.52)		(.52)
Household earnings (in $1,000s)	—	−.03***	—	−.02***†
		(.01)		(.01)
Education	—	−.44***†	—	.14*†
		(.08)		(.06)
Age	—	.40***†	—	.05†
		(.08)		(.08)
Age squared (/100)	—	−.44***†	—	−.15†
		(.08)		(.08)
Hours employed per week	—	−.17***†	—	−.08***†
		(.01)		(.01)
School enrollment (0 = no; 1 = yes)	—	−4.07**	—	−2.48**
		(.91)		(.82)
Disabled (0 = no; 1 = yes)	—	−5.34**	—	−2.96**
		(.86)		(.94)
Constant	36.67**	34.26**	17.83**	19.87**
	(.28)	(2.07)	(.25)	(2.08)
Root mean squared error	17.39	16.37	12.76	12.57
R^2	.08	.19	.02	.05
Number of cases	6,764	6,764	4,252	4,252

*$p < .05$ **$p < .01$ (two-tailed tests)

Note: Numbers in parentheses are standard errors. Equations in columns 2 and 4 include dummy variables for missing values on household earnings and disabled.

†Difference in coefficients for women and men is statistically significant at $p < .05$.

work. Although the difference between never-married men living in the parental home and married men becomes statistically nonsignificant when these variables are controlled, the absolute size of the decline (about 2.5 hours per week) is small. More important, with these controls the initially larger differences between married men and both divorced and widowed men actually increase.

Most of the explanatory variables have significant effects on time spent doing housework for either the men *or* the women, and many have significant effects for both sexes. Several variables have stronger effects among one sex than the other. The presence of children in the household creates more housework, especially for women, with pre-teenagers creating slightly more work than older children. The impact of children on housework hours tends to be significantly stronger for women than for men, a finding also found in studies limited to married couples (Bergen 1991; Rexroat and Shehan 1987). The presence in the household of the respondent's adult children also significantly affects housework hours, but the direction of the effect depends on both the sex of the adult children and the respondent. For female respondents, the presence of an adult male child increases housework hours, while for both female and male respondents the presence of an adult female child significantly reduces time allocated to housework. These findings are consistent with the view that men create housework, while women perform work men would otherwise do themselves (Hartmann 1981). Adults who are *not* children of the respondent do not add or subtract significantly, on average, from the respondent's housework time. This may be because the household is a heterogeneous group, including some roommates, siblings and other relatives, and elderly parents. Some household members may be helpful and others may be a burden, and their effects may cancel out.[7]

As expected, home ownership significantly increases housework time, and it appears to do so about equally for men and women. This may

be due to larger amounts of living space to be cleaned and to the increase in yard work and maintenance and repair chores among homeowners. Total household earnings reduce housework significantly more for women than for men, suggesting that purchased household services substitute more for women's than for men's domestic labor.[8] Among women, education is inversely associated with housework, while for men the association is positive and significant. Educated women and men tend to hold egalitarian attitudes, which may lead to greater symmetry in their housework patterns (Huber and Spitze 1983). The hypothesized curvilinear (bell-shaped) association between age and housework emerges for women, but not for men.

As indicated by the significant effects of employment and school enrollment on time spent doing housework, investing time in nonhousehold activities significantly reduces household labor. The impact of hours employed is significantly greater for women than for men, a finding consistent with prior research (Gershuny and Robinson 1988; Rexroat and Shehan 1987). This suggests that women have less discretionary time than men, so that increased expenditures of time outside the home must necessarily divert time away from housework.[9]

---

[7] While it is possible to separate persons in heterogeneous households into a number of categories and attempt to sort out those who tend to help and those who create more work, the small number of respondents with *any* other adult present suggests that this would not be a useful refinement to the analysis.

[8] The gender difference in the effect of household earnings on housework is complicated by the fact that, for couple households, wife's (or female cohabiting partner's) hours employed per week is controlled for in the women's equation, but not in the men's equation. If hours employed are deleted from both equations, the gender difference in the effect of household earnings becomes statistically nonsignificant. Hence, this difference, which is barely significant to begin with, should be interpreted cautiously.

[9] From the perspective of the New Home Economics, the amount of time allocated to housework and to paid labor are frequently considered to be jointly determined, and thus the inclusion of employment hours as a predictor of housework has been questioned (Godwin 1991). We believe that for most persons, and particularly persons in nonmarital households, decisions regarding the allocation of time to the paid labor force are made prior to decisions about housework time (especially given that our measure of housework excludes childcare), and thus that the treatment of paid employment as an explanatory variable is justified. In any event, omitting respondent's hours employed per week from the equations does not

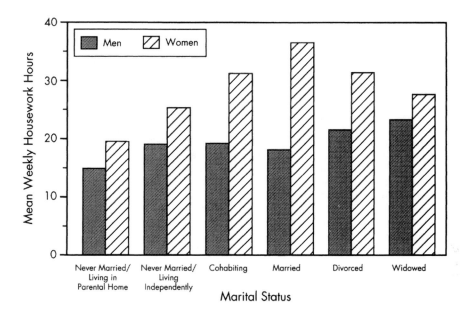

Figure 1. Mean Hours Spent Doing Housework Each Week, by Gender and Marital Status

Because the combined effects of gender and marital status are moderately complex, we present Figure 1 to help clarify the nature of their interaction. This figure graphs the (unadjusted) mean housework hours for men and women along the most common temporal sequence of marital statuses. In all marital statuses, women spend more hours than men on housework. The gender gap among never-married men and women living in the parental home is about 4 hours. Both never-married women and men who live independently do more housework than their counterparts who remain at home, but because the increase is slightly greater for women than for men (almost a 6-hour increase for women versus 4 hours for men), the gender difference in housework in this group grows to a little over 6 hours. Presumably, both men and women who live independently perform household tasks that previously had been done for them by their parents when the respondents resided in the parental homes.

appreciably alter the effects of marital status and gender that are the crux of our analysis, nor does the omission modify the impact of the other explanatory variables.

The gender difference in housework hours widens dramatically as one moves to the couple households—cohabiters and married persons. Cohabiting women do more housework than never-married women (regardless of the latter's living arrangements), while cohabiting men work about the same hours around the house as never-married men living independently. The result of these discrepant trajectories is that the gender difference among cohabiters increases to approximately 12 hours per week. The gender gap in housework hours reaches its zenith among married women and men, at approximately 19 hours per week. This disparity is primarily a consequence of married women doing substantially more housework than never-married and cohabiting women, although these differences diminish with controls, as shown in Table 2. Rather than simply maintaining a behavioral pattern established prior to forming a conjugal union, married and, to a lesser extent, cohabiting women appear to increase substantially the time they devote to housework. In contrast, the amount of housework done by married men is fairly similar to that done by never-married and cohabiting men. Hence, as the "gender perspective" would suggest, it is in marital and cohabiting

Table 3. Mean Hours Spent per Week in Various Household Tasks, by Marital Status and Gender: U.S. Men and Women, 1987 to 1988

Household Task[b]	Marital Status[a]					
	Never Married/ Living in Parental Home	Never Married/ Living Inde-pendently	Cohabiting	Married	Divorced	Widowed
Women						
Preparing meals	3.64	6.74	7.99	10.14	8.15	7.96
Washing dishes	3.92	4.38	5.51	6.11	5.14	4.73
Cleaning house	3.95	5.16	7.10	8.31	6.68	5.68
Washing/ironing	2.45	2.63	3.44	4.16	3.37	2.50
Outdoor maintenance	1.39	1.24	1.34	2.06	1.94	2.26
Shopping	1.72	2.28	2.69	2.86	2.67	2.40[ns]
Paying bills	.81[ns]	1.53	1.66	1.52	1.70	1.48[ns]
Car maintenance	.48	.42	.28	.16	.40	.20
Driving	.90[ns]	.65	1.10[ns]	1.34	1.30	.38[ns]
Total housework hours	19.26	25.04	31.12	36.67	31.37	27.59
Number of cases	383	649	248	3,838	829	817
Men						
Preparing meals	2.23	5.06	3.71	2.69	5.50	6.48
Washing dishes	1.92	2.77	2.63	2.15	3.24	3.87
Cleaning house	2.20	2.97	2.60	2.03	3.54	3.38
Washing/ironing	1.30	1.92	1.16	.70	1.75	1.67
Outdoor maintenance	3.56	1.56	3.18	4.94	2.60	3.38
Shopping	.83	1.92	1.73	1.58	1.93	2.14[ns]
Paying bills	.90[ns]	1.38	1.35	1.32	1.45	1.65[ns]
Car maintenance	1.23	.92	1.51	1.37	.99	.52
Driving	.75[ns]	.42	1.28[ns]	1.04	.57	.41[ns]
Total housework hours	14.93	18.92	19.16	17.83	21.56	23.49
Number of cases	477	476	181	2,668	323	127

[a] All associations between marital status and time spent on household tasks are significant at the $p < .05$ level.

[b] Within marital status and task type, all gender differences are significant at the $p < .05$ level with the following exceptions (marked ns): for never married in parental home—paying bills and driving; for cohabitors—driving; for widows—shopping, paying bills, and driving.

unions that gender differences in housework are most evident.

Among the formerly married, hours spent on housework by men and women begin to converge. Relative to their married counterparts, women who are divorced or widowed do less housework, while divorced or widowed men do more, with or without controlling for other variables. For women, this difference is perhaps best explained by a reduction in the total amount of housework required brought about by the absence of a husband in the household. For men, divorce and widowhood means doing household tasks previously done by a wife.

In general, then, patterns of time spent in housework across different marital statuses appear at least broadly consistent with the emerging "gender perspective." While there is a gender gap in housework in all marital statuses, this disparity varies dramatically and, as predicted, is widest for men and women in couple households (i.e., married or cohabiting relationships). However, to determine the extent to which these totals reflect behavior that becomes more gender-differentiated in couple households, we examine marital status differences in the completion of particular household tasks.

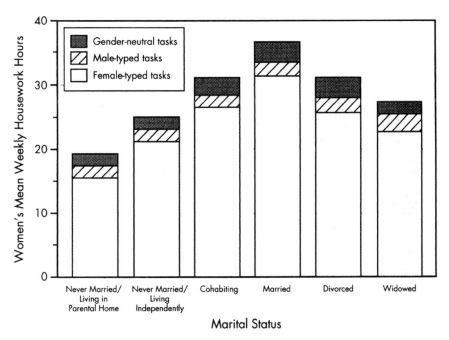

Figure 2. Mean Hours Spent by Women Doing Housework each Week, by Sex Type of Task and Marital Status

Accordingly, Table 3 presents the mean hours spent per week in each of the individual nine household tasks, disaggregated by gender and marital status. Figures 2 and 3 summarize the information in Table 3, graphing for women and men the (unadjusted) amounts of time spent in "female-typed" tasks (preparing meals, washing dishes, cleaning house, washing and ironing, and shopping), "male-typed" tasks (outdoor chores and automobile maintenance), and "gender-neutral" tasks (paying bills and driving other household members).[10] Among women, the marital status differences in *total* housework hours shown in Figure 1 are replicated for the female-typed tasks, which constitute in each marital status category the vast bulk of housework hours (see Figure 2). Of the female-typed tasks, the largest differences are in the number of hours spent prepar-

ing meals and cleaning house, although all five tasks consume more time for married women than for any of the other groups (Table 3). Because in each marital status the amount of time allocated to male-typed tasks is small, *differences* by marital status in these tasks are also slight. Married women do less car maintenance than do other women, but, with the exception of widows, spend slightly more time on outdoor maintenance. For women, then, marital status differences in total housework hours are largely a consequence of differences in hours spent on female-typed tasks.

Among men, however, marital status differences in gender-specific tasks do not always reflect those for housework as a whole. For example, as shown in Figure 3, although the difference in *total* housework hours between never-married men living independently and married men is small (about 1 hour), the difference is composed of several counterbalancing components. Never-married men living independently spend over 5 hours more per week than married men on female-typed tasks, but offset most of this difference by spending less time on male-typed tasks. Similarly, never-married men living independently spend al-

---

[10] This categorization is consistent with other analyses, including those by Ferree (1991b) and Aytac and Teachman (1992). Shelton (1992) shows shopping to be somewhat intermediate between female- and neutral-typed tasks, and others (e.g., Presser 1993) have treated it as a gender-neutral task.

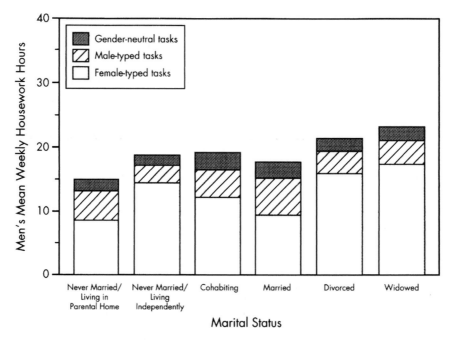

Figure 3. Mean Hours Spent by Men Doing Housework Each Week, by Sex Type of Task and Marital Status

most 3 hours per week more than cohabiting men in female-typed chores, but cohabiting men more than compensate for this difference by spending more time doing male-typed and gender-neutral tasks. Hence, to the extent that cohabiting men differ from never-married men living independently, they do so not by greater participation in female-typed chores, but by increasing their time doing stereotypically male tasks (e.g., automobile maintenance and outdoor chores) and gender-neutral tasks (e.g., driving other household members). On a smaller scale, the difference between cohabiting men and married men in total housework (about 1.3 hours per week) masks an important difference: Cohabiting men spend over 2.5 more hours per week than do married men on traditionally female chores, but married men make up over half of this difference by spending more time on outdoor maintenance. Like never-married men living independently, cohabiting men do more female-typed tasks than do married men, although they do not work on outdoor maintenance tasks to the same degree as their married counterparts.

The difference in total housework hours between married men and divorced men and be-

tween married men and widowed men is also composed of counterbalancing chores. Divorced and widowed men spend 6 to 8 hours more per week than married men on female-typed tasks, but the greater time expenditures by married men on outdoor and automobile maintenance partially offset this difference. In general, the distribution of housework hours by the sex-type of task appears consistent with the gender perspective: Married and cohabiting men spend less time on female-typed tasks and more time on male-typed tasks than do men in most other marital statuses.

DISCUSSION AND CONCLUSION

Doing housework is a significant part of many people's lives, yet few studies have explored housework patterns and determinants across household types. Indeed, because much prior research has been motivated by concerns about marital equity, the erroneous impression may exist that housework is performed only by members of married-couple families. Clearly, this is not the case.

Our results suggest that even never-married men, who might be expected to eschew housework, spend almost half as much time working around the home as they do in the paid labor force. Given prior studies suggesting little contribution by adult sons who live at home (Hartung and Moore 1992; Ward et al. 1992), the amount of housework reported being done by never-married men living in parental homes may seem surprisingly high — approximately 2 hours per day. However, the largest single component of this time (approximately one-quarter of it) is spent on outdoor maintenance, and outdoor and automobile maintenance together constitute one-third of the total time spent. Further, it is likely that much of the time spent in other chores, such as cooking, cleaning, or laundry, is directed more toward self-maintenance than to the well-being of the entire household (Hartung and Moore 1992). Thus, given this context, the amount of housework reported by never-married men living in the parental home appears reasonable.

The performance of housework by men is substantially similar across marital statuses. Differences in total housework hours among never-married, cohabiting, and married men are rather small and are partly attributable to differences in other social and economic characteristics. The most noteworthy differences among men in housework hours involve the appreciable differences between divorced and widowed men and the men in other marital statuses. The number of hours married women spend doing housework approaches a typical full-time work week and is termed the "second shift" by Hochschild (1989). But women in other living situations that do not include a male partner also spend 20 to 30 hours a week doing household chores. The gender gap in housework hours is highest in marriage, but is evident in other marital statuses as well. Although social and economic differences among women in various marital situations (especially the presence of children and hours spent in paid work) account for approximately half of these differences in housework hours, marital status differences in housework among women are generally greater than the corresponding differences among men.

From these patterns and from our detailed analysis of individual household tasks, we have concluded that there is suggestive evidence for the "gender perspective." Housework that women perform for and in the presence of men displays gender more so than the same work

performed with no man present. We find that the gender gap in housework time is greatest in married couple households relative to other households, and that much of this difference *cannot* be explained by the fact that marriage often brings children and reduced hours of paid work for women. Thus, we conclude that men and women must be "doing gender" when they live together. Moreover, relative to their unmarried counterparts, married men spend very little time in the traditionally female tasks of cooking and cleaning.

Of course, there are also significant gender gaps among persons in nonmarital households, implying that the dynamics of doing gender are not entirely absent in other household situations. However, we view our analysis and the patterns displayed in couple and noncouple households to be suggestive evidence that these dynamics operate differentially across household types. Perhaps our analysis and tentative interpretation will encourage those theorists working in the new gender perspective to further specify the conditions under which these processes operate so that future empirical tests can be more precise.

Analysis across household type and marital status may also have implications for the application of equity theory to the allocation of household labor. While most analyses of equity in household labor have used a comparison between husbands and wives as the implicit or explicit base for judging fairness, several recent discussions have raised the possibility that other standards may be used as well. Thompson (1991) discusses the issue of comparison referents and points out that husbands may compare themselves with *other husbands* and wives with *other wives*, while both Ferree (1990) and Kollock, Blumstein, and Schwartz (1988) present empirical comparisons between the predictive value of intracouple and intragender standards. To our knowledge, however, the idea that spouses may compare themselves to *their own past or projected experiences in another marital status*, or even to others who are not currently married, has not been discussed in the empirical literature on housework equity, although fear of divorce was certainly a potent factor in the ideological and behavioral choices of Hochschild's (1989) female respondents. Although this is necessarily speculative, we suggest that married men might use their experience prior to marriage as a reference point for both negotiating and evaluating their own contribution to household labor within marriage. People are spending increasing amounts of time in nonmarital statuses, particularly never-married, cohabiting, and divorced. During their lives, they often go through transitions which include a sequence from being never married to cohabiting to married to divorced or widowed. By examining the time men and women spend doing housework in each of these living situations we may be better able to understand what occurs when people negotiate how housework will be divided within marriage.

SCOTT J. SOUTH *is Associate Professor of Sociology at the State University of New York at Albany. His recent research focuses on the social demography of American families, with particular emphasis given to contextual influences on patterns of family formation and dissolution. He is Co-Editor (with Stewart E. Tolnay) of* The Changing American Family: Sociological and Demographic Perspectives *(Westview Press, 1992).*

GLENNA SPITZE *is Professor of Sociology and Women's Studies at the State University of New York at Albany. In addition to her research on household labor, she is working on a book with John R. Logan based on their research on family structure and intergenerational relations.*

REFERENCES

Aytac, Isik A. and Jay D. Teachman. 1992. "Occupational Sex Segregation, Marital Power, and Household Division of Labor." Paper presented at the meetings of the American Sociological Association, 20–24 Aug., Pittsburgh, PA.

Becker, Gary. 1981. *A Treatise on the Family.* Chicago, IL: University of Chicago.

Benin, Mary H. and Joan Agostinelli. 1988. "Husbands' and Wives' Satisfaction with the Division of Labor." *Journal of Marriage and the Family* 50:349–61.

Benin, Mary Holland and Debra A. Edwards. 1990. "Adolescents' Chores: The Difference Between Dual and Single-Earner Families." *Journal of Marriage and the Family* 52:361–73.

Bergen, Elizabeth. 1991. "The Economic Context of Labor Allocation." *Journal of Family Issues* 12:140–57.

Berk, Sarah Fenstermaker. 1985. *The Gender Factory.* New York: Plenum.

Blair, Sampson Lee. 1991. "The Sex-Typing of Children's Household Labor: Parental Influence on Daughters' and Sons' Housework." Paper presented at the meeting of the American Sociological Association, 23–27 Aug., Cincinnati, OH.

Blair, Sampson Lee and Michael P. Johnson. 1992. "Wives' Perceptions of the Fairness of the Division of Household Labor: The Intersection of Housework and Ideology." *Journal of Marriage and the Family* 54:570–81.

Blair, Sampson Lee and Daniel T. Lichter. 1991. "Measuring the Division of Household Labor: Gender Segregation Among American Couples." *Journal of Family Issues* 12:91–113.

Blood, Robert O. and Donald M. Wolfe. 1960. *Husbands and Wives.* New York: Free Press.

Blumberg, Rae Lesser and Marion Tolbert Coleman. 1989. "A Theoretical Look at the Gender Balance of Power in the American Couple." *Journal of Family Issues* 10:255–50.

Blumstein, Philip and Pepper Schwartz. 1983. *American Couples.* New York: William Morrow.

Coltrane, Scott and Masako Ishii-Kuntz. 1992. "Men's Housework: A Life-Course Perspective." *Journal of Marriage and the Family* 54:43–57.

Coverman, Shelley. 1985. "Explaining Husbands' Participation in Domestic Labor." *Sociological Quarterly* 26:81–97.

Coverman, Shelley and Joseph F. Sheley. 1986. "Changes in Men's Housework and Child-Care Time, 1965–1975." *Journal of Marriage and the Family* 48:413–22.

DeVault, Marjorie L. 1991. *Feeding the Family: The Social Organization of Caring as Gendered Work.* Chicago, IL: University of Chicago.

Ferree, Myra Marx. 1990. "Gender and Grievances in the Division of Household Labor: How Husbands and Wives Perceive Fairness." Paper presented at the meeting of the American Sociological Association, 11–15 Aug., Washington, DC.

———. 1991a. "Feminism and Family Research." Pp. 103–21 in *Contemporary Families,* edited by A. Booth. Minneapolis, MN: National Council on Family Relations.

———. 1991b. "The Gender Division of Labor in Two-Earner Marriages: Dimensions of Variability and Change." *Journal of Family Issues* 12:158–80.

Gershuny, Jonathan and John P. Robinson. 1988. "Historical Changes in the Household Division of Labor." *Demography* 25:537–52.

Gillespie, Dair L. 1971. "Who Has the Power: The Marital Struggle." *Journal of Marriage and the Family* 33:445–58.

Glazer-Malbin, Nona. 1976. "Housework." *Signs* 1:905–22.

Godwin, Deborah D. 1991. "Spouses' Time Allocation to Household Work: A Review and Critique." *Lifestyles: Family and Economic Issues* 12:253–94.

Goldscheider, Frances K. and Linda J. Waite. 1991. *New Families, No Families? The Transformation of the American Home.* Berkeley, CA: University of California.

Gove, Walter R. and Hee-Choon Shin. 1989. "The Psychological Well-Being of Divorced and Widowed Men and Women: An Empirical Analysis." *Journal of Family Issues* 10:122–44.

Grief, Geoffrey L. 1985. "Children and Housework in the Single Father Family." *Family Relations* 34:353–57.

Hartmann, Heidi I. 1981. "The Family as the Locus of Gender, Class, and Political Struggle: The Example of Housework." *Signs* 6:366–94.

Hartung, Beth and Helen A. Moore, 1992. "The Return of the 'Second Shift': Adult Children Who Return Home." Paper presented at the meeting of the American Sociological Association, 20–24 Aug., Pittsburgh, PA.

Hilton, Jeanne M. and Virginia A. Haldeman. 1991. "Gender Differences in the Performance of Household Tasks by Adults and Children in Single-Parent and Two-Parent, Two-Earner Families." *Journal of Family Issues* 12:114–30.

Hochschild, Arlie with Anne Machung. 1989. *The Second Shift: Working Parents and the Revolution at Home*. New York: Viking.

Huber, Joan and Glenna Spitze. 1983. *Sex Stratification: Children, Housework, and Jobs*. New York: Academic Press.

Ishii-Kuntz, Masako and Scott Coltrane. 1992. "Remarriage, Stepparenting, and Household Labor." *Journal of Family Issues* 13:215–33.

Kollock, Peter, Philip Blumstein, and Pepper Schwartz. 1988. "The Judgment of Equity in Intimate Relationships." Paper presented at the meeting of the American Sociological Association, 24–28 Aug., Atlanta, GA.

Oakley, Ann. 1974. *The Sociology of Housework*. New York: Pantheon.

Peterson, Larry R. and Judy L. Maynard. 1981. "Income, Equity, and Wives' Housekeeping Role Expectations." *Pacific Sociological Review* 24: 87–105.

Presser, Harriet B. 1993. "Gender, Work Schedules, and the Division of Family Labor." Paper presented at the meeting of the Population Association of American, 1–3 April, Cincinnati, OH.

Rexroat, Cynthia and Constance Shehan. 1987. "The Family Life Cycle and Spouses' Time in Housework." *Journal of Marriage and the Family* 49:737–50.

Rindfuss, Ronald R. and Audrey VandenHeuvel. 1992. "Cohabitation: A Precursor to Marriage or an Alternative to Being Single?" Pp. 118–42 in *The Changing American Family: Sociological and Demographic Perspectives*, edited by S. J. South and S. E. Tolnay. Boulder, CO: Westview Press.

Rodman, Hyman, 1967. "Marital Power in France, Greece, Yugoslavia, and the United States: A Cross-National Discussion." *Journal of Marriage and the Family* 29:320–24.

Ross, Catherine E. 1987. "The Division of Labor at Home." *Social Forces* 65:816–33.

Sanik, Margaret Mietus and Teresa Mauldin. 1986.

"Single Versus Two-Parent Families: A Comparison of Mothers' Time." *Family Relations* 35:53–56.

Shelton, Beth Anne. 1990. "The Distribution of Household Tasks: Does Wife's Employment Status Make a Difference?" *Journal of Family Issues* 11:115–35.

———. 1992. *Women, Men and Time*. New York: Greenwood Press.

Shelton, Beth Anne and Daphne John. 1993. "Does Marital Status Make a Difference? Housework Among Married and Cohabiting Men and Women." *Journal of Family Issues* 14:401–20.

Spitze, Glenna. 1986. "The Division of Task Responsibility in U.S. Households: Longitudinal Adjustments to Change." *Social Forces* 64:689–701.

———. 1988. "Women's Employment and Family Relations: A Review." *Journal of Marriage and the Family* 50:595–618.

Stafford, Rebecca, Elaine Backman, and Pamela Dibona. 1977. "The Division of Labor Among Cohabiting and Married Couples." *Journal of Marriage and the Family* 39:43–57.

Sweet, James, Larry Bumpass, and Vaughn Call. 1988. "The Design and Content of the National Survey of Families and Households." (Working Paper NSFH-1). Center for Demography and Ecology, University of Wisconsin, Madison, WI.

Sweet, James A. and Larry L. Bumpass. 1987. *American Families and Households*. New York: Russell Sage Foundation.

Thompson, Linda and Alexis J. Walker. 1991. "Gender in Families." Pp. 76–102 in *Contemporary Families*, edited by A. Booth. Minneapolis, MN: National Council on Family Relations.

Thompson, Linda. 1991. "Family Work: Women's Sense of Fairness." *Journal of Family Issues* 12: 181–96.

Timmer, Susan G., Jacquelynne Eccles, and Keith O'Brien. 1985. "How Children Use Time." Pp. 353–82 in *Time, Goods, and Well-Being*, edited by T. F. Juster and F. P. Stafford. Ann Arbor, MI: Institute for Social Research, University of Michigan.

Umberson, Debra, Camille B. Wortman, and Ronald C. Kessler. 1992. "Widowhood and Depression: Explaining Long-Term Gender Differences in Vulnerability." *Journal of Health and Social Behavior* 33:10–24.

Waite, Linda and Frances K. Goldscheider. 1992. "Work in the Home: The Productive Context of Family Relationships." Pp. 267–99 in *The Changing American Family: Sociological and Demographic Perspectives*, edited by S. J. South and S. E. Tolnay. Boulder, CO: Westview Press.

Walker, Kathryn E. and Margaret E. Woods. 1976. *Time Use: A Measure of Household Production of Goods and Services*. Washington, DC: American Home Economics Association.

Ward, Russell, John Logan, and Glenna Spitze. 1992. "The Influence of Parent and Child Needs on

Coresidence in Middle and Later Life." *Journal of Marriage and the Family* 54:209–21.

Warner, Rebecca A. 1986. "Alternative Strategies for Measuring Household Division of Labor: A Comparison." *Journal of Family Issues* 7:179–95.

Watkins, Susan Cotts, Jane A. Menken, and John Bongaarts. 1987. "Demographic Foundations of Family Change." *American Sociological Review* 52:346–58.

West, Candace and Don H. Zimmerman. 1981. "Doing Gender." *Gender and Society* 1:125–51.

White, Lynn K. and David B. Brinkerhoff. 1981a. "Children's Work in the Family: Its Significance and Meaning." *Journal of Marriage and the Family* 43:789–98.

———. 1981b. "The Sexual Division of Labor: Evidence from Childhood." *Social Forces* 60:170–81.

Appendix C

Table of Random Numbers

Line/Col.	(1)	(2)	(3)	(4)	(5)	(6)	(7)	(8)	(9)	(10)	(11)	(12)	(13)	(14)
1	10480	15011	01536	02011	81647	91646	69179	14194	62590	36207	20969	99570	91291	90700
2	22368	46573	25595	85393	30995	89198	27982	53402	93965	34095	52666	19174	39615	99505
3	24130	48360	22527	97265	76393	64809	15179	24830	49340	32081	30680	19655	63348	58629
4	42167	93093	06243	61680	07856	16376	39440	53537	71341	57004	00849	74917	97758	16379
5	37570	39975	81837	16656	06121	91782	60468	81305	49684	60672	14110	06927	01263	54613
6	77921	06907	11008	42751	27756	53498	18602	70659	90655	15053	21916	81825	44394	42880
7	99562	72905	56420	69994	98872	31016	71194	18738	44013	48840	63213	21069	10634	12952
8	96301	91977	05463	07972	18876	20922	94595	56869	69014	60045	18425	84903	42508	32307
9	89579	14342	63661	10281	17453	18103	57740	84378	25331	12566	58678	44947	05585	56941
10	85475	36857	43342	53988	53060	59533	38867	62300	08158	17983	16439	11458	18593	64952
11	28918	69578	88231	33276	70997	79936	56865	05859	90106	31595	01547	85590	91610	78188
12	63553	40961	48235	03427	49626	69445	18663	72695	52180	20847	12234	90511	33703	90322
13	09429	93969	52636	92737	88974	33488	36320	17617	30015	08272	84115	27156	30613	74952
14	10365	61129	87529	85689	48237	52267	67689	93394	01511	26358	85104	20285	29975	89868
15	07119	97336	71048	08178	77233	13916	47564	81056	97735	85977	29372	74461	28551	90707
16	51085	12765	51821	51259	77452	16308	60756	92144	49442	53900	70960	63990	75601	40719
17	02368	21382	52404	60268	89368	19885	55322	44819	01188	65255	64835	44919	05944	55157
18	01011	54092	33362	94904	31273	04146	18594	29852	71585	85030	51132	01915	92747	64951
19	52162	53916	46369	58586	23216	14513	83149	98736	23495	64350	94738	17752	35156	35749
20	07056	97628	33787	09998	42698	06691	76988	13602	51851	46104	88916	19509	25625	58104
21	48663	91245	85828	14346	09172	30168	90229	04734	59193	22178	30421	61666	99904	32812

Line/Col.	(1)	(2)	(3)	(4)	(5)	(6)	(7)	(8)	(9)	(10)	(11)	(12)	(13)	(14)
22	54164	58492	22421	74103	47070	25306	76468	26384	58151	06646	21524	15227	96909	44592
23	32639	32363	05597	24200	13363	38005	94342	28728	35806	06912	17012	64161	18296	22851
24	29334	27001	87637	87308	58731	00256	45834	15398	46557	41135	10367	07684	36188	18510
25	02488	33062	28834	07351	19731	92420	60952	61280	50001	67658	32586	86679	50720	94953
26	81525	72295	04839	96423	24878	82651	66566	14778	76797	14780	13300	87074	79666	95725
27	29676	20591	68086	26432	46901	20849	89768	81536	86645	12659	92259	57102	80428	25280
28	00742	57392	39064	66432	84673	40027	32832	61362	98947	96067	64760	64584	96096	98253
29	05366	04213	25669	26422	44407	44048	37937	63904	45766	66134	75470	66520	34693	90449
30	91921	26418	64117	94305	26766	25940	39972	22209	71500	64568	91402	42416	07844	69618
31	00582	04711	87917	77341	42206	35126	74087	99547	81817	42607	43808	76655	62028	76630
32	00725	69884	62797	56170	86324	88072	76222	36086	84637	93161	76038	65855	77919	88006
33	69011	65797	95876	55293	18988	27354	26575	08625	40801	59920	29841	80150	12777	48501
34	25976	57948	29888	88604	67917	48708	18912	82271	65424	69774	33611	54262	85963	03547
35	09763	83473	73577	12908	30883	18317	28290	35797	05998	41688	34952	37888	38917	88050
36	91567	42595	27958	30134	04024	86385	29880	99730	55536	84855	29080	09250	79656	73211
37	17955	56349	90999	49127	20044	59931	06115	20542	18059	02008	73708	83317	36103	42791
38	46503	18584	18845	49618	02304	51038	20655	58727	28168	15475	56942	53389	20562	87338
39	92157	89634	94824	78171	84610	82834	09922	25417	44137	48413	25555	21246	35509	20468
40	14577	62765	35605	81263	39667	47358	56873	56307	61607	49518	89656	20103	77490	18062
41	98427	07523	33362	64270	01638	92477	66969	98420	04880	45585	46565	04102	46880	45709

(Continued)

Line/Col.	(1)	(2)	(3)	(4)	(5)	(6)	(7)	(8)	(9)	(10)	(11)	(12)	(13)	(14)
42	34914	63976	88720	82765	34476	17032	87589	40836	32427	70002	70663	88863	77775	69348
43	70060	28277	39475	46473	23219	53416	94970	25832	69975	94884	19661	72828	00102	66794
44	53976	54914	06990	67245	68350	82948	11398	42878	80287	88267	47363	46634	06541	97809
45	76072	29515	40980	07391	58745	25774	22987	80059	39911	96189	41151	14222	60697	59583
46	90725	52210	83974	29992	65831	38857	50490	83765	55657	14361	31720	57375	56228	41546
47	64364	67412	33339	31926	14883	24413	59744	92351	97473	89286	35931	04110	23726	51900
48	08962	00358	31662	25388	61642	34072	81249	35648	56891	69352	48373	45578	78547	81788
49	95012	68379	93526	70765	10593	04542	76463	54328	02349	17247	28865	14777	62730	92277
50	15664	10493	20492	38391	91132	21999	59516	81652	27195	48223	46751	22923	32261	85653
51	16408	81899	04153	53381	79401	21438	83035	92350	36693	31238	59649	91754	72772	02338
52	18629	81953	05520	91962	04739	13092	97662	24822	94730	06496	35090	04822	86772	98289
53	73115	35101	47498	87637	99016	71060	88824	71013	18735	20286	23153	72924	35165	43040
54	57491	16703	23167	49323	45021	33132	12544	41035	80780	45393	44812	12515	98931	91202
55	30405	83946	23792	14422	15059	45799	22716	19792	09983	74353	68668	30429	70735	25499
56	16631	35006	85900	98275	32388	52390	16815	69298	82732	38480	73817	32523	41961	44437
57	96773	20206	42559	78985	05300	22164	24369	54224	35083	19687	11052	91491	60383	19746
58	38935	64202	14349	82674	66523	44133	00697	35552	35970	19124	63318	29686	03387	59846
59	31624	76384	17403	53363	44167	64486	64758	75366	76554	31601	12614	33072	60332	92325
60	78919	19474	23632	27889	47914	02584	37680	20801	72152	39339	34806	08930	85001	87820
61	03931	33309	57047	74211	63445	17361	62825	39908	05607	91284	68833	25570	38818	46920
62	74426	33278	43972	10119	89917	15665	52872	73823	73144	88662	88970	74492	51805	99378

Line/Col.	(1)	(2)	(3)	(4)	(5)	(6)	(7)	(8)	(9)	(10)	(11)	(12)	(13)	(14)
63	09066	00903	20795	95452	92648	45454	09552	88815	16553	51125	79375	97596	16296	66092
64	42238	12426	87025	14267	20979	04508	64535	31355	86064	29472	47689	05974	52468	16834
65	16153	08002	26504	41744	81959	65642	74240	56302	00033	67107	77510	70625	28725	34191
66	21457	40742	29820	96783	29400	21840	15035	34537	33310	06116	95240	15957	16572	06004
67	21581	57802	02050	89728	17937	37621	47075	42080	97403	48626	68995	43805	33386	21597
68	55612	78095	83197	33732	05810	24813	86902	60397	16489	03264	88525	42786	05269	92532
69	44657	66999	99324	51281	84463	60563	79312	93454	68876	25471	93911	25650	12682	73572
70	91340	84979	46949	81973	37949	61023	43997	15263	80644	43942	89203	71795	99533	50501
71	91227	21199	31935	27022	84067	05462	35216	14486	29891	68607	41867	14951	91696	85065
72	50001	38140	66321	19924	72163	09538	12151	06878	91903	18749	34405	56087	82790	70925
73	65390	05224	72958	28609	81406	39147	25549	48542	42627	45233	57202	94617	23772	07896
74	27504	96131	83944	41575	10573	08619	64482	73923	36152	05184	94142	25299	84387	34925
75	37169	94851	39117	89632	00959	16487	65536	49071	39782	17095	02330	74301	00275	48280
76	11508	70225	51111	38351	19444	66499	71945	05422	13442	78675	84081	66938	93654	59894
77	37449	30362	06694	54690	04052	53115	62757	95348	78662	11163	81651	50245	34971	52924
78	46515	70331	85922	38329	57015	15765	97161	17869	45349	61796	66345	81073	49106	79860
79	30986	81223	42416	58353	21532	30502	32305	86482	05174	07901	54339	58861	74818	46942
80	63798	64995	46583	09765	44160	78128	83991	42865	92520	83531	80377	35909	81250	54238
81	82486	84846	99254	67632	43218	50076	21361	64816	51202	88124	41870	52689	51275	83556
82	21885	32906	92431	09060	64297	51674	64126	62570	26123	05155	59194	52799	28225	85762
83	60336	98782	07408	53458	13564	59089	26445	29789	85205	41001	12535	12133	14645	23541

(Continued)

Presentation." Pp. 373–386 in *The Handbook of Social Research Ethics,* edited by Donna M. Mertens and Pauline E. Ginsberg. Thousand Oaks, CA: Sage.

Brown, Judith Belle. 1999. "The Use of Focus Groups in Clinical Research." Pp. 109–124 in *Doing Qualitative Research,* 2nd ed., edited by Benjamin F. Crabtree and William L. Miller. Thousand Oaks, CA: Sage.

Bruni, Frank. 2002. "Persistent Drop in Fertility Reshapes Europe's Future." *The New York Times,* December 26, pp. A1, A10.

Bryman, Alan. 2008. "The End of the Paradigm Wars?" Pp. 13–25 in *The Sage Handbook of Social Research Methods,* edited by Pertti Alasuutari, Leonard Bickman and Julia Brannen. Thousand Oaks, CA: Sage.

Bureau of Justice Statistics. 2011a. "Key Facts at a Glance: Four Measures of Serious Violent Crime." Washington, DC: Bureau of Justice Statistics, U.S. Department of Justice. Retrieved March 17, 2011, from http://bjs.ojp.usdoj.gov/content/glance/cv2.cfm

Bureau of Justice Statistics. 2011b. "Key Facts at a Glance: Drug Arrests by Age, 1970–2007." Washington, DC: Bureau of Justice Statistics, U.S. Department of Justice. Retrieved March 17, 2011, from http://bjs.ojp.usdoj.gov/content/glance/drug.cfm

Burke, Garance. 2000. "Mexico's Census Battles Perception of Corruption." *The Boston Globe,* March 2, p. A2.

Burt, Martha R. 1996. "Homelessness: Definitions and Counts." Pp. 15–23 in *Homelessness in America,* edited by Jim Baumohl. Phoenix, AZ: Oryx.

Bushman, Brad J., Roy F. Baumeister, and Angela D. Stack. 1999. "Catharsis, Aggression, and Persuasive Influence: Self-Fulfilling or Self-Defeating Prophecies?" *Journal of Personality and Social Psychology* 76:367–376.

Butler, Dore and Florence Geis. 1990. "Nonverbal Affect Responses to Male and Female Leaders: Implications for Leadership Evaluations." *Journal of Personality and Social Psychology* 58 (January):48–59.

Buttel, Frederick H. 2000. "World Society, the Nation-State, and Environmental Protection: Comment on Frank, Hironaka, and Schofer." *American Sociological Review* 65:117–121.

Buzawa, Eve S. and Carl G. Buzawa (Eds.). 1996. *Do Arrests and Restraining Orders Work?* Thousand Oaks, CA: Sage.

Cain, Carol. 1991. "Personal Stories: Identity Acquisition and Self-Understanding in Alcoholics Anonymous." *Ethos* 19:210–253.

California Penal Code. 1988. Street Terrorism Enforcement and Prevention Act, California Penal Code, sec. 186-22[f]. http://www.leginfo.ca.gov/cgi-bin/waisgate?WAISdocID=01537111441+2+0+0&WAISaction=retrieve. Downloaded September 2, 2011.

Campbell, Donald T. and M. Jean Russo. 1999. *Social Experimentation.* Thousand Oaks, CA: Sage.

Campbell, Donald T. and Julian C. Stanley. 1966. *Experimental and Quasi-Experimental Designs for Research.* Chicago: Rand McNally.

Campbell, Richard T. 1992. "Longitudinal Research." Pp. 1146–1158 in *Encyclopedia of Sociology,* edited by Edgar F. Borgatta and Marie L. Borgatta. New York: Macmillan.

Campbell, Wilson. 2002. "A Statement From the Governmental Accounting Standards Board and Performance Measurement Staff." American Society for Public Administration. Retrieved July 20, 2002, from www.aspanet.org/cap/forum_statement.html#top

Cannell, C. F., L. Oksenberg, G. Kalton, K. Bischoping, and Floyd J. Fowler. 1989. *New Techniques for Pretesting Survey Questions.* Final Report. Grant No. HS05616. National Center for Health Service Research, Health Care Technology Assessment.

Caplan, Brina, Russell K. Schutt, Winston M. Turner, Stephen M. Goldfinger, and Larry J. Seidman. 2006. "Change in Neurocognition by Housing Type and Substance Abuse Among Formerly Homeless Seriously Mentally Ill Individuals." *Schizophrenia Research* 83:77–86.

Carmines, Edward G. and Richard A. Zeller. 1979. *Reliability and Validity Assessment,* no. 17, *Quantitative Applications in the Social Sciences.* Beverly Hills, CA: Sage.

Cava, Anita, Reid Cushman, and Kenneth Goodman. 2007. "HIPAA and Human Subjects Research." *Collaborative Institutional Training Initiative.* Retrieved June 5, 2008, from https://www.citi program.org/members/learners

Cave, Emma and Soren Holm. 2003. "Milgram and Tuskegee—Paradigm Research Projects in Bioethics." *Health Care Analysis* 11:27–40.

Ceglowski, Deborah. 2002. "Research as Relationship." Pp. 5–27 in *The Qualitative Inquiry Reader,* edited by Norman Denzin and Yvonna S. Lincoln. Thousand Oaks, CA: Sage.

Center for Survey Research, University of Massachusetts at Boston. 1987. "Methodology: Designing Good Survey Questions." *Newsletter,* April, p. 3.

Centers for Disease Control and Prevention (CDC). 2009. "The Tuskegee Timeline." Atlanta, GA: National Center for HIV/AIDS, Viral Hepatitis, STD, and TB Prevention, Centers for Disease Control and Prevention. Retrieved March 16, 2011, from http://www.cc.gov/tuskegee/timeine.htm

Chalk, Rosemary and Joel H. Garner. 2001. "Evaluating Arrest for Intimate Partner Violence: Two Decades of Research and Reform." *New Directions for Evaluation* 90:9–23.

Chase-Dunn, Christopher and Thomas D. Hall. 1993. "Comparing World-Systems: Concepts and Working Hypotheses." *Social Forces* 71:851–886.

Chase-Lansdale, P. Lindsay, Robert A. Moffitt, Brenda J. Lohman, Andrew J. Cherlin, Rebekah Levine Coley, Laura D. Pittman, Jennifer Roff, and Elizabeth Votruba-Drzal. 2003. "Mothers' Transitions From Welfare to Work and the Well-Being of

Preschoolers and Adolescents." *Science* 299:1548–1552.

Chen, Huey-Tsyh. 1990. *Theory-Driven Evaluations.* Newbury Park, CA: Sage.

Chen, Huey-Tsyh and Peter H. Rossi. 1987. "The Theory-Driven Approach to Validity." *Evaluation and Program Planning* 10:95–103.

Christian, Leah, Scott Keeter, Kristen Purcell and Aaron Smith. 2010. "Assessing the Cell Phone Challenge to Survey Research in 2010." Washington, DC: Pew Research Center for the People & the Press and Pew Internet & American Life Project.

Church, Allan H. 1993. "Estimating the Effects of Incentives on Mail Survey Response Rates: A Meta-Analysis." *Public Opinion Quarterly* 57:62–79.

Church, A. Timothy. 2010. "Measurement Issues in Cross-Cultural Research." Pp. 151–175 in *The Sage Handbook of Measurement*, edited by Geoffrey Walford, Eric Tucker, and Madhu Viswanathan. Thousand Oaks, CA: Sage.

Clegg, Joshua and Brent D Slife. 2009. "Research Ethics in the Postmodern Context." Pp. 23–38 in *The Handbook of Social Research Ethics,* edited by Donna M. Mertens and Pauline E. Ginsberg. Thousand Oaks, CA: Sage.

Code of Federal Regulations. 2009. Title 45, Public Welfare. Department of Health and Human Services, Part 46, Protection of Human Subjects. http://www.hhs.gov/ohrp/human subjects/guidance/45cfr46.html#46.101 .Downloaded September 1, 2011.

Coffey, Amanda and Paul Atkinson. 1996. *Making Sense of Qualitative Data: Complementary Research Strategies.* Thousand Oaks, CA: Sage.

Cohen, S., R. Mermelstein, T. Kamarck, and H. M. Hoberman. 1985. "Measuring the Functional Components of Social Support." Pp. 73–94 in *Social Support: Theory, Research and Applications,* edited by I. G. Sarason and B. R. Sarason. The Hague, The Netherlands: Martinus Nijhoff.

Cohen, Susan G. and Gerald E. Ledford, Jr. 1994. "The Effectiveness of Self-Managing Teams: A Quasi-Experiment." *Human Relations* 47:13–43.

Coleman, James S. 1990. *Foundations of Social Theory.* Cambridge, MA: Harvard University Press.

Coleman, James S. and Thomas Hoffer. 1987. *Public and Private High Schools: The Impact of Communities.* New York: Basic Books.

Coleman, James S., Thomas Hoffer, and Sally Kilgore. 1982. *High School Achievement: Public, Catholic, and Private Schools Compared.* New York: Basic Books.

College Alcohol Study. 2008. "College Alcohol Study News." *Harvard School of Public Health.* Retrieved May 5, 2008, from www.hsph.harvard .edu/cas/Home.html

Collins, Patricia Hill. 2008. "Learning From the Outsider Within: The Sociological Significance of Black Feminist Thought." Pp. 308–310 in *Just Methods: An Interdisciplinary Reader,* edited by Alison M. Jaggar. Boulder, CO: Paradigm.

Collins, Randall. 1994. *Four Sociological Traditions.* New York: Oxford University Press.

Connolly, Francis J. and Charley Manning. 2001. "What 'Push Polling' Is and What It Isn't." *The Boston Globe,* August 16, p. A21.

Connors, Gerard J. and Robert J. Volk. 2004. "Self-Report Screening for Alcohol Problems Among Adults." Pp. 21–35 in *Assessing Alcohol Problems: A Guide for Clinicians and Researchers,* 2nd ed., edited by John P. Allen and Veronica B. Wilson. NIH Publication No. 03–3745, revised 2003. Bethesda, MD: National Institute on Alcohol Abuse and Alcoholism, National Institutes of Health. Retrieved October 6, 2005, from http://pubs.niaaa.nih.gov/publi cations/ Assesing%20Alcohol/index .htm#contents

Converse, Jean M. 1984. "Attitude Measurement in Psychology and Sociology: The Early Years." Pp. 3–40 in *Surveying Subjective Phenomena,* vol. 2, edited by Charles F. Turner and Elizabeth Martin. New York: Russell Sage Foundation.

Cook, Gareth. 2005. "Face It: Appearances Sway Voters." *The Boston Globe,* June 14, pp. E1, E4.

Cook, Thomas D. and Donald T. Campbell. 1979. *Quasi-Experimentation: Design*

and Analysis Issues for Field Settings. Chicago: Rand McNally.

Cook, Thomas D. and Vivian C. Wong. 2008. "Better Quasi-Experimental Practice." Pp. 134–165 in *The Sage Handbook of Social Research Methods,* edited by Pertti Alasuutari, Leonard Bickman, and Julia Brannen. Thousand Oaks, CA: Sage.

Cooper, Harris and Larry V. Hedges. 1994. "Research Synthesis as a Scientific Enterprise." Pp. 3–14 in *The Handbook of Research Synthesis,* edited by Harris Cooper and Larry V. Hedges. New York: Russell Sage Foundation.

Cooper, Kathleen B. and Michael D. Gallagher. 2004. *A Nation Online: Entering the Broadband Age.* Washington, DC: Economics and Statistics Administration and National Telecommunications and Information Administration, U.S. Department of Commerce. Retrieved June 15, 2005, from www.ntia.doc.gov/reports/anol/ index.html

Cooper, Kathleen B. and Nancy J. Victory. 2002a. "Foreword." P. iv in *A Nation Online: How Americans Are Expanding Their Use of the Internet.* Washington, DC: National Telecommunications and Information Administration, Economics and Statistics Administration, U.S. Department of Commerce.

Cooper, Kathleen B. and Nancy J. Victory. 2002b. *A Nation Online: How Americans Are Expanding Their Use of the Internet.* Washington, DC: National Telecommunications and Information Administration, Economics and Statistics Administration, U.S. Department of Commerce. Retrieved September 24, 2005, from www.ntia.doc.gov/ntiahome/dn/ana tiononline2.pdf

Core Institute. 1994. "Core Alcohol and Drug Survey: Long Form." Carbondale, IL: FIPSE Core Analysis Grantee Group, Core Institute, Student Health Programs, Southern Illinois University.

Correll, Joshua, Bernadette Park, Charles M. Judd, and Bernd Wittenbrink. 2002. "The Police Officer's Dilemma: Using Ethnicity to Disambiguate Potentially Threatening Individuals." *Journal of Personality and Social Psychology* 83:1314–1329.

Corse, Sara J., Nancy B. Hirschinger, and David Zanis. 1995. "The Use of the Addiction Severity Index With People With Severe Mental Illness." *Psychiatric Rehabilitation Journal* 19(1):9–18.

Costner, Herbert L. 1989. "The Validity of Conclusions in Evaluation Research: A Further Development of Chen and Rossi's Theory-Driven Approach." *Evaluation and Program Planning* 12:345–353.

Couper, Mick P. 2000. "Web Surveys: A Review of Issues and Approaches." *Public Opinion Quarterly* 64:464–494.

Couper, Mick P., Reginald P. Baker, Jelke Bethlehem, Cynthia Z. F. Clark, Jean Martin, William L. Nicholls II, and James M. O'Reilly (Eds.). 1998. *Computer-Assisted Survey Information Collection.* New York: Wiley.

Couper, Mick P. and Peter V. Miller. 2008. "Web Survey Methods: Introduction." *Public Opinion Quarterly* 72:831–835.

Couper, Mick P., Michael W. Traugott, and Mark J. Lamias. 2001. "Web Survey Design and Administration." *Public Opinion Quarterly* 65:230–253.

Cowan, Gloria. 2002. "Content Analysis of Visual Materials." Pp. 345–368 in *Handbook for Conducting Research on Human Sexuality,* edited by Michael W. Wiederman and Bernard E. Whitley, Jr. London: Lawrence Erlbaum.

Cress, Daniel M. and David A. Snow. 2000. "The Outcomes of Homeless Mobilization: The Influence of Organization, Disruption, Political Mediation, and Framing." *American Journal of Sociology* 4:1063–1104.

Creswell, John W. 2010. "Mapping the Developing Landscape of Mixed Methods Research." Pp. 45–68 in *Sage Handbook of Mixed Methods in Social & Behavioral Research,* 2nd ed., edited by Abbas Tashakkori and Charles Teddlie. Thousand Oaks, CA: Sage.

Cuba, Lee J. 2002. *A Short Guide to Writing About Social Science,* 4th ed. New York: Addison-Wesley.

Currie, Janet and Duncan Thomas. 1995. "Does Head Start Make a Difference?" *American Economic Review* 85:341–365.

Curtin, Richard, Stanley Presser, and Eleanor Singer. 2005. "Changes in Telephone Survey Nonresponse Over the Past Quarter Century." *Public Opinion Quarterly* 69:87–98.

Czopp, Alexander M., Margo J. Monteith, and Aimee Y. Mark. 2006. "Standing Up for a Change: Reducing Bias Through Interpersonal Confrontation." *Journal of Personality and Social Psychology* 90:784–803.

Dahlberg, Britt, Marsha N. Wittink, and Joseph J. Gallo. 2010. "Funding and Publishing Integrated Studies: Writing Effective Mixed Methods Manuscripts and Grant Proposals." Pp. 775–802 in *Sage Handbook of Mixed Methods in Social & Behavioral Research,* 2nd ed., edited by Abbas Tashakkori and Charles Teddlie. Thousand Oaks, CA: Sage.

Dale, Angela, Jo Wathan, and Vanessa Higgins. 2008. "Secondary Analysis of Quantitative Data Sources." Pp. 520–535 in *The Sage Handbook of Social Research Methods,* edited by Pertti Alasuutari, Leonard Bickman, and Julia Brannen. Thousand Oaks, CA: Sage.

D'Amico, Elizabeth J. and Kim Fromme. 2002. "Brief Prevention for Adolescent Risk-Taking Behavior." *Addiction* 97:563–574.

Dannefer, W. Dale and Russell K. Schutt. 1982. "Race and Juvenile Justice Processing in Court and Police Agencies." *American Journal of Sociology* 87(March):1113–1132.

D.A.R.E. 2008. *The "New" D.A.R.E. Program.* Retrieved May 31, 2008, from www.dare .com/newdare.asp

Davies, Philip, Anthony Petrosino, and Iain Chalmers. 1999. *Report and Papers From the Exploratory Meeting for the Campbell Collaboration.* London: School of Public Policy, University College.

Davis, James A. 1985. *The Logic of Causal Order.* Sage University Paper Series on Quantitative Applications in the Social Sciences, series No. 07–055. Beverly Hills, CA: Sage.

Davis, James A. and Tom W. Smith. 1992. *The NORC General Social Survey: A User's Guide.* Newbury Park, CA: Sage.

Davis, Ophera A. and Marie Land. 2007. "Southern Women Survivors Speak About Hurricane Katrina, the Children and What Needs to Happen Next." *Race, Gender & Class* 14:69–86.

Dawes, Robyn. 1995. "How Do You Formulate a Testable Exciting Hypothesis?" Pp. 93–96 in *How to Write a Successful Research Grant Application: A Guide for Social and Behavioral Scientists,* edited by Willo Pequegnat and Ellen Stover. New York: Plenum Press.

De Leeuw, Edith. 2008. "Self-Administered Questionnaires and Standardized Interviews." Pp. 311–327 in *The Sage Handbook of Social Research Methods,* edited by Pertti Alasuutari, Leonard Bickman, and Julia Brannen. Thousand Oaks, CA: Sage.

De Vaus, David. 2008. "Comparative and Cross-National Designs." Pp. 248–264 in *The Sage Handbook of Social Research Methods,* edited by Pertti Alasuutari, Leonard Bickman, and Julia Brannen. Thousand Oaks, CA: Sage.

Decker, Scott H. and Barrik Van Winkle. 1996. *Life in the Gang: Family, Friends, and Violence.* Cambridge, UK: Cambridge University Press.

Demos, John. 1998. "History Beyond Data Bits." *The New York Times,* December 30, p. A23.

Dentler, Robert A. 2002. *Practicing Sociology: Selected Fields.* Westport, CT: Praeger.

Denzin, Norman K. 2002. "The Interpretive Process." Pp. 349–368 in *The Qualitative Researcher's Companion,* edited by A. Michael Huberman and Matthew B. Miles. Thousand Oaks, CA: Sage.

Denzin, Norman and Yvonna S. Lincoln. 1994. "Introduction: Entering the Field of Qualitative Research." Pp. 1–28 in *The Handbook of Qualitative Research,* edited by Norman Denzin and Yvonna S. Lincoln. Thousand Oaks, CA: Sage.

Denzin, Norman and Yvonna S. Lincoln. 2000. "Introduction: The Discipline and Practice of Qualitative Research." Pp. 1–28 in *The Handbook of Qualitative Research,* 2nd ed., edited by Norman Denzin and Yvonna S. Lincoln. Thousand Oaks, CA: Sage.

Denzin, Norman K. and Yvonna S. Lincoln. 2008. *Strategies of Qualitative Inquiry,* 3rd ed. Thousand Oaks, CA: Sage Publications.

DeParle, Jason. 1999. "Project to Rescue Needy Stumbles Against the Persistence of Poverty." *The New York Times,* May 15, pp. A1, A10.

Department of Health, Education, and Welfare. 1979. *The Belmont Report: Ethical Principles and Guidelines for the Protection of Human Subjects of Research.* Washington, DC: The National Commission for the Protection of Human Subjects of Biomedical and Behavioral Research, Office of the Secretary, Department of Health, Education, and Welfare. Retrieved June 20, 2005, from www.hss.gov/ohrp/humansubjects/guidance/belmont.htm

Dewan, Shaila K. 2004a. "As Murders Fall, New Tactics Are Tried Against Remainder." *The New York Times,* December 31, pp. A24–A25.

Dewan, Shaila K. 2004b. "New York's Gospel of Policing by Data Spreads Across U.S." *The New York Times,* April 26, pp. A1, C16.

Diamond, Timothy. 1992. *Making Gray Gold: Narratives of Nursing Home Care.* Chicago: University of Chicago Press.

DiClemente, C. C., J. P. Carbonari, R. P. G. Montgomery, and S. O. Hughes. 1994. "The Alcohol Abstinence Self-Efficacy Scale." *Journal of Studies on Alcohol* 55:141–148.

Dill, Karen E. and Kathryn P. Thill. 2007. "Video Game Characters and the Socialization of Gender Roles: Young People's Perceptions Mirror Sexist Media Depictions." *Sex Roles* 57:851–864.

Dillman, Don A. 1978. *Mail and Telephone Surveys: The Total Design Method.* New York: Wiley.

Dillman, Don A. 1982. "Mail and Other Self-Administered Questionnaires." Chapter 12 in *Handbook of Survey Research,* edited by Peter Rossi, James Wright, and Andy Anderson. New York: Academic Press. As reprinted on pp. 637–638 in Delbert C. Miller, 1991. *Handbook of Research Design and Social Measurement,* 5th ed. Newbury Park, CA: Sage.

Dillman, Don A. 2000. *Mail and Internet Surveys: The Tailored Design Method,* 2nd ed. New York: Wiley.

Dillman, Don A. 2007. *Mail and Internet Surveys: The Tailored Design Method,* 2nd ed. Update with New Internet, Visual, and Mixed-Mode Guide. Hoboken, NJ: Wiley.

Dillman, Don A., James A. Christenson, Edwin H. Carpenter, and Ralph M. Brooks. 1974. "Increasing Mail Questionnaire Response: A Four-State Comparison." *American Sociological Review* 39 (October):744–756.

Dillman, Don A. and Leah Melani Christian. 2005. "Survey Mode as a Source of Instability in Responses Across Surveys." *Field Methods* 17:30–52.

Doucet, Andrea and Natasha Mauthner. 2008. "Qualitative Interviewing and Feminist Research." Pp. 328–343 in *The Sage Handbook of Social Research Methods,* edited by Pertti Alasuutari, Leonard Bickman, and Julia Brannen. Thousand Oaks, CA: Sage.

Douglas, Jack D. 1985. *Creative Interviewing.* Beverly Hills, CA: Sage.

Drake, Robert E., Gregory J. McHugo, and Jeremy C. Biesanz. 1995. "The Test-Retest Reliability of Standardized Instruments Among Homeless Persons With Substance Use Disorders." *Journal of Studies on Alcohol* 56(2):161–167.

Drew, Paul. 2005. "Conversation Analysis." Pp. 71–102 in *Handbook of Language and Social Interaction,* edited by Kristine L. Fitch and Robert E. Sanders. Mahwah, NJ: Erlbaum.

Duckworth, Kenneth, John H. Halpern, Russell K. Schutt, and Christopher Gillespie. 2003. "Use of Schizophrenia as a Metaphor in U.S. Newspapers." *Psychiatric Services* 54:1402–1404.

Duncombe, Jean and Julie Jessop. 2002. "'Doing Rapport' and the Ethics of 'Faking Friendship.'" Pp. 107–122 in *Ethics in Qualitative Research,* edited by Melanie Mauthner, Maxine Birch, Julie Jessop, and Tina Miller. Thousand Oaks, CA: Sage.

Durkheim, Emile. 1951. *Suicide.* New York: Free Press.

Durkheim, Emile. 1966. *Suicide: A Study in Sociology.* Translated by John A. Spaulding and George Simpson. New York: Free Press.

Durkheim, Emile. 1984. *The Division of Labor in Society.* Translated by W. D. Halls. New York: Free Press.

Dykema, Jennifer and Nora Cate Schaeffer. 2000. "Events, Instruments, and Reporting Errors." *American Sociological Review* 65:619–629.

Eckholm, Erik. 2006. "Report on Impact of Federal Benefits on Curbing Poverty Reignites a Debate." *The New York Times,* February 18, p. A8.

Elder, Keith, Sudha Xirasagar, Nancy Miller, Shelly Ann Bowen, Saundra Glover, and Crystal Piper. 2007. "African Americans' Decisions Not to Evacuate New Orleans Before Hurricane Katrina: A Qualitative Study." *American Journal of Public Health* 97(Suppl. 1):S124–S129.

Elliott, Jane, Janet Holland, and Rachel Thomson. 2008. "Longitudinal and Panel Studies." Pp. 228–248 in *The Sage Handbook of Social Research Methods,* edited by Pertti Alasuutari, Leonard Bickman, and Julia Brannen. Thousand Oaks, CA: Sage.

Ember, Carol R. and Melvin Ember. 2011. *A Basic Guide to Cross-Cultural Research.* New Haven, CT: Human Relations Area Files, Yale University. http://www.yale.edu/hraf/guides.htm. Downloaded August 31, 2011.

Emerson, Robert M. (Ed.). 1983. *Contemporary Field Research.* Prospect Heights, IL: Waveland.

Emerson, Robert M., Rachel I. Fretz, and Linda L. Shaw. 1995. *Writing Ethnographic Fieldnotes.* Chicago: University of Chicago Press.

Erikson, Kai T. 1966. *Wayward Puritans: A Study in the Sociology of Deviance.* New York: Wiley.

Erikson, Kai T. 1967. "A Comment on Disguised Observation in Sociology." *Social Problems* 12: 366–373.

Eurostat. 2003. *ECHP UDB Manual: European Community Household Panel Longitudinal Users' Database.* Brussels, Belgium: European Commission.

Exum, M. Lyn. 2002. "The Application and Robustness of the Rational Choice Perspective in the Study of Intoxicated and Angry Intentions to Aggress." *Criminology* 40:933–966.

Fears, Darryl. 2002. "For Latinos in U.S., Race Not Just Black or White." *The Boston Globe,* December 30, p. A3.

Fenno, Richard F., Jr. 1978. *Home Style: House Members in Their Districts.* Boston: Little, Brown.

Fenton, Steve. 1996. "Counting Ethnicity: Social Groups and Official Categories." Pp. 143–165 in *Interpreting Official Statistics,* edited by Ruth Levitas and Will Guy. New York: Routledge.

Fink, Arlene. 2005. *Conducting Research Literature Reviews: From the Internet to Paper,* 2nd ed. Thousand Oaks, CA: Sage.

Fischer, Constance T. and Frederick J. Wertz. 2002. "Empirical Phenomenological Analyses of Being Criminally Victimized." Pp. 275–304 in *The Qualitative Researcher's Companion,* edited by A. Michael Huberman and Matthew B. Miles. Thousand Oaks, CA: Sage.

Fisher, Celia B. and Andrea E. Anushko. 2008. "Research Ethics in Social Science." Pp. 94–109 in *The Sage Handbook of Social Research Methods,* edited by Pertti Alasuutari, Leonard Bickman, and Julia Brannen. Thousand Oaks, CA: Sage.

Forero, Juan. 2000a. "Census Takers Say Supervisors Fostered Filing of False Data." *The New York Times,* July 28, p. A21.

Forero, Juan. 2000b. "Census Takers Top '90 Efforts in New York City, With More to Go." *The New York Times,* June 12, p. A29.

Fowler, Floyd J. 1988. *Survey Research Methods,* Rev. ed. Newbury Park, CA: Sage.

Fowler, Floyd J. 1995. *Improving Survey Questions: Design and Evaluation.* Thousand Oaks, CA: Sage.

Fox, Nick and Chris Roberts. 1999. "GPs in Cyberspace: The Sociology of a 'Virtual Community.'" *The Sociological Review* 47:643–669.

Frank, David John, Ann Hironaka, and Evan Schofer. 2000. "The Nation-State and the Natural Environment Over the Twentieth Century." *American Sociological Review* 65:96–116.

Frankfort-Nachmias, Chava and Anna Leon-Guerrero. 2008. *Social Statistics for a Diverse Society,* 5th ed. Thousand Oaks, CA: Sage.

Franklin, Mark N. 1996. "Electoral Participation." Pp. 216–235 in *Comparing Democracies: Elections and Voting in Global Perspective,* edited by Lawrence LeDuc, Richard G. Niemi, and Pippa Norris. Thousand Oaks, CA: Sage.

Frohmann, Lisa. 2005. "The Framing Safety Project: Photographs and Narratives by Battered Women." *Violence Against Women* 11:1396–1419.

Frone, Michael R. 2008. "Are Work Stressors Related to Employee Substance Use? The Importance of Temporal Context in Assessments of Alcohol and Illicit Drug Use." *Journal of Applied Psychology* 93:199–206.

Gaiser, Ted J. and Anthony E. Schreiner. 2009. *A Guide to Conducting Online Research.* Thousand Oaks, CA: Sage.

Gall, Carlotta. 2003. "Armed with Pencils, Army of Census Workers Fans Out Into Afghan Outback." *The New York Times,* July 15:A4.

Gallup. 2011. *Election Polls—Accuracy Record in Presidential Elections.* Retrieved March 17, 2011, from http://www.gallup.com/poll/9442/Election-Polls-Accuracy-Record-Presidential-Elections.aspx?version=print

Gans, Curtis. 2008. "2008 Primary Turnout Falls Just Short of Record Nationally, Breaks Records in Most States." *AU News,* May 19. Washington, DC: Center for the Study of the American Electorate, American University. Retrieved June 2, 2008, from www.american.edu/media

Garfinkel, Harold. 1967. *Studies in Ethnomethodology.* Englewood Cliffs, NJ: Prentice Hall.

Gartrell, C. David and John W. Gartrell. 2002. "Positivism in Sociological Research: USA and UK (1966–1990)." *British Journal of Sociology* 53:639–657.

Geertz, Clifford. 1973. "Thick Description: Toward an Interpretive Theory of Culture." Pp. 3–30 in *The Interpretation of Cultures,* edited by Clifford Geertz. New York: Basic Books.

"'Get Tough' Youth Programs Are Ineffective, Panel Says." 2004. *The New York Times,* October 17, p. 25.

Gilchrist, Valerie J. and Robert L. Williams. 1999. "Key Informant Interviews." Pp. 71–88 in *Doing Qualitative Research,* 2nd ed., edited by Benjamin F. Crabtree and William L. Miller. Thousand Oaks, CA: Sage.

Gill, Hannah E. 2004. "Finding a Middle Ground Between Extremes: Notes on Researching Transnational Crime and Violence." *Anthropology Matters Journal* 6:1–9.

Gill, Richard T., Nathan Glazer, and Stephan A. Thernstrom. 1992. *Our Changing Population.* Englewood Cliffs, NJ: Prentice Hall.

Gilligan, Carol. 1988. "Adolescent Development Reconsidered." Pp. vii–xxxix in *Mapping the Moral Domain,* edited by Carol Gilligan, Janie Victoria Ward, and Jill McLean Taylor. Cambridge, MA: Harvard University Press.

Glaser, Barney G. and Anselm L. Strauss. 1967. *The Discovery of Grounded Theory: Strategies for Qualitative Research.* London: Weidenfeld and Nicholson.

Glavin, Paul, Scott Schieman, and Sarah Reid 2011. "Boundary-Spanning Work Demands and Their Consequences for Guilt and Psychological Distress." *Journal of Health and Social Behavior* 52:43–57.

Glover, Judith. 1996. "Epistemological and Methodological Considerations in Secondary Analysis." Pp. 28–38 in *Cross-National Research Methods in the Social Sciences,* edited by Linda Hantrais and Steen Mangen. New York: Pinter.

Glueck, Sheldon and Elenor Glueck. 1950. *Unraveling Juvenile Delinquency.* New York: Commonwealth Fund.

Gobo, Giampietro. 2008. "Re-Conceptualizing Generalization: Old Issues in a New Frame." Pp. 193–213 in *The Sage Handbook of Social Research Methods,* edited by Pertti Alasuutari, Leonard Bickman, and Julia Brannen. Thousand Oaks, CA: Sage.

Goertz, Gary. 2006. *Social Science Concepts: A User's Guide.* Princeton, NJ: Princeton University Press.

Goffman, Erving. 1961. *Asylums: Essays on the Social Situation of Mental Patients and Other Inmates.* Garden City, NY: Doubleday.

Goldfinger, Stephen M. and Russell K. Schutt. 1996. "Comparison of Clinicians' Housing Recommendations and Preferences of Homeless Mentally Ill Persons." *Psychiatric Services,* 47:413–415.

Goldfinger, Stephen M., Russell K. Schutt, Larry J. Seidman, Winston M. Turner, Walter E. Penk, and George S. Tolomiczenko. 1996. "Self-Report and Observer Measures of Substance Abuse Among Homeless Mentally Ill Persons in the Cross-Section and Over Time." *The Journal of Nervous and Mental Disease* 184(11):667–672.

Goleman, Daniel. 1993a. "Placebo Effect Is Shown to Be Twice as Powerful as Expected." *The New York Times,* August 17, p. C3.

Goleman, Daniel. 1993b. "Pollsters Enlist Psychologists in Quest for Unbiased Results." *The New York Times,* September 7, pp. C1, C11.

Goodnough, Abby. 2010. "A Wave of Addiction and Crime, with the Medicine Cabinet to Blame." *The New York Times,* September 22. Retrieved March 17, 2011, from www.nytimes.com

Gordon, Raymond. 1992. *Basic Interviewing Skills.* Itasca, IL: Peacock.

Graduate Program in Applied Sociology. 1990. *Handbook for Thesis Writers.* Boston: University of Massachusetts.

Grady, John. 1996. "The Scope of Visual Sociology." *Visual Sociology* 11:10–24.

Grant, Bridget F., Deborah A. Dawson, Frederick S. Stinson, S. Patricia Chou, Mary C. Dufour, and Roger P. Pickering. 2004. "The 12-Month Prevalence and Trends in DSM-IV Alcohol Abuse and Dependence: United States, 1991–1992 and 2001–2002." *Drug and Alcohol Dependence* 74:223–234.

Greenberg, David, Mark Shroder, and Mathew Onstott. 1999. "The Social Experiment Market." *Journal of Economic Perspectives* 13:157–172.

Grey, Robert J., Jr. 2005. "Jury Service: It's a Privilege." Retrieved June 18, 2005, from American Bar Association website, www.abanet.org/media/releases/opedjuror2.html

Griffin, Larry J. 1992. "Temporality, Events, and Explanation in Historical Sociology: An Introduction." *Sociological Methods & Research* 20:403–427.

Griffin, Larry J. 1993. "Narrative, Event-Structure Analysis, and Causal Interpretation in Historical Sociology." *American Journal of Sociology* 98(March):1094–1133.

Grinnell, Frederick. 1992. *The Scientific Attitude,* 2nd ed. New York: Guilford Press.

Grissom, Brandi. 2011. "Proposals Could Make It Harder to Leave Prison." *The New York Times,* March 12. Retrieved March 17, 2011, from www.nytimes.com

Grossman, Lev. 2010. "Mark Zuckerberg." *Time,* December 15. Retrieved March 14, 2011, from http://www.time.com/time/specials/packages/article/0,28804,2036683_2037183_2037185,00.html

Groves, Robert M. 1989. *Survey Errors and Survey Costs.* New York: Wiley.

Groves, Robert M. and Mick P. Couper. 1998. *Nonresponse in Household Interview Surveys.* New York: Wiley.

Groves, Robert M. and Robert L. Kahn. 1979. *Surveys by Telephone: A National Comparison With Personal Interviews.* New York: Academic Press. As adapted in Delbert C. Miller, 1991. *Handbook of Research Design and Social Measurement,* 5th ed. Newbury Park, CA: Sage.

Groves, Robert M., Eleanor Singer, and Amy Corning. 2000. "Leverage-Salience Theory of Survey Participation: Description and an Illustration." *Public Opinion Quarterly* 64:299–308.

Gruenewald, Paul J., Andrew J. Treno, Gail Taff, and Michael Klitzner. 1997. *Measuring Community Indicators: A Systems Approach to Drug and Alcohol Problems.* Thousand Oaks, CA: Sage.

Guba, Egon G. and Yvonna S. Lincoln. 1989. *Fourth Generation Evaluation.* Newbury Park, CA: Sage.

Guba, Egon G. and Yvonna S. Lincoln. 1994. "Competing Paradigms in Qualitative Research." Pp. 105–117 in *Handbook of Qualitative Research,* edited by Norman K. Denzin and Yvonna S. Lincoln. Thousand Oaks, CA: Sage.

Gubrium, Jaber F. and James A. Holstein. 1997. *The New Language of Qualitative Method.* New York: Oxford University Press.

Gubrium, Jaber F. and James A. Holstein. 2000. "Analyzing Interpretive Practice." Pp. 487–508 in *The Handbook of Qualitative Research,* 2nd ed., edited by Norman Denzin and Yvonna S. Lincoln. Thousand Oaks, CA: Sage.

Guterbock, Thomas M. 2008. *Strategies and Standards for Reaching Respondents in an Age of New Technology.* Presentation to the Harvard Program on Survey Research Spring Conference, New Technologies and Survey Research. Cambridge, MA: Institute of Quantitative Social Science, Harvard University, May 9.

Hacker, Karen, Jessica Collins, Leni Gross-Young, Stephanie Almeida, and Noreen Burke. 2008. "Coping with Youth Suicide and Overdose: One Community's Efforts to Investigate, Intervene, and Prevent Suicide Contagion." *Crisis* 29:86–95.

Hadaway, C. Kirk, Penny Long Marler, and Mark Chaves. 1993. "What the Polls Don't Show: A Closer Look at U.S. Church Attendance." *American Sociological Review* 58(December):741–752.

Hafner, Katie. 2004. "For Some, the Blogging Never Stops." *The New York Times,* May 27, pp. E1, E7.

Hage, Jerald and Barbara Foley Meeker. 1988. *Social Causality.* Boston: Unwin Hyman.

Hakim, Catherine. 1982. *Secondary Analysis in Social Research: A Guide to Data Sources and Methods With Examples.* London: George Allen & Unwin.

Hakimzadeh, Shirin and D'Vera Cohn. 2007. *English Usage Among Hispanics in the United States.* Washington, DC: Pew Hispanic Center. Retrieved May 24, 2008, from http://pewhispanic.org/ files/ reports/82.pdf

Halcón, Linda L. and Alan R. Lifson. 2004. "Prevalence and Predictors of Sexual Risks Among Homeless Youth." *Journal of Youth and Adolescence* 33:71–80.

Hallinan, Maureen T. 1997. "The Sociological Study of Social Change." *American Sociological Review* 62:1–11.

Hammersley, Martyn. 2008. "Assessing Validity in Social Research." Pp. 42–53 in *The Sage Handbook of Social Research Methods,* edited by Pertti Alasuutari, Leonard Bickman, and Julia Brannen. Thousand Oaks, CA: Sage.

Hampton, Keith N. 2003. "Grieving for a Lost Network: Collective Action in a Wired Suburb." *The Information Society* 19:417–428.

Hampton, Keith N. and Neeti Gupta. 2008. "Community and Social Interaction in the Wireless City: Wi-Fi Use in Public and Semi-public Spaces." *New Media & Society* 10(6):831–850.

Hampton, Keith, Oren Livio and Lauren Sessions Goulet. 2010. "The Social Life of Wireless Urban Spaces: Internet Use, Social Networks, and the Public Realm." *Journal of Communication* 60(4):701–722.

Hampton, Keith N. and Barry Wellman. 1999. "Netville On-line and Off-Line: Observing and Surveying a Wired Suburb." *American Behavioral Scientist* 43:475–492.

Hampton, Keith N. and Barry Wellman. 2000. "Examining Community in the Digital Neighborhood: Early Results From Canada's Wired Suburb." Pp. 475–492 in *Digital Cities: Technologies, Experiences, and Future Perspectives,* edited by Toru Ishida and Katherine Isbister. Berlin, Germany: Springer-Verlag.

Hampton, Keith N. and Barry Wellman. 2001. "Long Distance Community in the Network Society." *American Behavioral Scientist* 45:476–495.

Haney, C., C. Banks and Philip G. Zimbardo. 1973. "Interpersonal Dynamics in a Simulated Prison." *International Journal of Criminology and Penology* 1:69–97.

Hantrais, Linda and Steen Mangen. 1996. "Method and Management of Cross-National Social Research." Pp. 1–12 in *Cross-National Research Methods in the Social Sciences,* edited by Linda Hantrais and Steen Mangen. New York: Pinter.

Hard, Stephen F., James M. Conway, and Antonia C. Moran. 2006. "Faculty and College Student Beliefs About the Frequency of Student Academic Misconduct." *The Journal of Higher Education* 77:1058–1080.

Harding, David J. 2007. "Cultural Context, Sexual Behavior, and Romantic Relationships in Disadvantaged Neighborhoods." *American Sociological Review* 72:341–364.

Harris, David R. and Jeremiah Joseph Sim. 2002. "Who Is Multiracial? Assessing the Complexity of Lived Race." *American Sociological Review* 67:614–627.

Hart, Chris. 1998. *Doing a Literature Review: Releasing the Social Science Research Imagination.* London: Sage.

Hawkins, Donald N., Paul R. Amato, and Valarie King. 2007. "Nonresident Father Involvement and Adolescent Well-being: Father Effects or Child Effects?" *American Sociological Review* 72:990–1010.

Heath, Christian and Paul Luff. 2008. "Video and the Analysis of Work and Interaction." Pp. 493–505 in *The Sage Handbook of Social Research Methods,* edited by Pertti Alasuutari, Leonard Bickman, and Julia Brannen. Thousand Oaks, CA: Sage.

Heaton, Janet. 2008. "Secondary Analysis of Qualitative Data." Pp. 506–535 in *The Sage Handbook of Social Research Methods,* edited by Pertti Alasuutari, Leonard Bickman, and Julia Brannen. Thousand Oaks, CA: Sage.

Heckathorn, Douglas D. 1997. "Respondent-Driven Sampling: A New Approach to the Study of Hidden Populations." *Social Problems* 44:174–199.

Heckathorn, Douglas D. 2002. "Respondent-Driven Sampling II: Deriving Valid Population Estimates from Chain-Referral Samples of Hidden Populations." *Social Problems* 49:11–34.

Heckman, James, Neil Hohmann, and Jeffrey Smith. 2000. "Substitution and Dropout Bias in Social Experiments: A Study of an Influential Social Experiment." *The Quarterly Journal of Economics* 115:651–694.

Hedström, Peter and Richard Swedberg (Eds.). 1998. *Social Mechanisms: An Analytical Approach to Social Theory.* Cambridge, UK: Cambridge University Press.

Herek, Gregory. 1995. "Developing a Theoretical Framework and Rationale for a Research Proposal." Pp. 85–91 in *How to Write a Successful Research Grant Application: A Guide for Social and Behavioral Scientists,* edited by Willo Pequegnat and Ellen Stover. New York: Plenum Press.

Hesse-Biber, Sharlene Nagy, and Patricia Lina Leavy. 2007. *Feminist Research Practice: A Primer.* Thousand Oaks, CA: Sage.

Hesse-Biber, Sharon. 1989. "Eating Problems and Disorders in a College Population: Are College Women's Eating Problems a New Phenomenon?" *Sex Roles* 20:71–89.

Hirsch, Kathleen. 1989. *Songs From the Alley.* New York: Doubleday.

Hirschel, David, Eve Buzawa, Don Faggiani, and April Pattavina. 2008. "Domestic Violence and Mandatory Arrest Laws: To What Extent Do They Influence Police Arrest Decisions?" *Journal of Criminal Law & Criminology* 98(1):255–298.

Ho, D. Y. F. 1996. "Filial Piety and Its Psychological Consequences." Pp. 155–165 in *Handbook of Chinese Psychology,* edited by M. H. Bond. Hong Kong: Oxford University Press.

Holbrook, Allyson L., Melanie C. Green, and Jon A. Krosnick. 2003. "Telephone Versus Face-to-Face Interviewing of National Probability Samples With Long Questionnaires: Comparisons of Respondent Satisficing and Social Desirability Response Bias." *Public Opinion Quarterly* 60:58–88.

Hollingsworth, T. H. 1972. "The Importance of the Quality of Data in Historical Demography." Pp. 71–86 in *Population and Social Change,* edited by D. V. Glass and Roger Revelle. London: Edward Arnold.

Holmes, Steven A. 2000. "Stronger Response by Minorities Helps Improve Census Reply Rate." *The New York Times,* May 4, pp. A1, A22.

Holmes, Steven A. 2001a. "Census Officials Ponder Adjustments Crucial to Redistricting." *The New York Times,* February 12, p. A17.

Holmes, Steven A. 2001b. "The Confusion Over Who We Are." *The New York Times,* June 3, p. WK 1.

Horney, Julie, D. Wayne Osgood, and Ineke Haen Marshall. 1995. "Criminal Careers in the Short-Term: Intra-Individual Variability in Crime and Its Relation to Local Life Circumstances." *American Sociological Review* 60:655–673.

Horowitz, Carol R., Mimsie Robinson, and Sarena Seifer. 2009. "Community-Based Participatory Research From the Margin to the Mainstream: Are Researchers Prepared?" *Circulation* 119:2633–2642.

Howell, James C. 2003. *Preventing and Reducing Juvenile Delinquency: A Comprehensive Framework.* Thousand Oaks, CA: Sage.

Hoyle, Carolyn and Andrew Sanders. 2000. "Police Response to Domestic Violence: From Victim Choice to Victim Empowerment." *British Journal of Criminology* 40:14–26.

HRAF. 2005. *eHRAF Collection of Ethnography: Web.* New Haven, CT: Yale University. Retrieved July 3, 2005, from www.yale.edu/hraf/collections_body_ethnoweb.htm

Hrobjartsson, Asbjorn and Peter C. Gotzsche. 2001. "Is the Placebo Powerless? An Analysis of Clinical Trials Comparing Placebo With No Treatment." *New England Journal of Medicine* 344:1594–1602.

Huberman, A. Michael and Matthew B. Miles. 1994. "Data Management and Analysis Methods." Pp. 428–444 in *Handbook of Qualitative Research,* edited by Norman K. Denzin and Yvonna S. Lincoln. Thousand Oaks, CA: Sage.

Huff, Darrell. 1954. *How to Lie With Statistics.* New York: Norton.

Humphrey, Nicholas. 1992. *A History of the Mind: Evolution and the Birth of Consciousness.* New York: Simon & Schuster.

Humphreys, Laud. 1970. *Tearoom Trade: Impersonal Sex in Public Places.* Chicago: Aldine de Gruyter.

Humphries, Courtney. 2011. "Deeply Conflicted: How Can We Insulate Ourselves from Conflicts of Interest? The Most Popular Solution—Disclosing Them—Turns Out Not to Help." *Boston Globe,* May 15:K1, K3.

Hunt, Morton. 1985. *Profiles of Social Research: The Scientific Study of Human Interactions.* New York: Russell Sage Foundation.

Huston, Patricia and C. David Naylor. 1996. "Health Services Research: Reporting on Studies Using Secondary Data Sources." *Canadian Medical Association Journal* 155:1697–1702.

Hyvärinen, Matti. 2008. "Analyzing Narratives and Story-Telling." Pp. 447–460 in *The Sage Handbook of Social Research Methods,* edited by Pertti Alasuutari, Leonard Bickman, and Julia Brannen. Thousand Oaks, CA: Sage.

Iezzoni, Lisa I. 1997. "How Much Are We Willing to Pay for Information About Quality of Care?" *Annals of Internal Medicine* 126:391–393.

Innstrand, Siw Tone, Geir Arild Espries, and Reidar Mykletun. 2004. "Job Stress, Burnout and Job Satisfaction: An Intervention Study of Staff Working with People with Intellectual Disabilities." *Journal of Applied Research in Intellectual Disabilities* 17:119–126.

Internetworldstats.com. 2011. *Internet World Stats: Usage and Population Statistics.* Retrieved March 19, 2011, from http://www.internetworldstats.com/

Inter-University Consortium for Political and Social Research. 1996. *Guide to Resources and Services 1995–1996.* Ann Arbor, MI: ICPSR.

Irvine, Leslie. 1998. "Organizational Ethics and Fieldwork Realities: Negotiating Ethical Boundaries in Codependents Anonymous." Pp. 167–183 in *Doing Ethnographic Research: Fieldwork Settings.* Thousand Oaks, CA: Sage.

Jalbert, Sarah Kuck, William Rhodes, Christopher Flygare, and Michael Kane.

2010. "Testing Probation Outcomes in an Evidence-Based Practice Setting: Reduced Caseload Size and Intensive Supervision Effectiveness." *Journal of Offender Rehabilitation*, 49:233–253

James, Nalita and Hugh Busher. 2009. *Online Interviewing*. Thousand Oaks, CA: Sage.

Jarvis, Helen. 1997. "Housing, Labour Markets and Household Structure: Questioning the Role of Secondary Data Analysis in Sustaining the Polarization Debate." *Regional Studies* 31:521–531.

Jervis, Robert. 1996. "Counterfactuals, Causation, and Complexity." Pp. 309–316 in *Counterfactual Thought Experiments in World Politics: Logical, Methodological, and Psychological Perspectives*, edited by Philip E. Tetlock and Aaron Belkin. Princeton, NJ: Princeton University Press.

Jesnadum, Anick. 2000. Researchers Fear Privacy Breaches With Online Research." *Digital Mass*. Retrieved September 15, 2000, from www.digitalmass.com/ news/ daily/09/15/researchers.html

Johnson, Akilah. 2011. "From a Shelter to a Place Called Home." *The Boston Globe*, January 10:B1.

Johnson, David and Merry Bullock. 2009. "The Ethics of Data Archiving: Issues From Four Perspectives." Pp. 214–228 in *The Handbook of Social Research Ethics*, edited by Donna M. Mertens and Pauline E. Ginsberg. Thousand Oaks, CA: Sage.

Jones, Stephen R. G. 1992. "Was There a Hawthorne Effect?" *American Journal of Sociology* 98:451–468.

Kagay, Michael R. with Janet Elder. 1992. "Numbers Are No Problem for Pollsters. Words Are." *The New York Times*, October 9, p. E5.

Kale-Lostuvali, Elif. 2007. "Negotiating State Provision: State-Citizen Encounters in the Aftermath of the İzmit Earthquake." *The Sociological Quarterly* 48:745–767.

Kaplan, Fred. 2002. "NY Continues to See Plunge in Number of Felonies." *The Boston Globe*, April 15, p. A3.

Kaufman, Sharon R. 1986. *The Ageless Self: Sources of Meaning in Late Life*. Madison: University of Wisconsin Press.

Keeter, Scott. 2008. *Survey Research and Cell Phones: Is There a Problem?* Presentation to the Harvard Program on Survey Research Spring Conference, New Technologies and Survey Research. Cambridge, MA: Institute of Quantitative Social Science, Harvard University, May 9.

Kemmis, Stephen and Robin McTaggart. 2005. "Participatory Action Research: Communicative Action and the Public Sphere." Pp. 559–603 in *The Sage Handbook of Qualitative Research*, 3rd ed., edited by Norman K. Denzin and Yvonna S. Lincoln. Thousand Oaks, CA: Sage.

Kennedy, David M., Anthony M. Piehl and Anne A. Braga. 1996. "Youth Violence in Boston: gu n Markets, Serious Youth Offenders, and a Use-Reduction Strategy." *Law and Contemporary Problems* 59:147–196.

Kenney, Charles. 1987. "They've Got Your Number." *The Boston Globe Magazine*, August 30, pp. 12, 46–56, 60.

Kershaw, David and Jerilyn Fair. 1976. *The New Jersey Income-Maintenance Experiment*, vol. 1. New York: Academic Press.

Kershaw, Sarah. 2000. "In a Black Community, Mistrust of Government Hinders Census." *The New York Times*, May 16, p. A20.

Kifner, John. 1994. "Pollster Finds Error on Holocaust Doubts." *The New York Times*, May 20, p. A12.

Kincaid, Harold. 1996. *Philosophical Foundations of the Social Sciences: Analyzing Controversies in Social Research*. Cambridge, UK: Cambridge University Press.

King, Gary, Robert O. Keohane, and Sidney Verba. 1994. *Scientific Inference in Qualitative Research*. Princeton, NJ: Princeton University Press.

King, Miriam L. and Diana L. Magnuson. 1995. "Perspectives on Historical U.S. Census Undercounts." *Social Science History* 19:455–466.

King, Nigel and Christine Horrocks. 2010. *Interviews in Qualitative Research*. Thousand Oaks, CA: Sage.

Kiser, Edgar and Michael Hechter. 1991. "The Role of General Theory in Comparative-Historical Sociology." *American Journal of Sociology* 97:1–30.

Kitchener, Karen Strohm and Richard F. Kitchener. 2009. "Social Science Research Ethics: Historical and Philosophical Issues." Pp. 5–22 in *The Handbook of Social Research Ethics*, edited by Donna M. Mertens and Pauline E. Ginsberg. Thousand Oaks, CA: Sage.

Klein, Malcolm W. 1971. *Street Gangs and Street Workers*. Englewood Cliffs, NJ: Prentice Hall.

Klinenberg, Eric. 2002. *Heat Wave: A Social Autopsy of Disaster in Chicago*. Chicago: University of Chicago Press.

Knight, John R., Henry Wechsler, Meichun Kuo, Mark Seibring, E. R. Weitzman, and M. A. Schuckit. 2002. "Alcohol Abuse and Dependence Among U.S. College Students." *Journal of Studies on Alcohol* 63:263–270.

Koegel, Paul. 1987. *Ethnographic Perspectives on Homeless and Homeless Mentally Ill Women*. Washington, DC: Alcohol, Drug Abuse, and Mental Health Administration, Public Health Service, U.S. Department of Health and Human Services.

Kohn, Alfie. 2008. "Who's Cheating Whom?" *The Education Digest* 73:4–11.

Kohn, Melvin L. 1987. "Cross-National Research as an Analytic Strategy." *American Sociological Review* 52:713–731.

Kohut, Andrew. 1988. "Polling: Does More Information Lead to Better Understanding?" *The Boston Globe*, November 7, p. 25.

Kohut, Andrew. 2008. "Getting It Wrong." *The New York Times*, January 10, p. A27.

Kolbert, Elizabeth. 1992. "Test-Marketing a President." *The New York Times Magazine*, August 30, pp. 18–21, 60, 68, 72.

Kollock, Peter and Marc A. Smith. 1999. "Communities in Cyberspace." Pp. 3–25 in *Communities in Cyberspace,* edited by Peter Kollock and Marc A. Smith. New York: Routledge.

Korn, James H. 1997. *Illusions of Reality: A History of Deception in Social Psychology.* Albany: State University of New York Press.

Kotkin, Stephen. 2002. "A World War Among Professors." *The New York Times,* September 7, Arts pp. 1, 17.

Kozinets, Robert V. 2010. *Netnography: Doing Ethnographic Research Online.* Thousand Oaks, CA: Sage.

Kraemer, Helena Chmura and Sue Thiemann. 1987. *How Many Subjects? Statistical Power Analysis in Research.* Newbury Park, CA: Sage.

Kraut, Robert, Sara Kiesler, Bonka Boneva, Jonathon Cummings, Vicki Helgeson, and Anne Crawford. 2002. "Internet Paradox Revisited." *Journal of Social Issues* 58:49–74.

Kreuter, Frauke, Stanley Presser, and Roger Tourangeau. 2008. "Social Desirability Bias in CATI, IVR, and Web Surveys: The Effects of Mode and Question Sensitivity." *Public Opinion Quarterly* 72:847–865.

Krosnick, Jon A. 1999. "Survey Research." *Annual Review of Psychology* 50:537–567.

Krosnick, Jon A. 2006. *The Handbook of Questionnaire Design: Insights From Social and Cognitive Psychology.* Eric M. Mindich Encounters With Authors Symposium at Institute for Quantitative Social Science, Harvard University, Cambridge MA, January 19–21.

Krueger, Richard A. and Mary Anne Casey. 2009. *Focus Groups: A Practical Guide for Applied Research,* 4th ed. Thousand Oaks, CA: Sage.

Kubey, Robert. 1990. "Television and the Quality of Family Life." *Communication Quarterly* 38(Fall):312–324.

Kuhn, Thomas S. 1970. *The Structure of Scientific Revolutions*, 2nd ed. Chicago: University of Chicago Press.

Kuzel, Anton J. 1999. "Sampling in Qualitative Inquiry." Pp. 33–45 in *Doing Qualitative Research,* 2nd ed., edited by Benjamin F. Crabtree and William L. Miller. Thousand Oaks, CA: Sage.

Kvale, Steinar. 1996. *Interviews: An Introduction to Qualitative Research Interviewing.* Thousand Oaks, CA: Sage.

Kvale, Steinar. 2002. "The Social Construction of Validity." Pp. 299–325 in *The Qualitative Inquiry Reader,* edited by Norman K. Denzin and Yvonna S. Lincoln. Thousand Oaks, CA: Sage.

Labaw, Patricia J. 1980. *Advanced Questionnaire Design.* Cambridge, MA: ABT Books.

Langford, Terri. 2000. "Census Workers in Dallas Find the Well-Off Hard to Count." *The Boston Globe,* June 1, p. A24.

Larence, Eileen Regan. 2006. *Prevalence of Domestic Violence, Sexual Assault, Dating Violence, and Stalking. Letter to Congressional Committees.* Washington, D.C.: United States Government Accountability Office.

LaRossa, Ralph. Parenthood in Early Twentieth-Century America Project (PETCAP), 1900–1944 [Computer file]. Atlanta, GA: Georgia State University [producer], 1995. Ann Arbor, MI: Inter-University Consortium for Political and Social Research [distributor], 1997.

Larson, Calvin J. 1993. *Pure and Applied Sociological Theory: Problems and Issues.* New York: Harcourt Brace Jovanovich.

Lathrop, Barnes F. 1968. "History from the Census Returns." Pp. 79–101 in *Sociology and History: Methods,* edited by Seymour Martin Lipset and Richard Hofstadter. New York: Basic Books.

Latour, Francie. 2002. "Marching Orders: After 10 Years, State Closes Prison Boot Camp." *Boston Sunday Globe,* June 16, pp. B1, B7.

Latour, Francie. 2011. "The Bad Mother Complex: Why Are so Many Working Mothers Haunted by Constant Guilt?" *Boston Sunday Globe* March 13:K1, K2.

Lavin, Michael R. 1994. *Understanding the 1990 Census: A Guide for Marketers, Planners, Grant Writers and Other Data Users.* Kenmore, NY: Epoch Books.

Lavrakas, Paul J. 1987. *Telephone Survey Methods: Sampling, Selection, and Supervision.* Newbury Park, CA: Sage.

Layte, Richard and Christopher T. Whelan. 2003. "Moving In and Out of Poverty: The Impact of Welfare Regimes on Poverty Dynamics in the EU." *European Societies* 5:167–191.

Leakey, Tricia, Kevin B. Lunde, Karin Koga, and Karen Glanz. 2004. "Written Parental Consent and the Use of Incentives in a Youth Smoking Prevention Trial: A Case Study From Project SPLASH." *American Journal of Evaluation* 25:509–523.

LeDuc, Lawrence, Richard G. Niemi, and Pippa Norris (Eds.). 1996. *Comparing Democracies: Elections and Voting in Global Perspective.* Thousand Oaks, CA: Sage.

Lehrer, Jonah. 2010. "The Truth Wears Off: Is There Something Wrong with the Scientific Method?" *The New Yorker,* December 13:52-57.

Lempert, Richard. 1989. "Humility Is a Virtue: On the Publicization of Policy-Relevant Research." *Law & Society Review* 23: 146–161.

Lempert, Richard and Joseph Sanders. 1986. *An Invitation to Law and Social Science: Desert, Disputes, and Distribution.* New York: Longman.

Levine, James P. 1976. "The Potential for Crime Overreporting in Criminal Victimization Surveys." *Criminology* 14:307–330.

Levitas, Ruth and Will Guy. 1996. "Introduction." Pp. 1–6 in *Interpreting Official Statistics,* edited by Ruth Levitas and Will Guy. New York: Routledge.

Levitt, Heidi M., Rebecca Todd Swanger, and Jenny B. Butler. 2008. "Male Perpetrators' Perspectives on Intimate Partner Violence, Religion, and Masculinity." *Sex Roles* 58:435–448.

Levy, Paul S. and Stanley Lemeshow. 1999. *Sampling of Populations: Methods and Applications,* 3rd ed. New York: Wiley.

Lewin, Tamar. 2001a. "Income Education Is Found to Lower Risk of New Arrest." *The New York Times,* November 16, p. A18.

Lewin, Tamar. 2001b. "Surprising Result in Welfare-to-Work Studies." *The New York Times,* July 31, p. A16.

Lewis, Kevin, Nicholas Christakis, Marco Gonzalez, Jason Kaufman, and Andreas Wimmer. 2008. "Tastes, Ties, and Time: A New Social Network Dataset Using Facebook.com." *Social Networks* 30(4):330–342.

Lewis-Beck, Michael S., Alan Bryman, and Tim Futing Liao (Eds.). 2004. *The Sage Encyclopedia of Social Science Research Methods,* vol. 1. Thousand Oaks, CA: Sage.

Lieberson, Stanley. 1985. *Making It Count: The Improvement of Social Research and Theory.* Berkeley: University of California Press.

Lieberson, Stanley. 1991. "Small N's and Big Conclusions: An Examination of the Reasoning in Comparative Studies Based on a Small Number of Cases." *Social Forces* 70:307–320.

Lillard, Lee A. and Constantijn W. A. Panis. 1998. "Panel Attrition From the Panel Study of Income Dynamics: Household Income, Marital Status, and Mortality." *The Journal of Human Resources* 33:437–457.

Lincoln, Yvonna S. 2009. "Ethical Practices in Qualitative Research." Pp. 150–169 in *The Handbook of Social Research Ethics,* edited by Donna M. Mertens and Pauline E. Ginsberg. Thousand Oaks, CA: Sage.

Lindsay, Sally, Simon Smith, Frances Bell, and Paul Bellaby. 2007. "Tackling the Digital Divide: Exploring the Impact of ICT on Managing Heart Conditions in a Deprived Area." *Information, Communication & Society* 10:95–114.

Ling, Rich and Gitte Stald. 2010. "Mobile Communities: Are We Talking About a Village,

a Clan, or a Small Group?" *American Behavioral Scientist* 53(8):113–1147.

Link, Bruce G., Jo C. Phelan, Ann Stueve, Robert E. Moore, Michaeline Brenahan, and Elmer L. Struening. 1996. "Public Attitudes and Beliefs About Homeless People." Pp. 143–148 in *Homelessness in America,* edited by Jim Baumohl. Phoenix, AZ: Oryx.

Link, Michael. 2008. *Solving the Problems Cell Phones Create for Survey Research. Presentation to the Harvard Program on Survey Research Spring Conference, New Technologies and Survey Research.* Cambridge, MA: Institute of Quantitative Social Science, Harvard University, May 9.

Lipset, Seymour Martin. 1968. *Revolution and Counterrevolution.* New York: Basic Books.

Lipset, Seymour Martin. 1990. *Continental Divide: The Values and Institutions of the United States and Canada.* London: Routledge.

Lipsey, Mark W. and David B. Wilson. 2001. *Practical Meta-Analysis.* Thousand Oaks, CA: Sage.

Lipsky, Michael. 1980. *Street-Level Bureaucracy.* New York: Russell Sage Foundation.

Litwin, Mark S. 1995. *How to Measure Survey Reliability and Validity.* Thousand Oaks, CA: Sage.

Loader, Brian D., Steve Muncer, Roger Burrows, Nicolas Pleace, and Sarah Nettleton. 2002. "Medicine on the Line? Computer-Mediated Social Support and Advice for People With Diabetes." *International Journal of Social Welfare* 11:53–65.

Locke, Lawrence F., Stephen J. Silverman, and Waneen Wyrick Spirduso. 1998. *Reading and Understanding Research.* Thousand Oaks, CA: Sage.

Locke, Lawrence F., Waneen Wyrick Spirduso, and Stephen J. Silverman. 2000. *Proposals That Work: A Guide for Planning Dissertations and Grant Proposals,* 4th ed. Thousand Oaks, CA: Sage.

Lockwood, Daniel. *Violent Incidents Among Selected Public School Students in Two Large Cities of the South and the Southern Midwest, 1995:* [United States] [Computer file]. ICPSR version. Atlanta, GA: Clark Atlantic University [producer], 1996. Ann Arbor, MI: Inter-university Consortium for Political and Social Research [distributor], 1998.

Lofland, John and Lyn H. Lofland. 1984. *Analyzing Social Settings: A Guide to Qualitative Observation and Analysis,* 2nd ed. Belmont, CA: Wadsworth.

Lyall, Sarah. 2004. "Does Queen Get Her Mail on Time? You've Got to Wonder." *The New York Times,* May 28, p. A4.

Lynch, Michael and David Bogen. 1997. "Sociology's Asociological 'Core': An Examination of Textbook Sociology in Light of the Sociology of Scientific Knowledge." *American Sociological Review* 62:481–493.

Mabry, Linda. 2008. "Case Study in Social Research." Pp. 214–227 in *The Sage Handbook of Social Research Methods,* edited by Pertti Alasuutari, Leonard Bickman, and Julia Brannen. Thousand Oaks, CA: Sage.

Madden, Raymond. 2010. *Being Ethnographic: A Guide to the Theory and Practice of Ethnography.* Thousand Oaks, CA: Sage.

Mangione, Thomas W. 1995. *Mail Surveys: Improving the Quality.* Thousand Oaks, CA: Sage.

Manza, Jeff, Clem Brooks, and Michael Sauder. 2005. "Money, Participation, and Votes: Social Cleavages and Electoral Politics." Pp. 201–226 in *The Handbook of Political Sociology: States, Civil Societies, and Globalization,* edited by Thomas Janoski, Robert R. Alford, Alexander M. Hicks, and Mildred A. Schwartz. New York: Cambridge University Press.

Margolis, Eric. 2004. "Looking at Discipline, Looking at Labour: Photographic Representations of Indian Boarding Schools." *Visual Studies* 19:72–96.

Marini, Margaret Mooney and Burton Singer. 1988. "Causality in the Social Sciences." Pp. 347–409 in *Sociological Methodology,* vol. 18, edited by Clifford C. Clogg. Washington, DC: American Sociological Association.

Mark, Melvin M. and Chris Gamble. 2009. "Experiments, Quasi-Experiments, and Ethics." Pp. 198–213 in *The Handbook of Social Research Ethics,* edited by Donna M. Mertens and Pauline E. Ginsberg. Thousand Oaks, CA: Sage.

Markham, Annette N. 2008. "The Methods, Politics, and Ethics of Representation in Online Ethnography." Pp. 247–284 in *Collecting and Interpreting Qualitative Materials,* 3rd ed., edited by Norman Denzin and Yvonna S. Lincoln. Thousand Oaks, CA: Sage.

Markoff, John. 2005. "Transitions to Democracy." Pp. 384–403 in *The Handbook of Political Sociology: States, Civil Societies, and Globalization,* edited by Thomas Janoski, Robert R. Alford, Alexander M. Hicks, and Mildred A. Schwartz. New York: Cambridge University Press.

Marsden, Peter V. 1987. "Core Discussion Networks of Americans." *American Sociological Review* 52: 122–131.

Marshall, Gary D. and Philip G. Zimbardo. 1979. "Affective Consequences of Inadequately Explained Physiological Arousal." *Journal of Personality and Social Psychology* 37:970–988.

Martin, Lawrence L. and Peter M. Kettner. 1996. *Measuring the Performance of Human Service Programs.* Thousand Oaks, CA: Sage.

Martin, Linda G. and Kevin Kinsella. 1995. "Research on the Demography of Aging in Developing Countries." Pp. 356–403 in *Demography of Aging,* edited by Linda G. Martin and Samuel H. Preston. Washington, DC: National Academy Press.

Marx, Karl. 1967. *Capital: A Critique of Political Economy.* New York: International Publishers.

Marx, Karl and Friedrich Engels. 1961. "The Communist Manifesto." Pp. 13–44 in *Essential Works of Marxism,* edited by Arthur P. Mendel. New York: Bantam.

Matt, Georg E. and Thomas D. Cook. 1994. "Threats to the Validity of Research Syntheses." Pp. 503–520 in *The Handbook of Research Synthesis,* edited by Harris Cooper and Larry V. Hedges. New York: Russell Sage Foundation.

Maxwell, Joseph A. 1996. *Qualitative Research Design: An Interactive Approach.* Thousand Oaks, CA: Sage.

Maynard, Douglas W., Jeremy Freese, and Nora Cate Schaeffer. 2010. "Calling for Participation: Requests, Blocking Moves, and Rational (Inter)action in Survey Introductions." *American Sociological Review* 75:791-814.

Mayrl, Damon, Ben Moodie, Jon Norman, Jodi Short, Sarah Staveteig, and Cinzia Solari. 2004. "A Theory of Relativity." *Contexts* 3:10.

McCarty, John A. and L. J. Shrum. 2000. "The Measurement of Personal Values in Survey Research: A Test of Alternative Rating Procedures." *Public Opinion Quarterly* 64:271–298.

McIntyre, Alice. 2008. *Participatory Action Research.* Thousand Oaks, CA: Sage.

McLellan, A. Thomas, Lester Luborsky, John Cacciola, Jeffrey Griffith, Frederick Evans, Harriet L. Barr, and Charles P. O'Brien. 1985. "New Data from the Addiction Severity Index: Reliability and Validity in Three Centers." *The Journal of Nervous and Mental Disease* 173(7):412–423.

McPhail, Clark and John McCarthy. 2004. "Who Counts and How: Estimating the Size of Protests." *Contexts* 3:12–18.

McPherson, Miller, Lynn Smith-Lovin, and Matthew E. Brashears. 2006. "Social Isolation in America: Changes in Core Discussion Networks Over Two Decades." *American Sociological Review* 71: 353–375.

Meghani, S. H., Brooks, J., Gipson-Jones, T., Waite, R., Whitefield-Harris, L., Deatrick, J. (2009). Patient-provider Race-Concordance: Does it Matter in Improving Minority Patients' Health Outcomes. *Ethnicity & Health, 14(1),* 107–130.

Merton, Robert K., Marjorie Fiske, and Patricia L. Kendall. 1956. *The Focused Interview.* Glencoe, IL: Free Press.

Mieczkowski, Tom. 1997. "Hair Assays and Urinalysis Results for Juvenile Drug Offenders." *National Institute of Justice Research Preview.* Washington, DC: U.S. Department of Justice.

Milbrath, Lester and M. L. Goel. 1977. *Political Participation,* 2nd ed. Chicago: Rand McNally.

Miles, Matthew B. and A. Michael Huberman. 1994. *Qualitative Data Analysis,* 2nd ed. Thousand Oaks, CA: Sage.

Milgram, Stanley. 1963. "Behavioral Study of Obedience." *Journal of Abnormal and Social Psychology* 67:371–478.

Milgram, Stanley. 1964. "Issues in the Study of Obedience: A Reply to Baumrind." *American Psychologist* 19:848–852.

Milgram, Stanley. 1965. "Some Conditions of Obedience and Disobedience to Authority." *Human Relations* 18:57–76.

Milgram, S. 1974. *Obedience to Authority: An Experimental View.* New York: Harper & Row.

Milgram, S. 1977. "Subject Reaction: The Neglected Factor in the Ethics of Experimentation." *Hastings Law Review,* October, pp. 19–23 (as cited in Cave and Holm 2003:32).

Milgram, Stanley. 1992. *The Individual in a Social World: Essays and Experiments,* 2nd ed. New York: McGraw-Hill.

Mill, John Stuart. 1872. *A System of Logic: Ratiocinative and Inductive,* 8th ed., vol. 2. London: Longmans, Green, Reader, & Dyer.

Miller, Arthur G. 1986. *The Obedience Experiments: A Case Study of Controversy in Social Science.* New York: Praeger.

Miller, Delbert C. 1991. *Handbook of Research Design and Social Measurement,* 5th ed. Newbury Park, CA: Sage.

Miller, JoAnn. 2003. "An Arresting Experiment: Domestic Violence Victim Experiences and Perceptions." *Journal of Interpersonal Violence* 18:695–716.

Miller, Susan. 1999. *Gender and Community Policing: Walking the Talk.* Boston: Northeastern University Press.

Miller, Walter B. 1992. *Crime by Youth Gangs and Groups in the United States.* Washington, DC: Office of Juvenile Justice and Delinquency Prevention.

Miller, William L. and Benjamin F. Crabtree. 1999a. "Clinical Research: A Multimethod Typology and Qualitative Roadmap." Pp. 3–30 in *Doing Qualitative Research,* 2nd ed., edited by Benjamin F. Crabtree and William L. Miller. Thousand Oaks, CA: Sage.

Miller, William L. and Benjamin F. Crabtree. 1999b. "The Dance of Interpretation." Pp. 127–143 in *Doing Qualitative Research,* edited by Benjamin F. Crabtree and William L. Miller. Thousand Oaks, CA: Sage.

Miller, William L. and Benjamin F. Crabtree. 1999c. "Depth Interviewing." Pp. 89–107 in *Doing Qualitative Research,* 2nd ed., edited by Benjamin F. Crabtree and William L. Miller. Thousand Oaks, CA: Sage.

Mills, C. Wright. 1959. *The Sociological Imagination.* New York: Oxford University Press.

Minkler, Meredith. 2000. "Using Participatory Action Research to Build Healthy Communities." *Public Health Reports* 115:191–197.

Mirowsky, John. 1999. *Aging, Status, and the Sense of Control: Competing Continuation* (Proposal to the National Institute of Aging). Urbana: University of Illinois Press.

Mirowsky, John and Catherine E. Ross. 1999. "Economic Hardship Across the Life Course." *American Sociological Review* 64:548–569.

Mirowsky, John and Catherine E. Ross. 2001. *Aging, Status, and the Sense of Control (ASOC), 1995, 1998, 2001 [United States] Questionnaire* (ICPSR 3334). Ann Arbor, MI: Inter-University Consortium for Political and Social Research.

Mirowsky, John and Catherine E. Ross. 2003. *Education, Social Status, and Health.* New York: Aldine de Gruyter.

Mirowsky, John and Catherine Ross. n.d. *Informant Questionnaire.* Unpublished questionnaire.

Mitchell, Richard G., Jr. 1993. *Secrecy and Fieldwork.* Newbury Park, CA: Sage.

Moe, Angela M. 2007. "Silenced Voices and Structural Survival—Battered Women's Help Seeking." *Violence Against Women* 13(7):676–699.

Mohr, Lawrence B. 1992. *Impact Analysis for Program Evaluation.* Newbury Park, CA: Sage.

Monkkonen, Eric H. 1994. "Introduction." Pp. 1–8 in *Engaging the Past: The Uses of History Across the Social Sciences.* Durham, NC: Duke University Press.

Mooney, Christopher Z. and Mei Hsien Lee. 1995. "Legislating Morality in the American States: The Case of Abortion Regulation Reform." *American Journal of Political Science* 39:599–627.

Moore, Spencer, Mark Daniel, Laura Linnan, Marci Campbell, Salli Benedict, and Andrea Meier. 2004. "After Hurricane Floyd Passed: Investigating the Social Determinants of Disaster Preparedness and Recovery." *Family and Community Health* 27:204–217.

Morrill, Calvin, Christine Yalda, Madeleine Adelman, Michael Musheno, and Cindy Bejarano. 2000. "Telling Tales in School: Youth Culture and Conflict Narratives." *Law & Society Review* 34:521–565.

Mosher, Clayton J., Terance D. Miethe, and Dretha M. Phillips. 2002. *The Mismeasure of Crime.* Thousand Oaks, CA: Sage.

Muhr, Thomas and Susanne Friese. 2004. *User's Manual for ATLAS.ti 5.0,* 2nd ed. Berlin: Scientific Software Development.

Myers, Steven Lee. 2002. "Russia Takes Stock of a Nation's Transformation." *The New York Times,* September 29, p. 3.

Myers, Steven Lee. 2010. "Delays in a Head Count Keep Crucial Numbers a Matter of Guesswork." *The New York Times,* December 7:A10.

Nagourney, Adam. 2002. "Cellphones and Caller ID Are Making It Harder for Pollsters to Pick a Winner." *The New York Times,* November 5, p. A20.

Nakonezny, Paul A., Rebecca Reddick, and Joseph Lee Rodgers. 2004. "Did Divorces Decline After the Oklahoma City Bombing?" *Journal of Marriage and Family* 66:90–100.

Narayan, Sowmya and Jon A. Krosnick. 1996. "Education Moderates Some Response Effects in Attitude Measurement." *Public Opinion Quarterly* 60:58–88.

National Geographic Society. 2000. *Survey 2000.* Available at http://survey2000 .nationalgeographic.com

National Institute of Alcohol Abuse and Alcoholism. 1994. "Alcohol-Related Impairment." *Alcohol Alert* 25 (July):1–5.

National Institute of Alcohol Abuse and Alcoholism. 1997. "Alcohol Metabolism." *Alcohol Alert* 35 (January):1–4.

National Opinion Research Center. 2011. General Social Survey (GSS). Chicago: NORC at the University of Chicago. http://www .norc.org/Research/Projects/Pages/general-social-survey-aspx. Downloaded August 31, 2011.

National Technical Information Service, U.S. Department of Commerce. 1993. *Directory of U.S. Government Data Files for Mainframes and Microcomputers.* Washington, DC: Federal Computer Products Center,

National Technical Information Service, U.S. Department of Commerce.

Navarro, Mireya. 2003. "Going Beyond Black and White, Hispanics in Census Pick 'Other.'" *The New York Times,* November 9:1, 21.

Needleman, Carolyn. 1981. "Discrepant Assumptions in Empirical Research: The Case of Juvenile Court Screening." *Social Problems* 28 (February):247–262.

Neuendorf, Kimberly A. 2002. *The Content Analysis Guidebook.* Thousand Oaks, CA: Sage.

Newbury, Darren. 2005. "Editorial: The Challenge of Visual Studies." *Visual Studies* 20:1–3.

Newport, Frank. 1992. "Look at Polls as a Fever Chart of the Electorate." Letter to the Editor, *The New York Times,* November 6, p. A28.

Nie, Norman H. and Lutz Erbring. 2000. *Internet and Society: A Preliminary Report.* Palo Alto, CA: Stanford Institute for the Quantitative Study of Society.

Nie, Norman H. and Lutz Erbring. 2002. "Internet and Society: A Preliminary Report." *IT & Society* 1: 275–283.

NOAA (National Oceanic & Atmospheric Administration). 2005. *Summary of Hurricane Katrina.* Washington, DC: U.S. Department of Commerce. Retrieved May 25, 2008, from www.ncdc.noaa.gov/oa/climate/research/2005/katrina.html

Nordanger, Dag. 2007. "Discourses of Loss and Bereavement in Tigray, Ethiopia." *Culture, Medicine and Psychiatry* 31:173–194.

Norris, Pippa. 2004. "The Bridging and Bonding Role of Online Communities." Pp. 31–41 in *Society Online: The Internet in Context,* edited by Philip N. Howard and Steve Jones. Thousand Oaks, CA: Sage.

Nunberg, Geoffrey. 2002. "The Shifting Lexicon of Race." *The New York Times,* December 22, p. WK3.

Ogburn, William F. 1930. "The Folkways of a Scientific Sociology," *Scientific Monthly* 30:300–306.

Olzak, Susan, Suzanne Shanahan, and Elizabeth H. McEneaney. 1996. "Poverty, Segregation, and Race Riots: 1960 to 1993." *American Sociological Review* 61:590–613.

Onishi, Norimitsu. 2003. "Crime Rattles Japanese Calm, Attracting Politicians' Notice." *The New York Times,* September 7:A1, A4.

Orcutt, James D. and J. Blake Turner. 1993. "Shocking Numbers and Graphic Accounts: Quantified Images of Drug Problems in the Print Media." *Social Problems* 49(May):190–206.

Orr, Larry L. 1999. *Social Experiments: Evaluating Public Programs With Experimental Methods.* Thousand Oaks, CA: Sage.

Orshansky, Mollie. 1977. "Memorandum for Daniel P. Moynihan. Subject: History of the Poverty Line." Pp. 232–237 in *The Measure of Poverty. Technical Paper I: Documentation of Background Information and Rationale for Current Poverty Matrix,* edited by Mollie Orshansky. Washington, DC: U.S. Department of Health, Education, and Welfare.

Ousey, Graham C. and Matthew R. Lee. 2004. "Investigating the Connections Between Race, Illicit Drug Markets, and Legal Violence, 1984–1997." *Journal of Research in Crime and Delinquency* 41:352–383.

Orwin, Robert G. 1994. "Evaluating Coding Decisions." Pp. 138–162 in *The Handbook of Research Synthesis,* edited by Harris Cooper and Larry V. Hodges. New York: Russell Sage.

Pagnini, Deanna L. and S. Philip Morgan. 1996. "Racial Differences in Marriage and Childbearing: Oral History Evidence From the South in the Early Twentieth Century." *American Journal of Sociology* 101:1694–1715.

Paige, Jeffery M. 1999. "Conjuncture, Comparison, and Conditional Theory in Macrosocial Inquiry." *American Journal of Sociology* 105:781–800.

Panagopoulos, Costas. 2008. "Poll Accuracy in the 2008 Presidential Election." Retrieved March 17, 2011, from http://www.fordham.edu/images/academics/graduate_schools/gsas/elections_and_campaign_/poll%20accuracy%20in%20the%202008%20presidential%20election.pdf

Papineau, David. 1978. *For Science in the Social Sciences.* London: Macmillan.

Parks, Kathleen A., Ann M. Pardi, and Clara M. Bradizza. 2006. "Collecting Data on Alcohol Use and Alcohol-Related Victimization: A Comparison of Telephone and Web-Based Survey Methods." *Journal of Studies on Alcohol* 67:318–323.

Parlett, Malcolm and David Hamilton. 1976. "Evaluation as Illumination: A New Approach to the Study of Innovative Programmes." Pp. 140–157 in *Evaluation Studies Review Annual,* vol. 1, edited by G. Glass. Beverly Hills, CA: Sage.

Pate, Antony M. and Edwin E. Hamilton. 1992. "Formal and Informal Deterrents to Domestic Violence: The Dade County Spouse Assault Experiment." *American Sociological Review* 57(October):691–697.

Paternoster, Raymond, Robert Brame, Ronet Bachman, and Lawrence W. Sherman. 1997. "Do Fair Procedures Matter? The Effect of Procedural Justice on Spouse Assault." *Law & Society Review* 31(1):163–204.

Patterson, Orlando. 1997. "The Race Trap." *The New York Times,* July 11, p. A25.

Patton, Michael Quinn. 2002. *Qualitative Research & Evaluation Methods,* 3rd ed. Thousand Oaks, CA: Sage.

Paxton, Pamela. 2002. "Social Capital and Democracy: An Interdependent Relationship." *American Sociological Review* 67:254–277.

Paxton, Pamela. 2005. "Trust in Decline?" *Contexts* 4:40–46.

Pepinsky, Harold E. 1980. "A Sociologist on Police Patrol." Pp. 223–234 in *Fieldwork Experience: Qualitative Approaches to Social Research,* edited by William B. Shaffir, Robert A. Stebbins, and Allan Turowetz. New York: St. Martin's Press.

Peterson, Robert A. 2000. *Constructing Effective Questionnaires.* Thousand Oaks, CA: Sage.

Pew Hispanic Center. 2008. *Statistical Portrait of the Foreign-Born Population in the United States, 2006.* Statistical tables from 2000 Census and 2006 American Community Survey, Retrieved May 24, 2008, from http://pewhispanic.org/factsheets/factsheet.php?FactsheetID=36

Pew Internet & American Life Project. 2000. *Tracking Online Life: How Women Use the Internet to Cultivate Relationships With Family and Friends.* Washington, DC: Pew Internet & American Life Project. Available at www.pewinternet.org

Pew Internet & American Life Project. 2010. *Change in Internet Access by Age Group, 2000-2009.* Washington DC: Pew Internet & American Life Project. Retrieved May 15, 2011, from http://www.pewinternet.org/Infographics/2010/Internet-acess-by-age-group-over-time.aspx#

Phillips, David P. 1982. "The Impact of Fictional Television Stories on U.S. Adult Fatalities: New Evidence on the Effect of the Mass Media on Violence." *American Journal of Sociology* 87:1340–1359.

Piliavin, Jane Allyn and Irving M. Piliavin. 1972. "Effect of Blood on Reactions to a Victim." *Journal of Personality and Social Psychology* 23:353–361.

Plessy v. Ferguson, 163 U.S. 537 (1896).

Pollner, Melvin and Richard E. Adams. 1994. "The Interpersonal Context of Mental Health Interviews." *Journal of Health and Social Behavior* 35:283–290.

Porter, Stephen R. and Michael E. Whitcomb. 2003. "The Impact of Contact Type on Web Survey Response Rates." *Public Opinion Quarterly* 67:579–588.

Posavac, Emil J. and Raymond G. Carey. 1997. *Program Evaluation: Methods and Case Studies,* 5th ed. Upper Saddle River, NJ: Prentice Hall.

Presley, Cheryl A., Philip W. Meilman, and Rob Lyerla. 1994. "Development of the Core Alcohol and Drug Survey: Initial Findings and Future Directions." *Journal of American College Health* 42: 248–255.

Presser, Stanley and Johny Blair. 1994. "Survey Pretesting: Do Different Methods Produce Different Results?" *Sociological Methodology* 24:73-104.

Presser, Stanley, Mick P. Couper, Judith T. Lessler, Elizabeth Martin, Jean Martin, Jennifer M. Rothgeb, and Eleanor Singer. 2004. "Methods for Testing and Evaluating Survey Questions." *Public Opinion Quarterly* 68:109-130.

Price, Richard H., Michelle Van Ryn, and Amiram D. Vinokur. 1992. "Impact of a Preventive Job Search Intervention on the Likelihood of Depression Among the Unemployed." *Journal of Health and Social Behavior* 33 (June):158–167.

Punch, Maurice. 1994. "Politics and Ethics in Qualitative Research." Pp. 83–97 in *Handbook of Qualitative Research,* edited by Norman K. Denzin and Yvonna S. Lincoln. Thousand Oaks, CA: Sage.

Purdy, Matthew. 1994. "Bronx Mystery: 3rd-Rate Service for 1st-Class Mail." *The New York Times,* March 12, pp. 1, 3.

Putnam, Israel. 1977. "Poverty Thresholds: Their History and Future Development." Pp. 272–283 in *The Measure of Poverty. Technical Paper I: Documentation of Background Information and Rationale for Current Poverty Matrix,* edited by Mollie Orshansky. Washington, DC: U.S. Department of Health, Education, and Welfare.

Pyrczak, Fred. 2005. *Evaluating Research in Academic Journals: A Practical Guide to Realistic Evaluation,* 3rd ed. Glendale, CA: Pyrczak.

Radloff, Lenore. 1977. "The CES-D Scale: A Self-Report Depression Scale for Research in the General Population." *Applied Psychological Measurement* 1:385–401.

Ragin, Charles C. 1987. *The Comparative Method: Moving Beyond Qualitative and Quantitative Strategies.* Berkeley: University of California Press.

Ragin, Charles C. 1994. *Constructing Social Research.* Thousand Oaks, CA: Sage.

Ragin, Charles C. 2000. *Fuzzy-Set Social Science.* Chicago: University of Chicago Press.

Rainie, Lee and John Horrigan. 2005. A Decade of Adoption: How the Internet Has Woven Itself Into American Life. Pew Internet & American Life Project. Retrieved June 18, 2005, from the Pew Internet & American Life Project, www.pewinternet.org/PPF/r/148/report_display.asp (PDF version).

Rand, Michael R., James P. Lynch, and David Cantor. 1997. *Criminal Victimization, 1973–95.* Washington, DC: Office of Justice Programs, U.S. Department of Justice.

Rashbaum, William K. 2002. "Reasons for Crime Drop in New York Elude Many." *The New York Times,* November 29, p. A28.

Raudenbush, Stephen W. and Robert J. Sampson. 1999. "Ecometrics: Toward a Science of Assessing Ecological Settings, With Application to the Systematic Social Observation of Neighborhoods." *Sociological Methodology* 29:1–41.

Reinharz, Shulamit. 1992. *Feminist Methods in Social Research.* New York: Oxford University Press.

Reisman, David. [1950] 1969. *The Lonely Crowd: A Study of the Changing American Character.* New Haven, CT: Yale University Press.

Reiss, Albert J., Jr. 1971a. *The Police and the Public.* New Haven, CT: Yale University Press.

Reiss, Albert J., Jr. 1971b. "Systematic Observations of Natural Social Phenomena." Pp. 3–33 in *Sociological Methodology,* vol. 3, edited by Herbert Costner. San Francisco: Jossey-Bass.

Rele, J. R. 1993. "Demographic Rates: Birth, Death, Marital, and Migration." Pp. 2–1– 2–26 in *Readings in Population Research*

Methodology. vol. 1, *Basic Tools*, edited by Donald J. Bogue, Eduardo E. Arriaga, and Douglas L. Anderton. Chicago: Social Development Center, for the United National Population Fund.

Reverby, Susan M. 2011. "'Normal Exposure' and Inoculation Syphilis: A PHS 'Tuskegee' Doctor in Guatemala, 1946–1948." *Journal of Policy History* 23:6–28.

Rew, Lynn, Deborah Koniak-Griffin, Mary Ann Lewis, Margaret Miles, and Ann O'Sullivan. 2000. "Secondary Data Analysis: New Perspective for Adolescent Research." *Nursing Outlook* 48:223–229.

Reynolds, Paul Davidson. 1979. *Ethical Dilemmas and Social Science Research*. San Francisco: Jossey-Bass.

Richards, Thomas J. and Lyn Richards. 1994. "Using Computers in Qualitative Research." Pp. 445–462 in *Handbook of Qualitative Research*, edited by Norman K. Denzin and Yvonna S. Lincoln. Thousand Oaks, CA: Sage.

Richardson, Laurel. 1995. "Narrative and Sociology." Pp. 198–221 in *Representation in Ethnography*, edited by John Van Maanen. Thousand Oaks, CA: Sage.

Riedel, Marc. 2000. *Research Strategies for Secondary Data: A Perspective for Criminology and Criminal Justice*. Thousand Oaks, CA: Sage.

Riessman, Catherine Kohler. 2002. "Narrative Analysis." Pp. 217–270 in *The Qualitative Researcher's Companion*, edited by A. Michael Huberman and Matthew B. Miles. Thousand Oaks, CA: Sage.

Riessman, Catherine Kohler. 2008. *Narrative Methods for the Human Sciences*. Thousand Oaks, CA: Sage.

Ringwalt, Christopher L., Jody M. Greene, Susan T. Ennett, Ronaldo Iachan, Richard R. Clayton, and Carl G. Leukefeld. 1994. *Past and Future Directions of the D.A.R.E. Program: An Evaluation Review*. Research Triangle, NC: Research Triangle Institute.

Rives, Norfleet W., Jr. and William J. Serow. 1988. *Introduction to Applied Demography: Data Sources and Estimation Techniques*. Sage University Paper Series on Quantitative Applications in the Social Sciences, series No. 07–039. Thousand Oaks, CA: Sage.

Robertson, David Brian. 1993. "The Return to History and the New Institutionalism in American Political Science." *Social Science History* 17:1–36.

Robinson, Karen A. and Steven N. Goodman. 2011. "A Systematic Examination of the Citation of Prior Research in Reports of Randomized, Controlled Trials." *Annals of Internal Medicine* 154:50-55.

Rodríguez, Havidán, Joseph Trainor, and Enrico L. Quarantelli. 2006. "Rising to the Challenges of a Catastrophe: The Emergent and Prosocial Behavior Following Hurricane Katrina." *The Annals of the American Academy of Political and Social Science* 604:82–101.

Roman, Anthony. 2005. *Women's Health Network Client Survey: Field Report*. Unpublished report. Boston: Center for Survey Research, University of Massachusetts.

Rookey, Bryan D., Steve Hanway, and Don A. Dillman. 2008. "Does a Probability-Based Household Panel Benefit from Assignment to Postal Response as an Alternative to Internet-Only?" *Public Opinion Quarterly* 72:962–984.

Rosen, Lawrence. 1995. "The Creation of the Uniform Crime Report: The Role of Social Science." *Social Science History* 19:215–238.

Rosenbach, Margo, Carol Irvin, Angela Merrill, Shanna Shulman, John Czajka, Christopher Trenholm, Susan Williams, So Sasigant Limpa-Amara, and Anna Katz. 2007. National Evaluation of the State Children's Health Insurance Program: A Decade of Expanding Coverage and Improving Access: Final Report. Cambridge, MA: Mathematica Policy Research.

Rosenberg, Morris. 1968. *The Logic of Survey Analysis*. New York: Basic Books.

Rosenfeld, Richard. 2004. "The Case of the Unsolved Crime Decline." *Scientific American* 290:82–89.

Rosenthal, Elisabeth. 2000. "Rural Flouting of One-Child Policy Undercuts China's Census." *The New York Times*, April 14, p. A10.

Rosenthal, Rob. 1994. *Homeless in Paradise: A Map of the Terrain*. Philadelphia: Temple University Press.

Ross, Catherine E. 1990. *Work, Family, and the Sense of Control: Implications for the Psychological Well-Being of Women and Men*. Proposal submitted to the National Science Foundation. Urbana: University of Illinois.

Rossi, Peter H. 1989. *Down and Out in America: The Origins of Homelessness*. Chicago: University of Chicago Press.

Rossi, Peter H. 1999. "Half Truths with Real Consequences: Journalism, Research, and Public Policy. Three Encounters." *Contemporary Sociology* 28:1–5.

Rossi, Peter H. and Howard E. Freeman. 1989. *Evaluation: A Systematic Approach*, 4th ed. Newbury Park, CA: Sage.

Rossman, Gretchen B. and Sharon F. Rallis. 1998. *Learning in the Field: An Introduction to Qualitative Research*. Thousand Oaks, CA: Sage.

Roth, Dee. 1990. "Homelessness in Ohio: A Statewide Epidemiological Study." Pp. 145–163 in *Homeless in the United States, Vol. 1: State Surveys*, edited by Jamshid Momeni. New York: Greenwood.

Rotolo, Thomas and Charles R. Tittle. 2006. "Population Size, Change, and Crime in U.S. Cities." *Journal of Quantitative Criminology* 22:341–367.

Rubin, Herbert J. and Irene S. Rubin. 1995. *Qualitative Interviewing: The Art of Hearing Data*. Thousand Oaks, CA: Sage.

Rueschemeyer, Dietrich, Evelyne Huber Stephens, and John D. Stephens. 1992. *Capitalist Development and Democracy*. Chicago: University of Chicago Press.

Ruggles, Patricia. 1990. *Drawing the Line: Alternative Poverty Measures and Their Implications for Public Policy.* Washington, DC: Urban Institute Press.

Sacks, Stanley, Karen McKendrick, George DeLeon, Michael T. French, and Kathryn E. McCollister. 2002. "Benefit-Cost Analysis of a Modified Therapeutic Community for Mentally Ill Chemical Abusers." *Evaluation and Program Planning* 25:137–148.

Salisbury, Robert H. 1975. "Research on Political Participation." *American Journal of Political Science* 19(May):323–341.

Sampson, Robert J. 1987. "Urban Black Violence: The Effect of Male Joblessness and Family Disruption." *American Journal of Sociology* 93(September):348–382.

Sampson, Robert J. 2008. "Rethinking Crime and Immigration." *Contexts* 7:28–33.

Sampson, Robert J. and John H. Laub. 1990. "Crime and Deviance Over the Life Course: The Salience of Adult Social Bonds." *American Sociological Review* 55(October):609–627.

Sampson, Robert J. and John H. Laub. 1993. "Structural Variations in Juvenile Court Processing: Inequality, the Underclass, and Social Control." *Law & Society Review* 27(2):285–311.

Sampson, Robert J. and John H. Laub. 1994. "Urban Poverty and the Family Context of Delinquency: A New Look at Structure and Process in a Classic Study." *Child Development* 65:523–540.

Sampson, Robert J. and Stephen W. Raudenbush. 1999. "Systematic Social Observation of Public Spaces: A New Look at Disorder in Urban Neighborhoods." *American Journal of Sociology* 105:603–651.

Sampson, Robert J. and Stephen W. Raudenbush. 2001. "Disorder in Urban Neighborhoods—Does It Lead to Crime?" In *Research in Brief.* Washington, DC: National Institute of Justice, U.S. Department of Justice.

Sampson, Robert J., Stephen W. Raudenbush, and Felton Earls. 1997. "Neighborhoods and Violent Crime: A Multilevel Study of Collective Efficacy." *Science* 277:918–924.

Savage, Charlie. 2010. "Crime Rates Fell in '09 Despite Economy, F.B.I. Says." *The New York Times,* May 25:A15.

Schaeffer, Nora Cate and Stanley Presser. 2003. "The Science of Asking Questions." *Annual Review of Sociology* 29:65–88.

Schegloff, Emanuel A. 1996. "Issues of Relevance for Discourse Analysis: Contingency in Action, Interaction and Coparticipant Context." Pp. 3–35 in *Computational and Conversational Discourse: Burning Issues—An Interdisciplinary Account,* edited by Eduard H. Hovy and Donia R. Scott. New York: Springer.

Schleyer, Titus K. L. and Jane L. Forrest. 2000. "Methods for the Design and Administration of Web-Based Surveys." *Journal of the American Medical Informatics Association* 7:418–425.

Schober, Michael F. 1999. "Making Sense of Survey Questions." Pp. 77–94 in *Cognition and Survey Research,* edited by Monroe G. Sirken, Douglas J. Herrmann, Susan Schechter, Norbert Schwartz, Judith M. Tanur, and Roger Tourangeau. New York: Wiley.

Schofield, Janet Ward. 2002. "Increasing the Generalizability of Qualitative Research." Pp. 171–203 in *The Qualitative Researcher's Companion,* edited by A. Michael Huberman and Matthew B. Miles. Thousand Oaks, CA: Sage.

Schorr, Lisbeth B. and Daniel Yankelovich. 2000. "In Search of a Gold Standard for Social Programs." *The Boston Globe,* February 18, p. A19.

Schreck, Christopher J., Eric A. Steward, and Bonnie S. Fisher. 2006. "Self-control, Victimization, and their Influence on Risky Lifestyles: A Longitudinal Analysis Using Panel Data." *Journal of Quantitative Criminology* 22:319–340.

Schuman, Howard and Stanley Presser. 1981. *Questions and Answers in Attitude Surveys: Experiments on Question Form, Wording, and Context.* New York: Academic Press.

Schutt, Russell K. 1986. *Organization in a Changing Environment.* Albany: State University of New York Press.

Schutt, Russell K. (with the assistance of Tatjana Meschede). 1992. *The Perspectives of DMH Shelter Staff: Their Clients, Their Jobs, Their Shelters and the Service System.* Unpublished report to the Metro Boston Region of the Massachusetts Department of Mental Health. Boston: Department of Sociology, University of Massachusetts.

Schutt, Russell K. 2011. *Homelessness, Housing, and Mental Illness.* Cambridge, MA: Harvard University Press.

Schutt, Russell K. and W. Dale Dannefer. 1988. "Detention Decisions in Juvenile Cases: JINS, JDs and Gender." *Law & Society Review* 22(3):509–520.

Schutt, Russell K., Xiaogang Deng, Gerald R. Garrett, Stephanie Hartwell, Sylvia Mignon, Joseph Bebo, Matthew O'Neill, Mary Aruda, Pat Duynstee, Pam DiNapoli, and Helen Reiskin. 1996. *Substance Use and Abuse Among UMass Boston Students.* Boston: Department of Sociology, University of Massachusetts. Unpublished report.

Schutt, Russell K. and Jacqueline Fawcett. 2005. *Case Management in the Women's Health Network.* Boston: University of Massachusetts. Unpublished report.

Schutt, Russell K. and M. L. Fennell. 1992. "Shelter Staff Satisfaction With Services, the Service Network and Their Jobs." *Current Research on Occupations and Professions* 7:177–200.

Schutt, Russell K. and Stephen M. Goldfinger. 1996. "Housing Preferences and Perceptions of Health and Functioning Among

Homeless Mentally Ill Persons." *Psychiatric Services* 47:381–386.

Schutt, Russell K., Stephen M. Goldfinger, and Walter E. Penk. 1992. "The Structure and Sources of Residential Preferences Among Seriously Mentally Ill Homeless Adults." *Sociological Review* 3(3): 148–156.

Schutt, Russell K., Stephen M. Goldfinger, and Walter E. Penk. 1997. "Satisfaction With Residence and With Life: When Homeless Mentally Ill Persons Are Housed." *Evaluation and Program Planning* 20(2):185–194.

Schutt, Russell K., Suzanne Gunston, and John O'Brien. 1992a. "The Impact of AIDS Prevention Efforts on AIDS Knowledge and Behavior Among Sheltered Homeless Adults." *Sociological Practice Review* 3(1):1–7.

Schutt, Russell K., Walter E. Penk, Paul J. Barreira, Robert Lew, William H. Fisher, Angela Browne, and Elizabeth Irvine. 1992b. *Relapse Prevention for Dually Diagnosed Homeless.* Proposal to National Institute of Mental Health, for Mental Health Research on Homeless Persons, PA-91-60. Worcester: University of Massachusetts Medical School.

Schwandt, Thomas A. 1994. "Constructivist, Interpretivist Approaches to Human Inquiry." Pp. 118–137 in *Handbook of Qualitative Research,* edited by Norman K. Denzin and Yvonna S. Lincoln. Thousand Oaks, CA: Sage.

Schwartz, John. 2005. "Myths Run Wild in Blog Tsunami Debate." *The New York Times,* January 3, p. A9.

Schwarz, Norbert. 2010. "Measurement as Cooperative Communication: What Research Participants Learn from Questionnaires." Pp. 43–59 in *The SAGE handbook of measurement,* edited by Geoffrey Walford, Eric Tucker, and Madhu Viswanathan. Thousand Oaks, CA: Sage.

Scott, Janny. 2001. "A Nation by the Numbers, Smudged." *The New York Times,* July 1, pp. 21, 22.

Scriven, Michael. 1972a. "The Methodology of Evaluation." Pp. 123–136 in *Evaluating Action Programs: Readings in Social Action and Education,* edited by Carol H. Weiss. Boston: Allyn & Bacon.

Scriven, Michael. 1972b. "Prose and Cons About Goal-Free Evaluation." *Evaluation Comment* 3:1–7.

Scull, Andrew T. 1988. "Deviance and Social Control." Pp. 667–693 in *Handbook of Sociology,* edited by Neil J. Smelser. Newbury Park, CA: Sage.

Sechrest, Lee and Souraya Sidani. 1995. "Quantitative and Qualitative Methods: Is There an Alternative?" *Evaluation and Program Planning* 18:77–87.

Selm, Martine Van and Nicholas W. Jankowski. 2006. "Conducting Online Surveys." *Quality & Quantity* 40:435–456.

Services Research Review Committee, National Institute of Mental Health. 1992. *Summary Statement for Relapse Prevention for Dually Diagnosed Homeless.* Russell K. Schutt, Principal Investigator, application No. 1 R18 SM50851-01. Bethesda, MD: National Institute of Mental Health.

Shadish, William R., Thomas D. Cook, and Laura C. Leviton (Eds.). 1991. *Foundations of Program Evaluation: Theories of Practice.* Thousand Oaks, CA: Sage.

Sharma, Divya. 2009. "Research Ethics and Sensitive Behaviors: Underground Economy." Pp. 426–441 in *The Handbook of Social Research Ethics,* edited by Donna M. Mertens and Pauline E. Ginsberg. Thousand Oaks, CA: Sage.

Shepherd, Jane, David Hill, Joel Bristor, and Pat Montalvan. 1996. "Converting an Ongoing Health Study to CAPI: Findings From the National Health and Nutrition Study." Pp. 159–164 in *Health Survey Research Methods Conference Proceedings,* edited by Richard B. Warnecke. Hyattsville, MD: U.S. Department of Health and Human Services.

Sherman, Lawrence W. 1992. *Policing Domestic Violence: Experiments and Dilemmas.* New York: Free Press.

Sherman, Lawrence W. 1993. "Implications of a Failure to Read the Literature." *American Sociological Review* 58:888–889.

Sherman, Lawrence W. and Richard A. Berk. 1984. "The Specific Deterrent Effects of Arrest for Domestic Assault." *American Sociological Review* 49:261–272.

Sherman, Lawrence W. and Douglas A. Smith, with Janell D. Schmidt and Dennis P. Rogan. 1992. "Crime, Punishment, and Stake in Conformity." *American Sociological Review* 57:680–690.

Shermer, Michael. 1997. *Why People Believe Weird Things: Pseudoscience, Superstition, and Other Confusions of Our Time.* New York: W. H. Freeman.

Shrout, Patrick E. 2011. "Integrating Causal Analysis into Psychopathology Research." Pp. 3–24 in *Causality and Psychopathology: Finding the Determinants of Disorders and Their Cures,* edited by Patrick E. Shrout, Katherine M. Keyes, and Katherine Ornstein. New York: Oxford University Press.

Sieber, Joan E. 1992. *Planning Ethically Responsible Research: A Guide for Students and Internal Review Boards.* Thousand Oaks, CA: Sage.

Sjoberg, Gideon (Ed.). 1967. *Ethics, Politics, and Social Research.* Cambridge, MA: Schenkman.

Skinner, Harvey A. and Wen-Jenn Sheu. 1982. "Reliability of Alcohol Use Indices: The Lifetime Drinking History and the MAST." *Journal of Studies on Alcohol* 43(11):1157–1170.

Skocpol, Theda. 1979. *States and Social Revolutions: A Comparative Analysis of France, Russia, and China.* New York: Cambridge University Press.

Skocpol, Theda. 1984. "Emerging Agendas and Recurrent Strategies in Historical Sociology." Pp. 356–391 in *Vision and Method in Historical Sociology,* edited by Theda Skocpol. New York: Cambridge University Press.

Skocpol, Theda and Margaret Somers. 1979. "The Uses of Comparative History in Macrosocial Inquiry." Pp. 72–95 in *Social Revolutions in the Modern World,* edited by Theda Skocpol. New York: Cambridge University Press.

Smith, Joel. 1991. "A Methodology for Twenty-First Century Sociology." *Social Forces* 70:1–17.

Smith, Tom W. 1984. "Nonattitudes: A Review and Evaluation." Pp. 215–255 in *Surveying Subjective Phenomena,* vol. 2, edited by Charles F. Turner and Elizabeth Martin. New York: Russell Sage Foundation.

Smith-Lovin, Lynn. 2007. "Do We Need a Public Sociology? It Depends on What You Mean by 'Sociology.'" Pp. 124–134 in *Public Sociology: Fifteen Eminent Sociologists Debate Politics and the Profession in the Twenty-First Century,* edited by Dan Clawson, Robert Zussman, Joya Misra, Naomi Gerstel, and Randall Stokes. Berkeley: University of California Press.

Smithson, Janet. 2008. "Focus Groups." Pp. 357–370 in *The Sage Handbook of Social Research Methods,* edited by Pertti Alasuutari, Leonard Bickman, and Julia Brannen. Thousand Oaks, CA: Sage.

Smyth, Jolene D., Don A. Dillman, Leah Melani Christian, and Michael J. Stern. 2004. *How Visual Grouping Influences Answers to Internet Surveys.* Extended version of paper presented at the Annual Meeting of the American Association for Public Opinion Research, Phoenix, AZ, May 13. Retrieved July 5, 2005, from http://survey.sesrc.wsu.edu/ dillman/papers.htm

Snipp, C. Matthew. 2003. "Racial Measurement in the American Census: Past Practices and Implications for the Future." *Annual Review of Sociology* 29:563–588.

Snow, David L., Jacob Kraemer Tebes, and Michael W. Arthur. 1992. "Panel Attrition and External Validity in Adolescent Substance Use Research." *Journal of Consulting and Clinical Psychology* 60:804–807.

Sobell, Linda C., Mark B. Sobell, Diane M. Riley, Reinhard Schuller, D. Sigfrido Pavan, Anthony Cancilla, Felix Klajner, and Gloria I. Leo. 1988. "The Reliability of Alcohol Abusers' Self-Reports of Drinking and Life Events That Occurred in the Distant Past." *Journal of Studies on Alcohol* 49(2):225–232.

Sosin, Michael R., Paul Colson, and Susan Grossman. 1988. *Homelessness in Chicago: Poverty and Pathology, Social Institutions and Social Change.* Chicago: Chicago Community Trust.

South, Scott J. and Glenna Spitze. 1994. "Housework in Marital and Nonmarital Households." *American Sociological Review* 59:327–347.

Speiglman, Richard and Patricia Spear. 2009. "The Role of Institutional Review Boards: Ethics: Now You See Them, Now You Don't." Pp. 121–134 in *The Handbook of Social Research Ethics,* edited by Donna M. Mertens and Pauline E. Ginsberg. Thousand Oaks, CA: Sage.

Spretnak, Charlene. 1991. *States of Grace: The Recovery of Meaning in the Postmodern Age.* New York: HarperCollins.

St. Jean, Peter K. B. 2007. *Pockets of Crime: Broken Windows, Collective Efficacy, and the Criminal Point of View.* Chicago: University of Chicago Press.

St. Pierre, Robert G. and Peter H. Rossi. 2006. "Randomize Groups, Not Individuals: A Strategy for Improving Early Childhood Programs." *Evaluation Review* 30:656–685.

Stake, Robert E. 1995. *The Art of Case Study Research.* Thousand Oaks, CA: Sage.

Stake, Robert and Fazal Rizvi. 2009. "Research Ethics in Transnational Spaces." Pp. 521–536 in *The Handbook of Social Research Ethics,* edited by Donna M. Mertens and Pauline E. Ginsberg. Thousand Oaks, CA: Sage.

Stephen, Jason M., Michael F. Young, and Thomas Calabrese. 2007. "Does Moral Judgment Go Offline When Students Are Online? A Comparative Analysis of Undergraduates' Beliefs and Behaviors Related to Conventional and Digital Cheating." *Ethics & Behavior* 17:233–254.

Stewart, David W. and Michael A. Kamins. 1993. *Secondary Research: Information Sources and Methods,* 2nd ed. Thousand Oaks, CA: Sage.

Stille, Alexander. 2000. "A Happiness Index With a Long Reach: Beyond G.N.P. to Subtler Measures." *The New York Times,* May 20, pp. A17, A19.

Stokoe, Elizabeth. 2006. "On Ethnomethodology, Feminism, and the Analysis of Categorical Reference to Gender in Talk-in-Interaction." *The Sociological Review* 54:467–494.

Strauss, Anselm L. and Juliette Corbin. 1990. *The Basics of Qualitative Research: Grounded Theory Procedures and Techniques.* Newbury Park, CA: Sage.

Strickling, Lawrence E. 2010. *Digital Nation: 21st Century America's Progress Toward Universal Broadband Internet Access. An NTIA Research Preview.* Washington, DC: National Telecommunications and Information Administration, U.S. Department of Commerce.

Strunk, William, Jr. and E. B. White. 2000. *The Elements of Style,* 4th ed. New York: Allyn & Bacon.

Sudman, Seymour. 1976. *Applied Sampling.* New York: Academic Press.

Sulkunen, Pekka. 2008. "Social Research and Social Practice in Post-Positivist Society." Pp. 68–80 in *The Sage Handbook of Social Research Methods,* edited by Pertti Alasuutari, Leonard Bickman, and Julia Brannen. Thousand Oaks, CA: Sage.

Sunderland, Antonia. 2005. *Children, Families and Welfare Reform: A Three-City Study.* Princeton, NJ: The Robert Wood Johnson Foundation. Retrieved October 5, 2005, from www.rwjf.org/ reports/grr/037218.htm

"Survey on Adultery: 'I Do' Means 'I Don't.'" 1993. *The New York Times,* October 19, p. A20.

Survey Research Laboratory. 2008. *List of Academic and Not-for-Profit Survey Research Organizations (LANSRO)*. Chicago: College of Urban Planning and Public Affairs, University of Illinois at Chicago. Retrieved May 17, 2008, from www.srl.uic.edu/LANSRO.doc

Swarns, Rachel L. 2004. "Hispanics Debate Racial Grouping by Census." *The New York Times,* October 24, pp. A1, A18.

Tavernise, Sabrina. 2002. "How Many Russians? Let Us Weigh the Count, Cooperation or No." *The New York Times,* October 10, p. A13.

Taylor, Charles Lewis and David A. Jodice. 1986. *World Handbook of Political and Social Indicators III: 1948–1982*. File available from the Inter-University Consortium for Political and Social Research (ICPSR), Study #7761.

Taylor, Jerry. 1999. "DARE Gets Updated in Some Area Schools, Others Drop Program." *The Boston Sunday Globe,* May 16, pp. 1, 11.

Thomas, Neil. 2005. "Disaster Center Researchers to Study Katrina Response." *UDaily,* September 15. Retrieved May 26, 2008, from www.udel.edu/PR/UDaily/2005/mar/DRC091505.html

Thorne, Barrie. 1993. *Gender Play: Girls and Boys in School*. New Brunswick, NJ: Rutgers University Press.

Thrasher, Frederic M. 1927. *The Gang: A Study of 1,313 Gangs in Chicago*. Chicago: University of Chicago Press.

Timmer, Doug A., D. Stanley Eitzen, and Kathryn D. Talley. 1993. *Paths to Homelessness: Extreme Poverty and the Urban Housing Crisis*. Boulder, CO: Westview Press.

Tjaden, Patricia and Nancy Thoennes. 2000. *Extent, Nature, and Consequences of Intimate Partner Violence: Findings From the National Violence Against Women Survey, NCJ 181867*. Washington, DC: Office of Justice Programs, National Institute of Justice and the Centers for Disease Control and Prevention.

Toby, Jackson. 1957. "Social Disorganization and Stake in Conformity: Complementary Factors in the Predatory Behavior of Hoodlums." *Journal of Criminal Law, Criminology and Police Science* 48:12–17.

Toppo, Greg. 2002. "Antidrug Program Backed by Study." *The Boston Globe,* October 29, p. A10.

Tourangeau, Roger. 1999. "Context Effects." Pp. 111–132 in *Cognition and Survey Research,* edited by Monroe G. Sirken, Douglas J. Herrmann, Susan Schechter, Norbert Schwarz, Judith M. Tanur, and Roger Tourangeau. New York: Wiley.

Tourangeau, Roger. 2004. "Survey Research and Societal Change." *Annual Review of Psychology* 55:775–801.

Townsend, Meg, Dana Hunt, Caity Baxter, and Peter Finn. 2005. *Interim Report: Evaluability Assessment of the President's Family Justice Center Initiative*. Cambridge, MA: Abt Associates *Retrieved March 20, 2011,* from www.ncjrs.gov/pdffiles1/nij/grants/212278.pdf

Tufte, Edward R. 1983. *The Visual Display of Quantitative Information*. Cheshire, CT: Graphics Press.

Turabian, Kate L. 1996. *A Manual for Writers of Term Papers, Theses, and Dissertations,* 6th ed. Chicago: University of Chicago Press.

Turkle. Sherry. 2011. *Alone Together: Why We Expect More from Technology and Less from Each Other*. New York: Basic Books.

Turner, Charles F. and Elizabeth Martin (Eds.). 1984. *Surveying Subjective Phenomena,* vols. 1 and 2. New York: Russell Sage Foundation.

Turner, Jonathan H., Leonard Beeghley, and Charles H. Powers. 1995. *The Emergence of Sociological Theory,* 3rd ed. Belmont, CA: Wadsworth.

Turner, Stephen P. 1980. *Sociological Explanation as Translation*. Cambridge, UK: Cambridge University Press.

Tyler, Tom R. 1990. "The Social Psychology of Authority: Why Do People Obey an Order to Harm Others?" *Law & Society Review* 24:1089–1102.

Uchitelle, Louis. 1997. "Measuring Inflation: Can't Do It, Can't Stop Trying." *The New York Times,* March 16, p. 4.

Uchitelle, Louis. 1999. "Devising New Math to Define Poverty." *The New York Times,* October 16, pp. A1, A14.

UCLA Center for Communication Policy. 2001. *The UCLA Internet Report 2001: Surveying the Digital Future*. Los Angeles: UCLA Center for Communication Policy.

UCLA Center for Communication Policy. 2003. *The UCLA Internet Report: Surveying the Digital Future, Year Three*. Los Angeles: UCLA Center for Communication Policy. Retrieved June 15, 2005, from www.digitalcenter.org/pdf/InternetReportYearThree.pdf

Udry, J. Richard. 1988. "Biological Predispositions and Social Control in Adolescent Sexual Behavior." *American Sociological Review* 53:709–722.

Universities of Essex and Manchester. 2011. *ESDS Qualidata*. Retrieved May 28, 2011, from http://www.esds.ac.uk/qualidata/

U.S. Bureau of Economic Analysis. 2004. *Customer Satisfaction Survey Report, FY 2004*. Retrieved October 4, 2005, from www.bea.gov/ bea/about/cssr_2004_complete.pdf

U.S. Bureau of Labor Statistics, Department of Labor. 1991. *Major Programs of the Bureau of Labor Statistics*. Washington, DC: U.S. Bureau of Labor Statistics, Department of Labor.

U.S. Bureau of Labor Statistics, Department of Labor. 1997a. *Employment and Earnings*. Washington, DC: U.S. Bureau of Labor Statistics, Department of Labor.

U.S. Bureau of Labor Statistics, Department of Labor. 1997b. *Handbook of Methods*. Washington, DC: U.S. Bureau of Labor Statistics, Department of Labor.

U.S. Bureau of the Census. 1981. *Section 1, Vital Statistics. Statistical Abstract of the United States, 1981, 102nd edition*. Washington,

DC: U.S. Department of Commerce, Bureau of the Census.

U.S. Bureau of the Census. 1994. *Census Catalog and Guide, 1994.* Washington, DC: Department of Commerce, U.S. Bureau of the Census.

U.S. Bureau of the Census. 1999. *United States Census 2000, Updated Summary: Census 2000 Operational Plan.* Washington, DC: U.S. Department of Commerce, Bureau of the Census, February.

U.S. Bureau of the Census. 2001. "Statement by William G. Barron Jr. on the Current Status of Results of Census 2000 Accuracy and Coverage Evaluation Survey." *United States Department of Commerce News,* July 13. Retrieved January 19, 2003, from www.census.gov/PressRelease/www/2001/cb01cs06.html

U.S. Bureau of the Census. 2003. "Census Bureau to Test Changes in Questionnaire, New Response Technology." *United States Department of Commerce News,* January 16. Retrieved January 19, 2003, from www.census.gov/Press-Release/www/2003/cb03cn02.html

U.S. Bureau of the Census. 2006. *Census Bureau Guideline: Language Translation of Data Collection Instruments and Supporting Materials.* Washington, DC: U.S. Census Bureau, Census Advisory Committees. Retrieved May 24, 2008, from www.census.gov/cac/www/007585.html

U.S. Bureau of the Census. 2010a. "$1.6 Billion in 2010 Census Savings Returned." *United States Department of Commerce News,* August 10. Retrieved March 17, 2011, from http://2010.census.gov/news/releases/operations/

U.S. Bureau of the Census. 2010b. "Door-to-Door Visits Begin for 2010 Census." *United States Department of Commerce News,* April 30. Retrieved March 17, 2011, from http://2010.census.gov/news/releases/operations/

U.S. Bureau of the Census. 2010c. "2010 Census Forms Arrive in 120 Million Mailboxes Across Nation." *United States Department of Commerce News,* March 15. Retrieved March 17, 2011, from http://2010.census.gov/news/releases/operations/

U.S. Bureau of the Census. 2010d. *Questions and Answers: Real People, Real Questions, Real Answers.* Washington, DC: U.S. Department of Commerce, Bureau of the Census. Retrieved March 17, 2011, from http://2010.census.gov/2010census/about/answers.php

U.S. Bureau of the Census. 2010e. *Internet Use in the United States: October 2009.* Washington, DC: U.S. Department of Commerce, Bureau of the Census. Retrieved March 18, 2011, from http://www.census.gov/population/www/socdemo/computer/2009.html

U.S. Department of Health and Human Services, Substance Abuse and Mental Health Services Administration, Center for Mental Health Services. 1995. *Client-Level Evaluation Procedure Manual.* Washington, DC: U.S. Department of Health and Human Services.

U.S. Department of Health and Human Services. 2011. *Fact Sheet on the 1946–48 U.S. Public Health Service Sexually Transmitted Diseases Inoculation Study.* Retrieved March 16, 2011, from http://www.hhs.gov/1946inoculationstudy/factsheet.html

U.S. Government Accountability Office. 2006. *Prevalence of Domestic Violence, Sexual Assault, Dating Violence, and Stalking: Briefing to Congressional Committees.* GAO 07–148R Washington, DC: GAO.

U.S. Government Accounting Office. 2001. *Health and Human Services: Status of Achieving Key Outcomes and Addressing Major Management Challenges June.* Retrieved April 8, 2003, from www.gao.gov/new.items/d01748.pdf

U.S. Office of Management and Budget. 2002. *Government and Performance Results Act of 1993.* Washington, DC: U.S. Office of Management and Budget, Executive Office of the President.

Vaessen, Martin. 1993. "Evaluation of Population Data: Errors and Deficiencies." Pp. 4–1–4–69 in *Readings in Population Research Methodology,* vol. 1, *Basic Tools,* edited by Donald J. Bogue, Eduardo E. Arriaga, and Douglas L. Anderton. Chicago: Social Development Center, for the United Nations Population Fund.

Vaillant, George E. 1995. *The Natural History of Alcoholism Revisited.* Cambridge, MA: Harvard University Press.

van de Vijver, Fons and Kwok Leung. 1997. *Methods and Data Analysis for Cross-Cultural Research.* Thousand Oaks, CA: Sage.

Van Hoye, Greet and Filip Lievens. 2003. "The Effects of Sexual Orientation on Hirability Ratings: An Experimental Study." *Journal of Business and Psychology* 18:15–30.

Van Maanen, John. 1982. "Fieldwork on the Beat." Pp. 103–151 in *Varieties of Qualitative Research,* edited by John Van Maanen, James M. Dabbs, Jr., and Robert R. Faulkner. Beverly Hills, CA: Sage.

Van Maanen, John. 1995. "An End to Innocence: The Ethnography of Ethnography." Pp. 1–35 in *Representation in Ethnography,* edited by John Van Maanen. Thousand Oaks, CA: Sage.

Ventimiglia, Vincent J. 2006. *Letter of Eileen R. Larence, GAO.* Washington, DC: U.S. Department of Health & Human Services.

Verba, Sidney and Norman Nie. 1972. *Political Participation: Political Democracy and Social Equality.* New York: Harper & Row.

Verba, Sidney, Norman Nie, and Jae-On Kim. 1978. *Participation and Political Equality: A Seven-Nation Comparison.* New York: Cambridge University Press.

Vernez, Georges, M. Audrey Burnam, Elizabeth A. McGlynn, Sally Trude, and Brian S. Mittman. 1988. *Review of California's Program for the Homeless Mentally Disabled.* Santa Monica, CA: RAND.

Vidich, Arthur J. and Stanford M. Lyman. 2004. "Qualitative Methods: Their History in Sociology and Anthropology." Pp. 27–84 in *The Handbook of Qualitative Research,* 2nd ed., edited by Norman Denzin and Yvonna S. Lincoln. Thousand Oaks, CA: Sage.

Villarreal, Andrés. 2010. "Stratification by Skin Color in Contemporary Mexico." *American Sociological Review* 75:652-678.

Vincus, Amy A., Chris Ringwalt, Melissa S. Harris, Stephen R. Shamblen. 2010. "A Short-Term, Quasi-Experimental Evaluation of D.A.R.E.'s Revised Elementary School Curriculum." *Journal of Drug Education* 40:37–49.

Viswanathan, Madhu. 2005. *Measurement Error and Research Design.* Thousand Oaks, CA: Sage.

Wageman, Ruth. 1995. "Interdependence and Group Effectiveness." *Administrative Science Quarterly* 40:145–180.

Wagner, William E., III. 2011. *Using IBM SPSS Statistics for Social Statistics and Research Methods,* 3rd ed. Thousand Oaks, CA: Sage.

Walker, Robert, Mark Tomlinson, and Glenn Williams. 2010. "The Problem with Poverty: Definition, measurement and Interpretation." Pp. 353–376 in *The Sage Handbook of Measurement,* edited by Geoffrey Walford, Eric Tucker, and Madhu Viswanathan. Thousand Oaks, CA: Sage Publications.

Walford, Geoffrey, Eric Tucker, and Madhu Viswanathan. eds. 2010. *The Sage Handbook of Measurement.* Thousand Oaks, CA: Sage.

Wallace, Walter L. 1983. *Principles of Scientific Sociology.* New York: Aldine.

Wallgren, Anders, Britt Wallgren, Rolf Persson, Ulf Jorner, and Jan-Aage Haaland. 1996. *Graphing Statistics and Data: Creating Better Charts.* Thousand Oaks, CA: Sage.

Walters, Pamela Barnhouse, David R. James, and Holly J. McCammon. 1997. "Citizenship and Public Schools: Accounting for Racial Inequality in Education for the Pre- and Post-Disfranchisement South." *American Sociological Review* 62:34–52.

Ward, Lester Frank. 1897. *Dynamic Sociology,* 2nd ed. New York: D. Appleton.

Warr, Mark. 1995. "The Polls-Poll Trends: Public Opinion on Crime and Punishment." *Public Opinion Quarterly* 59:296–310.

Watson, Charles G., Curt Tilleskjor, E. A. Hoodecheck-Schow, John Pucel, and Lyle Jacobs. 1984. "Do Alcoholics Give Valid Self-Reports?" *Journal of Studies on Alcohol* 45(4):344–348.

Weatherby, Norman L., Richard Needle, Helen Cesari, Robert Booth, Clyde B. McCoy, John K. Waters, et al. 1994. "Validity of Self-Reported Drug Use Among Injection Drug Users and Crack Cocaine Users Recruited Through Street Outreach." *Evaluation and Program Planning* 17(4):347–355.

Webb, Eugene J., Donald T. Campbell, Richard D. Schwartz, and Lee Sechrest. 2000. *Unobtrusive Measures,* rev. ed. Thousand Oaks, CA: Sage.

Weber, Max. 1949. *The Methodology of the Social Sciences.* Translated and edited by Edward A. Shils and Henry A. Finch. New York: Free Press.

Weber, Robert Philip. 1990. *Basic Content Analysis,* 2nd ed. Newbury Park, CA: Sage.

Wechsler, Henry, Jae Eun Lee, Meichun Kuo, Mark Seibring, Toben F. Nelson, and Hang Lee. 2002. "Trends in College Binge Drinking During a Period of Increased Prevention Efforts." *Journal of American College Health* 50:203–217.

Weinberg, Darin. 2000. "'Out There': The Ecology of Addiction in Drug Abuse Treatment Discourse." *Social Problems* 47:606–621.

Weisburd, David, Cynthis M. Lum, and Anthony Petrosino. 2001. "Does Research Design Affect Study Outcomes in Criminal Justice?" *Annals of the American Academy of Political and Social Science* 578:50-70.

Weiss, Carol H. 1993. "Where Politics and Evaluation Research Meet." *Evaluation Practice* 14:93–106.

Wellman, Barry. 2004. "Connecting Communities: On and Offline." *Contexts* 3:22–28.

Wellman, Barry, Anabel Quan Haase, James Witte, and Keith Hampton. 2001. "Does the Internet Increase, Decrease, or Supplement Social Capital? Social Networks, Participation, and Community Commitment." *American Behavioral Scientist,* 45:436–455.

Wellman, Barry and Keith Hampton. 1999. "Living Networked in a Wired World." *Comparative Sociology* 28:1–12.

Wells, L. Edward and Joseph H. Rankin. 1991. "Families and Delinquency: A Meta-Analysis of the Impact of Broken Homes." *Social Problems* 38(February):71–93.

Wengraf, Tom. 2001. *Qualitative Research Interviewing: Biographic Narrative and Semi-Structured Methods.* Thousand Oaks, CA: Sage.

West, Steven L. and Ken K. O'Neal. 2004. "Project D.A.R.E. Outcome Effectiveness Revisited." *American Journal of Public Health* 94:1027–1029.

Wheeler, Peter M. 1995. *Social Security Programs Throughout the World—1995.* Research Report #64, SSA Publication No. 13–11805. Washington, DC: Office of Research and Statistics, Social Security Administration.

White, Aaron M., Courtney L. Kraus, Lindsey A. McCracken, and H. Scott Swartzwelder. 2003. "Do College Students Drink More Than They Think? Use of a Free-Pour Paradigm to Determine How College Students Define Standard Drinks." *Alcoholism: Clinical and Experimental Research* 27:1750–1756.

White, Michael J. 1993. "Measurement of Population Size, Composition, and Distribution." Pp. 1–1 – 1–29 in *Readings in Population Research Methodology,* vol. 1, *Basic Tools,* edited by Donald J. Bogue, Eduardo E. Arriaga, and Douglas L. Anderton. Chicago: Social Development Center, for the United Nations Population Fund.

Whyte, William Foote. 1955. *Street Corner Society.* Chicago: University of Chicago Press.

Whyte, William Foote. 1991. *Social Theory for Social Action: How Individuals and*

Organizations Learn to Change. Newbury Park, CA: Sage.

Whyte, William H., Jr. 1956. *The Organization Man.* New York: Simon & Schuster.

Williams, Kirk R. and Richard Hawkins. 1986. "Perceptual Research on General Deterrence: A Critical Review." *Law and Society Review* 20:545–572.

Wilson, William Julius. 1987. *The Truly Disadvantaged: The Inner City, the Underclass, and Public Policy.* Chicago: University of Chicago Press.

Wilson, William Julius. 1998. "Engaging Publics in Sociological Dialogue Through the Media." *Contemporary Sociology* 27:435–438.

Witkin, Belle Ruth and James W. Altschuld. 1995. *Planning and Conducting Needs Assessments: A Practical Guide.* Thousand Oaks, CA: Sage.

Witte, James C., Lisa M. Amoroso, and Philip E. N. Howard. 2000. "Research Methodology: Method and Representation in Internet-Based Survey Tools—Mobility, Community, and Cultural Identity in Survey 2000." *Social Science Computer Review* 18:179–195.

Wolcott, Harry F. 1995. *The Art of Fieldwork.* Walnut Creek, CA: AltaMira Press.

Wolf, Amanda, David Turner, and Kathleen Toms. 2009. "Ethical Perspectives in Program Evaluation." Pp. 170–184 in *The Handbook of Social Research Ethics*, edited by Donna M. Mertens and Pauline E. Ginsberg. Thousand Oaks, CA: Sage.

World Health Organization. 2005. *WHO Multi-Country Study on Women's Health and Domestic Violence Against Women: Summary Report of Initial Results on Prevalence, Health Outcomes and Women's Responses.* Geneva: World Health Organization.

World Health Organization. n.d. *Process of Translation and Adaptation of Instruments.* Retrieved January 25, 2011, from the World Health Organization, http://www.who.int/substance_abuse/research_tools/translation/en/

World Medical Association (WMA). 2011. *WMA International Code of Medical Ethics.* http://www.wma.net/en/30publications/10policies/c8/index.htm. Downloaded August 31, 2011.

Wunsch, Guillaume J. and Marc G. Termote. 1978. *Introduction to Demographic Analysis: Principles and Methods.* New York: Plenum Press.

Xu, Yili, Mora L. Fiedler and Karl H. Flaming. 2005. "Citizens' Judgment Discovering the Impact of Community Policing: The Broken Windows Thesis, Collective Efficacy, and Citizens' Judgment." *Journal of Research in Crime and Delinquency* 42:147–186.

Yi, Hsiao-ye, Gerald D. Williams, and Barbara A. Smothers. 2004. *Trends in Alcohol-Related Fatal Traffic Crashes, United States, 1977–2002.* Surveillance Report #69. Bethesda, MD: Division of Epidemiology and Prevention Research, National Institute on Alcohol Abuse and Alcoholism.

Young, Robert L. 1992. "Religious Orientation, Race and Support for the Death Penalty." *Journal for the Scientific Study of Religion* 31:76–87.

Zaret, David. 1996. "Petitions and the 'Invention' of Public Opinion in the English Revolution." *American Journal of Sociology* 101:1497–1555.

Zielbauer, Paul. 2000. "2 Cities Lag Far Behind the U.S. in Heeding the Call of the Census." *The New York Times,* April 21, p. A21.

Zimbardo, Philip G. 2004. "A Situationist Perspective on the Psychology of Evil: Understanding How Good People Are Transformed Into Perpetrators." Pp. 21–50 in *The Social Psychology of Good and Evil: Understanding Our Capacity for Kindness and Cruelty,* edited by Arthur G. Miller. New York: Guilford Press.

Zimbardo, Philip. 2007. *The Lucifer Effect: Understanding How Good People Turn Evil.* New York: Random House.

Zitner, Aaron. 1996. "A Cloudy Gaze Into a Crystal Ball." *The New York Times,* September 26, pp. D1, D14.

Credits

Preface

Exhibit P.1: Glavin et al. (2011:51).

Chapter 1

Exhibit 1.1: U.S. Department of Commerce. 2010. *Digital Nation: 21st Century America's Progress Toward Universal Broadband Internet Access: An NTIA Research Preview.* Washington DC: US Department of Commerce, National Telecommunications and Information Administration.

Exhibit 1.2: McPherson et al. (2006:358).

Exhibit 1.3: U.S. Bureau of the Census (2010e).

Exhibit 1.6: Shermer (1997:26).

Exhibit 1.8: Hampton and Wellman (2000). Reprinted with permission.

Chapter 2

Exhibit 2.1: World Health Organization (2005).

Exhibit 2.3: Pate and Hamilton (1992:695). Reprinted with permission.

Exhibit 2.9: Data from Sherman and Berk (1984:267).

Exhibit 2.12: Based on Hirschel et al. (2008:275).

Exhibit 2.15: Schutt et al. (1992b).

Chapter 3

Exhibit 3.1, 3.2: From the film OBEDIENCE. Copyright © 1968 by Stanley Milgram, © Renewed 1993 by Alexandra Milgram, and distributed by Penn State, Media Sales.

Exhibit 3.3: Tuskegee Syphilis Study Administrative Records. Records of the Centers for Disease Control and Prevention. National Archives—Southeast Region (Atlanta).

Exhibit 3.5: From *The Lucifer Effect* by Philip G. Zimbardo. Copyright © 2007 by Philip G. Zimbardo, Inc. Used by permission of Random House, Inc., and Random House Group Ltd. .

Exhibit 3.8: http://www.hhs.gov/ohrp/humansubjects/guidance/decisioncharts.htm

Exhibit 3.9: Adapted from Miller (2003:704).

Exhibit 3.10: Guba and Lincoln (1989).

Chapter 4

Exhibit 4.1: Based on Howell (2003:76).

Exhibit 4.3: Absolute Standard image: ©iStockphoto.com/michellegibson; Subjective Standard image: A combination of ©iStockphoto.com/kevinruss, ©iStockphoto.com/GordonsLife, ©iStockphoto.com/clifflocalinicom, ©iStockphoto.com/mevans; Relative Standard graph: U.S. Census Bureau, CPS, 2010 ASEC.

Exhibit 4.4: Adam Henerey.

Exhibit 4.5: Based on Black (1976).

Exhibit 4.6: Adapted from Viswantathan (2005:7).

Exhibit 4.7: Reprinted with permission from the *Annual Review of Sociology,* Volume 29 ©2003 by Annual Reviews www.annualreviews.org

Exhibit 4.9: Data from Hadaway et al. (1993:744–746).

Exhibit 4.11: Arthur et al. (2005). Favorable Attitudes Toward Antisocial Behavior. Center for Substance Abuse Prevention (CSAP) website, http://www.activeguidellc.com/

cmi/menu_frameset.htm (choose CMI Viewer, Individual/Peer, Antisocial attitudes).

Exhibit 4.12: Core Institute (1994:3). Copyright 1994; reproduced with permission.

Exhibit 4.15: Hawkins, Amato, and King (2007:1007).

Exhibit 4.16: Reprinted with permission from the *Diagnostic and Statistical Manual of Mental Disorders, Fourth Edition, Text Revision* (Copyright 2000). American Psychiatric Association.

Exhibit 4.18: Based on Schutt (1988:7–10, 15, 16). Results reported in Schutt and Fennell (1992).

Chapter 5

Exhibit 5.1: Based on information from Levy and Lemeshow (1999).

Exhibit 5.3: Gallup (2011); Panagopoulos (2008).

Exhibit 5.4: Based on Keeter (2008).

Exhibit 5.10: Rossi (1989:225). Reprinted with permission from the University of Chicago Press.

Exhibit 5.13: Based on Heckathorn (1997:178).

Exhibit 5.14: Reprinted from Heckathorn (2002:11–34), by permission. Copyright 2002 by the Society for the Study of Social Problems.

Exhibit 5.15, 5.17: Data from General Social Survey (1996).

Exhibit 5.18: Retrieved June 26, 2005, from http://www.statsoft.comtextbook/stpowan.html Copyright © Statsoft, Inc 1984–2003, STATISTICA is a trademark of StatSoft, inc.

Chapter 6

Exhibit 6.1: Bureau of Justice Statistics (2011a).

Exhibit 6.3, 6.4: Based on Sampson and Raudenbush (1999).

Exhibit 6.6: Schreck et al. (2006).

Exhibit 6.8, 6.9: Exum (2002).

Exhibit 6.10: Sampson and Laub (1990).

Exhibit 6.14: Based on Sampson and Raudenbush (1999).

Chapter 7

Exhibit 7.1: Czopp et al. (2006:797).

Exhibit 7.2: Based on Czopp et al. (2006: 795–796).

Exhibit 7.5: Innstrand, Espries, and Mykletun (2004:124).

Exhibit 7.7: Wageman (1995:170). Reprinted with permission from *Administrative Science Quarterly.*

Exhibit 7.8: Adapted from Philips (1982:1347). Reprinted with permission from the University of Chicago Press.

Exhibit 7.9: Nakonezny et al. (2004:94). Reprinted with permission.

Exhibit 7.10: Based on Cohen & Ledford (1994).

Exhibit 7.11: Van Hoye & Lievens (2003:22, Table 2). Reprinted with permission by Springer.

Chapter 8

Exhibit 8.1: Link (2008); Christian et al. (2010).

Exhibit 8.3: Based on Schuman and Presser (1981: 121).

Exhibit 8.4: Based on Mirowsky and Ross (2001:9).

Exhibit 8.6: Radloff (1977:387). Copyright 1977 by West Publishing Company/Applied Psychological Measurement, Inc.; reproduced by permission.

Exhibit 8.7: Reprinted with permission from *Journal of Studies on Alcohol,* Volume 55, pp. 141–148, 1994. Center for Alcohol Studies, Rutgers.

Exhibit 8.8: Ross (1990). Reprinted with permission of the author.

Exhibit 8.11: Ross (1990:7).

Exhibit 8.12: Curtin, Presser, and Singer (2005:91).

Exhibit 8.13: Keeter (2008).

Exhibit 8.14: Ross (1990).

Exhibit 8.17: Adapted from Dillman (1978: 74–75). *Mail and Telephone Surveys: The Total Design Method.* Reprinted by permission of John Wiley & Sons, Inc.

Chapter 9

Exhibit 9.1: © Rick Wilking/Reuters/Corbis.

Exhibit 9.2: Miller and Crabtree (1999a:16). Reprinted with permission from SAGE Publications, Inc.

Exhibit 9.4: Adapted from Marshall and Rossman (1999:75–76). Reprinted with permission from SAGE Publications, Inc.

Exhibit 9.6: Field Notes from an ECH made available by Norma Ware, unpublished ethnographic notes, 1991.

Exhibit 9.7: Source: Raudenbush and Sampson (1999:15).

Exhibit 9.8: ©Peter K. B. St. Jean. Reprinted with permission.

Exhibit 9.9: Horowitz, Carol R., Mimsie Robinson, and Sarena Seifer. 2009. "Community-Based Participatory Research from the Margin to the Mainstream: Are Researchers Prepared?" *Circulation* May 19, 2009. Table 1, Characteristics of CBPR, p. 2634. Reprinted with permission of Wolters Kluwer Health, LWW.

Chapter 10

Exhibit 10.1: Miller and Crabtree (1999b:139, Figure 7.1, based on Addison 1999). Reprinted with permission from SAGE Publications, Inc.

Exhibit 10.2–10.4: Miles and Huberman (1994). Reprinted with permission from SAGE Publications, Inc.

Exhibit 10.5: Patton (2002). Reprinted with permission from SAGE Publications, Inc.

Exhibit 10.6: Miles and Huberman (1994). Reprinted with permission from SAGE Publications, Inc.

Exhibit 10.7: Stokoe (2006:479–480).

Exhibit 10.8: Morrill et al. (2000:551, Table 1). Copyright 2000. Reprinted with permission of Blackwell Publishing Ltd.

Exhibit 10.10: Levitt et al. (2008:439).

Exhibit 10.11: Cress & Snow (2000:1097, Table 6,). Reprinted with permission from the University of Chicago Press.

Exhibit 10.12: Margolis (2004:78).

Exhibit 10.13: Frohmann (2005).

Exhibit 10.14: Needleman (1981:248–256).

Exhibit 10.15: From *Homelessness, Housing, and Mental Illness* by Russell K. Schutt, with Stephen M. Goldfinger, p. 135. Cambridge, Mass.: Harvard University Press. Copyright © 2011 by the President and Fellows of Harvard College. Reprinted by permission of the publisher.

Exhibit 10.16: Muhr and Friese (2004:29).

Chapter 11

Exhibit 11.1: Greenberg, Shroder, and Onstott (1999:159). Reprinted with permission of Urban Institute Press.

Exhibit 11.2: Adapted from Martin and Kettner (1996).

Exhibit 11.3: Based on Goldfinger and Schutt (1996).

Exhibit 11.4: From Meg Townsend, Dana Hunt, Caity Baxter, and Peter Finn. 2005. *Interim Report: Evaluability Assessment of the President's Family Justice Center Initiative*. Cambridge, MA: Abt Associates Inc. Reprinted with permission.

Exhibit 11.5: Ringwalt et al. (1994:58).

Exhibit 11.6: Chen (1990:210). Reprinted with permission from SAGE Publications, Inc.

Exhibit 11.7: Based on D'Amico & Fromme (2002:569).

Exhibit 11.8: From Margo Rosenbach, Carol Irvin, Angela Merrill, Shanna Shulman, John Czajka, Christopher Trenholm, Susan Williams, So Sasigant Limpa-Amara, and Anna Katz. 2007. *National Evaluation of the State Children's Health Insurance Program: A Decade of Expanding Coverage and Improving Access: Final Report*. Cambridge, MA: Mathematica Policy Research, Inc.

Exhibit 11.9: Orr (1992:224, Table 6.5). Reprinted with permission from SAGE Publications, Inc.

Exhibit 11.12: Leakey et al. (2004).

Chapter 12

Exhibit 12.2: Adapted from Griffin (1993:1110). Reprinted with permission from the University of Chicago Press.

Exhibit 12.3, 12.4: From "The Nation-State and the Natural Environment over the Twentieth Century" David John Frank, Ann Hironaka, and Evan Schofer. February 2000: Vol. 65, No. 1, pp. 96–116. *American Sociological Review*.

Exhibit 12.5: Brooks, Clem and Jeff Manza. June, 2006. "Social Policy Responsiveness in Developed Democracies." *American Sociological Review* 71(3):474–494.

Exhibit 12.6: Bail, Christopher A. February, 2008. "The Configuration of Symbolic Boundaries against Immigrants in Europe." *American Sociological Review* 73(1):37–59.

Exhibit 12.7: Reproduced by permission of International IDEA from Turnout in the World—Country by Country Performance (1945–1998). From Voter Turnout: A Global Survey (http://www. int/vt/survey/voter_turnout_pop2–2 .cfm). © International Institute for Democracy and Electoral Assistance.

Exhibit 12.9: Rueschemeyer et al. (1002:161). Reprinted with permission from the University of Chicago Press.

Exhibit 12.10: Adapted from Skocpol (1984:379).

Exhibit 12.11: Adapted from Sanders (1994:514–517) and Skocpol and Somers (1979:80).

Exhibit 12.12: Skocpol (1979).

Exhibit 12.13: Bogue, Arriaga, and Anderton (1993).

Chapter 13

Exhibit 13.1: U.S. Bureau of the Census (1994:8).

Exhibit 13.8: From *Detroit Area Study, 1997: Social Change in Religion and Child Rearing.* Inter-University Consortium for Politcal and Social Research. Reprinted with permission.

Exhibit 13.9: From Detroit Area Study, 1997: Social Change in Religion and Child Rearing. Inter-University Consortium for Politcal and Social Research. Reprinted with permission.

Exhibit 13.10: Neuendor (2002).

Exhibit 13.11: Reprinted by permission from *Organization in a Changing Environment: Unionization of Welfare Employees,* edited by Russell K. Schutt, the State University of New York Press © 1986, State University of New York. All rights reserved

Exhibit 13.12: Neuendorf (2002).

Chapter 14

Exhibit 14.2: Center for the Study of the American Electorate, American University, Preliminary Primary Turnout Report.

Exhibit 14.3: General Social Survey, National Opinion Research Center (2010).

Exhibit 14.4: U.S. Bureau of Economic Analysis (2004:14).

Exhibit 14.5–14.7: General Social Survey, National Opinion Research Center (2010).

Exhibit 14.8: Adapted from Orcutt and Turner (1993). Copyright 1993 by the Society for the Study of Social Problems. Reprinted by permission.

Exhibit 14.9–14.14: General Social Survey, National Opinion Research Center (2010).

Exhibit 14.23–14.25: General Social Survey, National Opinion Research Center (2010).

Exhibit 14.26: Rand et al. (1997).

Exhibit 14.27, 14.28: General Social Survey, National Opinion Research Center (2010).

Exhibit 14.29: Frankfort-Nachmias and Leon-Guerrero (2006:230). Reprinted with permission from SAGE Publications, Inc.

Exhibit 14.30–14.32: General Social Survey, National Opinion Research Center (2010).

Exhibit 14.34–14.40: General Social Survey, National Opinion Research Center (2010).

Chapter 15

Exhibit 15.2–15.4: Meghani et al. (2009).

Exhibit 15.5: Adapted from Wells and Rankin (1991:80).

Exhibit 15.6: Aseltine and Kessler (1993:237–251).

Exhibit 15.7: Vernez et al. (1988). Reprinted with permission.

Exhibit 15.8: Schutt et al. (1996).

Index